Communications in Computer and Information Science 1911

Editorial Board Members

Joaquim Filipe, *Polytechnic Institute of Setúbal, Setúbal, Portugal*
Ashish Ghosh, *Indian Statistical Institute, Kolkata, India*
Raquel Oliveira Prates, *Federal University of Minas Gerais (UFMG), Belo Horizonte, Brazil*
Lizhu Zhou, *Tsinghua University, Beijing, China*

Rationale

The CCIS series is devoted to the publication of proceedings of computer science conferences. Its aim is to efficiently disseminate original research results in informatics in printed and electronic form. While the focus is on publication of peer-reviewed full papers presenting mature work, inclusion of reviewed short papers reporting on work in progress is welcome, too. Besides globally relevant meetings with internationally representative program committees guaranteeing a strict peer-reviewing and paper selection process, conferences run by societies or of high regional or national relevance are also considered for publication.

Topics

The topical scope of CCIS spans the entire spectrum of informatics ranging from foundational topics in the theory of computing to information and communications science and technology and a broad variety of interdisciplinary application fields.

Information for Volume Editors and Authors

Publication in CCIS is free of charge. No royalties are paid, however, we offer registered conference participants temporary free access to the online version of the conference proceedings on SpringerLink (http://link.springer.com) by means of an http referrer from the conference website and/or a number of complimentary printed copies, as specified in the official acceptance email of the event.

CCIS proceedings can be published in time for distribution at conferences or as post-proceedings, and delivered in the form of printed books and/or electronically as USBs and/or e-content licenses for accessing proceedings at SpringerLink. Furthermore, CCIS proceedings are included in the CCIS electronic book series hosted in the SpringerLink digital library at http://link.springer.com/bookseries/7899. Conferences publishing in CCIS are allowed to use Online Conference Service (OCS) for managing the whole proceedings lifecycle (from submission and reviewing to preparing for publication) free of charge.

Publication process

The language of publication is exclusively English. Authors publishing in CCIS have to sign the Springer CCIS copyright transfer form, however, they are free to use their material published in CCIS for substantially changed, more elaborate subsequent publications elsewhere. For the preparation of the camera-ready papers/files, authors have to strictly adhere to the Springer CCIS Authors' Instructions and are strongly encouraged to use the CCIS LaTeX style files or templates.

Abstracting/Indexing

CCIS is abstracted/indexed in DBLP, Google Scholar, EI-Compendex, Mathematical Reviews, SCImago, Scopus. CCIS volumes are also submitted for the inclusion in ISI Proceedings.

How to start

To start the evaluation of your proposal for inclusion in the CCIS series, please send an e-mail to ccis@springer.com.

Fazilah Hassan · Noorhazirah Sunar ·
Mohd Ariffanan Mohd Basri ·
Mohd Saiful Azimi Mahmud ·
Mohamad Hafis Izran Ishak ·
Mohamed Sultan Mohamed Ali
Editors

Methods and Applications for Modeling and Simulation of Complex Systems

22nd Asia Simulation Conference, AsiaSim 2023
Langkawi, Malaysia, October 25–26, 2023
Proceedings, Part I

Editors
Fazilah Hassan
Universiti Teknologi Malaysia
Johor, Malaysia

Noorhazirah Sunar
Universiti Teknologi Malaysia
Johor, Malaysia

Mohd Ariffanan Mohd Basri
Universiti Teknologi Malaysia
Johor, Malaysia

Mohd Saiful Azimi Mahmud
Universiti Teknologi Malaysia
Johor, Malaysia

Mohamad Hafis Izran Ishak
Universiti Teknologi Malaysia
Johor, Malaysia

Mohamed Sultan Mohamed Ali
Universiti Teknologi Malaysia
Johor, Malaysia

ISSN 1865-0929 ISSN 1865-0937 (electronic)
Communications in Computer and Information Science
ISBN 978-981-99-7239-5 ISBN 978-981-99-7240-1 (eBook)
https://doi.org/10.1007/978-981-99-7240-1

© The Editor(s) (if applicable) and The Author(s), under exclusive license
to Springer Nature Singapore Pte Ltd. 2024

This work is subject to copyright. All rights are reserved by the Publisher, whether the whole or part of the material is concerned, specifically the rights of translation, reprinting, reuse of illustrations, recitation, broadcasting, reproduction on microfilms or in any other physical way, and transmission or information storage and retrieval, electronic adaptation, computer software, or by similar or dissimilar methodology now known or hereafter developed.
The use of general descriptive names, registered names, trademarks, service marks, etc. in this publication does not imply, even in the absence of a specific statement, that such names are exempt from the relevant protective laws and regulations and therefore free for general use.
The publisher, the authors, and the editors are safe to assume that the advice and information in this book are believed to be true and accurate at the date of publication. Neither the publisher nor the authors or the editors give a warranty, expressed or implied, with respect to the material contained herein or for any errors or omissions that may have been made. The publisher remains neutral with regard to jurisdictional claims in published maps and institutional affiliations.

This Springer imprint is published by the registered company Springer Nature Singapore Pte Ltd.
The registered company address is: 152 Beach Road, #21-01/04 Gateway East, Singapore 189721, Singapore

Paper in this product is recyclable.

Preface

The 22nd Asia Simulation Conference (AsiaSim 2023) was held on Langkawi Island, Malaysia for two days on 25–26th October 2023. This Asia Simulation Conference is annually organized by the following Asian Simulation societies: the Korea Society for Simulation (KSS), the China Simulation Federation (CSF), the Japan Society for Simulation Technology (JSST), the Society of Simulation and Gaming of Singapore, and the Malaysia Simulation Society (MSS). This conference provides a platform for scientists, academicians, and professionals from around the world to present and discuss their work and new emerging simulation technologies with each other. AsiaSim 2023 aimed to serve as the primary venue for academic and industrial researchers and practitioners to exchange ideas, research findings, and experiences. This would encourage both domestic and international research and technical innovation in these sectors.

Bringing the concept of 'Experience the Power of Simulation', the papers presented at this conference have been compiled in this volume of the series Communications in Computer and Information Science. The papers in these proceedings tackle complex modelling and simulation problems in a variety of domains. The total of submitted papers to the conference was 164 papers. All submitted papers were reviewed by at least three (3) reviewers based on single-blind peer review. The papers presented in AsiaSim 2023 are included in two (2) volumes (1911 and 1912) containing 77 full papers divided into seven (7) main tracks, namely: Artificial Intelligence and Simulation, Digital Twins (Modelling), Simulation and Gaming, Simulation for Engineering, Simulation for Industry 4.0, Simulation for Sustainable Development, and Simulation for Social Sciences.

As editorial members of the AsiaSim 2023 conference, we would like to express our gratitude to all the authors who chose this conference as a venue for their publications and to all reviewers for providing professional reviews and comments. We are also very grateful for the support and tremendous effort of the Program and Organizing Committee members for soliciting and selecting research papers with a balance of high quality and new ideas and applications. Thank you from us.

Fazilah Hassan
Noorhazirah Sunar
Mohd Ariffanan Mohd Basri
Mohd Saiful Azimi Mahmud
Mohamad Hafis Izran Ishak
Mohamed Sultan Mohamed Ali

Preface

Organization

Advisor

Yahaya Md. Sam	Universiti Teknologi Malaysia, Malaysia

Chair

Zaharuddin bin Mohamed	Universiti Teknologi Malaysia, Malaysia

Deputy Chair

Norhaliza Abdul Wahab	Universiti Teknologi Malaysia, Malaysia

Secretariats

Mohd Saiful Azimi Mahmud	Universiti Teknologi Malaysia, Malaysia
Nurulaqilla Khamis	Universiti Teknologi Malaysia, Malaysia

Treasurer

Noorhazirah Sunar	Universiti Teknologi Malaysia, Malaysia

Technical and Publication

Mohamed Sultan Mohamed Ali	Universiti Teknologi Malaysia, Malaysia
Mohd. Ridzuan Ahmad	Universiti Teknologi Malaysia, Malaysia
Fazilah Hassan	Universiti Teknologi Malaysia, Malaysia

Program

Salinda Buyamin — Universiti Teknologi Malaysia, Malaysia
Mohd Ariffanan Mohd Basri — Universiti Teknologi Malaysia, Malaysia

Local Arrangement

Anita Ahmad — Universiti Teknologi Malaysia, Malaysia
Nurul Adilla Mohd Subha — Universiti Teknologi Malaysia, Malaysia
Mohamad Hafis Izran Ishak — Universiti Teknologi Malaysia, Malaysia

Website and Publicity

Herman Wahid — Universiti Teknologi Malaysia, Malaysia
Hazriq Izzuan Jaafar — Universiti Teknikal Malaysia Melaka, Malaysia

Sponsorship

Abdul Rashid Husain — Universiti Teknologi Malaysia, Malaysia
Leow Pei Ling — Universiti Teknologi Malaysia, Malaysia

International Program Committee

Abul K. M. Azad — Northern Illinois University, USA
Andi Andriansyah — Universitas Mercu Buana, Indonesia
Axel Lehmann — Universitat de Bundeswehr München, Germany
Bidyadhar Subudhi — National Institute of Technology Rourkela, India
Bohu Li — Beijing University of Aeronautics and Astronautics, China
Byeong-Yun Chang — Ajou University, South Korea
Cai Wentong — National University of Singapore, Singapore
Gary Tan — National University of Singapore, Singapore
He Chen — Hebei University, China
Jie Huang — Beijing Institute of Technology, China
Le Anh Tuan — Vietnam Maritime University, Vietnam
Liang Li — Ritsumeikan University, Japan
Lin Zhang — Beihang University, China
Mehmet Önder Efe — Hacettepe University, Turkey

Mohammad Hasan Shaheed	Queen Mary University of London, UK
Muhammad Abid	COMSATS University Islamabad, Pakistan
Mukarramah Yusuf	Universitas Hasanuddin, Indonesia
Ning Sun	Nankai University, China
Ramon Vilanova Arbos	University of Barcelona, Spain
Sarvat M. Ahmad	King Fahd University of Petroleum and Minerals, Saudi Arabia
Satoshi Tanaka	Ritsumeikan University, Japan
Sumeet S. Aphale	University of Aberdeen, UK
Teo Yong Meng	National University of Singapore, Singapore
Walid Aniss	Aswan University, Egypt
Xiao Song	Beihang University, China
Yahaya Md Sam	Universiti Teknologi Malaysia, Malaysia
Yuanjun Laili	Beihang University, China
Yun Bae Kim	Sungkyunkwan University, South Korea

Contents – Part I

Contents – Part II

FPGA Based Accelerator for Image Steganography

Liqaa N. Sabeeh and Mohammed A. Al-Ibadi(✉)

Computer Engineering Department, University of Basrah, Basrah, Iraq
{pgs.liqaa.nabeel,mohammed.joudah}@uobasrah.edu.iq

Abstract. In this study, a novel approach is proposed for enhancing the speed and efficiency of steganography system by designing a hardware model that adopts the image-in-image (grey-in-color) method and utilizes the Least Significant Bit (LSB) algorithm. The sizes of secret and cover images are chosen so that all secret bits can be embedded inside the LSB bits of the pixels for the three RBG matrices for the cover image and produce the color stego-image. On the receiving part, these embedded secret bits will be extracted from stego-image and reform the original secret image. Using XSG and Vivado design suite, a successful FPGA-based steganography system was developed as an accelerator tool with high-speed processing that reaches 250 times faster than the software-based system. The quality of the stego-image and the extracted secret image was calculated by three metrics, which are the Mean Square Error (MSE), Peak Signal to Noise Ratio (PSNR), and Cross Correlation (CCR). The comparison metrics values affirm a high level of trust in the generated stego-image, and the extracted secret image being identical to the original. The FPGA chip utilized in the implemented system consumes only 1% of the hardware resources, and a remarkable speedup factor of 250 is achieved, indicating that expanding the system for larger image sizes becomes highly feasible. Moreover, this expansion can be accomplished economically by utilizing low-cost field-programmable gate array (FPGA) chips.

Keywords: Steganography · LSB algorithm · FPGA · XSG

1 Introduction

The transmission of data through networks today requires the use of digital communication methods. There are always significant dangers to safe data transfer due to networking and digital communication advancements [1]. A hidden image can be concealed within a digital image using various data-concealing techniques. One of the most commonly accepted methods of data concealing is image steganography. Steganography aims to transmit a secret image while concealing its presence [2]. The operating principles of steganography are different from those of cryptography. It provides a high level of security in communication while concealing the existence of data, as opposed to cryptography, which is a security technology that conceals data to preserve its confidentiality and integrity. The ability to integrate a hidden image into a cover image without leaving

© The Author(s), under exclusive license to Springer Nature Singapore Pte Ltd. 2024
F. Hassan et al. (Eds.): AsiaSim 2023, CCIS 1911, pp. 1–12, 2024.
https://doi.org/10.1007/978-981-99-7240-1_1

a noticeable trace is both an art and a science [1]. A "stego-image" is created when the cover image has a secret image buried inside it [3]. Steganography began as a kind of art in the past, but it has since evolved into one of the most specialized fields in cybersecurity [4]. The secret data, the cover file (the media file into which the secret data is incorporated), and the stego-file (the cover file after the secret data has been hidden within) are the three basic components of a steganographic approach. Three crucial characteristics of a powerful steganographic technique are robustness (the resistance to the secret data being removed from the stego-file), imperceptibility (the property that makes the stego-file and cover file indiscernible from one another), and capacity (the maximum amount of secret data that the cover file can conceal). The frequency and spatial domains are the two domains where steganography could be used [5]. Steganographic techniques modify the pixels of the images in the spatial domain to hide the secret data in. Steganographic methods employ the modification of data in the frequency domain to conceal sensitive data by altering the image frequencies. The secret data is hidden using steganographic techniques that alter the frequencies of the images in the frequency domain [5]. The system uses a spatial domain method called least significant bit (LSB) steganography. The LSB-based image steganography algorithm consists of two primary phases: the embedding phase and the extraction phase. The algorithm analyzes the cover image and inserts the secret image into it such that a human eye cannot see it. The resultant image, called a "stego-image," is then obtained and sent across a communication channel. The LSBs of the stego-image are used to extract the concealed image during the extraction stage [3]. This project aims to employ FPGA as an accelerator for an image steganography system, aiming to address the extended execution time associated with the conventional software version.

2 LSB Algorithm for Steganography

Steganography is a technique used for hiding important data within video, audio, or image files before transmitting them [6]. A trustworthy piece of information can have concealed data sent without anybody being aware of it. Since steganography is hidden within a carrier and travels through it, it prohibits unauthorized parties from accessing the news [7]. In image steganography, the carrier for the hidden image is an image that either reveals or hides it. Figure 1 displays the fundamental image steganography diagram. The term "cover image" refers to the image that will carry the hidden data, and the "secret image" refers to the data that need to be hidden. In the reality, the "embedding technique" refers to the process or algorithm used to conceal the "secret image" behind the "cover image," or "stego-image." In a manner similar to embedding, extraction of the hidden information is also possible, with the "extraction technique" being the method of recovering the "secret image" from the "stego-image" [8].

The Least Significant Bit (LSB) algorithm is one of the core and traditional techniques that can conceal more significant hidden information in a cover image without obvious visual distortions is steganography [9]. It basically works by substituting secret message bits for the low-order bits (LSBs) of randomly (or) carefully chosen pixels in the cover image. A stego may choose the pixels to use or the embedding order. As time went on, steganographic techniques used various iterations of LSB's pixel or bit-planes. Due

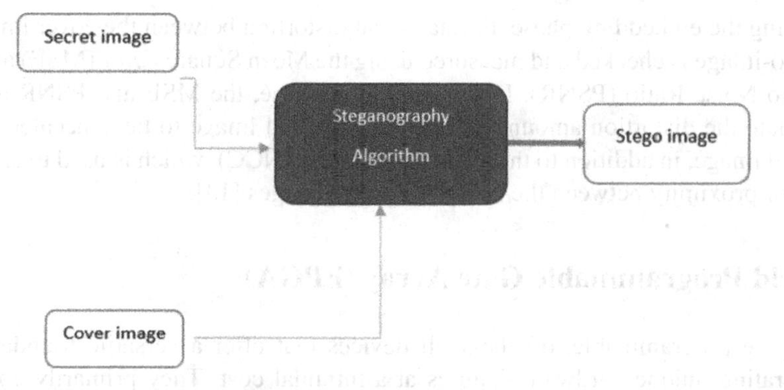

Fig. 1. Image steganography technique.

to the ease of LSB embedding, many techniques have been developed in an effort to optimize payload while enhancing visual quality and undetectability. To estimate the number of LSB bits for data embedding, some of them use adaptive LSB replacement depending on the edges, texture, intensity level, and brightness of the cover images [10].

The spatial domain operations are straightforward and do not require any kind of modification before being applied to the pixels [11]. LSB algorithm is the one that is most frequently employed in the spatial domain. LSB is a quick and easy method that turns any image into a grayscale image, regardless of its type. Each pixel is represented by 1 byte, with the least important information being contained in the final bit or the rightmost bit [12].

This bit is swapped out by the LSB algorithm with one of the secret data bits, which will be concealed inside the image. There will not be any visible visual changes in the image because the swapped bit is the least important one. The block diagram of the LSB embedding process is shown in Fig. 2, where the new LSB value has changed from 1 to 0.

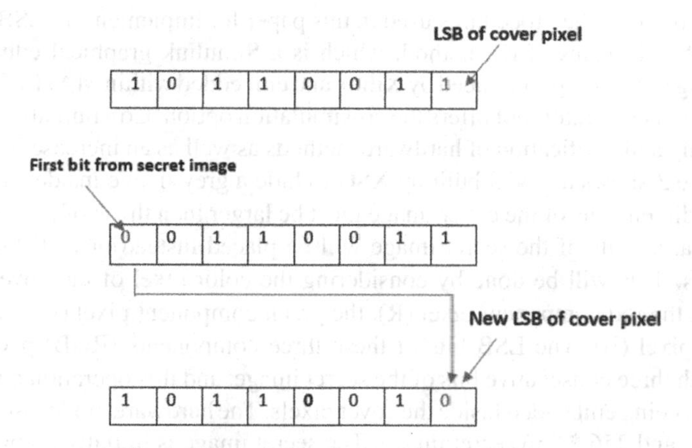

Fig. 2. Process of replacing LSB cover image.

During the embedding phase, the statistical distortion between the cover image and the stego-image is checked and measured using the Mean Square Error (MSE) and Peak Signal to Noise Ratio (PSNR). In the extraction phase, the MSE and PSNR are used to estimate the distortion amount between the original image to be concealed and the extracted image, in addition to the Cross Correlation (NCC), which is used to assess the degree of proximity between the cover and stego-images [13].

3 Field Programmable Gate Array (FPGA)

FPGAs are programmable, off-the-shelf devices that offer a versatile foundation for incorporating unique hardware features at a minimal cost. They primarily consist of configurable logic blocks (CLBs) [14], which are a collection of programmable logic cells, a programmable interconnection network, and a collection of programmable input and output cells positioned everywhere around the device [15]. Additionally, they have a wide range of embedded components, including block RAMs (BRAMs), look-up tables (LUTs), flip-flops (FFs), clock management units, high-speed I/O links, and others. DSP blocks are used to perform arithmetic-intensive operations like multiply-and-accumulate. FPGAs are frequently used as accelerators for computationally intensive applications because they provide models with highly flexible fine-grained parallelism and associative operations, such as broadcast and collective response [16]. FPGAs are being employed in a wide range of applications, including robotics, telecommunications, medical care, and image processing. However, programming FPGAs using conventional design approaches like VHDL is time-consuming; therefore, new programming tools such as XSG are adopted for fast-developing FPGA-based systems [17].

4 FPGA-Based LSB Algorithm Implementation

4.1 Hiding System

This section outlines the procedures used in this paper for implementing LSB algorithm inside FPGA chip using XSG method, which is a Simulink graphical editor tool for programming FPGA chips produced by Xilinx and embedded within MATLAB software.

The System Generator tool offers the co-simulation option. Co-simulation allows for the processing and verification of hardware methods as well as an increase in simulation speed. Figure 3 shows a model built by XSG to hide a grey image inside another color image. The dimensions of the cover image must be larger than those of the secret image to ensure that all bits of the secret image will be placed instead of LSB bits of cover image pixels. This will be done by considering the color pixel of the cover image as three pixels, the red component pixel (R), the green component pixel (G), and the blue component pixel (B). The LSB bits of these three components (RGB) pixels will be replaced with three consecutive bits of the secret image, and this operation continues till all secret bits being embedded inside the cover pixels. The hardware model uses 384 * 512 cover image, and 256 * 256 secret image. The secret image is initially preprocessed by converting it into a bit stream using a parallel-to-serial block. In the other hand, the 7 MSB bits of each pixel of the cover image will be separately decomposed, as illustrated in Fig. 4. At the same time, one bit of the secret image will be placed in the LSB position (by concat block) and produce the new pixel of the stego-image. This operation is done sequentially until all secret bits are involved inside the cover image pixels (one by one) and introduce the total stego-image. To align the secret image bits with the other seven bits of the cover image pixel, the latency to block (p_to_s) must be 1. The serial sample format is transformed by the gateway-in into an unsigned fixed-point format with WL = 8 and FL = 0 (width length and fractional length, respectively).

The periods of the Simulink system and the gateway-in block should both be set to value (1/8) in order to trace the replacement of LSB bit correctly, as illustrated in Figs. 5 and 6. Finally, the stego-image produced from the embedded secret image inside the cover image using the proposed system is presented in Fig. 7.

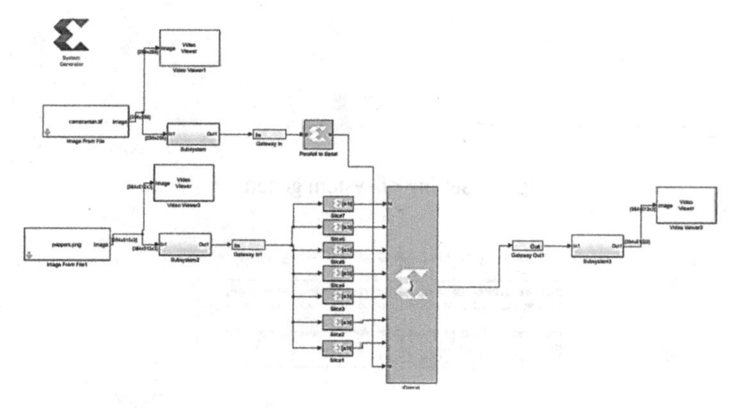

Fig. 3. FPGA implementation of image steganography with LSB algorithm.

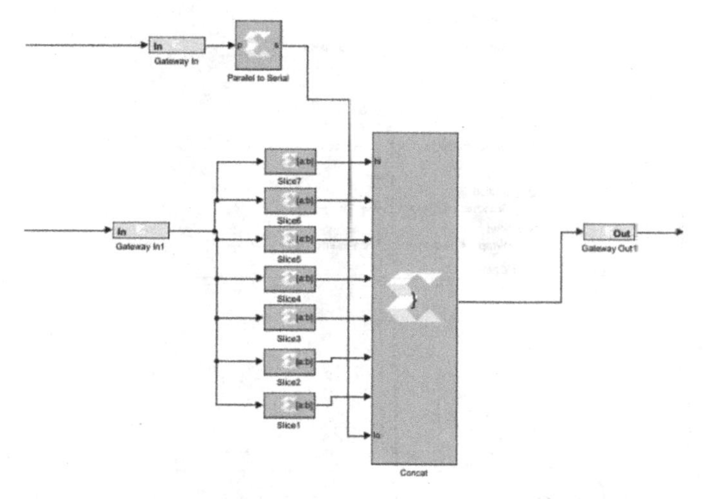

Fig. 4. LSB Algorithm for image hidden system.

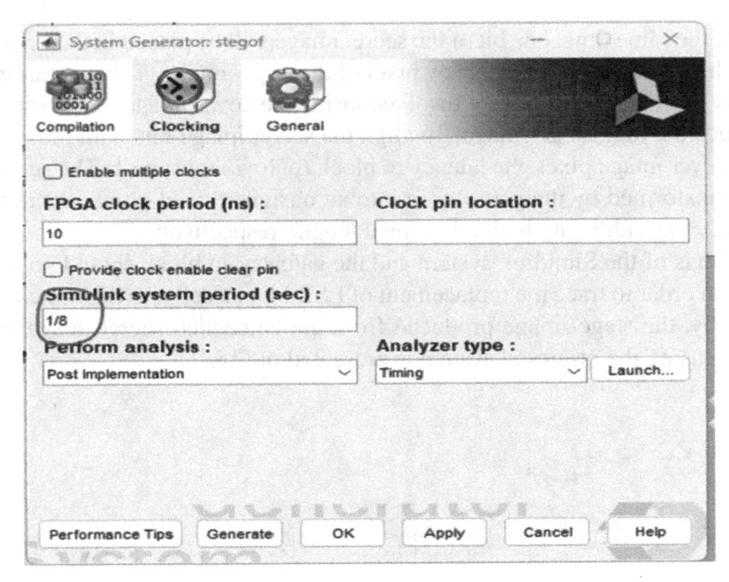

Fig. 5. Setting of system generator.

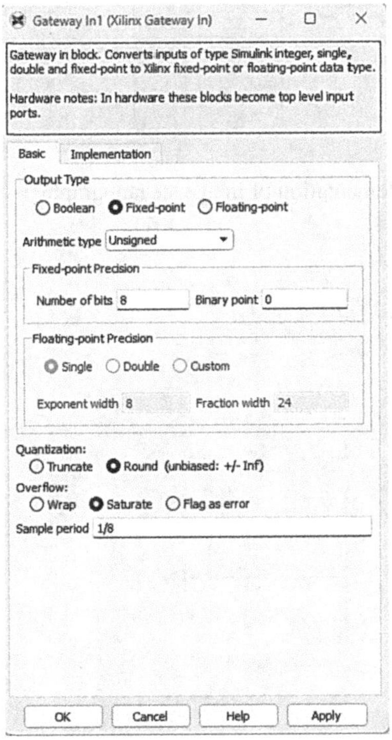

Gateway In1 (Xilinx Gateway In)

Gateway in block. Converts inputs of type Simulink integer, single, double and fixed-point to Xilinx fixed-point or floating-point data type.

Hardware notes: In hardware these blocks become top level input ports.

Basic Implementation

Output Type

Boolean Fixed-point Floating-point

Arithmetic type Unsigned

Fixed-point Precision

Number of bits 8 Binary point 0

Floating-point Precision

Single Double Custom

Exponent width 8 Fraction width 24

Quantization:

Truncate Round (unbiased: +/- Inf)

Overflow:

Wrap Saturate Flag as error

Sample period 1/8

OK Cancel Help Apply

Fig. 6. Gateway in for the one pixel of the cover image.

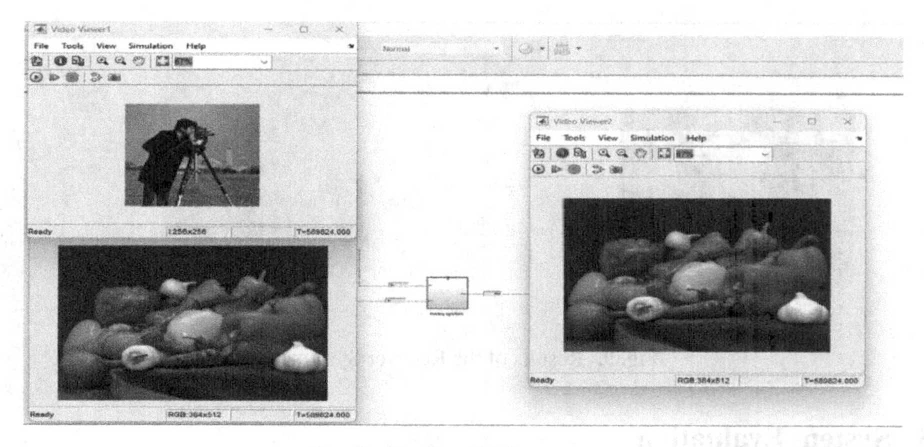

Fig. 7. Result of hiding system.

4.2 Recovering System

After the stego-image is generated, it will be sent to the target place through the communication media. Just if the receiver catches the stego-image, the secret image may be extracted. This operation is done simply by reverse operation of hiding operation. The LSB bits of the stego-image pixels (RGB) will be extracted and reformed as groups of 8 bits in sequence in order to regenerate secret image pixels and then rearrange the extracted pixels in the original two dimensions to produce the original secret image.

Figure 8 illustrates the hardware recovering system. The stego-image enters the system as pixel by pixel. LSB bit of each pixel will pass the concat block, which works as a demultiplexer for this purpose. Then, the extracted LSB bits will be accumulated as groups of 8 bits that are extracted sequentially for recomposing the original secret pixels. This operation continues till all secret pixels are reformed. The slice8 block and S-to-P block are set appropriately (lower bit location and width = 1) for exact secret image reformation. Figure 9 shows recovering the secret image from the stego-image using the proposed hardware system.

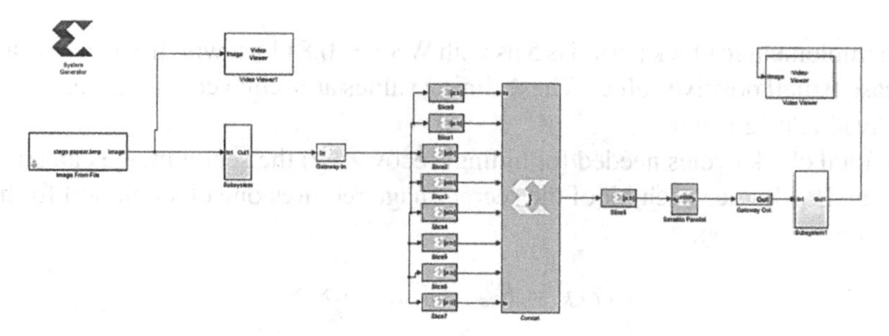

Fig. 8. Recover image from LSB steganography algorithm.

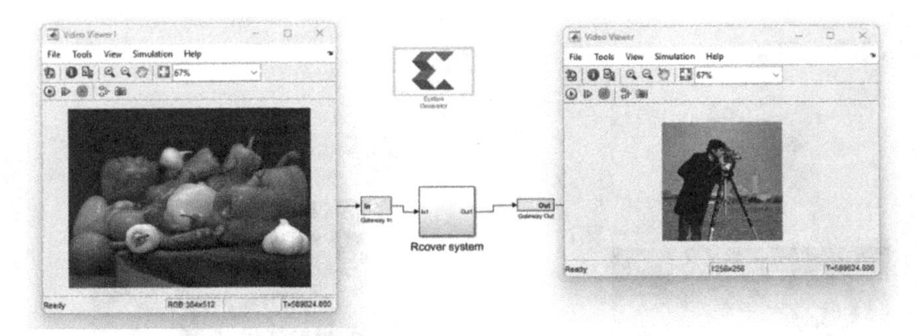

Fig. 9. Results of the Recover System.

5 System Evaluation

The system has been implemented in the ARTY A7 board, including the xc7a100t-lcsg324 FPGA chip. The XSG tool and Vivado design suite were used for implementing the proposed hiding/recovering systems. The area usage, clock period, and consumption power for the FPGA chip after implementation are listed in Table 1.

Table 1. FPGA implementation of the project

Resource	Used	Available	Utilization
LUT	589	63400	%0.93
LUTRAM	5	19000	%0.03
FF	998	126800	%0.79
BRAM	2	135	%1.48
Implemented Power on chip			0.29 W
Minimum clock (Ts)			5 ns
Worst negative slack (WNS)			0.811 ns

The implemented clock period is 5 ns with WSN = 0.811 ns, which can be considered a best case (small positive value). These timing values are achieved via timing constrains using Vivado design suite.

The total clock cycles needed for hiding (recovering) the secret image can be calculated using Eq. 1 since each bit of the secret image requires one clock period for hiding or extracting operation:

$$T_{CLK} = R_{sec} \times C_{sec} \cdots \times 8 \qquad (1)$$

where T_{CLK} is the total clock cycle for embedding (extracting) all bits of the secret image, R_{sec} is the number of rows of the secret image, and C_{sec} is the number of columns of the secret image. The term 8 in Eq. 1 represents the 8 bits of one pixel of the secret image (gray level image).

For the implemented work, the secret image dimension is 255 × 255 Gy level image, so the $T_{CLK} = 255 \times 255 \times 8 = 520{,}200$ clock cycles. Because the proposed system's clock period is 5 ns, the complete time for achieving the secret image hiding is 2,621,440 ns, the same time period for extracting the complete secret image in the recovering stage.

To calculate the speed gained in the proposed systems for the hiding and recovering operations, the standard software version of the LSB steganography algorithm was utilized with the same set of secret and cover images. Using MATLAB running on PC with Windows 11 Pro (64-bit operating system, x64-based processor), core i7-8550U CPU and 8 GB RAM, the minimum time required for completing the hiding and recovering operations is approximately 0.5 s. To obtain the speedup factor for hardware version against the software version of LSB steganography algorithms, Eq. 2 can be used.

$$\text{speedup factor} = \frac{\text{Time for software version}}{\text{Time for hardware version}} = \frac{0.5}{0.002} = 250 \tag{2}$$

The speedup value reflex that the accelerator (hardware) version is fast as 250 times as the conventional (software) version for this steganography implementation.

The stego-image produced by LSB algorithm should not have any visible differences when compared with the original cover image, where it noticed during the transmission stage. To ensure that, many evaluation metrics are used for this purpose, and the most used in literature are the Mean Square Error (MSE), Peak Signal-to-Noise Ratio (PSNR), and Cross Correlation (CCR). The MSE and PSNR are two error metrics. The MSE is the sum of the squared errors between the stego-image and the cover image, whereas the PSNR measures the peak error. The MSE value approaches zero, indicating that the similarity approaches identicality. The PSNR value belonging to the [50 dB–60 dB] is generally regarded as the favorable value in scenarios involving lossy conditions. The MSE error can be calculated using the following equation [18, 19]:

$$\text{MSE} = \frac{1}{\text{L}} \sum_{\text{i}=1}^{\text{N}-1} \sum_{\text{j}=1}^{\text{M}-1} [\text{Is(i, j)} - (\text{Ic(i, j)}]^2 \tag{3}$$

where *Is, Ic, M, N, L* represent the stego-image, cover image, number of rows, number of columns, and total number of cover image, respectively. For color image, the MSE_{RBG} is calculated as the average of the MSE for the three colors (R, B, G) component images as follows:

$$MSE_{RBG} = \left[\frac{MSE_R + MSE_B + MSE_G}{3} \right] \tag{4}$$

where MSE_R, MSE_B, and MSE_G are the MSE for the color image's red, blue, and green image matrices.

The other error metric, PSNR, can be determined for color images using the following equation [20]:

$$\text{PSNR} = 10 \times \log_{10}\left(\frac{cmax^2}{MSE_{RGB}}\right) \tag{5}$$

where cmax represents the highest possible pixel value from the original color image.

The CCR is one of the most techniques employed for determining how closely the cover image matches the stego-image. The cross-correlation result falls within the range of -1 to $+1$. A significantly positive value indicates a robust positive correlation, whereas a notably negative value indicates a strong negative correlation. When the value approaches zero, it suggests no correlation. Equation 6 can be used to determine the CCR [20].

$$\mathrm{CCR} = \frac{\sum_{i=0}^{N-1} \sum_{j=0}^{M-1} Is(i,j) * Ic(i,j)}{\sum_{i=0}^{N-1} \sum_{j=0}^{M-1} Ic^2(i,j)} \tag{6}$$

where *Is, Ic, M, and N* represent the stego-image, cover image, number of rows, and number of columns, respectively.

Table 2 summarizes the values of the three metrics between the original cover image and the stego-image produced by the proposed FPGA-based hardware system. These values absolutely demonstrate that the stego-image of the proposed hiding system closely matches the original cover image.

Table 2. MSE, PSNR, and CCR values of the proposed HIDING SYSTEM

MSE	0.499
PSNR	51.145
CCR	0.99

On the other hand, Table 3 illustrates the identicality of the original secret image with that extracted by the proposed FPGA-based recovering system.

Table 3. MSE, PSNR, and CCR values of the proposed RECOVERING system.

MSE	0.00
PSNR	∞
CCR	1.00

Where PSNR $= \infty$ implies that there is no discernible difference between the original secret image and the reconstructed secret image.

6 Conclusions

This paper presents the implementation of the LSB algorithm for steganography within the FPGA chip, aiming to achieve a high-speed steganography system. By taking advantage of the parallel and efficient processing capabilities of the FPGA chip's hardware components, the proposed system achieved a significant acceleration in hiding and recovering processes, surpassing that of the software versions of the same LSB algorithm. The achieved result demonstrates that the proposed FPGA-based steganography system exhibits a superior performance, surpassing the conventional software version by 250 times. Using the proposed system, the generated stego-image has good comparison metrics with the original cover image. Furthermore, the hardware recovering system successfully extracts the secret image, ensuring its exact copy is identical to the original secret image before it was embedded within the cover image.

On the other hand, besides using Vivado design suite, Xilinx system generator is a very effective tool for developing FPGA-based systems. It might be characterized as practical and faster than the other developing tools such as VHDL.

The resource utilization of the suggested system uses approximately 1% of the targeted FPGA chip. It is clear that the proposed system can be extended for larger size images and also can be modified for hiding color-in color images.

References

1. Emad, E., Safey, A., Refaat, A., Osama, Z., Sayed, E., Mohamed, E.: A secure image steganography algorithm based on least significant bit and integer wavelet transform. J. Syst. Eng. Electron. **29**(3), 639–649 (2018)
2. Al-Korbi, H.A., Al-Ataby, A., Al-Taee, M.A., Al-Nuaimy, W.: High-capacity: image steganography based on Haar DWT for hiding miscellaneous data. In: 2015 IEEE Jordan Conference on Applied Electrical Engineering and Computing Technologies (AEECT), pp. 1–6 (2015)
3. Safey, A., Zahran, O., Kordy, M.: FPGA implementation of robust image steganography technique based on Least Significant Bit (LSB) in spatial domain. Int. J. Comput. Appl. **145**, 43–52 (2016)
4. Shet, K.S., Aswath, A.R., Hanumantharaju, M.C., Gao, X.-Z.: Novel high-speed reconfigurable FPGA architectures for EMD-based image steganography. Multimed. Tools Appl. **78**(13), 18309–18338 (2019)
5. Abed, S., Almutairi, M., Alwatyan, A., Almutairi, O., AlEnizy, W., Al-Noori, A.: An automated security approach of video steganography based LSB using FPGA implementation. J. Circuits Syst. Comput. **28**, 1950083 (27 pages) (2019)
6. Hussain, M., Wahid, A., Idris, M., Ho, A., Jung, K.-H.: Image steganography in spatial domain: a survey. Signal Process. Image Commun. **65**, 46–66 (2018)
7. Kothari, L., Thakkar, R., Khara, S.: Data hiding on web using combination of Steganography and Cryptography. In: 2017 International Conference on Computer, Communications and Electronics (Comptelix), pp. 448–452 (2017)
8. Yahaya, M., Ajibola, A.: Cryptosystem for secure data transmission using Advance Encryption Standard (AES) and Steganography. Int. J. Sci. Res. Comput. Sci. Eng. Inf. Technol. **5**, 317–322 (2019)
9. Devi, A.G., Thota, A., Nithya, G., Majji, S., Gopatoti, A., Dhavamani, L.: Advancement of digital image steganography using deep convolutional neural networks. In: 2022 International Interdisciplinary Humanitarian Conference for Sustainability (IIHC), pp. 250–254 (2022)
10. Wahed, M.A., Nyeem, H.: Efficient LSB substitution for interpolation based reversible data hiding scheme. In: 2017 20th International Conference of Computer and Information Technology (ICCIT), pp. 1–6, August 2017

11. Rustad, S., Setiadi, D.R.I.M., Syukur, A., Andono, P.N.: Inverted LSB image steganography using adaptive pattern to improve imperceptibility. J. King Saud Univ. - Comput. Inf. Sci. **34**(6, Part B), 3559–3568 (2022). https://doi.org/10.1016/j.jksuci.2020.12.017
12. Andono, P.N., Setiadi, D.R.I.M.: Quantization selection based on characteristic of cover image for PVD Steganography to optimize imperceptibility and capacity. Multimed. Tools Appl. **82**(3), 3561–3580 (2023)
13. Sun, H., Luo, H., Wu, T.-Y., Obaidat, M.: A PSNR-controllable data hiding algorithm based on LSBs substitution, pp. 1–7 (2015)
14. Ji, Y.: FPSA: a full system stack solution for reconfigurable ReRAM-based NN accelerator architecture (2019)
15. Shawahna, A., Sait, S.M., El-Maleh, A.: FPGA-based accelerators of deep learning networks for learning and classification: a review. IEEE Access **7**, 7823–7859 (2019)
16. Al-Ibadi, M.A.: Hardware implementation for high-speed parallel adder for QSD 2D data arrays. In: 2020 International Conference on Electrical, Communication, and Computer Engineering (ICECCE), pp. 1–5 (2020)
17. Meyer, M., Kenter, T., Plessl, C.: In-depth FPGA accelerator performance evaluation with single node benchmarks from the HPC challenge benchmark suite for Intel and Xilinx FPGAs using OpenCL. J. Parallel Distrib. Comput. **160**, 79–89 (2022). https://doi.org/10.1016/j.jpdc.2021.10.007
18. Li, L., Luo, B., Li, Q., Fang, X.: A color images steganography method by multiple embedding strategy based on Sobel operator. In: 2009 International Conference on Multimedia Information Networking and Security, pp. 118–121.187 (2009)
19. Mishra, M., Routray, A.R., Kumar, S.: High security image steganography with modified Arnold cat map. arXiv, abs/1408.3838 (2012)
20. Rao, Y.: Application of normalized cross correlation to image registration. Int. J. Res. Eng. Technol. **03**, 12–16 (2014)

Research on Data Acquisition and Real-Time Communication for Intelligent Manufacturing Training Equipment Based on Model of Things and Intranet Penetration

Nengwen Wan[1], Wenjun Zhang[1], Tan Li[1](✉), Meng Chen[1], Weining Song[2], Honglin Chen[1], Jingyu Zhu[1], Huifeng Cao[1], Nanjiang Chen[1], Qi Wang[3], Yanwen Lin[3], and Runqiang Li[3]

[1] Nanchang University, Nanchang Jiangxi, China
someone8584@sina.com
[2] East China University of Technology, Nanchang, Jiangxi, China
[3] QuickTech Co., Ltd., Haidian, Beijing, China

Abstract. Intelligent Manufacturing Training Equipment (IMTE) are featured with high-value shareable training property and high security requirement on human-machine interaction, which prompt urgent demands on equipment sharing and virtual training of IMTEs based on Digital Twin (DT). Enabling the effective data acquisition and real-time interaction for the DT system construction for IMTEs, Data Acquisition Interface Definition and Cross-Network Real-Time Communication technologies are researched, including the universal IMTE data acquisition interface definition method based on Model of Things (MoT) and the real-time cross-network communication technology based on Intranet Penetration. A prototype IMTE DT system is built for the effectiveness verification on interface definition, acquisition and transferring of IMTE data, where the MoT of IMTE facilitates the PLC data collection and Intranet Penetration accelerates the data transfer, thus drive the real-time interaction between IMTE and the DT system, realizing the simplification of data acquisition and real-time cross-network communication for IMTEs.

Keywords: Intelligent Manufacturing · Training equipment · IOT · Intranet Penetration · Digital Twin · Model of Things

1 Introduction

Currently, intelligent manufacturing becomes commanding height for the global industry competition. USA, Germany, Japan and other industrial powers have put forward their own national development strategies for intelligent manufacturing, as well as "Made in China 2025" for China to optimize and upgrade industrial structure and achieve high-quality economic development [1]. With the gradual advancement of the development strategy, intelligent manufacturing has promoted profound changes in the national economic structure, including the optimization of industrial structure and human resource

© The Author(s), under exclusive license to Springer Nature Singapore Pte Ltd. 2024
F. Hassan et al. (Eds.): AsiaSim 2023, CCIS 1911, pp. 13–27, 2024.
https://doi.org/10.1007/978-981-99-7240-1_2

structure, which requires the adjustment and innovation of the talent training system, and trained more professionals with skills about intelligent manufacturing.

The intelligent manufacturing strategies require a large number of professionals with manufacturing expertise, however, there is currently an imbalance between knowledge level and practical ability in traditional education. Especially in the field of intelligent manufacturing, many research and problems are new, and often come from the practice of the industry. Intelligent manufacturing training is a practical training that simulates the real working environment and relies on IMTE to train practitioners to master intelligent manufacturing technology and work flow, which can play an effective role in solving this dilemma. As the key support for intelligent manufacturing training, IMTE is different from c-conventional industrial equipment and experimental equipment. Table 1 [2] compares and analyzes the characteristics of the three types of equipment.

Table 1. Equipment comparison (★refers to the relative level among the three types)

		experimental instruments and equipment	**industrial site equipment**	**IMTE**
precision		★★★	★★	★
		Highest precision to complete science experiments	Lower precision but assure product quality	Lowest precision, just meet the training needs
stability		★★★	★★★	★
		High stability over time to ensure the precision	High stability to ensure consistent production quality	General functions, providing repeatable training course
efficiency	**high efficiency**	★	★★★	★
		Precision is much more important than efficiency	Rapidest operation response while production	Smooth operation and easy learning for the trainee
	Usage efficiency	★	★★★	★
		5*8, low utilization ratio based on the research task	Normally uninterrupted work through 7*24	5*8, even less on vacation
security	**machine security**	★	★★	★★★
		Complex operation, experience required for security	Specific equipment security protection mechanism	Full protection mechanism to avoid misoperation damage
	operator security	★★	★★★	★★★
		Laboratories may work at high risk in certain tasks	Operators may work at risk in certain environment	Trainee Security more important than equipment

Comparison of the figure above can be summarized as high security and low utilization rate of IMTE, and the following is an elaboration of the characteristics and required functions of IMTE:

High Security

Since the equipment is intended for training users who have little or no experience in operating the equipment, IMTE for equipment safety protection and personal safety requires high. Therefore, students can be completely in the simulation environment to familiarize themselves with the equipment, to be improved by the students with the operating level, operating experience and then the actual operation, so the need for IMTE to achieve the establishment of the object model and holographic uploading of equipment data to support the virtual simulation.

Low Utilization Rate

Within the university, IMTE is generally used during the students' class period, i.e. 5 * 8 working hours arrangement, and the rest of the time is completely idle except for the working day time. Moreover, due to the lack of sharing awareness, although the equipment belongs to the fixed assets of the school, the rights of use and maintenance are basically in the hands of each subject group. In order to avoid the trouble of responsibility sharing and cost management that exists in the sharing process, even schools are not willing to share equipment internally [3]. At the same time, many small and medium-sized enterprises also need a large number of practical training equipment to train their employees, but the price of IMTE is not cheap and maintenance for small and medium-sized enterprises is a big burden.

Therefore, three demands are derived from the status quo and characteristics of IMTE: virtual simulation, effective management and utilization of equipment in schools, and a large number of industrial enterprises have demands for efficient IMTE. For these needs, online sharing of equipment supported by digital twin technology can provide good solutions. However, at the same time, it also puts forward a higher demand for efficient data acquisition, access and remote real-time communication of equipment, especially IMTE.

In the application of industrial equipment data acquisition, Li Congbo et al. used power sensors based on Modbus protocol to collect energy consumption data, collected machining parameters based on TCP/IP protocol, and interacted with MES to obtain machining tasks through intelligent terminals, realizing digital twin based multi-level energy efficiency monitoring of machining workshop. The digital twin-based multilevel energy efficiency monitoring of the machine shop is realized [4]. Cao Wei et al. proposed an RFID-based real-time data collection method for discrete manufacturing workshops, which realizes RFID configuration and data acquisition around WIP processes, process flows, batches and lots, and then realizes multi-level visualization and monitoring of the manufacturing process [5]. Jiao Yuyang et al. proposed an OPC UA unified architecture for forging workshop energy consumption data collection and supervision system, for the traditional forging workshop equipment is scattered between the plant area is too large, built for the forging workshop energy consumption data collection and supervision system overall network architecture, to achieve the data collection, storage and interaction [6].

In terms of remote real-time communication, Tong Wang et al. introduce a mosaic antenna consisting of spatially distributed small transceiver units to realize long-distance communication in order to replace high-power amplifiers and large directional antennas to form a cross-domain communication network that dynamically adapts to losses and sudden threats in confrontational environments in order to achieve battlefield-stabilized long-distance communication [7]. Xu Feilong et al. use RD120-6W communication board, GT6805 industrial control board, ULN2003 driver chip and on-board dip switches to complete the construction of the communication machine, and realize the connection with the communication machine through the Ethernet interface RS485 to provide communication support for the remote teaching management system [8]. Yang Yalian et al. used PIC18F4580 with built-in CAN controller module and industrial-grade DTU-MD610 in the vehicle terminal, and transmitted the data and GPS information to the remote control terminal through GPRS (general packet radio service) and Internet network to support the remote communication of real-time monitoring and control system of the hybrid vehicle [9].

In summary, many scholars have carried out a lot of research and practice around the issue of industrial efficient data acquisition and remote real-time communication. But in the actual research project of intellectual manufacturing training, it is found that the lack of data acquisition interface definition of IMTE will be inconvenient for the data access from the physical end to the simulation end, and IMTE is often faced with the problem of cross-network communication in the remote communication. However, there is relatively little research and practice on the efficient data acquisition with interface definition and cross-network real-time communication. In view of this, this paper describes the training device based on the object model, i.e., the interface definition of the training device, followed by the real-time data collection of the training device through the data acquisition tool. Finally, the data is transmitted remotely through the intranet penetration and the data interface definition is conveniently accessed to the data, in order to drive the real-time collaborative digital twin of the Unity 3D simulation model, and the feasibility is verified through an actual example.

2 Methods for Data Acquisition and Interface Definition of Training Equipment Based on Model of Things

2.1 Overview of Model of Things

Internet of Things (IoT) is built based on the Internet environment and combines various technologies such as information sensors, RFID technology to real-time collect information from objects or processes. The information is accessed through various network protocols, enabling ubiquitous connections between objects and people and serving as a carrier for interconnected information. With the implementation of the Made in China 2025 strategy, the manufacturing industry has an urgent need for digital transformation and cost reduction [10]. In this context, industrial IoT has flourished, with its industrial applications and use cases continuously expanding, but it has also brought forth challenges. In industrial applications, there is often a massive amount of heterogeneous data that needs to be integrated into the IoT. How to efficiently parse and assign semantic information to this data to achieve unified management of diverse devices becomes a constraint on the further development of industrial IoT.

In industrial IoT scenarios, the massive heterogeneity of physical entities results in a vast amount of heterogeneous data. To connect these entities to an IoT platform, it is necessary to digitally model the physical entities. Without unified modeling specifications and languages, this process becomes extremely challenging. Different devices and data formats make data parsing, processing, and storage difficult, greatly increasing the difficulty of entity integration into the network [11]. To address this issue, the concept of a "thing model" emerges.

A thing model is a digital representation model of real-world objects in a virtual space, used in IoT platforms to abstract and generalize physical objects. In simple terms, a thing model provides a standardized interface for connecting real-world objects to the IoT. It defines the characteristics and behaviors of objects and describes and represents the attributes, methods, events, etc., of the connected objects. Through Model of Things, an IoT platform can better manage, interact with, and control the connected IoT devices. From a business perspective based on IoT platforms, the existence of Model of Things greatly facilitates device management, data exchange, remote monitoring and control, data analysis, and application development. Furthermore, Model of Things promote interoperability among IoT devices, enabling the standardization and normalization of connected entities and simplifying the utilization of data. They establish a one-to-one correspondence between physical entities and virtual models, bridging communication barriers between different vendors and device types, enabling device coordination, and storing data from connected entities in a unified format in the cloud. This creates standardized service libraries and databases, streamlines data parsing, and greatly simplifies the development process.

2.2 MoT Framework of IMTE

The Thing Model of the intelligent manufacturing training platform serves as an integrated system comprising various components, such as robots and their controllers, diverse sensors, a range of actuators and controllers, and a human-machine interaction panel. As a result, the Thing Model of the training platform needs to include individual sub-model units representing each component. Following this hierarchical architecture, a designated Thing Model, illustrated in Fig. 1, can effectively describe and represent IMTE that requires seamless integration into IoT.

Due to the diverse objectives of different training tasks, there is a wide variety of training platforms available. Even for similar training objectives, different companies may design their training platforms with significant differences. Therefore, when constructing an IoT platform based on training devices, it is essential to have a suitable Thing Model to integrate various training platforms into the IoT, enabling remote monitoring, remote control, real-time data analysis, and more. In response to these requirements, this Thing Model divides the training platform into five major components: robots, actuators, sensors, controllers, and supporting structures. Each component can have multiple similar objects. These five components cover the primary subsystems utilized in training platforms, making this Thing Model capable of describing the majority of intelligent manufacturing training platforms comprehensively and accurately.

For the Thing Model, the geometric properties are mostly described using engineering models or engineering design drawings, while the basic information and work information are extracted from detailed specifications provided by equipment manufacturers. Operational data, on the other hand, needs to be collected through data acquisition methods. The static information of the training equipment, such as basic information, work information, and geometric properties, will be described in the format of static properties as shown in Fig. 1.

The various types of data generated by the operation of each component of the training equipment will be recorded in the format shown in Fig. 1 and stored in the running data properties of the training platform in JSON format. The various types of data generated by the operation of each component of the training equipment will be recorded in the format shown in Fig. 1 and stored in the running data properties of the training platform in JSON format.

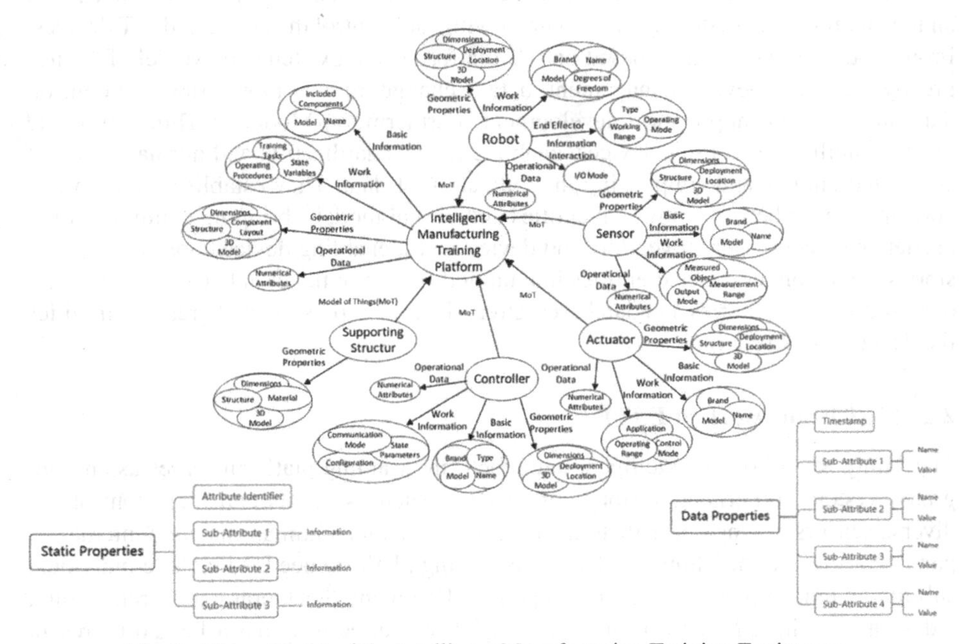

Fig. 1. Thing Model of the Intelligent Manufacturing Training Equipment

3 Real-Time Cross-Network Communication Based on Intranet Penetration

3.1 Overview of Frp Intranet Penetration

When communicating remotely, the restriction from the intranet to the extranet will hinder the communication, and the intranet penetration is a technology that can provide the local server's intranet IP port to the extranet to realize the extranet connection access. Intranet penetration is widely used in various scenarios, such as remote access, server management, remote monitoring of IoT devices, etc. Frp, peanut shells, ngrok, natapp, etc. are common intranet penetration tools. Frp has the advantages of supporting multiple protocols, simple configuration, TLS encryption and authentication, etc. compared to other intranet penetration tools.

Frp is based on Go language development, support for multi-platform deployment of a high-performance fast reverse proxy for intranet penetration, can be behind the firewall or NAT local services in a safe and convenient way through the public IP nodes with the transit exposed to the Internet. Frp intranet penetration can be understood as using a server with a publicly accessible IP address as an intermediary to establish a “connection” between the connected computer or other terminals and the computer currently in operation. In other words, the frp program connects two computers located in two different LANs through a server with a publicly accessible IP address. Therefore, the method of frp intranet penetration to achieve remote communication is to map the intranet service to the public network, thus enabling external network access.

Overall, frp intranet penetration is a convenient and practical tool to solve the problem that intranet environments can not be directly accessed, so that public network users can safely and conveniently access intranet services.

3.2 Cross-Network Communication Solution for IMTE

Aiming at the demand for remote real-time communication of IMTE, based on the frp intranet penetration technology to support the cross-network real-time communication of the intellectual manufacturing training equipment, as shown in the technical structure diagram in Fig. 2. Frp intranet penetration uses the client-server architecture. In the intranet, frp client frpc establishes a connection with frp server frps and sends requests for intranet services to the server. The server side is responsible for receiving the request and forwarding it to the service port on the intranet.

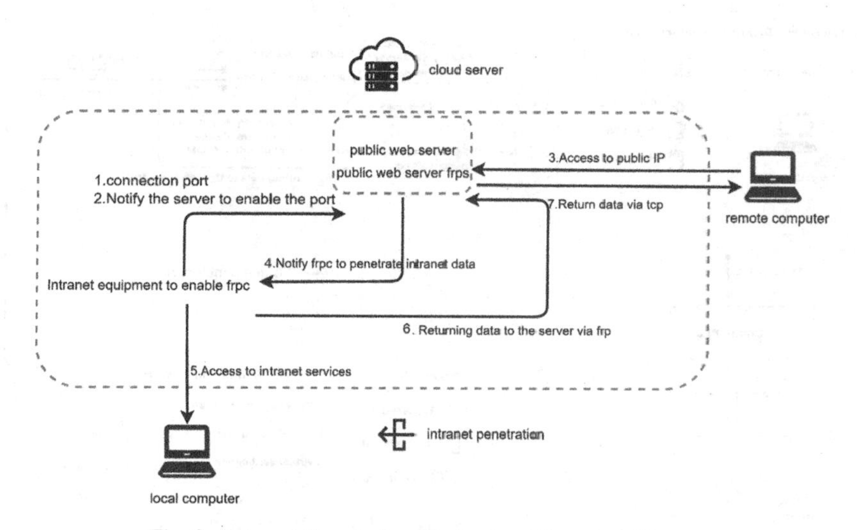

Fig. 2. Intranet Penetration Technology Structure Diagram

The specific process of realizing intranet penetration is as follows: first, the frpc client sends information such as the extranet address and port, as well as the address and port of the intranet service that needs to be mapped, to the frps server. At the same time, the server assigns a unique token to the client for verifying the client’s identity and for subsequent communication. Next, the client initiates a tunnel request to the server, in which the client specifies the address and port number of the intranet service that needs to be mapped to the public network. After the server receives the tunnel request, it assigns the extranet address and port number and sends that address and port number to the client. The client receives it and informs the service in the intranet about the address and port number, so that the external network can access the service in the intranet through this address and port number. When the external network accesses the external address and port number assigned by the server, the request is forwarded to the client, which in turn forwards the request to the service in the intranet. Finally, when the service in the intranet has a response, the client sends the response to the server, which in turn forwards the response to the external network. By the above way, a communication tunnel is established between the public network and the intranet, mapping the intranet services to the public network.

4 A Real-Time Communication System Designed for Data Acquisition, Interface Definition and Cross-Network

4.1 The Overall System Framework

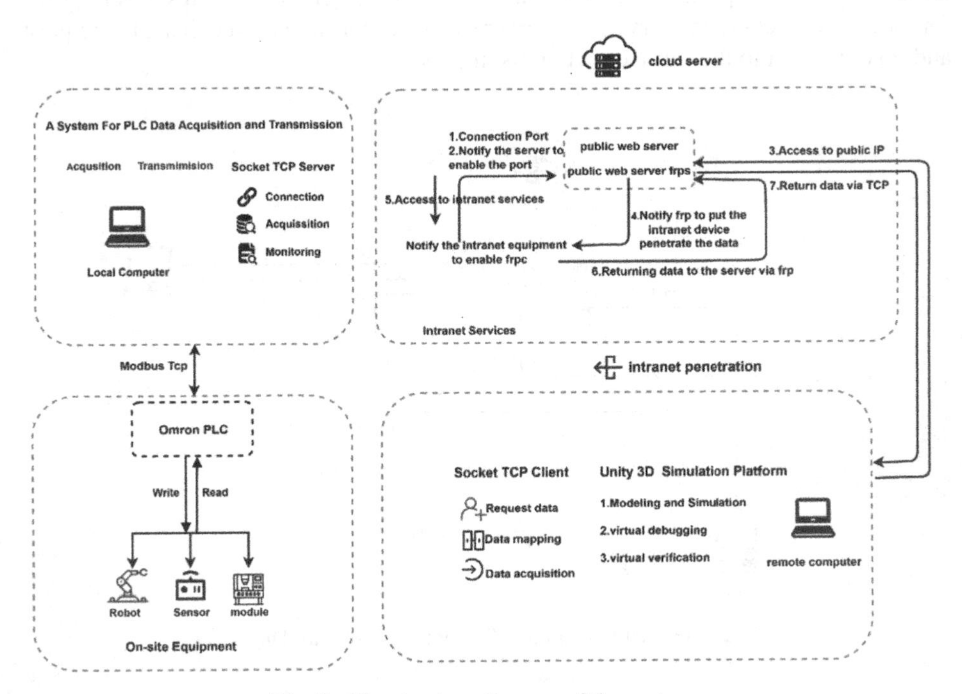

Fig. 3. The structure diagram of the system

As shown in Fig. 3 for the data acquisition, interface definition and cross-network real-time communication system structure. The data of IMTE mainly comes from the robot, actuators, sensors, and controllers, in which the actuators and contact type sensors, non-contact ordinary type sensors and so on are summarized as the training module. Firstly, the PLC reads the data of IMTE, then the TCP-based Socket communication module connects and collects the PLC data. Then the data is converted into Json format and sent to the Unity 3D simulation platform that integrates the Socket client, and in the process of designing, it is found that the data access is cumbersome and inefficient, as well as the remote cross-network communication due to the limitations of the intranet to the extranet. Thus ultimately, it is designed an object model-based data acquisition and interface definition method for training equipment and a real-time cross-network communication system based on intranet penetration.

4.2 PLC Reads Data from Training Equipment

Figure 3 contains the PLC communication module, which mainly demonstrates that the PLC communicates with the other parts of the IMTE through three communication methods, namely, communicating with the robot through Modbus TCP, collecting digital sensors data based on Modbus RTU through the RS485 serial port, and monitoring and controlling the training module through the entity I/O.

Socket is an abstraction layer that provides an interface between an application program and network communications Socket combines the Socket programming interface and the TCP protocol to provide a convenient way to develop communication applications. It allows applications to create sockets to establish a connection and transfer data using the TCP protocol. The general communication flow between a socket server and a client is shown in Fig. 4, which closes the socket and moves to the listening and waiting for client connection state when communication is complete, the other party initiates a close request, or there is a network failure, etc.

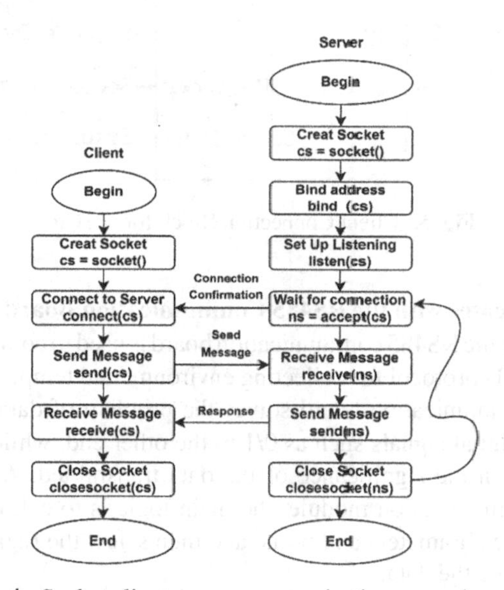

Fig. 4. Socket client-server communication general process

PLC and Robot Communication. As shown in Fig. 5 is the connection program block of Modbus TCP communication library of Omron NX/NJ series PLC user-defined package. The communication library can be used for PLC control of other controllers or for data transfer, in this case used to connect the robot controller, read the robot axes, gripper and other information and can automatically or manually control the robot action, the connection is made in accordance with the steps of the socket connection in Table 2.

Table 2. The PLC communicates with the robot based on TCP Socket

Program structure	Framework code
Circular judgment	Cycle:=Timer…// Loop to determine if it is needed

Socket	IF NOT Busy AND Cycle …// Conditional judgment is true
	TCP_Status_Init…// Initialization state
	IF Connect AND NOT Connected// Wait for connection
	TCP_Connect_Init…// Apply for a handshake
	IF (TCP_Connect_Init.Done)…// Whether to shake hands three times
	ELSEIF (TCP_Connect_Init.Error)…// More than three times
	IF (Connected AND NOT Connect)…// Connected
	TCP_Close_Init…// End

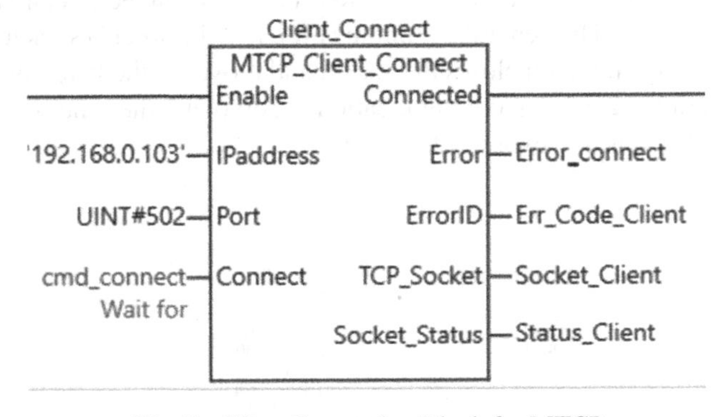

Fig. 5. Client Connection Block for MTCP

The PLC Communicates with the RS485 Communication Board Based on Modbus RTU Protocol. Here the RS485 communication board is used to communicate with PLC based on Modbus RTU protocol for collecting environmental temperature and humidity data. The RS485 communication board solves the problem of data transmission, that is, how to transmit digital signals such as 0/1 to the other end, while the Modbus RTU protocol is to agree on the significance of the data transmitted. As Fig. 6 shows the RS485 and PLC communication module, the main logic is to call the communication board to configure the parameters and ports, and then select the registers as well as the address to read or write the data.

The PLC Monitors the Training Module Via I/O Mapping. PLC's I/O Mapping refers to mapping the input and output signals of external devices to the PLC's input

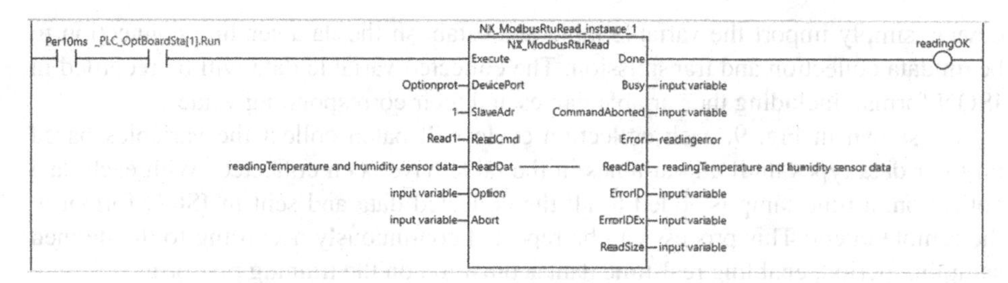

Fig. 6. RS485 and PLC communication program module

and output modules for monitoring and controlling the external devices. I/O Mapping defines the connection relationship between the PLC's input and output points and the external devices. In a PLC system, input modules are used to receive signal inputs from external devices, such as the status of sensors, push button presses, and so on. The output modules are used to control external devices, such as driving actuators, relays, etc. This system uses an Omron series PLC. When programming the PLC, first specify the external device and signal type corresponding to each input/output point, such as Bool type, and then map it to the appropriate input/output module. By correctly configuring the I/O mapping, the PLC can monitor the status of external devices in real time and control the output signals as needed.

4.3 Data Acquisition and Communication Platform Based on Modbus TCP and Socket

This data acquisition and communication platform, based on Modbus TCP communication protocol, is designed as a Socket client to connect and access PLC using Modbus TCP. It reads data from holding registers and converts the data into JSON format. Finally, the data is transmitted to the Unity 3D client over the Internet of Things interface through Socket server using local network tunneling.

First, the Modbus TCP protocol is used as the communication protocol between the data acquisition platform and the controller (such as PLC, robot control cabinet, computer, etc. In this case, it is PLC). Then, based on the data access methods described earlier, the data acquisition process is simplified. The entire data acquisition workflow is shown in Fig. 7.

In this workflow, the data address mapping table is imported into the data acquisition tool. The data acquisition tool then automatically configures the settings based on the imported table, allowing the data acquisition process to begin. As shown in Fig. 8, the specific implementation involves creating a table for the variables to be collected, specifying the variable names, data types, and communication protocol addresses. The controller determines the internal variables to be collected and maps the variable data to the communication protocol addresses using the relevant Modbus TCP communication program. With these preparations, the data acquisition process can be initiated on the training platform.

To facilitate the data acquisition process, a developed data acquisition tool can be used for real-time batch collection of the required variables. After connecting to the device, simply import the variable table and establish the data sending connection to begin data collection and transmission. The collected variable data will be recorded in JSON format, including the variable names and their corresponding values.

As shown in Fig. 9, each collection cycle will batch collect the variables based on their data types until all variables in the table have been collected. With each data collection, a timestamp is added to all the collected data and sent in JSON format to the remote client. This process can be repeated continuously according to the defined sampling period, enabling real-time data acquisition on the training platform.

Based on the above logic and principles, the data acquisition and communication platform, as shown in Fig. 10, can be constructed. By combining this platform with local network tunneling and data access methods, data acquisition and remote cross-network transmission can be achieved.

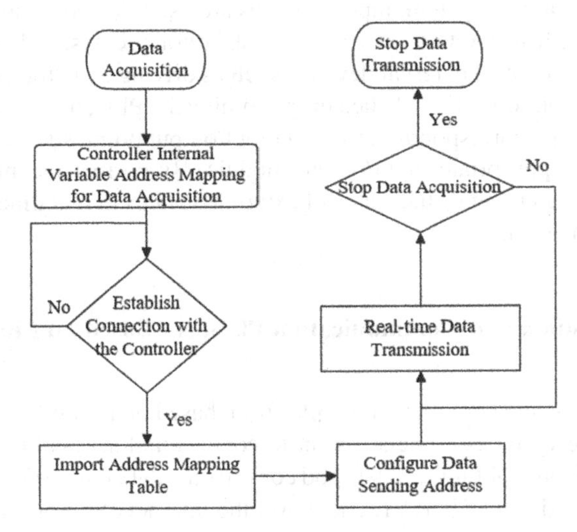

Fig. 7. Data Acquisition Workflow

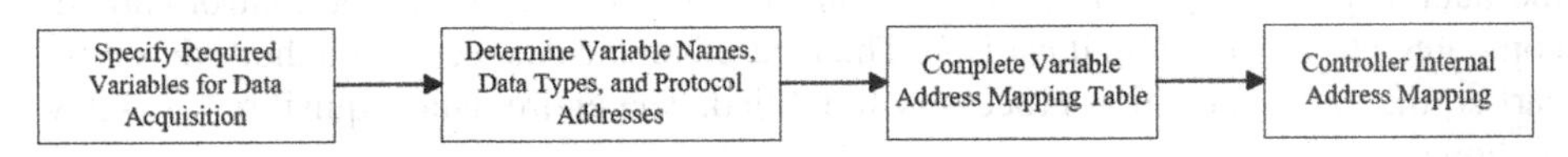

Fig. 8. Controller Preparation in Data Acquisition

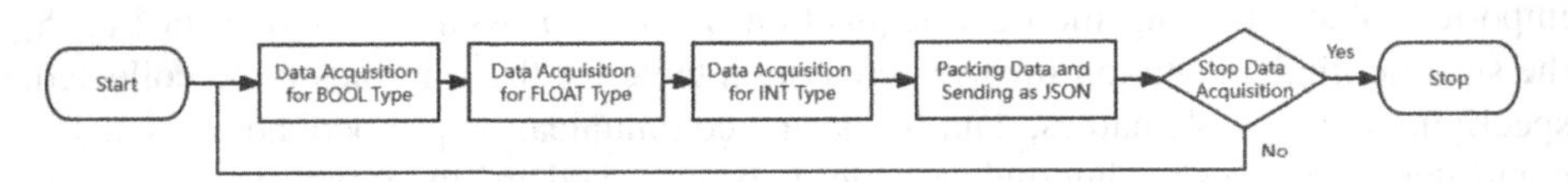

Fig. 9. Variable Data Acquisition Process

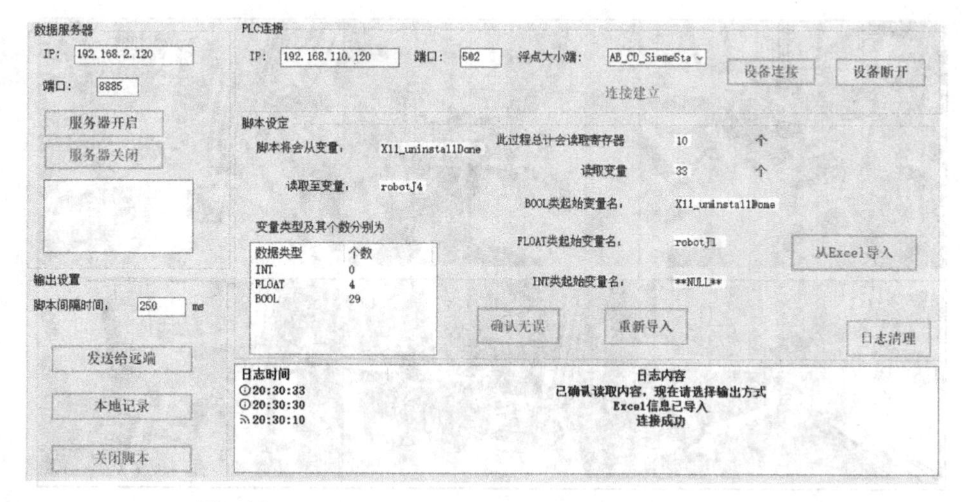

Fig. 10. The data acquisition and communication platform

4.4 Implementation and Validation of the System

After the data collection is completed, it is remotely transmitted to Unity 3D via intranet penetration to support its remote real-time simulation, as shown in Fig. 11, which shows the top half of the figure is the running image of IMTE in Jiangxi Province, and the bottom half of the figure is the image of the digital twin in Hubei Province. The IMTE modeled after the industrial assembly of lithium battery production line, the left image is 1 m 24 s AUBO series of six-axis robotic arm clamping and placement of lithium battery cores, the middle image is 5 m 16 s robotic arm assembly of lithium batteries, the right image is 10 m 4 s robotic arm disassembly of lithium batteries. Table 3 is a data table comparing the data of each axis of the actual robotic arm, PLC reads data and real-time simulation robotic arm data, in which the data of each axis of the actual robotic arm is monitored by the AOBO supporting host computer.

From the comparison of the above figure and table, it can be verified that the accuracy of PLC's data acquisition of the robotic arm as well as the digital twin robotic arm basically meets the requirements in real-time, and the delay time is less than 1 s in the case of the basic stability of the network in the testing process, the specific method is to read the PLC data at the same time the host computer reads the time and sends it to Unity when the network is stable, Unity receives and saves the data when the receiving time is displayed, and then randomly take 100 groups to get the difference and then take the average. Thus it is considered that the systems have effectively solved the problem of the tedious and inefficient access to data and the problem of remote cross-network communications due to the restriction of the intranet to the extranet.

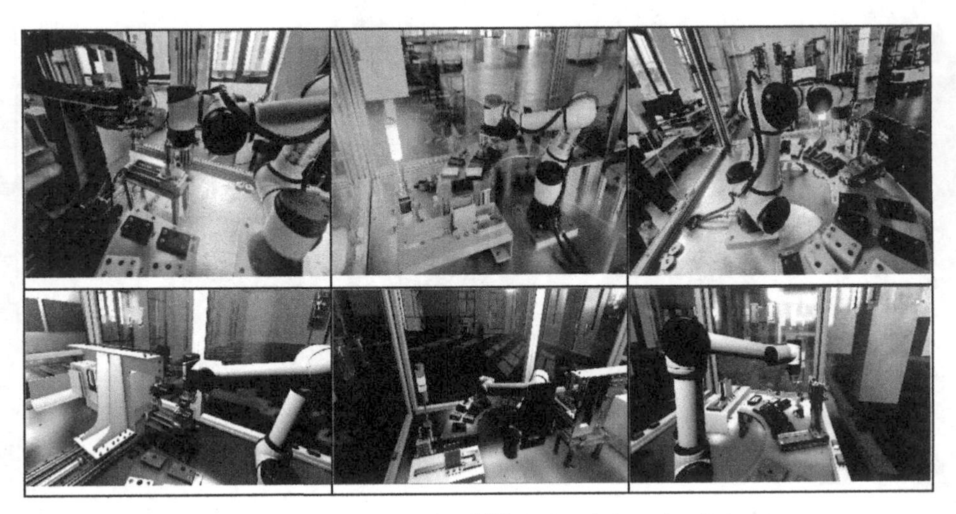

Fig. 11. Comparison of real life and real time simulation

Table 3. Data comparison

Axle	Physical data/deg			PLC data/deg			Simulation data/deg		
J1	−162.895	−40.630	−49.365	−162.768	−40.754	−49.476	−162.678	−40.9876	−49.2679
J2	11.095	17.162	−16.394	11.258	17.265	−16.438	11.3765	17.8977	−16.7626
J3	109.332	120.073	82.980	109.319	119.658	82.916	109.8764	89.9754	82.0862
J4	7.401	15.112	10.605	7.298	15.168	10.596	7.5647	5.7875	10.8972
J5	90.978	86.561	90.372	90.879	86.489	90.597	84.6598	82.8765	80.9756
J6	−12.502	−16.806	−31.180	−12.575	−16.769	−31.685	−8.9866	−16.9761	−31.9865

5 Conclusion

Enabling the effective data acquisition and real-time interaction for the DT system construction for IMTEs, Data Acquisition Interface Definition and Cross-Network Real-Time Communication technologies are researched, including the universal IMTE data acquisition interface definition method based on MoT and the real-time cross-network communication technology based on Intranet Penetration. A prototype IMTE DT system is built for the effectiveness verification on interface definition, acquisition and transferring of IMTE data, where the MoT of IMTE facilitates the PLC data collection and Intranet Penetration accelerates the data transfer, thus drive the real-time interaction between IMTE and the DT system, realizing the simplification of data acquisition and real-time cross-network communication for IMTEs. Real-time cross-network communication and simplified of data acquisition and access is realized, in which the delay time is less than 1 s in the case of basic network stability during the test. This system can be universally applied to IMTEs, providing support for intelligent manufacturing practical training and intelligent manufacturing. The next step of the research will further optimize the compatibility and universality of the data interface definition, so that it can realize the convenient data collection and access of different brands of PLC equipment, in addition to optimizing the intranet penetration of multiple platforms, thus improving the utilization rate of IMTEs.

References

1. Qin, J., Lu, X., Si, H., et al.: New features of welding materials based on intelligent manufacturing. Mater. Guide **37**(11), 128–134 (2023)
2. Luo, J.X., Lan, T.H., Li, T., et al.: Survey on sharing technology and applications of intelligent manufacturing training equipment based on industrial internet and man-in-loop simulation. In: Fan, W., Zhang, L., Li, N., Song, X. (eds) Methods and Applications for Modeling and Simulation of Complex Systems. AsiaSim 2022. CCIS, vol. 1713. Springer, Singapore (2022). https://doi.org/10.1007/978-981-19-9195-0_48
3. Zhu, X., Zhang, G.: Literature review on management of research instruments and equipment in universities. Lab. Res. Explor. **38**(11), 274–277 (2019)
4. Li, C., Cao, B., Wu, Y., et al.: Multi-level energy efficiency monitoring in machine shops based on digital twin. Comput. Integr. Syst.1–24 (2023). http://kns.cnki.net/kcms/detail/11.5946.TP.20230607.0958.002.html
5. Cao, W., Jiang, P., Jiang, K., et al.: Real-time data collection and visualization monitoring method for discrete manufacturing workshop based on RFID technology. Comput. Integr. Manuf. Syst. **23**(02), 273–284 (2017). https://doi.org/10.13196/j.cims.2017.02.006
6. Jiao, Y., Li, L., Nie, H., et al.: Energy consumption data collection and supervision system of forging workshop using OPC unified architecture. China Mech. Eng. **32**(20), 2492–2500 (2021)
7. Wang, T., Li, P., Jiang, Q., et al.: Flexible networking distributed mosaic communication project realizes new mode of remote communication. Tactical Missile Technol. **01**, 107–114 (2021). https://doi.org/10.16358/j.issn.1009-1300.2021.1.507
8. Xu, F., Lu, X.: Research on JSP technology based on the design of telecommunication teaching management system. Mod. Electron. Technol. **43**(17), 130–133 (2020). https://doi.org/10.16652/j.issn.1004-373x.2020.17.030
9. Yang, Y., Zheng, Y., Song, A., et al.: Development of real-time remote monitoring system for hybrid vehicles. J. Chongqing Univ. **35**(06), 1–8 (2012)
10. He, J.: Design and realization of IoT data platform based on object model and rule engine. Metall. Autom. Res. Des. Inst. (2022). https://doi.org/10.27448/d.cnki.gyzyy.2022.000001
11. Sha, M., Liu, C., Chen, L.: Design and realization of intelligent facility management platform based on object model. Electron. Technol. Softw. Eng. **2021**(02), 170–171 (2021)
12. Yu, G.: Internet of Things communication experiment based on Frp intranet penetration and Flask framework. Inf. Technol. Informatization **2021**(08), 197–199 (2021)

Research on Intelligent Manufacturing Training System Based on Industrial Metaverse

Yuqi Zhou[1], Tan Li[1](✉), Bohu Li[2,3], Gang Wu[1], Xianghui Meng[1], Jin Guo[1], Nengwen Wan[1], Jingyu Zhu[1], Shimei Li[1], Weining Song[4], Chunhui Su[2], Nanjiang Chen[1], Yalan Xing[3], Qi Wang[5], Yanwen Lin[5], and Runqiang Li[5]

[1] Nanchang University, Nanchang, Jiangxi, China
someone8584@sina.com
[2] China Aerospace Science and Industry Co., Ltd, Haidian, Beijing, China
[3] Beihang University, Haidian, Beijing, China
[4] East China University of Technology, Nanchang, Jiangxi, China
[5] QuickTech Co. Ltd., Haidian, Beijing, China

Abstract. Skilled human resource becomes an essential resource for implementing intelligent manufacturing in the new era, prompting high demands on Intelligent Manufacturing Training (IMT). Empowering the effective IMT, the new mode of IMT based on Industrial Metaverse is proposed as well as detailed comparison with traditional training modes. The layered technical architecture is discussed as a guidance for training system construction, as well as specific solutions for the six key technologies based on primary research, including rapid modeling, natural interaction, real-time communication, industrial avatar/agent, industrial tools access, industrial AIGC, etc. Verifying the effectiveness of Industrial Metaverse based IMT, a prototype system "TrAiN" for industrial internet skill training is built, constructing a private Industrial Metaverse based on specific industrial equipment and fields in certain factory, facilitating the virtual training. Future research hotspots on Industrial Metaverse based IMT are prospected at the end based on the primary research and application.

Keywords: Industrial Metaverse · Training · Intelligent manufacturing

1 Introduction

According to the report "The Future of Jobs in the Post-Pandemic Era" released by the MGI, the COVID-19 pandemic will accelerate the reshaping of the employment structure of major global economies, including China, over the next decade [1]. Currently, digital transformation and technological progress have put forward higher requirements for the skills quality of the manufacturing talent team. The phenomenon of training not meeting the demands of economic and social development is intensifying. How to adjust the content of education and skills training and promote the transformation of the skills training system to adapt to the demands of the industrial economy era on workers, has become an issue urgently considered by all countries.

© The Author(s), under exclusive license to Springer Nature Singapore Pte Ltd. 2024
F. Hassan et al. (Eds.): AsiaSim 2023, CCIS 1911, pp. 28–43, 2024.
https://doi.org/10.1007/978-981-99-7240-1_3

This chapter starts with the analysis of the development status at home and abroad to reveal that the "Metaverse and Training" is a new paradigm of the current training model, and then by comparing training modes 1.0, 2.0 and 3.0 to illustrate that "Metaverse and Training" is a new trend of the future training mode. Making the basis for the proposed Intelligent Manufacturing Training System based on Industrial Metaverse.

1.1 Global Development on IMT

Intelligent Manufacturing Training (IMT) is a practical training and education method based on intelligent manufacturing technology. The aim is to simulate the real industrial production environment, train students/workers to master knowledge and skills related to intelligent manufacturing and enable them to solve practical engineering problems. At present, the Industrial Metaverse, as a new type of industrial digital space, a novel industrial intelligence interconnected system, and a new carrier for the integrated development of the digital economy and the real economy [2], is receiving wide-spread attention from the scientific and technological community in the industrial field. The question of how to combine the Industrial Metaverse with Intelligent Manufacturing Practice Training to create a new form of "Intelligence+" training that embodies "virtual-real mapping, virtual-real interaction, virtual-real integration, enhance reality through virtuality, and promoting reality through virtuality" is becoming a new trend in the intelligent manufacturing training field. Many top technology teams globally have invested a considerable amount of manpower and funds in the development and research of "Metaverse and Training".

Abroad, Dominic Gorecky and others proposed as early as 2017 to introduce virtual training in future factories to meet the constantly changing technology trends [3]. Upadhyay A K and others explored using the Metaverse as a training ecosystem and the research results indicate that immersive training has great potential [4]. Danylec A and others crated a driver training framework based on the metaverse, and the effectiveness of the developed model was confirmed through a case study of loco-motive vehicles [5]. In addition, in 2020, Mercedes-Benz USA reached a digital trans-formation cooperation agreement with Microsoft and launched the Mercedes-Benz Virtual Remote Support—the first mixed-reality car maintenance system to remotely support technicians in dealing with complex maintenance issues.

In China, Chen Guo and others have developed and implemented a coal mine drilling machine virtual assembly training system from a first-person perspective, providing trainees with a novel and vivid teaching experience through friendly human-computer interaction [6]. Xie Gang and others, through the combination of gesture recognition and VR technology, designed a highly similar and flexible human-computer interaction virtual simulation assembly training system based on the existing assembly teaching course, exploring the innovative development model of the new engineering assembly teaching system [7]. China National Offshore Oil Corporation has built and put into use the country's first well control intelligent scenario training system, which uses holographic projection and 3D modeling technology to simulate different types of drilling platforms and various emergencies, effectively helping workers in the drilling industry to undergo efficient training and emergency training to deal with challenging situations.

Hence, it can be seen that in the field of education and training, the metaverse is bringing about significant innovation to the education and training industry of the 21st century through its high interactivity and strong immersion scenario empowerment.

1.2 Comparative Analysis of Training Modes

In the context of the new era, the quick and diversified trends of society have put forward higher requirements for talent cultivation. As an important link in the construction of a high-quality talent team, education and training must keep pace with the times to adapt to the rapidly changing social development. This paper makes a detailed comparison of Training 1.0, 2.0 and 3.0 (see Fig. 1) [8].

Training Methods	Training 1.0	Training 2.0	Training 3.0
Main Features	Face-to-face training, transfer of knowledge and basic skills development are the main focus	Web-based media such as multimedia, online courses and social platforms	Immersive virtual space, diversified hands-on training
Learning State	Learners are passive recipients of knowledge	Strong learner initiative and participation	Immersive learning experience for participants
Learning Resources	Limited paper books and handwritten notes	Rich online courses and multimedia resources	Diversified hands-on training and sharing of global educational resources
Learning Motivation	Reliance on traditional classrooms and teacher supervision	Strong learner initiative and participation	Immersive experience and fun teaching motivate learning
Space&Time	Greater constraints of time, space and teaching environment	Freedom to choose the time and place of study	Breaking down time and space constraints, global learning interactions
Practical Training	The practice is more effective, but there are some operational risks	Dependence on online media, limited opportunities for practice	Provide immersive virtual practice and full hands-on training opportunities
Individualised Teaching	Limited personalised teaching	Students choose their own learning resources	Personalised instructional content based on learning analytics and data mining
Educational Resources	Resource limitations within traditional classrooms	Rich online resources and social networks for learning	Cross-cultural communication and sharing of global resources

Fig. 1. Comparison of Training 1.0, 2.0 and 3.0.

Training 1.0 normally adopts face-to-face training, written materials, classroom teaching etc. to conduct training in physical classrooms, which conforms to the learning habits of most people. Training 2.0 relies on media such as multimedia, online courses, and social platforms to provide training, which emphasizes the proactivity and participation of learners.

In recent years, technologies such as digital twins and the metaverse have provided new possibilities for the future virtualization of education and training, gradually forming a virtual-real interactive Training 3.0. Training 3.0 can use artificial intelligence, virtual reality, and other metaverse-related technologies to construct an immersive virtual space that maps to the real world, providing learners with diversified practical training and immersive learning experiences. In addition, this mode of training can not only carry out a deep analysis of the learner's learning process and results through learning analysis and data mining technology, to develop personalized learning content and paths according to the learner's learning needs and characteristics; but also break geographical and time zone restrictions, establish social networks in the virtual world, and allow learners to interact, collaborate and discuss in real-time with teachers and learners worldwide, promoting the sharing of education resources and the formation of cross-cultural exchanges.

2 Technical Architecture

According to the mapping between real space and virtual space, this paper proposes a technical architecture of the IMT based on Industrial Metaverse (see Fig. 2). The architecture mainly includes the following parts.

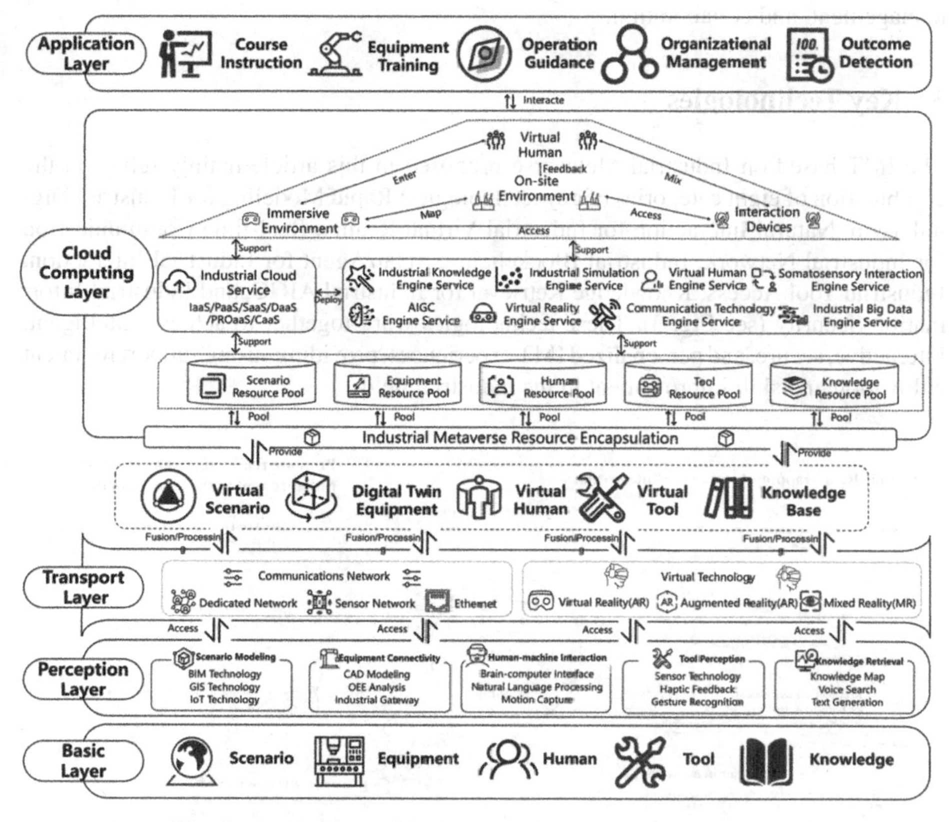

Fig. 2. Architecture of the IMT based on Industrial Metaverse

Basic Layer: This layer uses the five elements of scenario, equipment, human, tool, and knowledge as the primary objects for construction, providing physical space support for the IMT based on Industrial Metaverse.

Perception Layer: This layer realizes the simulation and sensory access of the five elements through five major types of technologies.

Transport Layer: This layer is composed of a communications network and virtual technology and plays the role of a mediator. Based on the completion of sensory access to the five elements, on one hand it relies on the communication network to realize real-time communication and transmission of information; on the other hand, it relies on VR, AR, and MR technologies to map the virtual space to the real world.

Cloud Computing Layer: The cloud computing layer can pool elements into the industrial metaverse. Through industrial cloud services and various engine services, it realizes the virtual-real integration of people, machines, and environment.

Application Layer: This layer mainly includes application scenarios in the intelligent manufacturing training process, such as: course instruction, equipment training, operation guidance, organization management, and result detection, etc.

Overall, these layers cooperate with each other to jointly construct a complete system architecture. This allows the IMT system to effectively carry out tasks such as training, management, and collaboration.

3 Key Technologies

The IMT based on Industrial Metaverse proposed in this article mainly relies on the combination of eight categories of key technologies: Rapid Modeling for Industrial Digital Twin, Natural Interaction for Industrial Virtual Reality, Real-time Communication for Industrial Network, Industrial Blockchain, Avatar/Agent for Industrial Interaction, Industrial Tool Access, Knowledge Retrieval for Industrial AIGC, and Industrial Information Security (see Fig. 3). These technologies work together to achieve intelligent, interactive, secure, and personalized IMT experiences, providing strong support for talent cultivation and skill improvement in the industrial field.

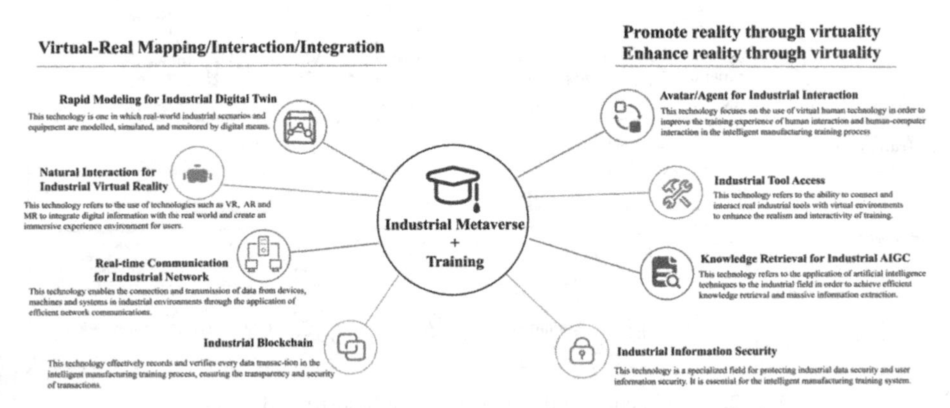

Fig. 3. Key Technologies of IMT based on Industrial Metaverse

3.1 Rapid Modeling Technology for Industrial Equipment and Scenarios Based on CAD and BIM

Accurate 3D equipment models and realistic industrial scenario models are the basis for IMT. This paper proposes a rapid modeling technology for industrial equipment and scenarios based on CAD and BIM. It is an effective means to integrate the design, construction, operation and maintenance information of industrial equipment, buildings, infrastructure, and scenarios into digital models, and to convert geometry, material, texture, and animation data into data that can be recognized and processed by virtual engines. This technology includes several aspects, such as model generation, model optimization, and so on.

In terms of model generation, for equipment modeling based on CAD, object dimensional data can first be obtained using accurate measurement tools, and make them become standardized CAD models. Secondly, to import these equipment models into the virtual environment, it is necessary to use 3D model conversion tools to convert them into specific formats. For industrial scenario modeling based on BIM, the modeling process is generally similar to the equipment modeling process based on CAD. The differences are that in BIM, it usually starts from predefined "objects" with certain attributes, and these attributes are modified to quickly create and modify designs. In addition, because BIM models contain "nD information", it is necessary to validate the model, such as conflict detection, energy simulation, etc.

In terms of model optimization, whether it's the models can sometimes be very detailed, with a high polygon count that can slow down rendering performance. To solve these problems, this technology can simplify geometry by removing unnecessary details, merging repetitive elements, reducing the number of faces, and low polygon model usage. Moreover, optimize performance and increase rendering speed by reducing texture resolution, using compressed texture formats, and setting material properties appropriately (Fig. 4).

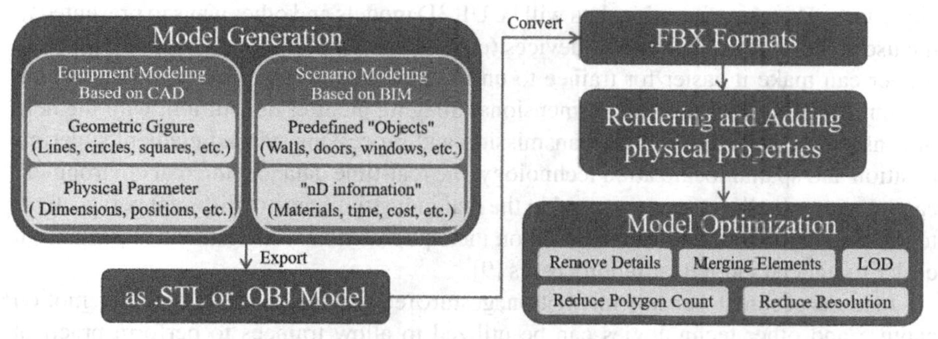

Fig. 4. Architecture of Rapid modeling architecture of industrial entities and scenarios based on CAD and BIM

3.2 Industrial Data Visualization and Somatosensory Interaction Technology Based on AR/MR

AR technology is an advanced technology that combines virtual information with the real world, and it is a link between industrial metaverse and intelligent manufacturing training. In order to provide a more intuitive, richer, and immersive practical training environment, this paper proposes an industrial data visualization and somatosensory interaction technology based on AR/MR (Fig. 5).

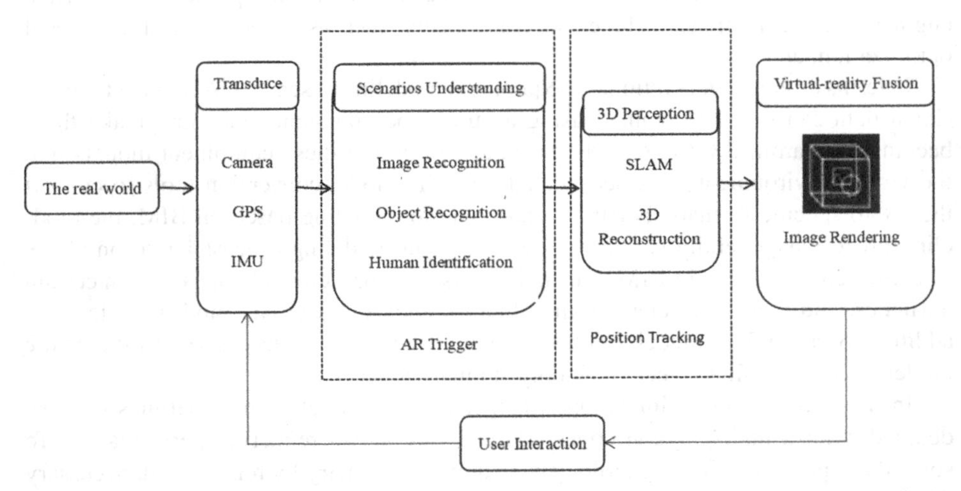

Fig. 5. Architecture of Industrial Data Visualization and Somatosensory Interaction Technology Based on AR/MR

In terms of industrial data visualization, this technology use AR/MR development toolkits (e.g., Unity, ARCore, ARKit, etc.) to build applications for specific educational objectives. The manufacturing data will be UI, 3D models and other ways to presented to the users. By wearing AR or MR devices (e.g., Microsoft HoloLens, Magic Leap, etc.), trainer can make it easier for trainee to understand the intuitive virtual devices which present in the form of the three-dimensional imaging picture. In addition, with the help of sensor data acquisition, data transmission and processing, virtual element superimposition and spatial localization technology, the real-time data and the real environment can integrate. It allows users to stand in the real industrial scene in different perspectives to see the real-time data superimposed on the equipment or workstations, also makes it easier to understand industrial processes [9].

In terms of somatosensory interaction, gesture recognition, voice recognition, motion capture and other technologies can be utilized to allow trainees to perform practical operations in a safe virtual scenario. Trainers can perform the corresponding operation of the virtual equipment, and the trainees can follow the operation at the same time, and the virtual operation is basically the same as that of the real world, which will help the trainees to familiarize themselves with the equipment, understand the operation process.

3.3 The Technology of Real-Time Industrial Data Communication and Sharing Based on Intranet Penetration

Remote real-time communication is an important guarantee for industrial enterprises to use IMT equipment efficiently. So this paper proposes a real-time communication and sharing technology of industrial data based on frp intranet penetration to support remote communication of IMT equipment.

Frp intranet penetration can be understood as using a server with a publicly accessible IP address as an intermediary to establish a "connection" between the connected computer or other terminals and the computer currently in operation. As shown in the technical structure diagram in Fig. 6. Frp intranet penetration uses the client-server architecture. In the intranet, frp client frpc establishes a connection with frp server frps and sends requests for intranet services to the server. The server side is responsible for receiving the request and forwarding it to the service port on the intranet.

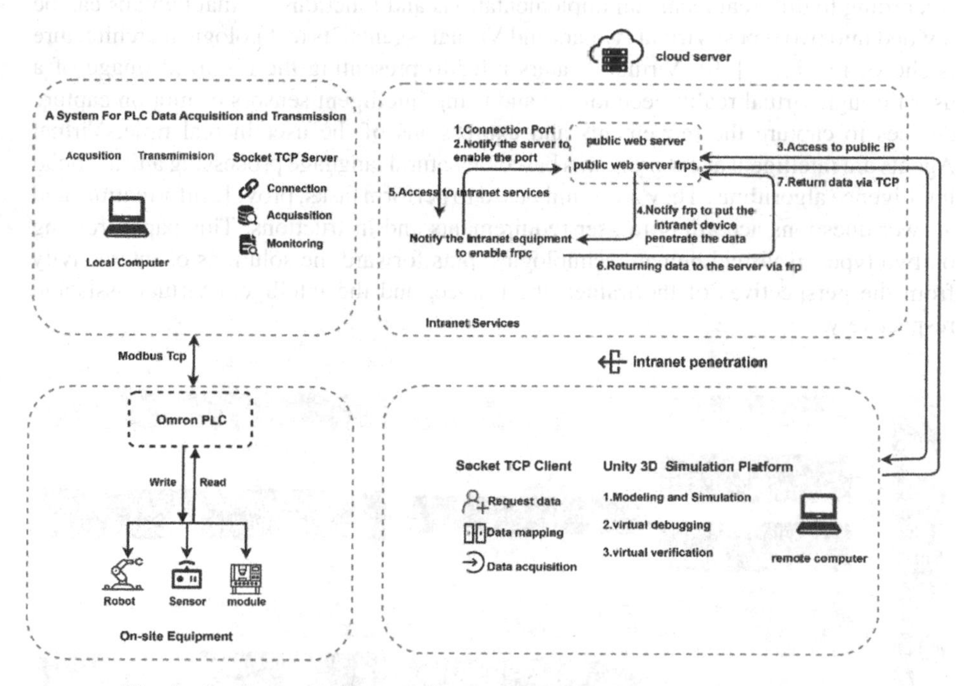

Fig. 6. Architecture of Intranet Penetration Technology Structure Diagram

The specific process of realizing intranet penetration is as follows: first, the frpc client sends information such as the extranet address and port, as well as the address and port of the intranet service that needs to be mapped, to the frps server. At the same time, the server assigns a unique token to the client for verifying the client's identity and for subsequent communication. Next, the client initiates a tunnel request to the server, in which the client specifies the address and port number of the intranet service that needs to be mapped to the public network. After the server receives the tunnel request, it assigns the extranet address and port number and sends that address and port number to the client. The client receives it and informs the service in the intranet about the address and port number, so that the external network can access the service in the intranet through this address and port number. When the external network accesses the external address and port number assigned by the server, the request is forwarded to the client, which in turn forwards the request to the service in the intranet. Finally, when the service in the intranet has a response, the client sends the response to the server, which in turn forwards the response to the external network. By the above way, a communication tunnel is established between the public network and the intranet, mapping the intranet services to the public network, so that the external network can access the services in the intranet.

3.4 Interactive Training Teaching Technology Based on Virtual Humans

According to different technical implementations and functions, virtual humans can be divided into two types: Virtual Avatars and Virtual Agents. Its technological architecture is shown in Fig. 7 [10]. Virtual Avatars refer to presenting the digitized image of a user through virtual reality technology and using intelligent sensors or motion capture devices to capture the movements and expressions of the user in real time. Virtual Agents are intelligent agent programs based on natural language processing and artificial intelligence algorithms. They are mainly used to perform tasks, provide information, and answer questions according to user requirements and instructions. This paper, relying on two types of virtual human technologies, puts forward the solutions of interactivity from the perspectives of the trainer, the trainee, and the intelligent virtual assistant, respectively.

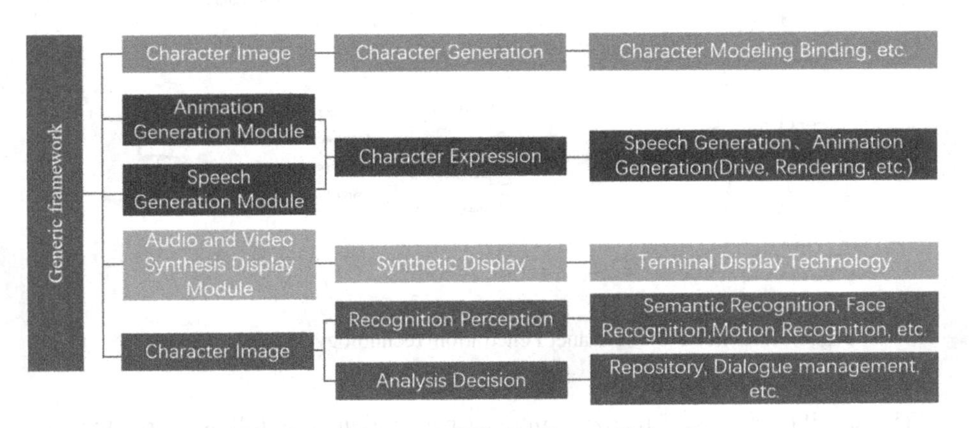

Fig. 7. Architecture of Virtual Human Technology

In the IMT based on Industrial Metaverse, virtual avatar technology can be used to construct virtual images of trainers and trainees. From the trainer's perspective, trainer's avatar can overcome time and space constraints to enter the virtual scenario, demonstrate specific operation details and provide theoretical instruction on debugging and troubleshooting for trainees. Meanwhile, these high-quality knowledge and experiences can be fully preserved in the system and be widely disseminated and shared for the benefit of a wide range of students, enterprises, and organizations.

From the trainee's perspective, they can create their own virtual avatars and carry out operation training in virtual scenarios, deepening their understanding of actual process flows. Accessing the system in a more realistic virtual avatar form, it's as if trainee could face-to-face collaborate with trainees from other regions, making interactions become more real and intimate.

From the perspective of intelligent virtual assistants, this type of virtual human usually exists in the form of virtual agents, serving as auxiliary tools for teaching, providing help and support to users. For example, intelligent virtual assistants can provide navigation functions to guide trainees to find the required work area, equipment, or resources quickly and accurately in the virtual scenario. They also can provide 24-h online detailed explanations of equipment or process flows, introduce relevant knowledge and specific steps of training, answer trainee' questions. In addition, by collecting and analyzing student data, intelligent virtual assistants help monitor trainees' learning progress in real-time and create personalized learning support according to trainees' individual differences and learning needs.

3.5 Industrial Tool Perception Access Technology Based on Inertial Motion Capture

The deployment of industrial tools is an integral component of industrial skills training. Among the existing IMT, the use of VR handles is the most prevalent. However, the sense of authenticity is insufficient. To solve the above problems, this paper proposes an industrial tool perception access technology based on inertial motion capture (its technological architecture is shown in Fig. 8). It is a technology that transforms real tools into virtual interactive tools. Through the modeling, tracking, and capturing of real tools and interaction design, tool appearance information and operation functions are integrated into the virtual training system, so that users can manipulate the tool model in the virtual world through the real tools to carry out the practical training operations.

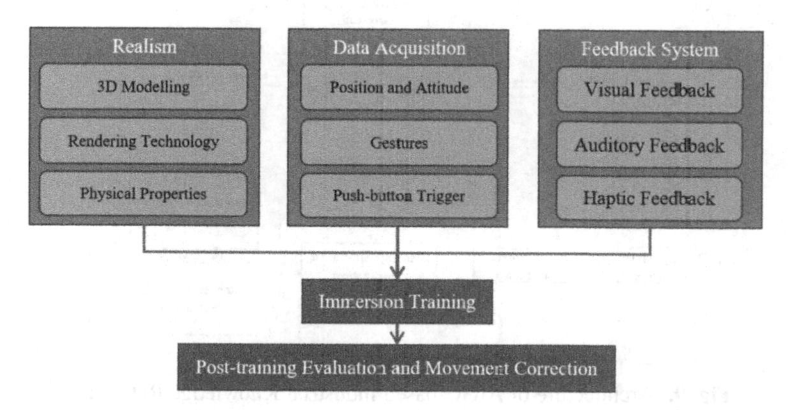

Fig. 8. Architecture of Industrial Tool Perception Access Process

To augment the realism of operational tools, this technology begins with the utilization of high-precision 3D scanning and measurement technology to acquire detailed information about the appearance of industrial tools. Subsequently, professional 3D modeling software and rendering technology can be applied to create realistic models. Furthermore, utilizing a physics engine to establish the model's physical characteristics, such as collision detection, gravity, and friction, aids in restoring more detailed information about the tool.

To ensure real-time data acquisition and analysis, the user's operation information of the tool is obtained by rigidly binding the tracker to the real tool, to restore the position and movement trajectory of the tool. Based on the collected data, non-compliant operations can be identified, and operational suggestions can be provided to the user simultaneously.

To enhance the interaction between humans and tools, on one hand the interface provided by the tracker can be developed into a data interaction channel between the real tool and the virtual tool. On the other hand, in the visual and auditory aspects, the same animation effects (e.g., tool rotation, telescopic movement) and sound effects can be added, in terms of haptic experience, a variety of haptic feedback devices such as vibration motors and temperature controllers are integrated into the tools for the effects of using different tools.

3.6 Industrial Knowledge Retrieval and Quiz Generation Technology Based on AIGC

In the IMT, the application scenarios of industrial knowledge retrieval and quiz generation technology based on AIGC mainly lie in effectively integrating natural language processing and knowledge graph technology to retrieve issues quickly and accurately in the industrial field and obtain related information. Furthermore, this technology enhances user experience through real-time voice interaction, and realizes the automatic question generation function for text knowledge to meet diverse needs. So this paper proposes an industrial knowledge retrieval and quiz generation technology based on AIGC. Its technological architecture is shown in Fig. 9

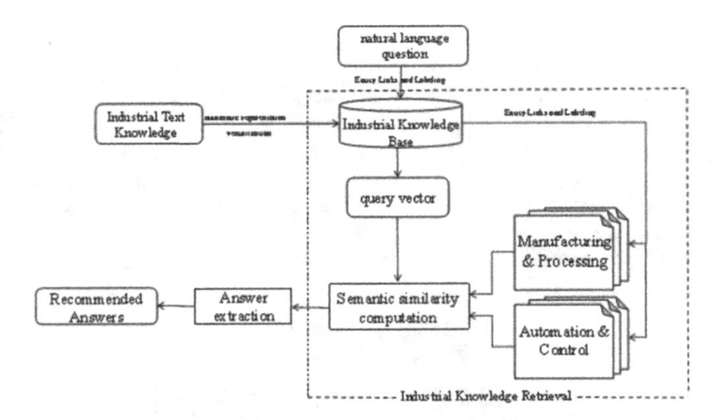

Fig. 9. Architecture of AIGC-based Industrial Knowledge Retrieval

The main method of AIGC-based knowledge retrieval technology is to use information retrieval technology to extract relevant answers in Web documents, they cannot get more satisfactory results in actual Q&A systems in specific domains such as industry. This is mainly due to the domain-specific answer acquisition process with the help of more domain-related resources and domain text features. To address this problem, domain knowledge base can be introduced to automatically annotate the questions and domain documents, mapping entities or concepts in the text to the knowledge base, so as to find out more semantic information (image cited here). For example, chatGLM can automatically segment and vectorize the textual knowledge of a specific domain, and other operations, to construct the corresponding knowledge base of the domain. For the IMT, users can provide textual knowledge specialized in the domain, and the system can generate a knowledge base specific to the domain, thus obtaining more satisfactory results. Moreover, questions can be automatically generated based on the generated knowledge base to improve teaching efficiency and user learning experience. To realize this technology, a large amount of data can be used as a driver to generate questions using machine learning or deep learning techniques, in which a question template can be defined and new questions can be generated by replacing some variables or parameters in the template to generate questions of different types and difficulty ranges.

4 Prototype Construction

Verifying the effectiveness of Industrial Metaverse based IMT, this paper builds the "TrAiN" prototype system, a combination of "Treatment", "AI", and "Network". "Treatment" refers to the execution and control processes in the production manufacturing field, including Industrial Robots, Mechatronics, Industrial Control, Fault Diagnosis and Maintenance, etc. "AI" refers to industrial intelligence driven by a mix of models and data, covering technologies such as Big Data and Machine Learning, Industrial Virtual Reality with AR/VR/MR, and large models like AIGC. "Network" refers to the cloud-physical-mobile integrated industrial network, involving technologies such as the

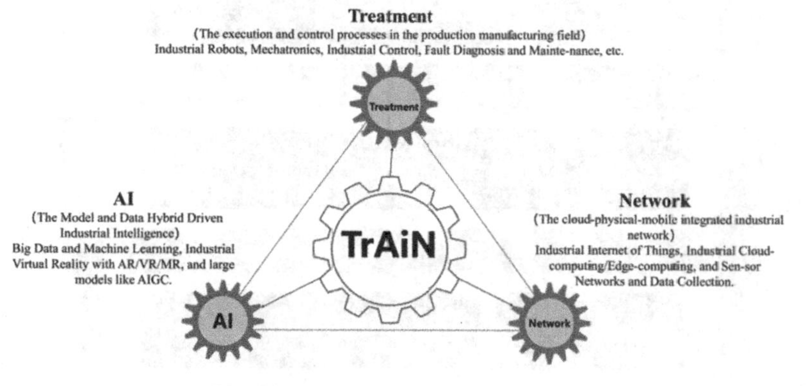

Fig. 10. "TrAiN" System Architecture

Industrial Internet of Things, Industrial Cloud-computing/Edge-computing, and Sensor Networks and Data Collection (Fig. 10).

Based on the layered technical architecture presented in Sect. 2 and six key technologies mentioned in Sect. 3, this paper, proposes a simple architecture model as shown in Fig. 11, and constructs a private Industrial Metaverse based on specific industrial equipment and fields in certain factory. In this architecture model, the six elements of intelligent manufacturing training (i.e., scenarios, equipment, human, tools, and knowledge) are abstracted into virtual models (i.e., virtual scenarios, digital twin equipment, virtual human, virtual tools, and knowledge base) through the Rapid modeling technology for industrial equipment and scenarios based on CAD and BIM. Then users can access the intelligent manufacturing training system based on the industrial metaverse through Industrial data visualization and somatosensory interaction technology based on AR/MR. In the industrial metaverse, Interactive training teaching technology based on Virtual Humans, Industrial tool perception access technology based on Inertial Motion Capture, and Industrial knowledge retrieval and quiz generation technology based on AIGC can provide strong support for the organic integration of human, immersive space, knowledge, and reality. Above all, four types of applications for the private Industrial

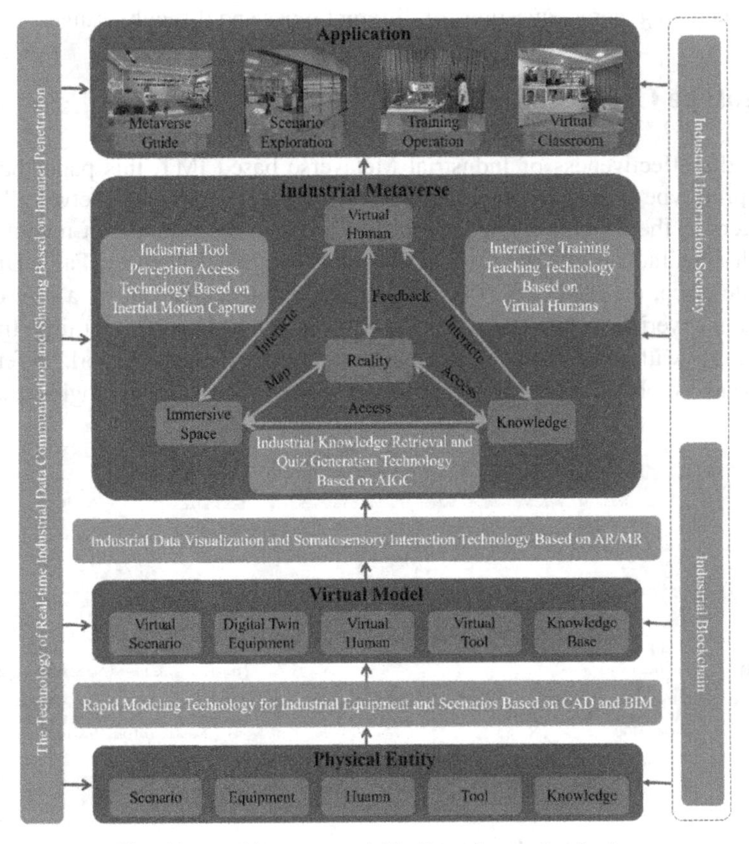

Fig. 11. Architecture model built on key technologies

Metaverse have been empowered. It is worth noting that Real-time industrial data communication and sharing technology based on Intranet Penetration serves as a safeguard for the realization of virtual-real mapping in the IMT.

Firstly, when the user puts on AR equipment and enters the private Industrial Metaverse, a metaverse guide will appear to interact with the user in real time. Relying on virtual human technology, the metaverse guide will not be a static interface element, but a dynamic interactive entity. It can provide intelligent guidance for users through voice interaction technology, answer common questions, and provide practical suggestions to help users quickly familiarize and adapt to the IMT environment (Fig. 12).

Fig. 12. Metaverse Guide

Secondly, users can explore the IMT scenarios according to freely move and rotate their perspectives to observe the details of the laboratory environment and equipment from different angles. Through industrial data visualization and somatosensory interaction technology based on AR/MR, users can also click or hover over the corresponding physical equipment to obtain its detailed information, such as the name, function, working principle, and current operating parameters of the equipment, to understand the specific indicators and operation methods of the equipment (Fig. 13).

Fig. 13. Scenario Exploration

Moreover, through industrial tool sensing technology, users can use the IMT tools similar to real industrial tools to enter the metaverse space for safe skill training. This technology, while retaining the advantages of virtual practical training such as high safety, low cost, and no spatial constraints, maximally replicates the real feel of using industrial tools, enhancing the immersion of students in metaverse space training, and can significantly improve the familiarity with the tools (Fig. 14).

Fig. 14. Training Operation

Lastly, users can enter the metaverse immersive virtual classroom for learning. In the classroom, the metaverse digital teacher can not only provide an interactive and diverse teaching experience but also provide real-time feedback on the student's learning situation and timely adjust the teaching content to meet the personalized needs of the students. In addition, by integrating ChatGLM and AIGC, the knowledge base can provide users with access to a wide range of information and create a highly personalized learning experience for each learner, helping them to better understand and retain the information they have learned in the training program (Fig. 15).

Fig. 15. Virtual Classroom

5 Conclusion and Perspectives

As the culmination of digital technology, the intelligent manufacturing training system based on the industrial metaverse promotes the transformation of education and training models to some extent. It can not only achieve seamless integration of virtual and reality, cross-organizational collaborative work, safety guarantees for training equipment and personnel, and personalized customization of learning plans, but also achieve real-time sharing of data knowledge and cost reduction and efficiency improvement of corporate training.

However, this research only built the intelligent manufacturing training system based on the industrial metaverse at the experimental stage and has not yet been applied on a large scale. In addition, this research only provides six types of technical solutions, and these solutions are only preliminary explorations that still need to be further improved and verified. Future research will further consider introducing technologies in the fields of industrial blockchain and industrial information security to further enhance the function and security of the system and provide more comprehensive support for practical applications.

References

1. https://www.mckinsey.com/featured-insights/future-of-work/the-future-of-work-after-cov id-19
2. Zheng, Z., et al.: Industrial metaverse: connotation, features, technologies, applications and challenges. In: Fan, W., Zhang, L., Li, N., Song, X. (eds.) Methods and Applications for Modeling and Simulation of Complex Systems. AsiaSim 2022. Communications in Computer and Information Science, vol. 1712, pp. 239–263 (2022). https://doi.org/10.1007/978-981-19-9198-1_19
3. Gorecky, D., Khamis, M., Mura, K.: Introduction and establishment of virtual training in the fac-tory of the future. Int. J. Comput. Integr. Manuf. **30**(1), 182–190 (2017)
4. Upadhyay, A.K., Khandelwal, K.: Metaverse: the future of immersive training. Strateg. HR Rev. **21**(3), 83–86 (2022)
5. Danylec, A., Shahabadkar, K., Dia, H., et al.: Cognitive implementation of metaverse embedded learning and training framework for drivers in rolling stock. Machines **10**(10), 926 (2022)
6. Chen, G., Liu, G.: Development of a virtual assembly training system for coal mine drilling rigs based on unity3D. Packag. Eng. Art Des. **43**(12), 106–112 (2022)
7. Xie, G., Qi, P., Xiang, Y., et al.: Research on assembly virtual simulation system based on 3D interaction. Industrial Control Computer (2022)
8. Luo, J.X., et al.: Survey on sharing technology and applications of intelligent manufacturing training equipment based on industrial internet and man-in-loop simulation. In: Fan, W., Zhang, L., Li, N., Song, X. (eds.) Methods and Applications for Modeling and Simulation of Complex Systems. AsiaSim 2022. Communications in Computer and Information Science, vol. 1713, pp. 593–610. Springer, Singapore (2022). https://doi.org/10.1007/978-981-19-9195-0_48
9. Fang, W., Wu, Y., Chen, C.: Application of AR/MR technology in intelligent wearable devices for industrial inspection. Digit. Technol. Appl. **39**(07), 150–152 (2021)
10. Deep Report Series on the Metaverse in the Media Industry: Digital Virtual Humans, the Intersection of Technology and Humanities, Empowering Industries

Convolutional Transformer Network: Future Pedestrian Location in First-Person Videos Using Depth Map and 3D Pose

Kai Chen[1](✉), Yujie Huang[1], and Xiao Song[2]

[1] Nanjing University of Aeronautics and Astronautics, Nanjing 210016, China
chen_kai@nuaa.edu.cn

[2] Beijing University of Aeronautics and Astronautics, Beijing 102206, China

Abstract. Future pedestrian trajectory prediction in first-person videos (egocentric videos) offers great prospects for autonomous vehicles and social robots. Given a first-person video stream, we aim to predict that person's location and depth (distance between the observed person and the camera) in future frames. To locate the future trajectory of the person, we mainly consider the following three key factors: a) The image in the video sequence is the mapping of the actual 3D space scene on the 2D plane. We restore the spatial distribution of pedestrians in two-dimensional images to three-dimensional space. The distance of the pedestrian from the camera, which can be represented by the depth of the image, is the third dimension of information that is lost. b) First-person videos can utilize people's 3D poses to represent intention interactions among people. c) The rules governing a pedestrian's historical trajectory are very important for the prediction of pedestrian's future trajectory. We incorporate these three factors into a multi-channel tensor to represent a deployment of the scene in three-dimensional space. We put this tensor into an end-to-end fully convolutional framework based on transformer architecture. Experimental results reveal our method to be effective on public benchmark MOT16.

Keywords: Convolutional Neural Network · Attention · 3D pose · Image depth

1 Introduction

With the development of intelligent autonomous driving technology and intelligent robot technology, accurate pedestrian trajectory prediction has become an important part of machine intelligence and also receives more and more attention in the computer vision community [1–4]. However, most existing social-interaction methods [1, 2, 5, 6] are only applicable to monitoring systems and are difficult to apply to autonomous vehicles and social robots which observe the scene in first-person view. Predicting the future trajectories of pedestrians in the video taken by the camera with first-person view is a complex task because it needs to deal with the following three factors simultaneously:

© The Author(s), under exclusive license to Springer Nature Singapore Pte Ltd. 2024
F. Hassan et al. (Eds.): AsiaSim 2023, CCIS 1911, pp. 44–59, 2024.
https://doi.org/10.1007/978-981-99-7240-1_4

Fig. 1. RGB image frame and its depth confidence map.

Social Spatial Dependencies: When people navigate in public places, they usually socialize with other pedestrians around them. In practical applications, for the first-person videos captured by autonomous vehicles and social robots, extracting the relative spatial positions of the crowd from these two-dimensional image data is a key factor. As shown in Fig. 1, (b) is the depth confidence map of image (a); the darkness level of (b) indicates the distance of an object in the graphic from the camera. If the depth of the image is ignored, the spatial distance of the four pedestrians in the scene is very small. But in the actual scene, they still have a certain distance between them according to the depth of the image, which represents the distance of pedestrians relative to the camera. At the same time, when predicting the future trajectories of pedestrians, we need to predict not only the positions but also the depth of pedestrians in the image, which is very important for the visual assistance of autonomous vehicles and social robots.

Social Interaction Dependencies with Neighbors' Intentions: Every pedestrian in a social place has his/her own intention, which may be adjusted according to the intentions of other pedestrians around. Therefore, accurately modelling a pedestrian's intention is an important part of pedestrian trajectory prediction modeling.

Social Temporal Dependencies: In crowded scenes, people tend to move at random speeds and in random directions. Pedestrian movement changes greatly with time, so the effective mining of pedestrian historical trajectory rules is essential for the prediction of pedestrian trajectory.

Based on the above three key factors, we propose a multi-task learning model that predicts not only the future locations but also the future depths in the scene for the subject pedestrian. During the data modeling, we use the depth of the image and the locations of all pedestrians in the scene to model the three-dimensional spatial structure of the scene, and use the postures of all pedestrians to model their intention interactions. The spatial structure of the scene and pedestrian intentions will be integrated into a three-dimensional tensor. When constructing the deep neural network model, in order to ensure the spatial invariance of the input tensor, we propose a fully convolutional transformer (Conv-Transformer) based on the fully connected transformer (FC-Transformer) to establish the global dependence of the pedestrians' historical and future trajectories.

2 Background

In recent years, data-driven neural network models have proved superior to traditional mathematical statistical models. LSTM-based series models and GAN models based on generative adversarial networks are the most representative. Helbing [11] and Pellegriniet [4] successfully demonstrate the benefits of establishing a social interaction model, but they require manual rules that cannot be generalized to new scenarios. Alahi [1] uses a recurrent architecture to consider multiple time steps for pedestrian behavior, but does not consider the physical cues of the scene. In social-scene methods [5, 12, 13], scene context supplements social features to improve the accuracy of human trajectory prediction. These methods assume that people tend to have the same walking pattern in similar scene layouts. In order to extract scene features, [5, 12] use convolutional neural networks (CNN) [15, 16] that have also been successfully used in image classification [14]. However, these methods are not designed to handle first-person videos.

The latest research [3, 17] notes that people usually walk in public places with a specific intention. Humans can read the body language of others to predict whether they will cross the road or continue walking along the sidewalk. Kooij [18] modeled the faces of pedestrians to build an intention model using a dynamic Bayesian network for driving recorder video to predict whether they would cross the road, but pedestrians' faces are unavailable in most scenes and will also be affected by image resolution. In [3], human 2D key point features and convolutional neural networks are used to predict the future path in first-person videos. However, in [3], the scale of the pedestrian is used to represent the distance of the pedestrian relative to the camera, which will cause a large error because the sizes of the various people are different and the camera is not fixed. The latest research [19] proposes a method for predicting dense depth in the case of a monocular camera with people in the scene moving freely.

Existing methods [20, 25, 26] also define pedestrian trajectory prediction as a sequence-to-sequence model, using the encoder-decoder structure to build a network model. In the encoder-decoder configuration, sequence prediction is based on a complex RNN or CNN, and the best performing model also needs to connect the encoder and decoder through an attention mechanism. [9] proposed a new type of simple network architecture transformer, which is completely based on the attention mechanism and abandons recurrence and convolution. Although a transformer can also be used to solve pedestrian space-time sequence prediction problems, the existing model uses a fully connected transformer (FC-Transformer) that does not consider spatial correlation. In this article, we propose a new convolutional transformer (Conv-Transformer) network for pedestrian trajectory prediction in first-person videos.

3 Approach

In practical applications, in order to assist autonomous vehicles and social robots' navigation, we propose the task of predicting the pedestrian's future trajectory in a first-person video. People usually navigate in space with a specific intention. Effectively capturing the intention interactions of the subject pedestrian and surrounding persons can greatly orient the subject pedestrian's future trajectory. Unlike a video in perspective view, it is obviously unreasonable in a first-person video to simply use the relative coordinate distance of the pedestrian in the two-dimensional image to express interaction between pedestrians. What we need to consider more is the distance of the pedestrian in terms of image depth, that is, the distance of the pedestrian from the camera. At the same time, autonomous vehicles and robots, in order to avoid collisions with pedestrians, not only need to predict the positions of pedestrians in the image but also need to predict the distance between pedestrians and themselves. This motivates us to study future path prediction jointly with depth.

3.1 Network Architecture

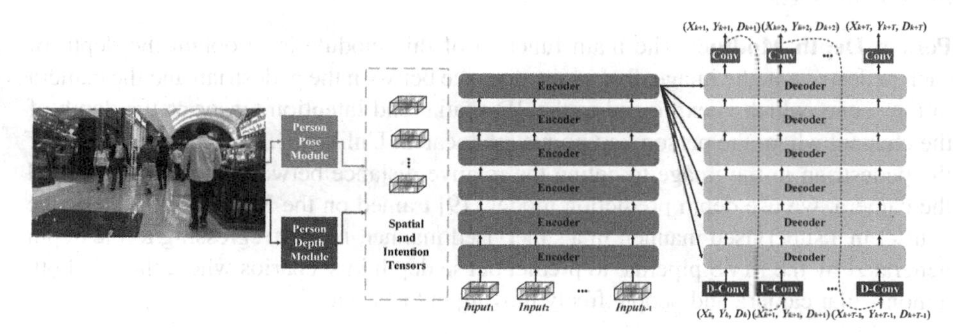

Fig. 2. The overview network architecture of our model.

Figure 2 shows the overall network architecture of our model. Given a sequence of frames containing the person whose trajectory is to be predicted, our data module utilizes a person pose module and person depth module to model a frame in a spatial and intention tensor. Our data module uses a novel spatial and intention tensor that takes into account person-person relations in a 3D space for joint depth (D) and location (X, Y) prediction. Unlike most existing work that oversimplifies a person as a spatial point, our data module uses two modules (person pose and depth module) to integrate the intentions and interaction of all pedestrians in a 3D scene into a 3D tensor. To construct a deep neural network, we discarded the LSTM structure commonly used in existing methods and replaced it with a fully convolutional transformer encoder-decoder structure.

3.2 Data Module

As shown in Fig. 3, the data module integrates the intention and spatial distribution of each person in the scene to a 3D spatial and intention tensor (SIT) base on person pose module and person depth module. SIT contains three aspects of information: the behavior of all pedestrians in the scene (the intention of the pedestrians); the location information of the pedestrian in the two-dimensional image and the distance between the pedestrian and the camera (the depth of the image), and the latter two constitute the pedestrians' 3D spatial distribution information. Our main purpose is to map the pedestrian distribution information (behavior, position and depth) in a 2D image into a 3D tensor.

Person Pose Module. This module encodes the visual information about every individual in a scene. As opposed to oversimplifying a person as a point in space, we model the person's the appearance and body movement. To capture human motion, we base on the pedestrian 3D pose and shape estimation model MVSPIN [21] and detection model YOLO [24] to extract pedestrian 3D key point information. As shown in Fig. 3, unlike most existing methods, our data module not only models the body movement of the subject pedestrian but first models all visible pedestrians' movement change at different times in the scene.

Person Depth Module. The main function of this module is to obtain the depth of each pedestrian in the image, that is, the distance between the pedestrian and the camera in the scene, which is expressed in the 3D spatial and intention tensor as the depth of the channel where the person key-points are located. Unlike [3], which uses the size of the pedestrian in the image to define the relative distance between the pedestrian and the camera, we use depth prediction model [19] trained on the Mannequin- Challenge dataset in a supervised manner, in a supervised manner, i.e., by regressing to the depth generated by the MVS pipeline to predict dense depth in scenarios where there is both a monocular camera and people freely moving in the scene.

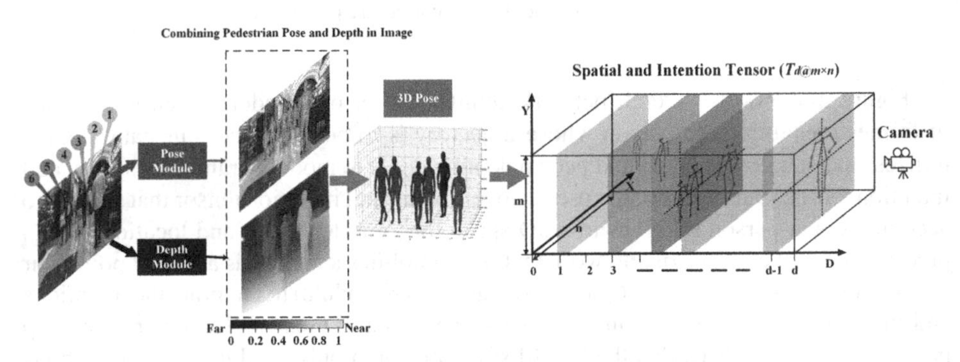

Fig. 3. Data module structure frame diagram.

With the above two submodules, the data module is able to integrate a person's intention and depth into a spatial and intention tensor (Fig. 4).

Algorithm 1 Generating Spatial and Intention Tensor $T_{d@m\times n}$

1: **Input:**
2: Initial a zero tensor $T_{d@m\times n}$
3: Given each person 3D pose key point location set L_j, (x_i, y_i, z_i), i=1, ..., nL_j, j=1, ... N;
4: Given image depth matrix $D_{m\times n}$
5: Given subject pedestrian bounding box coordinates (x_0, y_0, w, h)
6: **Output:** Spatial and Intention Tensor $T_{d@m\times n}$
7: **for** $j = 1 : N$ **do**
8: Initial a zero person pose matrix $P_{m\times n}$
9: **for** i = 1 : nL_j **do**
10: **if** (x_i, y_i, z_i) in (x_0, y_0, w, h) **then**
11: $P_{x_i y_i z_i}$=1
12: **Else**
13: $P_{x_i y_i z_i}$=-1
14: **end for**
15: $d_j = KMeans(D_{x_1 y_1 z_1}, D_{x_2 y_2 z_2}, \ldots, D_{x_i y_i z_i})$
16: $T_{d_j@m\times n}$= $P_{m\times n}$
17: **end for**

Fig. 4.Some samples of spatial and intention tensors generated by our data module.

Fig. 4. Some samples of spatial and intention tensors generated by our data module.

Spatial and Intention Tensor Build Strategy. As shown in Fig. 3, the data module combines the posture, position, and depth information of all visible pedestrians in the scene to form a spatial and intention tensor $T_{d@m\times n}$ of size $d@m \times n$, where d is the channel of $T_{d@m \times n}$ and represents the depth of the pedestrian in the image, and m and n are the height and width of each channel matrix $M_{m \times n}$, respectively. The $T_{d@m \times n}$ generating algorithm is summarized in Algorithm (1). N is the total number of visible pedestrians in the scene. L_j is the set of position coordinates for 18 key pose points for each pedestrian. n L_j is the number of coordinates contained in L_j, because not all 18 3D key points can be detected, for example, when a pedestrian faces away from the camera, the key points of his face are undetectable, so n $L_j \leq 18$. $D_{m \times n}$ is the image depth matrix generated by the depth module, and the value of each element is within [0, 1]. In this paper, we define $d = 32$ in $T_{d@m \times n}$. Therefore, we map the values of the elements in $D_{m \times n}$ to [1,32]. When obtaining the depth d_j of each pedestrian, due to the inevitable error in the depth module, some of the n L_j key points of the pedestrian will be at the wrong depth. KMean [22] algorithm is used to discard these abnormal depths. The subject pedestrian's bounding box coordinates (x_0, y_0, w, h) distinguish the target pedestrian from other pedestrians in the scene. The experimental part uses the first-person videos with 1080 * 1920 pixels, so we define $m = 54$, $n = 96$ in $T_{d@m \times n}$. Figure 2 shows some samples of the spatial and intention tensors generated by our data module.

3.3 Encoder and Decoder Stacks

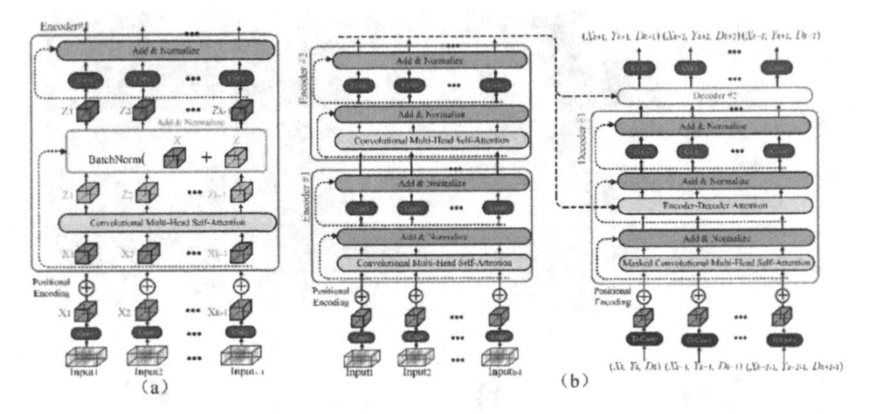

Fig. 5. The structure of a single encoder(a) and decoder(b).

Encoder. As shown in Fig. 2, the encoder is composed of 6 stacks of the same layer. As shown in Fig. 5(a), the tensor sequence generated by the data module first changes the size of the input tensor to 32 @ 32 $\times$ 32 through a convolution network, then adds the position information through position encoding to finally generate the tensor as input for a single encoder. A single encoder layer has two sublayers; the first is a multi-head convolutional self-attention mechanism, and the second is a simple convolutional network. We use residual connections around each of the two sublayers and then batch normalize. That is to say, the output of each sublayer is BatchNorm(x + Sublayer(x)), where Sublayer(x) is a function realized by the sublayer itself. To facilitate these residual connections, all sublayers in the model and the convolutional network layer produce a tensor output of size 32 @ 32 × 32.

Positional Encoding Tensor. Unlike the LSTM network, the transformer model does not contain a recurrent structure, so in order to learn the sequence order, we must label the input information with its relative or absolute position in the sequence. To do this, we add "position coding" to the input embedded at the bottom of the encoder and decoder stack. The dimension of the position coding is the same as the dimension inputted into the encoder, so the two can be added. In this paper, we refer to [9] and use the sine and cosine functions of different frequencies to construct positional encodings.

$$PE_{(pos,2i)} = sin(pos/10000^{2i/d_{model}}) \tag{1}$$

$$PE_{(pos,2i+1)} = cos(pos/10000^{2i/d_{model}}) \tag{2}$$

In formula (1) and (2), where pos is the position and i is the dimension. That is, each dimension of the positional encoding corresponds to a sinusoid. The wavelengths form a geometric progression from 2π to 10000•2π. We chose this function because we hypothesized it would allow the model to easily learn to attend to relative positions, since for any fixed offset k, PE_{pos+k} can be represented as a linear function of PE_{pos}. d_{model} is the length of the positional encoding; in this paper, the dimension of the positional encoding is 32@32 × 32, so $d_{model} = 32 \times 32 \times 32 = 32768$. . Thus, the dimension of vector PE_{pos} is 32768, and we can then transpose it into a tensor of size 32@32 × 32 as the terminal position encoding tensor.

Decoder. Decoder is composed of 6 stacks of the same layer. As shown in Fig. 5(b), comparing a single encoder and decoder, it can be seen that in addition to the two sub-layers in each encoder layer, the decoder also inserts a third sub-layer encoder-decoder attention layer that performs multi-head attention on the output of the encoder stack. Similar to the encoder, residual connections are used around each sub-layer, and then batch normalization is performed. We also modify the convolutional multi-head self-attention sub-layer in the decoder stack to prevent positions from attending to subsequent positions. This masking, combined with the fact that the output embeddings are offset by one position, ensures that the predictions for position i can depend only on the known outputs at positions before i.

3.4 Convolutional Multi-head Self-attention

As shown in Fig. 6, the input of the convolutional multi-head attention module is a tensor sequence X_{sqe} $(x_1, x_2, \ldots, x_h)$ with position coding added; h is the length of X_{sqe}. X_{sqe} generates A_{sqe} $(a_1, a_2, \ldots, a_h)$ through the convolution kernel K_a; A_{sqe} and X_{sqe} are of the same size. A_{sqe} generates the query sequence Q_{sqe} $(q_1, q_2, \ldots, q_h)$, keys sequence K_{sqe} $(k_1, k_2, \ldots, k_h)$, and value sequence V_{sqe} $(v_1, v_2, \ldots, v_h)$ through three convolution kernels, K_q, K_k, K_v. As for the multi-head attention mechanism, take q_1, k_1, v_1 as examples: we divide them into header sequences Q_1^{head} $(q_1^1, q_1^2, \ldots, q_1^n)$, K_1^{head}

$(k_1^1, k_1^2, \ldots, k_1^n)$, V_1^{head} $(v_1^1, v_1^2, \ldots, v_1^n)$ in the channel dimension; n is the number of heads. Q_{seq}^{head} $(Q_1^{head}, Q_2^{head}, \ldots, Q_h^{head})$ is convoluted by kernel K_{seq}^{head} $(K_1^{head}, K_2^{head}, \ldots, K_h^{head})$ to get a correlation coefficient sequence α_{seq} $(\alpha_{i,1}^l, \alpha_{i,2}^l, \ldots, \alpha_{i,h}^l)$, which generates attention scores $\widehat{\alpha}_{seq}$ $(\widehat{\alpha}_{i,1}^l, \widehat{\alpha}_{i,2}^l, \ldots, \widehat{\alpha}_{i,h}^l)$ by *SoftMax*. For the production process of the attention tensor sequence B_{sqe} $(b_1, b_2, \ldots, b_h)$, which consists of the outputs of the convolutional multi-head self-attention module, taking b_1 $(b_1^1, b_1^2, \ldots, b_1^n)$ as an example, $b_1^1 = \sum_{m=1}^{n} \alpha_{1,j}^1 \cdot v_1^m$. Formulas (3) - (9) are the overall calculation process of the convolutional multi-head attention module from the input tensor sequence X_{sqe} to the attention tensor sequence B_{sqe}, where '$*$' denotes the convolution operator and '$\cdot$' denotes the Hadamard product:

$$A_{sqe} = X_{sqe} * K_a \tag{3}$$

$$Q_{sqe} = A_{sqe} * K_q \tag{4}$$

$$K_{sqe} = A_{sqe} * K_k \tag{5}$$

$$V_{sqe} = A_{sqe} * K_v \tag{6}$$

$$\alpha_{seq} = K_{sqe} * V_{sqe} \tag{7}$$

$$\widehat{\alpha}_{seq} = softMax(\alpha_{seq}) \tag{8}$$

$$B_{sqe} = \Sigma \widehat{\alpha}_{seq} \cdot V_{sqe} \tag{9}$$

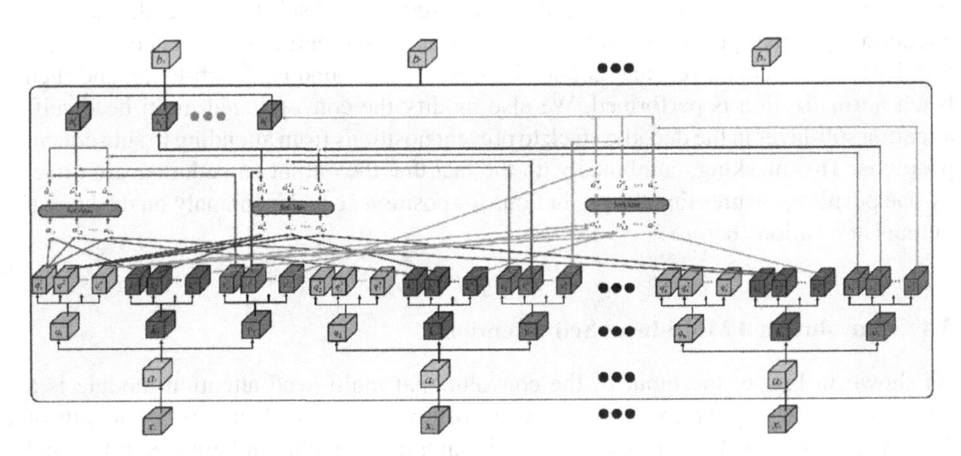

Fig. 6. The internal structure of the convolutional multi-head self-attention module.

As for the masked convolutional multi-head self-attention layers in the decoder, which allow each position in the decoder to attend to all positions in the decoder up to and including that position, we need to prevent leftward information flow in the decoder to preserve the auto-regressive property. We refer to [9] to mask out (set to $-\infty$) all values in the input of the softMax that correspond to illegal connections.

4 Experiments

This work evaluates the proposed method using the public benchmark: Multiple Object Tracking Benchmark 2016 (MOT16) [10]. Instead of ETH [7] or UCY [8], taken from a bird's-eye view, the MOT datasets consist of more practical videos from a first-person view or tilted camera view. We compare our method with state-of-the-art approaches in terms of several standard metrics.

4.1 Datasets and Metrics

Datasets. MOT16 contains 14 challenging video sequences in unconstrained environments filmed with both static and moving cameras. All sequences have been annotated using the MOT benchmark with high accuracy, strictly following a well-defined protocol. Following [1, 2], our method observes 3.2 s (8 frames) for each person and predicts the next 4.8 s (12 frames) of that person's trajectory.

The video sequences of MOT16 have two shooting angles: bird's-eye view and side view, in this paper, we mainly study the pedestrian trajectory prediction in the first-view video sequence, so we choose 5 of the first-person view video sequences (MOT16-02, MOT16-09, MOT16-10, MOT16-11, MOT16-13) as training and test datasets (11324 available pedestrian trajectories). We downsample the videos to 2.5 fps. Since we do not have a homographic matrix, we adopt the pixel values for the trajectory coordinates, as in [3].

Metrics. We evaluate our model using the following three metrics:

Average Displacement Error (ADE). The average Euclidean distance between the ground truth coordinates and the predicted coordinates over all time instants.

$$ADE = \frac{\sum_{i=1}^{N}\sum_{t=1}^{T_{pred}}||\tilde{Y}_t^l - Y_t^i||_2}{N * T_{pred}} \tag{10}$$

Final Displacement Error (FDE). The Euclidean distance between the predicted points and the ground truth point at the final predicted time instant T_{pred}.

$$FDE = \frac{\sum_{i=1}^{N}||\tilde{Y}_{T_{pred}}^l - Y_{T_{pred}}^i||_2}{N} \tag{11}$$

Average Depth Error (A-Depth-E). The average Euclidean distance between the ground truth depths and the predicted depths over all time instants. The depth value is normalized to [0, 1].

$$A - Depth - E = \frac{\sum_{i=1}^{N}\sum_{t=1}^{T_{pred}}||\tilde{D}_t^l - D_t^i||_2}{N * D_{pred}} \tag{12}$$

4.2 Implementation Details

Our convolutional transformer (Conv-Transformer) is trained with an Adam optimizer [23], an extension to stochastic gradient descent, to update network weights during the training process. The learning rate is 0.001 and the dropout value is 0.2. We adopt a gradient clipping of 10 and weight decay of 0.0001. The model is trained on a GTX1080Ti GPU. We refer to [1, 2] to train and test Conv-Transformer and compare it with existing methods.

Training. We refer to [1, 2] to train and test Conv-Transformer and compare it with existing methods. The training phase is divided into two stages:

Stage 1. We select four out of five video sequences (MOT16-02, MOT16-09, MOT16-10, MOT16-11 and MOT16-13) for training and verification. We select 80% of each video sequence as the training set and 20% as the verification set. The best model parameters during the training are those that make the ADE of the verification set the lowest. The remaining, fifth data set is left for use in Stage 2. The above process is called "leave-one-out." For the five video sequences, we repeat the above process five times to train/verify the four videos in Stage 1 and obtain the five sets of model weight parameters used in Stage 2 for each remaining screen sequence that is not model-learned. We execute 100 epochs in each training session.

Stage 2. The model parameters trained and validated by Stage 1 are further trained on the fifth, un-modeled video sequence, and the weight parameters are further updated to obtain the best performing model. At this stage, 50% of the fifth video sequence is used to train 10 epochs. We obtain the best model parameters and adopt the remaining 50% of the data in the fifth video sequence for testing.

Testing. The model is initialized with the final set of weights from Stage 2. The testing process is repeated at each frame by observing eight frames (3.2 s) and then predicting the next 12 frames (4.8 s) in a sliding window manner.

Loss Function. For the fully convolutional network, we apply the Smooth-L1 loss function in (13), where N is the length of the model prediction output sequence, k is the length of the model input sequence, and $Y_{T_{pred}}$ and Y_T are the predicted and true values of the time T.

$$\mathrm{L} = \begin{cases} \frac{1}{N}\sum_{T=k+1}^{k+N} 0.5 * (Y_{T_{pred}} - Y_T)^2, \left|Y_{T_{pred}} - Y_T\right| \leq 1 \\ \frac{1}{N}\sum_{t=k+1}^{k+N} \left|Y_{T_{pred}} - Y_T\right| - 0.5, \left|Y_{T_{pred}} - Y_T\right| > 1 \end{cases} \tag{13}$$

4.3 Comparison with State-of-the-Art Methods

We implement the same training and testing procedures for three kinds of state-of-the-art methods, including social-interaction models (Social-LSTM [1], Social-GAN [2]), a first-person video prediction model (CNN [3]), and an encoder-decoder model base on convolutional LSTM (Conv-LSTM)) on MOT16.

Social-LSTM [1]: We evaluated Social LSTM (one of the latest methods of human trajectory prediction) and made some minor modifications to better handle first-person video. We train the Social-LSTM model to directly predict trajectory coordinates instead of Gaussian parameters. We added the image depth information to the input. The method of obtaining the depth information is the same as the data module in this article in two steps, first detecting the 3D key points of the pedestrian, and then using the KMean [22] algorithm to find the depth of most key points as the depth information of the pedestrian.

SocialGan [2]: SocialGan added adversarial training on Social-LSTM to improve performance. We train the model in the paper base on the released code from SocialGan (https://github.com/agrimgupta92/sgan/). We also added the image depth information to the model input as the above Social-LSTM.

CNN [3]: A new method of pedestrian trajectory prediction in first-person video, this method uses the size of the pedestrian in the image to define the relative distance between the pedestrian and the camera, and only focus on the posture of the subject pedestrian and ignores the intentions of other pedestrians in the scene. We train two model variants (CNN & CNN-D) detailed in the paper. CNN uses the pedestrian size in [3] to represent the distance between the pedestrian and the camera, and CNN-D uses the image depth information in our paper to indicate the distance between the pedestrian and the camera.

Conv-LSTM: In this paper, we replace the convolutional multi-head self-attention module in Fig. 5 with a convolutional LSTM, and the rest of the structure remains unchanged to build an end-to-end sequence-to-sequence model. Its purpose is to compare the effectiveness of LSTM and convolutional multi-head self-attention modules in predicting trajectories.

Quantitative Results. All methods predict human trajectories in 12 frames, using 8 observed frames. Each score describes the ADE and FDE in pixels with respect to the frame size of 1080 × 1920-pixels. The A-Depth-E is normalized to [0, 1]. The testing results for each sequence are calculated on the last 50% of the data of each video sequence. The quantitative results in Table 1 show that our Conv-Transformer outperforms the other methods with respect to two metrics (ADE and FDE) for all video sequences. Better ADE and FDE than with other methods are achieved in all sequences. The results in Table 1 demonstrate that Conv-Transformer is more efficient in predicting the pedestrian's final position, because Conv-Transformer successfully learns their advancement intentions from the 3D key points of their pose. MOT16–02 and MOT16–09 were taken by a fixed first-person view camera, MOT16–10, MOT-11, and MOT16–13 were taken by a mobile first-person view camera. The results in Table 1 show that Conv-Transformer performs stably in videos with both fixed and mobile first-person view camera. Since walking speeds and directions of people were quite diverse and changed dynamically over time, naive baselines like CNN and SocialGan did not perform well. Moreover, we found that Social-LSTM [1] performed poorly. Compared with CNN and CNN-D, it is not difficult to find that in the first-person video, the depth information of the pedestrian (the distance between the pedestrian and the camera) can slightly improve the prediction performance of the model. When Conv-LSTM and Conv-Transformer use the same sequence tensors as the training and test sets, the results show

Table 1. Quantization results of related methods in 5 different video sequences from MOT16.

Metrics	Sequences	Social-LSTM	Social Gan	CNN	CNN-D	Conv-LSTM	Our
ADE	MOT16-02	105.64	95.36	72.32	69.54	63.32	49.31
	MOT16-09	107.39	98.31	70.51	68.33	65.21	38.88
	MOT16-10	120.54	99.98	79.36	72.28	69.55	45.21
	MOT16-11	124.69	100.31	80.97	77.98	72.91	47.22
	MOT16-13	123.47	105.76	89.31	83.57	75.88	43.78
	Average	116.35	99.94	78.49	74.34	69.37	**44.88**
FDE	MOT16-02	110.34	100.66	75.39	73.21	70.13	50.65
	MOT16-09	111.28	105.87	76.21	75.66	71.47	49.47
	MOT16-10	129.47	112.04	85.97	79.95	75.32	51.71
	MOT16-11	129.39	111.07	86.38	84.22	77.87	49.97
	MOT16-13	134.22	120.64	94.36	88.69	80.14	58.64
	Average	122.94	110.06	83.66	80.35	74.99	**52.01**
A-Depth-E	**Average**	–	–	–	–	–	0.074

that the convolutional multi-head self-attention module can mine the input data more effectively than LSTM to reduce the error of trajectory prediction. As for A-Depth-E, our Conv-Transformer performed well, with a small error.

Qualitative Evaluation. Figure 7 presents several visual examples of how well each method worked. The red line represents our Conv-Transformer, the green line represents the ground truth, and the brightness of the line represents the depth of the pedestrian in the image. Examples (a) and (b) with a fixed first-person view camera show that Conv-Transformer more accurately predicts the trajectories of pedestrians and successfully captures their intentions in a simple scene. Examples (c) and (d) with a mobile first-person view camera show that only Conv-Transformer can decrease prediction error. Example (e) is a video of a moving bus. In Example (e), our Conv-Transformer accurately predict that the subject pedestrian crosses the road, as well as their final position. Conv-Transformer successfully performed in this case because it could capture the depth and postural changes of the subject pedestrian for prediction.

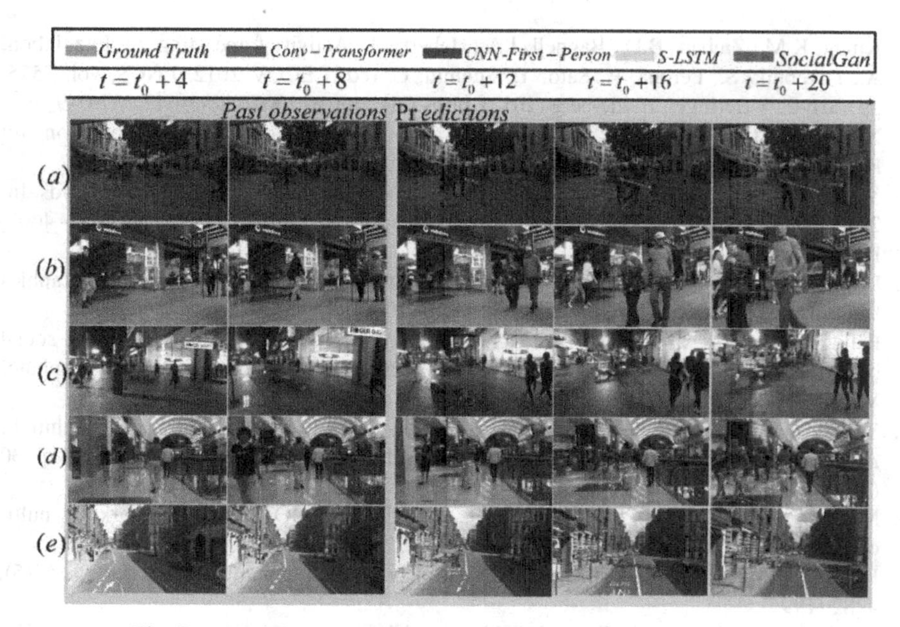

Fig. 7. Visual Examples of Future Person Localization and Depth.

5 Conclusion

This work comprehensively considers the social spatial, temporal, and intention interaction factors of pedestrian trajectory prediction in first-person videos. It first uses the depth of an image to model the relative distance between the subject pedestrian and the camera and to predict this depth in the future. Then, it considers the 3D key pose points of all people in the scene to represent the social intention interaction among these people. Finally, we integrate people's spatial positions (*xyz*-coordinates and depths) and the 3D key pose points into a multi-channel tensor to represent the practical 3D scene data. As for the deep neural network, we construct an end-to-end fully convolutional model based on a transformer that solely uses attention mechanisms, dispensing with recurrence entirely. The experimental results on public benchmark MOT16 demonstrate that the proposed Conv-Transformer outperforms state-of-the-art methods.

References

1. Alahi, A., Goel, K., Ramanathan, V., Robicquet, A., Fei-Fei, L., Savarese, S.: Social LSTM: human trajectory prediction in crowded spaces. In: Proceedings of the IEEE Conference on Computer Vision and Pattern Recognition, pp. 961–971 (2016)
2. Gupta, A., Johnson, J., Fei-Fei, L., Savarese, S., Alahi, A.: Social GAN: socially acceptable trajectories with generative adversarial networks. In: Proceedings of the IEEE Conference on Computer Vision and Pattern Recognition, pp. 2255–2264 (2018)
3. Yagi, T., Mangalam, K., Yonetani, R., Sato, Y.: Future person localization in first-person videos. In: Proceedings of the IEEE Conference on Computer Vision and Pattern Recognition, pp. 7593–7602 (2018)

4. Kitani, K.M., Ziebart, B.D., Bagnell, J.A., Hebert, M.: Activity forecasting. In: Fitzgibbon, A., Lazebnik, S., Perona, P., Sato, Y., Schmid, C. (eds.) ECCV 2012. LNCS, vol. 7575, pp. 201–214. Springer, Heidelberg (2012). https://doi.org/10.1007/978-3-642-33765-9_15
5. Nikhil, N., Tran Morris, B.: Convolutional neural network for trajectory prediction. In: Proceedings of the European Conference on Computer Vision (ECCV), (2018)
6. Vemula, A., Muelling, K., Oh, J.: Social attention: modeling attention in human crowds. In: 2018 IEEE international Conference on Robotics and Automation (ICRA), pp. 4601–4607. IEEE (2018)
7. Lerner, A., Chrysanthou, Y., Lischinski, D.: Crowds by example. In: Computer graphics forum, vol. 26, pp. 655–664. Blackwell, Oxford (2007)
8. Pellegrini, S., Ess, A., Schindler, K., Van Gool, L.: You'll never walk alone: Modeling social behavior for multi-target tracking. In: 2009 IEEE 12th International Conference on Computer Vision, pp. 261–268. IEEE (2009)
9. Vaswani, A., Shazeer, N., Parmar, N., Uszkoreit, J., Jones, L., Gomez, A. N., Polosukhin, I.: Attention is all you need. In: Advances in Neural Information Processing Systems, vol. 30 (2017)
10. Milan, A., Leal-Taixé, L., Reid, I., Roth, S., Schindler, K.: MOT16: a benchmark for multi-object tracking. arXiv preprint arXiv:1603.00831 (2016)
11. Helbing, D., Molnar, P.: Social force model for pedestrian dynamics. Phys. Rev. E **51**(5), 4282 (1995)
12. Manh, H., Alaghband, G.: Scene-LSTM: a model for human trajectory prediction. arXiv preprint arXiv:1808.04018 (2018)
13. Jaipuria, N., Habibi, G., How, J. P.: A transferable pedestrian motion prediction model for intersections with different geometries. arXiv preprint arXiv:1806.09444 (2018)
14. Haralick, R.M., Shanmugam, K., Dinstein, I.H.: Textural features for image classification. IEEE Trans. Syst. Man Cybern. **6**, 610–621 (1973)
15. Lian, J., Ren, W., Li, L., Zhou, Y., Zhou, B.: PTP-STGCN: pedestrian trajectory prediction based on a spatio-temporal graph convolutional neural network. Appl. Intell. **53**(3), 2862–2878 (2023)
16. Simonyan, K., Zisserman, A.: Very deep convolutional networks for large-scale image recognition. arXiv preprint arXiv:1409.1556 (2014)
17. Liang, J., Jiang, L., Niebles, J.C., Hauptmann, A.G., Fei-Fei, L.: Peeking into the future: Predicting future person activities and locations in videos. In: Proceedings of the IEEE/CVF Conference on Computer Vision and Pattern Recognition, pp. 5725–5734 (2019)
18. Kooij, J.F.P., Schneider, N., Flohr, F., Gavrila, D.M.: Context-based pedestrian path prediction. In: Fleet, D., Pajdla, T., Schiele, B., Tuytelaars, T. (eds) Computer Vision–ECCV 2014. ECCV 2014. Lecture Notes in Computer Science, vol. 8694, pp. 618–633. Springer, Cham (2014). https://doi.org/10.1007/978-3-319-10599-4_40
19. Li, Z., et al: Learning the depths of moving people by watching frozen people. In: Proceedings of the IEEE/CVF Conference on Computer Vision and Pattern Recognition, pp. 4521–4530 (2019)
20. Li, Y.: Pedestrian path forecasting in crowd: a deep spatio-temporal perspective. In: Proceedings of the 25th ACM International Conference on Multimedia, pp. 235–243 (2017)
21. Li, Z., Oskarsson, M., Heyden, A.: 3D human pose and shape estimation through collaborative learning and multi-view model-fitting. In Proceedings of the IEEE/CVF Winter Conference on Applications of Computer Vision, pp. 1888–1897 (2021)
22. Krishna, K., Murty, M.N.: Genetic K-means algorithm. IEEE Trans. Syst. Man Cybern. Part B (Cybern.) **29**(3), 433–439 (1999)
23. Kingma, D.P., Ba, J.: Adam: a method for stochastic optimization. arXiv preprint arXiv:1412.6980 (2014)

24. Bochkovskiy, A., Wang, C.Y., Liao, H.Y.M.: YOLOv4: optimal speed and accuracy of object detection. arXiv preprint arXiv:2004.10934 (2020)
25. Fu, J., Zhao, X.: Action-aware encoder-decoder network for pedestrian trajectory prediction. J. Shanghai Jiaotong Univ. (Sci.) **28**(1), 20–27 (2023)
26. Huang, X., Liu, Q., Yang, Y.: Triple GNN: a pedestrian-scene-object joint model for pedestrian trajectory prediction. In: Fang, L., Povey, D., Zhai, G., Mei, T., Wang, R. (eds.) Artificial Intelligence. CICAI 2022. Lecture Notes in Computer Science, vol. 13604, pp. 67–79. Springer, Cham (2022). https://doi.org/10.1007/978-3-031-20497-5_6

Conceive-Design-Implement-Operate (CDIO) Approach in Producing Wiring Projects for Domestic Electrical Wiring Course

Hanifah Jambari[1](✉), Nurul Farahin Ismail[2], Ishak Taman[3], Mohamad Rasidi Pairin[1], and Muhammad Fathullah Hamzah[1]

[1] School of Education, Faculty of Social Sciences and Humanities, Universiti Teknologi Malaysia, Johor Bahru, Malaysia
hanifahjambari@gmail.com

[2] Sekolah Menengah Kebangsaan Sagil, Muar Johor, Malaysia

[3] Politeknik Ibrahim Sultan, Pasir Gudang, Johor, Malaysia

Abstract. This descriptive study aims to evaluate the Conceive-Design-Implement-Operate (CDIO) approach in producing wiring projects for the Domestic Electrical Wiring course among students from the Technical with Education courses at Universiti Teknologi Malaysia. This study is based on four main objectives, which are to evaluate the elements of "conceive, design, implement and operate" among students completing wiring projects in the Domestic Electrical Wiring course. The methodology used in this study is a quantitative. A set of questionnaires was used as a research instrument for data collection by distributing a Google form to respondents based on the research questions that have been constructed. The questionnaire used in this study was divided into five parts, namely Part A about demographics, Part B about the "conceive" element in electrical wiring, Part C about the "design" element in electrical wiring, Part D about the "implement" element in electrical wiring and Part E about the "operate" element in electrical wiring. The selection of respondents is random. The respondents of this study consisted of 110 students in the fourth and third year of electric and electronics, and life skills education programme. The results of the study were analysed using the Statistical Package for Social Science (SPSS) software version 26.0. Overall, the results of the study showed that students can design their work process according to the actual CDIO standards. In conclusion, this study is very important for students to complete their project properly on time, especially for students at the Institute of Technical and Vocational Education, Engineering and other related fields, as well as for teaching staff at the institutes related to the production of a project that is implemented.

Keywords: Conceive · Design · Implement · Operate · Domestic Electrical Wiring Course

© The Author(s), under exclusive license to Springer Nature Singapore Pte Ltd. 2024
F. Hassan et al. (Eds.): AsiaSim 2023, CCIS 1911, pp. 60–70, 2024.
https://doi.org/10.1007/978-981-99-7240-1_5

1 Introduction

Technical knowledge and skills comprise a large part of the educational curriculum that is essential for both graduates and the industry. In particular, new types of graduates who have skills that go beyond technical mastery (Duderstadt, 2008); (Alessandro Testa, 2015) in which they possess skill processes such as leadership skills, teamwork, and communication, are in high demand from industry. Therefore, CDIO which means "conceive", "design", "implement", and "operate" is one of the initiatives to design the work process of students in a more systematic way, which is of higher quality (Hanifah et.al 2019).

This standard has been widely adopted for engineering courses around the world and can also be used for courses that require processes or working steps in their learning. This is an innovative educational framework to produce a generation of quality leaders in the future (Hanifah et.al 2019). Industry will benefit because CDIO produces graduates with the knowledge, talent and experience it specifically needs. Educators are interested in this approach because the CDIO syllabus forms the basis for curriculum planning and outcomes-based assessment that can be universally adapted for all technical schools. In addition, students will be more enthusiastic because they can meet all needs with various personal, interpersonal, and system development experiences that allow them to excel in the real world, and produce new products and systems. This is essential to maintain productivity, innovation, and excellence in an environment based on increasingly complex technological systems. In recent years, conflicts have arisen between technical education pedagogies, and to resolve this contemporary technical education conflict, education must have a new concept and a new vision must be developed.

The CDIO initiative envisages education that emphasises the principles set in the context of practising, planning, implementing and operating product systems (Yuxin et al. 2021). However, in addition to emphasising technical fundamentals, it must prepare students to play a successful role in developing product systems. The curriculum is structured according to CDIO standards in a very simple manner. The standards are mutually supportive and interactive. The programme must be rich in student projects complemented by experience in industry, and feature active learning groups, experiences and groups in classrooms and workspaces or modern learning laboratories, connected to the outside world. It is constantly improved through a comprehensive assessment and evaluation process.

To improve the pedagogy of the CDIO initiative, it must consider what educators know about student experiences and their impact on learning (Crawley et al. 2014). To address these and other learning needs, the CDIO initiative sets improvements in four basic areas: increased active and hands-on learning, emphasis on problem formulation and solving, increased emphasis on conceptual learning, and increased learning feedback mechanisms. Educational research confirms that active learning techniques improve student learning dramatically. CDIO's emphasis on active learning encourages students to take a more active role in themselves while studying. Hands-on and team-based learning are important examples of active learning, but techniques can be used to increase student activity even in conventional classroom settings. Problem solving is an important skill in the technical field. The CDIO initiative supports learning in problem formulation, estimation, modelling and solution. Related efforts in concept learning serve to ensure that students master tools and skills as well as deeper basic concepts. The CDIO initiative is a new way to get feedback and decisions on the way students learn today.

CDIO which stands for 'conceive-design-implement-operate' is one of the educational frameworks for systems and products in real industry. The CDIO syllabus consists of: technical knowledge and reasoning, personal and professional skills and interpersonal skills. This framework allows education students to emphasize the basics of wiring according to the standards in the context of 'conceive, design, implement and operate'. The purpose of CDIO at Universiti Teknologi Malaysia (UTM) is to create projects or learning based on familiar problems. In the application of this approach, learning and teaching in the field of engineering aims to promote innovative thinking so as to equip students with problem-solving skills to a high level in a way to build theory while improving efficiency and practical abilities. Thus, this research was conducted to identify the implementation of the CDIO in domestic electrical wiring practise for technical and vocational education students with the appropriated knowledge and skills in the specified time.

1.1 Research Problem

Electrical education students should have technical knowledge and skills in wiring before graduated and working into the real world. However, during wiring in class, disorganized such done their work resulting in less neat wiring and pressure to complete the practice on time.

In addition, to sketch the wiring diagram and calculating various factors may took a lot of time and has an impact on completing the wiring. This is because of time management by students who always delay to start sketching the wiring diagram. Student should have sufficient knowledge in terms of all aspects such as how to use equipment and materials, knowledge of safety in the workshop and technical work that requires certain techniques to prevent unwanted things from happening. The attitude of doing work at the last minute and also relying too much on workshop assistants is also often a problem for students in the preparation of materials in the workshop such as vessels, 2 mm cables, 4 mm cables, hand tools and so on. Most of the time when students want to start practical in the workshop, they have been confused about what needs to be done first and that will take up their time to complete the project as required.

As a solution to these problems, planning the framework before starting the project will produce a higher quality product when students take their respective roles during active learning. Thus, the CDIO approach was implemented during wiring project to overcome the problem as stated earlier. Furthermore, the use of this approach, learning and teaching in the field of education aims to promote innovative thinking in order to equip students with high-level problem-solving skills by building theory while improving efficiency and practical abilities.

1.2 Research Objective

The objective of this study is to evaluate the elements of; conceive, design, implement and operate among students while completing wiring projects in the domestic electrical wiring course that follow the CDIO rule.

2 Methodology

This quantitative study was carried out and involved a total of 110 students from electric and electronics and life skills education programme at Universiti Teknologi Malaysia. The results of data collection were analysed using the Statistical Package for Social Science (SPSS) software version 26.0 to obtain frequency values, percentages and mean scores. The research instrument used in this study was a questionnaire, which was distributed using Google form to collect data. The items in this questionnaire are divided into five sections, namely Section A regarding the respondent's demographic data (3 items), Section B regarding the conceive element (8 items), Section C regarding the design element (7 items), Section D regarding the implement element (7 items) and Section E regarding the operating elements (7 items). The researcher analysed each element of CDIO applied in the learning of the domestic electrical wiring course. Table 1 below shows the constructs of the items presented in the questionnaire for this study.

Table 1. Construct and sub-construct of items in the questionnaire.

Section	Main Construct	Sub-Construct	No. of Items
A	Demographic Data	Course Gender Age	3
B	Conceive Element in Electrical Wiring	Conceive	8
C	Design Element in Electrical Wiring	Design	7
D	Implement Element in Electrical Wiring	Implement	7
E	Operate Element in Electrical Wiring	Operate	7

The items developed in Part B, C, D and E use a Likert-type scale that contains five scale options. This scale is also often used and is familiar to respondents, making it easier for respondents to understand how to answer this questionnaire. The measurement level for the Likert scale is shown in Table 2.

Table 2. Likert Scale.

Measurement Scale	Scale
Strongly Disagree	1
Disagree (D)	2
Not Sure (N)	3
Agree (A)	4
Strongly Agree (SA)	5

Descriptive statistics is a summary of data obtained from a data processing system whose results can be seen in the form of information, tables, diagrams or graphs (Mohd Majid, 1990). The data analysis in this study only takes the mean score value to interpret the data obtained. The interpretation of the mean score is shown in Table 3 below.

Table 3. Mean Score scale and interpretation.

Mean Score	Mean Score Interpretation
1.00 – 2.33	Low
2.34 – 3.66	Medium
3.67 – 5.00	High

3 Findings

The acquisition of data about this CDIO approach involved 110 students of electric and electronics and life skills education at Universiti Teknologi Malaysia who learned the domestic electrical wiring course. Table 4 shows the demographic distribution of the respondents involved in this study.

Table 4. Demographic Distribution of Study Respondents.

Background	Information	Frequency	Percentage
Gender	Male	60	54.5
	Female	50	45.5
Age	21 – 23 years	54	49.1
	24 – 26 years	48	43.6
	27 years and above	8	7.3
Course	4SPPR	14	6.4
	4SPPH	46	48.9
	3SPPR	22	23.4
	3SPPH	15	2.1
	2SPPR	13	19.1

3.1 Descriptive Findings

The mean scores of all the study constructs involving the elements of conceive, design, implement and operate are shown in Table 5 below.

Table 5. Mean Values of Study Constructs.

No	Item Construct	Mean
1	Conceive Element	4.36
2	Design Element	4.18
3	Implement Element	4.27
4	Operate Element	4.21

3.2 "Conceive" Element

Table 6 shows that each item of the knowledge aspect is at a high level. The highest mean value is 4.64, for the statement "I can identify the safety elements in the workshop". This item has a high mean value in the "conceive" element because students understand and have been exposed to safety practices since they were in school. This statement is supported by (Hashim, 2005) where students are made aware and understand about safety practices during the workshop. Hence, each individual has the value of awareness and common sense in identifying the elements of safety in the workshop, while the lowest mean value is 4.17, for the statement "I really understand the theories taught and can put them into practice in the workshop". Through the questionnaire items for this "conceive" section, we can identify where the students' weaknesses are before doing the wiring and what causes them to be slow in the learning process.

Table 6. Mean Scores for Conceive Element.

No	Item	Mean
1	I understand the learning concept taught by the lecturer	4.32
2	I can imagine the steps to take next	4.20
3	I really understand the theories taught and can put them into practice in the workshop	4.17
4	I can identify the needs of users in domestic electrical wiring	4.38
5	I can meet the needs of users according to the IEE rules that have been set	4.26
6	I can learn the hand tools needed in domestic electrical wiring	4.40
7	I can identify the safety elements in the workshop	4.64
8	I can identify how to use the hand tools in the workshop	4.51
Overall mean score		4.36

3.3 "Design" Element

Table 7 shows that the highest mean value is 4.45, for the statement "which is in reality I can arrange safety measures before starting domestic electrical wiring". Meanwhile, the lowest mean value is 3.85, for the statement "I can calculate various factors before sketching the electrical wiring diagram". Students understand how to plan wiring projects and understand all safety measures in the workshop where they follow CDIO standards for "design" element. These data are supported by (Lee et al. 2018), where design thinking is a human-centred approach to designing products, processes, systems and services.

Table 7. Mean Scores for Design Element

No	Item	Mean
1	I can arrange safety measures before starting domestic electrical wiring	4.45
2	I can organise the material requirements in electrical wiring practice	4.36
3	I can strategies before doing the wiring	4.36
4	I can draw a domestic electrical wiring diagram according to the established IEE regulations	4.09
5	I can calculate various factors before sketching the electrical wiring diagram	3.85
6	I can identify the compatibility of components before sketching the wiring diagram	4.02
7	I can identify the load requirements after calculating various factors	4.15
Overall mean score		4.18

3.4 "Implement" Element

Table 8 shows the mean score values of the CDIO approach for the Implement element. The highest mean value is 4.43, for the statement "I can test the functionality of electrical wiring correctly" while the lowest mean value is 4.06, for the statement "I can apply domestic electrical wiring according to IEE regulations". The activities and learning outcomes applied in these two courses have the same objective and follow the CDIO standards from NALI UTM. In general, most students understand the use and importance of CDIO in domestic electrical wiring. Students are confident and understand how to test the functionality of their wiring by implementing existing skills. CDIO challenges students to be more competitive. This statement is supported by (Batdorj et al. 2018), where students carry out projects every year and they get the opportunity to use their knowledge and skills that have been acquired previously while working in teams to design and develop certain products, systems or processes.

Table 8. Mean Scores for Implement Element

No	Item	Mean
1	I can apply domestic electrical wiring according to IEE regulations	4.06
2	I can identify the needs and suitability of the load required by the user	4.32
3	I can identify the needs and suitability on the load required by the user	4.30
4	I can determine the size of MCB and ELCB required according to user requirements based on IEE requirements	4.32
5	I can test the functionality of the electrical wiring correctly	4.43
6	I can perform domestic electrical wiring correctly	4.28
7	I can calculate the various factors correctly based on the IEE table	4.15
Overall mean score		4.27

3.5 "Operate" Element

Table 9 shows the mean score values for the operate element. The highest mean value is 4.36, for the statement "I refer to the safety procedure guide when doing the testing", while the lowest mean value is 4.00, for the statement "I know how to repair the wiring circuit if the circuit does not work". Students learn to solve problems and complete

Table 9. Mean Scores for Operate Element

No	Item	Mean
1	I can identify ways to perform dead circuit testing	4.30
2	I can identify ways to perform live circuit testing	4.32
3	I know how to repair a wiring circuit if the circuit does not work	4.00
4	I refer to the safety procedure guide when doing the testing	4.36
5	I can identify all types of electrical tools in the workshop well	4.34
6	I can make circuit repair steps if it does not work. I can correctly use all types of electrical wiring testing tools	4.06
7	I know the maintenance required after the wiring is operational	4.06
Overall mean score		4.21

projects according to CDIO standards. In this context, lecturers should play a role not only as teaching staff but also act as facilitators when doing electrical wiring by monitoring and guiding students where according to (Lan and Vu, 2018), 74 is the basis that students go through the learning process and the role of lecturers as a facilitator, is to guide students by presenting some ideas, methods, and tools for teaching.

4 Discussion

Subsequent to improve the teaching and learning of technology students, as well as electric and electronics and life skills education at UTM, CDIO aims to make four important improvements in which the programme must increase active and practical learning, emphasise formulation and problem solving, and explore carefully the basic concepts of tools or components and wiring techniques, and finally collect feedback in an innovative way to interest students. This improvement aims to fully engage educational students in all stages of life productivity. In addition, another importance of the CDIO approach in wiring is that CDIO promotes project-based and purposeful goal-oriented learning. Learning outcomes also need to be clearly stated to students before they do any wiring work or before instructions are given. It also has prompted curriculum reform that includes design and construction projects to coordinate and link other subjects in electrical wiring. This statement is supported by (Batdorj et al., 2018) where students implement a project every year and they get the opportunity to use their knowledge and skills that have been acquired previously while working in a team to design and develop a specific product, system or process.

Also, because of the innovative teaching style, initiative is needed in the alternative assessment process. It is important to continue making comprehensive assessments of the students and the programme itself. CDIO is open and available to all programs in the University in adapting themselves to specific needs especially in electrical education and life skills education courses at UTM.

The difference between the current approach and the CDIO approach is that the current approach is heavy on learning outcomes where lecturers only know what they think students are capable of doing. But it is slightly different from using the CDIO approach where the learning outcomes are determined by what graduates expect in the future world of work such as competence in work. According to (Piironen and Padley, 2018), students learn to solve problems and complete projects according to CDIO standards. In this context, the lecturer should play a role not only as an instructor but also as a facilitator when doing electrical wiring by monitoring and guiding students; according to (Lan and Vu, 2018), it is fundamental that students go through the learning process, and the role of the lecturer as a facilitator, is to guide students by presenting some ideas, methods, and tools for teaching. Therefore, students will be more competitive in doing their electrical wiring.

5 Conclusion

To improve the teaching and learning of technology students, as well as electric and electronics and life skills education at UTM, CDIO aims to make four important improvements in which the programme must increase active and practical learning, emphasise formulation and problem solving, and explore carefully the basic concepts of tools or components and wiring techniques, and finally collect feedback in an innovative way to interest students. This improvement aims to fully engage educational students in all stages of life productivity. In addition, another importance of the CDIO approach in wiring is that CDIO promotes project-based and purposeful goal-oriented learning. Learning outcomes also need to be clearly stated to students before they do any wiring work or before instructions are given. It has also prompted curriculum reform that includes design and construction projects to coordinate and link other subjects in electrical wiring. This aspires to create a challenging experience where students design, build and operate electrical wiring. Also, because of the innovative teaching style, initiative is needed in the alternative assessment process. CDIO is open and available to all programmes in the University in adapting themselves to specific needs especially in electric and electronics and life skills education courses at UTM. The difference between the current approach and the CDIO approach is that the current approach is heavy on learning outcomes where lecturers only know what they think students are capable of. But it is slightly different from using the CDIO approach where the learning outcomes are determined by what graduates expect in the future world of work such as competence in work. Using the CDIO approach, CDIO elements can be evaluated mainly on the use of knowledge such as processes, rubrics and others.

Acknowledgement. The authors would like to acknowledge the financial support from the Ministry of Higher Education under Fundamental Research Grant Scheme (FRGS) with vote number, FRGS/1/2020/SS0/UTM/02/6 and Universiti Teknologi Malaysia for the funding under UTM Encouragement Research (UTMER) with vote number Q. J130000.3853.20J05.

References

Testa, A., Cinque, M., Coronato, A., De Pietro, G., Augusto, J.S.: Heuristic strategies for assessing wireless sensor network resiliency: an event-based formal approach. J. Heuristics **21**, 145–175 (2015)

Rashid, A., Salleh, A.N., Halim, M. A. R.: Inovasi dan Teknologi dalam P&P: Pengesan Korosakan Pendawaian Elektrik Domestik Satu Fasa. Persidangan Kebangsaan Penyelidikan Dan Inovasi Dalam Sistem Pendidikan Dan Latihan Teknikal Dan Vokasional., pp. 473–483 (2012)

Bachtiar, Y.: Penerapan Konsep CDIO pada Praktikum Pemprograman Komputer Sebagai Media Pembelajaran Kreatif dan Aplikasi. Gastrointest. Endosc. **10**(1), 279–288 (2018)

Batdorj, T., Purevsuren, N., Purevdorj, U., Tungalag, U., Gonchigsumlaa, K.: Experience of Developing Students' CDIO Skills Using (2018)

Chua, P.: CDIO Experience For New Faculty: Integrating CDIO Skills into A Statistics Module (2011)

Crawley, E.F., Lucas, W.A.: The CDIO Syllabus V2. 0 An Updated Statement of Goals for Engineering Education (2011)

Crawley, E., Malmqvist, J., Östlund, S., Brodeur, D., Edström, K.: Rethinking Engineering Education: The CDIO Approach, 2nd edn. Springer Verlag, New York (2014)

Jambari, H.: Impacts of conceive-design-implement-operate knowledge and skills for innovative capstone project. Inter. J. Online Biomed. Eng. (iJOE) – **15**(10), 146–153 (2019)

Hanif, A.S., Azman, M.N., Pratama, H., Nazirah, N., Imam, M.: Kit pemantauan penyambungan litar elektrik: satu kajian efikasi alat bantu mengajar. Geografia : Malaysian J. Soc. Space **12**(3), 69–78 (2016)
Hafiz, H., Mohd, Y.S.: e-Learning Pendawaian Domestik (2013)
Hashim, A.B.H.: Statistical Package for Social Sciences (SPSS) (2005)
Lan, T., Vu, A.: Building CDIO Approach training programmes against challenges of industrial revolution 4. 0 for engineering and technology. Development **11**(7), 1129–1148 (2018)
Lee, C., Lee, L., Kuptasthien, N.: Design Thinking for CDIO Curriculum Development (June 2018). http://ds.libol.fpt.edu.vn/handle/123456789/2437
Piironen, A.K., Padley, A.: Student-centered Learning in CDIO Framework.Tenaga, Pendawaian Elektrik Di Bangunan Kediaman, pp. 1–64 (2018).
Wang, Y., Gao, S., Liu, Y., Fu, Y.: . Design and implementation of project-oriented CDIO approach of instrumental analysis experiment course at Northeast agricultural university. Educ. Chem. Eng. **34**, 47–56 (2021)
Zawawi, S.S.A.: transformasi PTV: kesediaan guru-guru vokasional terhadap pelaksanaan koleh vokasional KPM dari aspek tahap kemahiran. Persidangan Kebangsaan Penyelidikan Dan Inovasi Dalam Pendidikan Dan Latihan Teknik Dan Vokasional, vol. 10 (2011)
Zubaidah, S.: SitiZubaidah-STKIPSintang-10Des2016. Seminar Nasional Pendidikan **2**, 1–17 (2016)

Students' Digital Readiness in Vocational College for Industrial Revolution 4.0

Nurul Azlynie Rosly[1], Hanifah Jambari[2(✉)], Muhammad Fathullah Hamzah[2], Umi Salmah Mihad[3], Sharifah Osman[2], and Mohamad Rasidi Pairin[2]

[1] Electrical Engineer at Layar Identiti Sdn. Bhd, Johor Bahru Sentral, Johor Bahru, Malaysia
[2] School of Education, Faculty of Social Sciences and Humanities, Universiti Teknologi Malaysia, Johor Bahru, Malaysia
hanifahjambari@gmail.com
[3] School of Professional and Continuing Education, Universiti Teknologi Malayisa, Kuala Lumpur, Malaysia

Abstract. Students often encounter challenges in developing their digital technology skills, primarily due to a lack of proficiency in using computer applications, internet-based apps, and similar tools. The failure of graduates in mastering the digital technology has a negative effect on their ability to get a job related to Industrial Revolution 4.0 (IR4.0). Thus, the objective of this study is to identify students' digital readiness for IR4.0 in the aspects of attitude, knowledge and technical skills at one of Vocational College in the Southern Zone in Malaysia. A total of 306 respondents were selected using the stratified random sampling. This study is quantitative method and the research instrument consists of a set of questionnaires containing 28 question items divided into Sections A, B, C and D. All the data collected through the questionnaires were analysed descriptively, using the Statistical Package for Social Science Version 20.0 (SPSS), to obtain frequency values, percentages, means and standard deviations. The results of the study reveal that the level of the students' digital readiness for IR4.0 from the aspects of attitude, knowledge and technical skills is at a satisfactory level. The mean average value obtained for all the research questions is 3.34, and it is at a moderate score level. Therefore, vocational colleges need to pay more attention to students' digital readiness for IR4.0 from the aspect of attitude, knowledge and technical skills, so that the graduates can meet the needs of the industry now, and propel the nation towards IR4.0.

Keywords: Digital Skills · Knowledge · Attitude · Technical Skills

1 Introduction

In the current era of globalisation, skills are important for the country to continue developing and preparing for the 21st century. The marketability of a technical graduate after completing Technical and Vocational Education and Training (TVET), especially in preparation for Industrial Revolution 4.0, should have strong digital skills. Technical graduates should be aware of this revolutionary change, and they need to master the

© The Author(s), under exclusive license to Springer Nature Singapore Pte Ltd. 2024
F. Hassan et al. (Eds.): AsiaSim 2023, CCIS 1911, pp. 71–80, 2024.
https://doi.org/10.1007/978-981-99-7240-1_6

digital skills related to the subjects they are studying, which can then be used in the world of work later. According to Zubaidah (2018), the country needs to focus on training of skilled workers who are creative, highly proficient, and knowledgeable, in line with the present era of globalisation. Therefore, the government has taken an approach by empowering schools with digital technology, including Technical and Vocational Education Institutions providing courses related to technical skills, so that Malaysia can produce highly skilled youth, especially in digital competence, which will blend in well with the country's transformation towards the digital economy.

In this era of Industrial Revolution 4.0, digital information is essential, particularly in the industrial sector. Industrial Revolution 4.0 was introduced in 2016; it is the third version of upgrading advanced automation in the traditional industrial sector that uses modern technology. Digital 4.0 is a big change, especially in the industrial sector which applies a combination of cyber technology and automation. In Digital 4.0, industrial technology has entered the latest automation that entails data exchange. These changes greatly affect people's lifestyles, especially in the sectors of economy, social activities and way of living. Hence, this Digital 4.0 can be integrated with the human social sphere so that life activities are easier and faster. The three technological changes introduced are physical technology, digital technology and biological technology through nine main pillars in the industry, which include these domains: horizontal and vertical system integration; cybersecurity; Internet of Things (IoT); public computing; simulation and virtual reality; power data analysis; supply chain; robot automation; and additive manufacturing.

In the 21st century, digital skills for IR4.0 are crucial for graduates so that they can apply these skills confidently after entering the working world, especially in the industrial sector. In facing the challenges of this Industry 4.0, Technical and Vocational Education plays a role in training these technical graduates to master the digital technology skills of IR4.0 before they venture into the industrial sector. According to Nur Fatin (2006), Vocational Education is an organisation's effort to produce students who are highly skilled in the technical field and prepare them to be a competent workforce in the industrial sector one day. A Technical and Vocational Education institution is one that provides specific training courses according to technical fields; for example, Electrical and Electronic Technology Courses, Mechanical Technology, Building Construction Technology and others.

2 Problem Statement

The purpose of this study is to identify the level of students' digital readiness for IR4.0 among Vocational College students in Melaka for future industry needs. Provision of digital skills in accordance to IR4.0 needs to be emphasised, especially in relation to technical and vocational students. In education, the core of a Technical and Vocational Education institution is providing solid technical skills to attain the goal of developing highly competent students in a particular field. The courses offered by the Ministry at various Vocational Colleges have all been recognised with the award of Malaysian Skills Certificate (SKM) at level 3, and the Malaysian Vocational Diploma (DVM) at level 4. Vocational College students are able to obtain these certificates if they complete all tasks given to them according to those levels. In addition, Vocational College students also need to take academic subjects such as Malay, Mathematics, Science, English, Islamic Education, Morals and History in order to obtain the Malaysian Vocational Certificate (SVM), which is in line with the requirement of the Malaysian Certificate of Education (SPM). It is at this stage that students are exposed to digital-based technical skills, which are in line with the industrial demand in today's revolutionary era.

It is necessary to brief students on the revolutionary changes to today's industrialisation, so that Vocational College students do not fall behind, and face difficulties in the industrialized world in the future. In this Industrial Revolution 4.0, digital skills ought to be widely disseminated to improve the existing skills and even acquire new skills by Vocational College students. This will enable the country to produce highly skilled individuals in technical and vocational fields for the future industry needs.

In Malaysia, employers' feedback shows that graduates fail to meet the industry's requirements, and many of them face the difficulties of getting a job (Seetha, 2014). Although Malaysia is now developing rapidly, inevitable issues crop up alongside the latest technological advances. Graduates are the youth who will be the country's future administrators and leaders. According to Ismail (2011), unemployment is a known issue among graduates, which has become the talk of the town. The community perceives that the country's higher education system has failed in producing employable graduates. According to a report by the Department of Statistics Malaysia (2013), 31.4 percent of the community aged 15–64 years comprise retirees, housewives, students and so on.

3 Objectives

The objectives of this study are as follows:

i. Identify the level of Vocational College students' digital readiness for IR4.0 from the attitude aspect.
ii. Identify the level of Vocational College students' digital readiness for IR4.0 from the knowledge aspect.
iii. Identify the level of Vocational College students' digital readiness for IR4.0 from the technical skills aspect.

4 Methodology

For the purpose of this study, a non-experimental approach was used to identify the level of Vocational College students' digital readiness for IR4.0 from the attitude, knowledge and technical skills aspects. The quantitative data survey method was conducted using a 4-point Likert scale questionnaire to collect respondents' data among the students. The online survey was undertaken using Google Forms specifically for the students of this vocational college, which focused on 306 students from year 1 to year 4, who were selected from these courses: Electrical & Electronic Technology, Building Construction Technology, and Mechanical Technology. The first pilot test was carried out among 30 students randomly selected from the business management course to answer the questionnaire. The number of pilot studies is sufficient to analyse validity and reliability before conducting the actual study. According to Johanson and Brooks (2010), the minimum number of respondents for pilot study is thirty. According to Cooper and Schindler (2011), the appropriate number for a pilot study is between 25 and 100 respondents. These students were chosen because they were not related to the actual respondents. Cronbach's Alpha reliability values were then calculated. The results obtained from the SPSS software show that the overall Cronbach's Alpha value is 0.71, indicating that the instrument has a high level of reliability.

5 Findings

The quantitative survey method using a 4-point Likert scale questionnaire was used in this study. The table below shows the mean values calculated using SPSS20.0 to identify the items most agreed and least agreed with by the students. The results and findings of this study are described in the sub-sections as follows;

5.1 Part B Analysis: Vocational College Students' Digital Readiness for Ir4.0 From the Attitude Aspect

Table 1 shows the results of the descriptive analysis of the items in Part B.

Table 1. Mean scores from the attitude aspect

Item	Statement	F	SDA	DA	AG	SA	Mean	Standard Deviation
1	I am interested in learning computer software as a subject at school	F	–	24	163	119	3.31	0.61
		%		7.8	53.3	38.9		
2	Learning computer software in the subject helps me to complete the assignments given	F	–	14	136	156	3.47	0.58
		%		4.4	44.4	51.1		
3	I like learning computer software because it encourages me to think	F	–	55	153	98	3.14	0.70
		%		17.8	50.0	32.2		
4	I easily learn computer software with the help of a teacher	F	–	27	126	153	3.41	0.65
		%		8.9	41.1	50.0		
5	Learning computer software has a positive impact on my life	F	–	48	146	112	3.21	0.69
		%		15.6	47.8	36.7		
6	Learning computer software stimulates my interest to learn more	F	–	58	136	112	3.18	0.73
		%		18.9	44.4	36.7		

(continued)

Table 1. (*continued*)

Item	Statement	F	SDA	DA	AG	SA	Mean	Standard Deviation
7	I agree with the use of simulation methods using appropriate computer software in engineering subjects	F	–	34	122	150	3.30	0.68
		%		11.1	40	48.9		
8	I agree that learning of computer software can be applied in everyday life	F	–	40	133	133	3.30	0.70
		%		13.3	43.3	43.3		
Overall Average							3.40	0.59

Based on the findings of the study, the overall mean value for the attitude aspect is 3.40, indicating that the level of student readiness from the attitude aspect is at a moderate level 156 respondents (51.1%) strongly agreed that 'learning computer software in the subject helps me to complete the assignments given', which achieved the highest overall mean value of 3.47 in the attitude item. Meanwhile, Item 3 shows the lowest mean value of 3.14. A total of 98 respondents (32.2%) strongly agreed with Item 3; a total of 153 respondents (50.0%) agreed with Item 3, a total of 55 respondents (17.8%) disagreed with 'learning computer software because it encourages me to think'.

5.2 Part C Analysis: Vocational College Students' Digital Readiness for Ir4.0 From The Knowledge Aspect

Table 2 shows the results of the descriptive analysis of the items in Part C.

Table 2. Mean scores from the knowledge aspect

Item	Statement	F	SDA	DA	AG	SA	Mean	Standard Deviation
1	I know the Internet of Things (IoT)/ Virtual reality (VLE FROG)/ Basics of robotics	F	3	55	180	68	3.02	0.67
		%	1.1	17.8	58.9	22.2		
2	The internet of Things (IoT) learning improves mastery of computer software knowledge in engineering subjects	F	3	17	204	82	3.19	0.58
		%	1.1	5.6	66.7	26.7		

(*continued*)

Table 2. (*continued*)

Item	Statement	F	SDA	DA	AG	SA	Mean	Standard Deviation
3	I know how to use computer software to improve knowledge	F	–	24	166	116	3.30	0.61
		%		7.8	54.4	37.8		
4	I can improve my knowledge through learning this computer software	F	–	10	170	126	3.38	0.55
		%		3.3	55.6	41.1		
5	I know how to submit assignments over the Internet	F	–	10	150	146	3.44	0.56
		%		3.3	48.9	47.8		
6	I understand computer software with the help of a teacher	F	–	24	143	139	3.38	0.63
		%		7.8	46.7	45.6		
7	I know how to access digital technology myself to increase knowledge	F	4	14	163	125	3.34	0.62
		%	1.1	4.4	53.3	41.1		
8	I know the basic knowledge related to computer software from home	F	–	24	166	116	3.30	0.61
		%		7.8	54.4	37.8		
Overall Average							3.29	0.60

Table 2 shows the findings of the level of students' digital readiness for IR4.0 among vocational college students from the knowledge aspect. The overall mean value is 3.29, which shows that the students' digital readiness for IR4.0 is at a moderate level. Item 5 records the highest mean value of 3.44, with 146 respondents (47.8%) strongly agreed, 150 respondents (48.9%) agreed, while 10 respondents (3.3%) disagreed with the statement 'I know how to submit assignments over the Internet'. Meanwhile, Item 1 has the lowest mean value of 3.02 with 68 respondents (22.2%) strongly agreed, 180 respondents (58.9%) agreed, 55 respondents (17.8%) disagreed, and 3 respondents (1.1%) strongly disagreed with the statement 'I know the Internet of Things (IoT)/ Virtual reality (VLE FROG)/ Basics of robotics'.

5.3 Part D Analysis: Vocational College Students' Digital Readiness for Ir4.0 From The Technical Skills Aspect

Table 3 shows the results of the descriptive analysis of the items in Part D.

Table 3. Mean scores from the technical skills aspect

Item	Statement	F	SDA	DA	AG	SA	Mean	Standard Deviation
1	I am good at accessing information using the Internet	F	–	17	149	140	3.40	0.60
		%		5.6	48.9	45.9		
2	I am good at uploading pictures on the Internet	F	–	31	140	135	3.34	0.66
		%		10.0	45.6	44.4		
3	I am good at downloading pictures from the Internet	F	–	7	153	146	43.3	0.54
		%		2.2	50.0	47.8		
4	I am good at uploading videos on the Internet	F	–	24	149	133	3.36	0.62
		%		7.8	48.9	43.3		
5	I am good at downloading videos on Ihe internet	F	–	17	153	135	3.39	0.59
		%		5.6	50.0	44.4		
6	I am good at creating simulations using applications on the Internet	F	–	34	98	174	3.22	0.65
		%		11.1	32.2	56.7		
7	I am proficient in using Google Drive to store information	F	–	34	156	116	3.27	0.65
		%		11.1	51.1	37.8		
8	I am proficient in using Microsoft Word interactively	F	4	27	153	122	3.29	0.67
		%	1.1	8.9	50.0	40.0		
9	I am proficient in using Microsoft PowerPoint interactively	F	–	24	160	122	3.32	0.62
		%		7.8	52.2	40.0		
10	I am proficient in using Microsoft Excel interactively	F	4	51	163	88	3.10	0.70
		%	1.1	16.7	53.3	28.9		
11	I am proficient in using Google Meet during online classes	F	–	10	143	153	3.47	0.57
		%		3.3	46.7	50.0		
12	The use of digital in engineering subjects improves my technical skills	F	–	17	163	126	3.36	0.59
		%		5.6	53.3	41.1		
Overall Average							3.33	0.62

Table 3 shows the results of the findings about the level of students' digital readiness for IR4.0 among Vocational College students from the technical skills aspect. The overall mean value is 3.33, which shows that the students' digital readiness for IR4.0 is at a moderate level. Item 11 records the highest mean value of 3.47, with 153 respondents (50.0%) strongly agreed, 143 respondents (46.7%) agreed, while 10 respondents (3.3%) disagreed with the statement 'I am proficient in using Google Meet during online classes. Meanwhile, Item 10 has the lowest mean value of 3.10, with 88 respondents (28.9%)

strongly agreed, 163 respondents (53.3%) agreed, 51 respondents (16.7%) disagreed and 4 respondents (1.1%) strongly disagreed with the statement 'I am proficient in using Microsoft Excel interactively'.

6 Discussion

The results in Table 1 show that the level of students' digital readiness for IR4.0 from the attitude aspect is at a moderate level. It is evident that there are still students who lack interest in digital tools such as the use of computer software in class subjects. This is attributed to a lack of exposure to the use of digital skills in schools. Therefore, the Education sector needs to take actions by holding digital awareness programmes for IR4.0. According to Abd Kadir, Aziz, Hassan, Rahman, & Sidek (2019), there are suggestions for the Education sector to hold programmes in schools such as conducting courses, workshops, guidance briefing, and campaigns, as well as providing facilities and giving exposure in the curriculum related to digital skills for IR4. 0. This matter is also supported by Lai, Chundra, & Lee (2020) that in this era of Industrial Revolution 4.0, every individual needs to equip himself or herself in the best ability with digital skills for IR4.0.

The analysis results for Table 2 show that the level of vocational college students' digital readiness for IR4.0 from the knowledge aspect is at a moderate level. This means that the level of preparedness in terms of vocational college students' knowledge of digital skills for IR4.0 is low, and does not meet the requirements set by the Education sector, which aspires to make digitalisation technology changes among students. (Zulnaidi and Majid, 2020). Azmi, et al. (2018) said that knowledge related to digital skills for IR4.0 among students is still at a low level even though the Industrial Revolution is rapidly expanding. Exposure to the use of digital skills needs to be emphasised because it can increase knowledge related to new technological information such as the Internet of Things (IoT)/Virtual Reality (VLE FROG)/ Basic Robotics among vocational college students.

The analysis results for Table 3 show that the level of vocational college students' digital readiness for IR4.0 from the technical skills aspects is at a moderate level. This research question shows the student's capability level in the digital 4.0 usage, where skills are abilities and techniques acquired through specific training or experience involving body parts (Reader's Digest Universal Dictionary). From the aspect of technical skills, actions are performed by the different body parts to implement digital skills for IR4.0. These technical skills are important to students, because with the skills, it is easier for the students to carry out tasks given by the teacher such as accessing information using the Internet.

7 Conclusion

This study has proven that the overall level of students' digital readiness for IR4.0 is only at a moderate level. It requires a lot of improvement and initiative from the responsible parties to think about the actions that will be taken to upgrade the knowledge and technical skills of vocational college students in the use of digital skills. The ultimate goal is to produce graduates who are competent in digital skills for IR4.0 in order to move towards IR 4.0, which is a common feature in the global market. Further research in this area should be done so that improvements can be implemented from time to time, and various parties, especially the education sector, need to play their role well to deal with the current issues. Similar studies can help the top management take immediate action. Therefore, students' digital readiness for IR4.0 needs to be emphasised to achieve IR 4.0.

Acknowledgement. The authors would like to acknowledge the financial support from the Ministry of Higher Education under Fundamental Research Grant Scheme (FRGS) with vote number, FRGS/1/2020/SS0/UTM/02/6 and Universiti Teknologi Malaysia for the funding under UTM Encouragement Research (UTMER) with vote number Q. J130000.3853.20J05.

References

Abd Kadir, R.B., Aziz, M.A.B.A., Hassan, M.K.B., Rahman, N.B.A., Sidek, M.A.B.: Tahap Pengetahuan Dan Tahap Kesediaan Guru Pelatih Institut Pendidikan Guru Kampus Pendidikan Teknik (Ipgkpt) Terhadap Revolusi Industri 4.0 (IR 4.0). Jurnal Penyelidikan Teknokrat II (Jilid xxi), 27–37 (2019)

Rahim, A.N.: Penggunaan Mobile Learning (m-Learning) untuk tujuan pembelajaran dalam kalangan pelajar kejuruteraan UTHM (Doctoral dissertation, Universiti Tun Hussein Onn Malaysia) (2013)

Nabil, A.: Kerangka pengajaran amali kursus teknologi elektronik di Kolej Vokasional Malaysia. Thesis Ph.D (Teknikal dan Vokasional) - Universiti Teknologi Malaysia (2016)

Cooper, D.R., Schindler, P.S.: Business research methods, 11th edn. McGraw-Hill/Irwin, New York (2011)

Hanafi, S.: Kesediaan pelajar dari aspek kemahiran teknikal terhadap pembentukan kebolehkerjaan di Kolej Vokasional Wilayah Selatan. Universiti Tun Hussein Onn Malaysia (2015)

Ilias, K., &Ladin, C.A.: Pengetahuan dan kesediaan Revolusi Industri 4.0 dalam kalangan pelajar Institut Pendidikan Guru Kampus Ipoh. O-JIE: Online J. Islamic Educ. **6**(2), 18–26 (2018)

Ishak, R., Mansor, M.: The relationship between knowledge management and organizational learning with academic staff readiness for education 4.0. Eurasian J. Educ. Res. **85**, 169–184 (2020)

Ismail, N.A.: Graduates characteristics and unemployment: a study among malaysian graduates. Int. J. Bus. Soc. Sci. **2**, 1 (2011)

Perangkaan, J., Buruh, P.T.: Malaysia, Mac 2013. Dicapai pada Mei 25 dari (2013). http://www.statistics.gov.my/portal/images/stories/files/LatestReleases/employment/2

Johanson, G.A., Brooks, G.P.: Initial scale development: sample size for pilot studies. Educ. Psychol. Measur.Measur. **70**(3), 394–400 (2010)

Tinggi, K.P.: Punca Graduan Sukar Dapat Kerja. Dicapai pada Mei 20, 2014, dari (2012). http://blog.mohe.gov.my/2012/11/punca-graduan-sukar-dapat-

Krejcie, R.V., Morgan, D.W.: Determining sample size for research activities. Educ. Psychol. Measur. **30**, 607–610 (1970)

Lai, C.S., Chundra, U., Lee, M.F.: Teaching and Learning Based on IR 4.0: Readiness of Attitude among Polytechnics Lecturers. J. Phys. Conf. Ser. **1529**(3), 032105 (2020)

Muhali, M.: Arah Pengembangan Pendidikan Masa Kini Menurut Perspektif Revolusi Industri 4.0. In: Prosiding Seminar Nasional Lembaga Penelitian dan Pendidikan (LPP) Mandala (2018)

Sauffie, N.F.M.: Technical and vocational education transformation in Malaysia: shaping the future leaders. J. Educ. Practice **6**(22) (2015). www.iiste.org

Abdullah, Q.A., Humaidi, N., Shahrom, M.: Industry revolution 4.0: the readiness of graduates of higher education institutions for fulfilling job demands. Romanian J. Inform. Technol. Autom. Control **30**(2), 15–26 (2020)

Ramli, M.A., Mustapha, R., Abd Rahman, R.: Hubungan Kemahiran Kebolehkerjaan Pelajar Kolej Vokasional Pertanian Dengan Kesediaan Menghadapi Revolusi Industri 4.0. Politeknik & Kolej Komuniti Journal of Life Long Learn. **2**(1), 1–15 (2018)

Zubaidah, S.: Mengenal 4C: Learning and innovation skills untuk menghadapi era revolusi industri 4.0. In: 2nd Science Education National Conference, pp. 1–18 (2018).

Zulnaidi, H., Majid, M.Z.A.: Readiness and understanding of technical vocational education and training (TVET) lecturers in the integration of industrial revolution 4.0. Inter. J. Innovation, Creativity Change **10**(10), 31–43 (2020)

Implementing Blockchain Technology for Accreditation and Degree Verification

Az Mukhlis Iskandar Azli[1], Nur Haliza Abdul Wahab[1(✉)], DaYong Zhang[1], Khairunnisa A. Kadir[1], and Noorhazirah Sunar[2]

[1] Faculty of Computing, Universiti Teknologi Malaysia, UTM Johor, Johor Bahru, Malaysia
nur.haliza@utm.my

[2] Faculty of Electrical Engineering, Universiti Teknologi Malaysia, UTM Johor, Johor Bahru, Malaysia

Abstract. In the evolving academic landscape, the integrity of educational certificates is paramount. These documents serve as formal attestations of one's educational attainment, however, the increasing prevalence of document fraud undermines their credibility. To combat this, we introduce a robust, web-based certificate validation system Universiti Teknologi Malaysia's Blockchain-Based Accreditation and Verification System (UTM-BADVES). Built upon the transparent and immutable infrastructure of Blockchain technology, complemented with a diverse array of advanced technologies, UTM-BADVES offers a secure, efficient, and intuitive solution for real-time validation of academic credentials. The system focuses on data privacy, enabling transcript verification, selective data dissemination, and efficient credential revocation. By doing so, it significantly reduces the potential for academic credential fraud. Consequently, UTM-BADVES safeguards the integrity of educational qualifications, while simultaneously providing immense benefits for educational institutions, employers, and society as a whole. This paper delves into the design and application of UTM-BADVES, elucidating the crucial role it plays in maintaining the sanctity of academic certifications in the digital age.

Keywords: Blockchain · Blockchain-based Verification · Decentralized Applications

1 Introduction

Education in Malaysia receives significant government expenditure, accounting for 16.8% of the total budget in 2020 [1]. This allocation reflects the importance of education as a fundamental pillar of society, enabling social mobility, economic advancement, and political stability. However, the value attached to education also makes it vulnerable to manipulation and corruption. Transparency International's study in 2013 highlighted various forms of corruption in the education sector, including falsified credentials, ghost schools, bribery, and misuse of funds [2]. Falsified academic credentials, in particular, pose significant harm by devaluing the credibility of higher education systems, discouraging genuine investment in education, and leading to workplace incompetence and potential dangers in certain fields.

© The Author(s), under exclusive license to Springer Nature Singapore Pte Ltd. 2024
F. Hassan et al. (Eds.): AsiaSim 2023, CCIS 1911, pp. 81–95, 2024.
https://doi.org/10.1007/978-981-99-7240-1_7

In traditional practices, the verification of academic credentials has been marred by costly procedures, excessive time consumption, and bureaucratic red tape. To rectify these prevailing challenges, our project introduces a comprehensive solution steeped in the revolutionary capabilities of Blockchain technology. By harnessing the inherent transparency and immutability of Blockchain, we aspire to establish a resilient, future-proof verification system that effectively eradicates risks related to credential falsification. This avant-garde approach greatly simplifies the verification process while concurrently preserving the integrity of educational qualifications. Consequently, it authenticates the legitimacy of individual accomplishments and safeguards workplaces from potential damages resulting from inadequately skilled personnel. The revolutionary, decentralised nature of blockchain technology is what makes digital currencies like Bitcoin possible, but its ramifications go well beyond financial ones. A blockchain may be thought of as a digital ledger that records transactions across a network of computers in a decentralised and secure manner [3]. The blockchain network functions without a trusted third party thanks to a consensus process that ensures all users accept the legitimacy of all transactions. Each block of transactions includes a link to the prior block, forming an unbreakable chain of data. Using cryptography, the data is protected from prying eyes and cannot be altered in any way. All blockchain transactions are publicly viewable by all nodes in the network, which increases transparency and promotes trust and accountability [3].

Academic credential fraud is driven by individuals seeking quick success and employment without the necessary qualifications. It involves intentionally misrepresenting academic achievements, posing risks to other parties [4]. Falsifying or enhancing credentials has become prevalent due to the pressure to outperform peers in competitive job markets [5]. This practice not only compromises ethics but also poses significant risks, such as appointing individuals with fake degrees to important positions, potentially hindering a country's educational development. Furthermore, the traditional paper-based format of academic credentials is vulnerable to forgery, loss, and inefficiencies, leading to costly and time-consuming verification processes [5]. To address these challenges, a web-based certificate system utilizing Blockchain technology, UTM-BADVES, is proposed. It provides a secure database accessible by issuing institutions, degree holders, and third-party verifiers, enabling quick and reliable verification of students' credentials. The paper will focus on:

- To design and develop the proposed UTM Blockchain-Based Accreditation and Degree Verification System (UTM-BADVES).
- To test and evaluate the developed UTM-BADVES for reliability, data immutability and privacy, integrity, and the credential revocation mechanism.

2 Literature Review

A Blockchain is fundamentally a decentralized ledger, documenting transactions within a peer-to-peer network [6, 7]. Conceptualized as a chain of blocks, each housing data and references to preceding blocks, a Blockchain manifests itself in three primary architectures: public, private, and consortium. Public Blockchains, typified by Bitcoin and Ethereum, permit universal participation and data accessibility to anyone with internet connectivity [8]. In contrast, private Blockchains necessitate explicit invitations and are governed by particular authorities or organizations [9]. Consortium Blockchains are collaborative, encompassing nodes from multiple institutions.

The structural bedrock of Blockchain architecture rests on several key com-ponents: blocks, nodes, transactions, chains, miners, and a consensus mechanism [10]. Upon the initiation of a new transaction, a fresh block is generated and appended to the Blockchain. Each block incorporates data and alludes to its antecedent block, thus preserving the continuity and integrity of the chain. Block verification is realized through a consensus mechanism amongst the network's nodes.

Blockchain technology ushers in a multitude of advantages, such as cost efficiency, as the nodes eliminate the necessity for physical servers, thereby functioning as a distributed database [11]. The Blockchain acts akin to a global archive or library, facilitating instantaneous, worldwide access to transaction verification. The incorporation of hashing in Blockchain amplifies data integrity and security, albeit with potential implications for performance speed [12]. On the whole, Blockchain establishes a resilient and secure framework for transaction processing and data management.

In terms of degree verification systems, UTM has an existing system in place to ensure the authenticity of degrees issued by the university, although the specific details and name of the system were not disclosed for safety reasons. However, it can be inferred that UTM does not currently utilize a QR code or web-based verification service. On the other hand, LuxTag's e-Scroll initiative, developed in collaboration with a consortium of Malaysian public universities including UTM, offers a verification feature through scanning a unique QR code on the received degree [13]. This system utilizes the Catapult NEM Blockchain to authenticate the degrees and verify the legitimacy of the registrar's department signature and co-signer. LuxTag has successfully secured the certifications of thousands of graduates from multiple countries [14]. Another notable Blockchain-based certif-icate ecosystem is Blockcerts, a collaboration between MIT Media Lab and Learning Machine. Blockcerts enables the creation, issuance, inspection, and validation of certificates on the Bitcoin Blockchain, adhering to open-source standards such as IMS Open Badges and W3C linked data signatures [15]. It supports certificate revocation and addresses scalability through batch issuance using Merkle trees, although it requires recipients and verifiers to maintain crypto-graphic credentials for participation in the ecosystem and verification service [16].

Table 1 shows provides a summary of the existing systems discussed earlier, highlighting their constraints and properties. LuxTag's e-Scroll is labeled as "not document-ed" for the issuance feature, as it relies on the university to issue academic credentials to graduates [17]. The current issuance and verification systems of UTM and LuxTag's e-Scroll are not documented. Blockcerts has a documented revocation feature, but it is labeled as "partially" because the revocation listings are centralized and susceptible to potential hacking [18]. None of the systems have the feature of selective data disclosure, where graduates can share specific data items without revealing all the original information. Additionally, UTM and LuxTag's e-Scroll do not require graduates to maintain cryptographic credentials, unlike Blockcerts.

To address these constraints, this project introduces UTM-BADVES. UTM-BADVES offers scalability through batch issuance and can serve as a solution at the university, consortium, or national level. It implements smart contracts and on-chain and multi-party schemes for a more robust revocation mechanism. UTM-BADVES prioritizes data protection and gives students the option to selectively disclose verification information based on their needs. It offers enhanced usability by using QR codes for verification, eliminating the need for users to manage digital identities or cryptographic credentials. QR codes can be included on resumes or publicly published, enabling verification without the physical certificate.

Table 1. Summary comparison of existing systems

Features	UTM (Current)	e-Scroll	Blockcerts	UTM-BADVES
Issuance	Yes	Not documented	Yes	Yes
Verification	Yes	Yes	Yes	Yes
Revocation	Not documented	Not documented	Partially	Yes
Selective Disclosure	Not documented	No	No	Yes
No Cryptographic Keys	Yes	Yes	No	Yes

3 System Development Methodology

For this project, the Rapid Application Development (RAD) methodology was chosen as the preferred approach. RAD is a subset of agile methodology known for its quick response to requirements and delivery of high-quality results [19, 20]. Unlike the Waterfall methodology that focuses on design, RAD emphasizes on processes and iterative development.

The development of our system relies on a range of technologies tailored for various tasks in the project. To construct the fundamental structure of the web-site, HTML (HyperText Markup Language) is utilized, a standard language for designing web pages. Complementing HTML, CSS (Cascading Style Sheets) is employed to stylize the website, enhancing its aesthetics and user-interface by controlling the layout of multiple web pages simultaneously.

For scripting purposes, TypeScript, a strongly typed superset of JavaScript, is adopted to ensure enhanced scalability and maintainability of the web application. Additionally, we employ React, a renowned JavaScript framework, to optimize the process of web development by providing reusable UI components.

On the Blockchain front, the Algorand network is integrated into our system. Algorand is a scalable and secure Blockchain network designed to speed up transactions and enhance security. To interact seamlessly with the Algorand Blockchain network, the Algo Software Development Kit (SDK) is utilized, thereby simplifying the development and execution of Blockchain-based applications.

Microsoft Visual Studio Code serves as the primary source code editor, thanks to its comprehensive suite of powerful features, including support for debugging, embedded Git control, and extensive functionality extensions. As a runtime environment for executing JavaScript, NodeJS is implemented to build the server-side of the application. Lastly, for database needs, we make use of Firebase, a Google-backed platform, offering a NoSQL cloud database for storing and synchronizing data in real-time, contributing to a robust and responsive application. PyTeal: PyTeal is a Python language binding for Algorand Smart Contracts (ASC) that provides a higher-level, more pythonic interface for creating smart contracts for the Algorand blockchain. In the UTM-BADVES system, PyTeal is utilized to develop smart contracts, forming the backbone of the certificate validation process.

4 Requirement Analysis and Design

This section introduces the design phase of the UTM-BADVES system. The chapter highlights the importance of carefully implementing this phase to meet user specifications. It outlines the key aspects of the design, including the modules, entity relationships, system workflow, and the database structure. UML techniques, such as use case diagrams and sequence diagrams, will be used to explain the processes involved. Additionally, the chapter mentions that the requirement analysis phase is essential for identifying user expectations. It identifies four main actors involved in the system: UTM, the graduates, and the Blockchain users.

4.1 Use Case Diagram

Figure 1 presents a Use Case diagram, a visual representation using Unified Modelling Language (UML), which illustrates the actors involved in the UTM-BADVES system and their corresponding user stories. The use cases depicted in the diagram represent the various interactions and functionalities that the system pro-vides to fulfil the requirements of the actors. The diagram serves as a visual tool to understand the roles and responsibilities of each actor and their involvement in the system's processes.

The UTM-BADVES system offers an innovative solution for managing academic credentials, with three principal user types: Admin, Blockchain Users, and Graduates. Each user type has distinct abilities and responsibilities within the system.

Admin holds the highest level of control in the system, responsible for connecting to and disconnecting from the wallet for secure blockchain transactions, verifying the validity of certificates, creating and revoking academic credentials as required.

Graduates are the recipients of the certificates, capable of interacting with their blockchain wallet, checking the validity of their digital certificates, viewing and sharing their academic information securely, and logging in or out of their accounts as needed.

Blockchain users, interested in verifying the authenticity of the credentials is-sued, can interact with their blockchain wallet, verify the validity of the certificates, and view the academic credentials of a graduate with necessary permissions.

These use cases underscore the broad functionality of the UTM-BADVES system, showcasing its ability to provide a secure, transparent, and efficient platform for managing and verifying academic credentials.

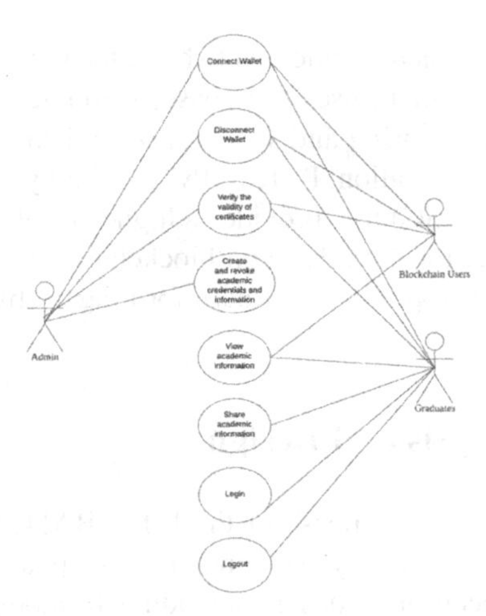

Fig. 1. Use case of the system

4.2 System Architecture

Figure 2 presents a clear visualization of the system architecture for the certificate verification program, providing a roadmap for the life cycle of certificates, right from their inception to their validation. Occupying the apex of the architecture is the Algorand Blockchain Network, a decentralized, open-source, public blockchain platform employing a pure proof-of-stake (PPoS) consensus mechanism. This network, overseen by the University, plays a dual role. Firstly, it acts as a secure digital vault, storing the digitally issued certificates. Secondly, it serves as a reference point for the web application, aiding in the validation of these certificates.

The next layer of the architecture houses Firebase, a flexible, scalable cloud-based database. Firebase is responsible for storing user-specific data within the web application. Users are a diverse set of stakeholders encompassing university staff, registrars, and students. Each user group interacts with the system differently; thus their data handling and storage needs vary, and Firebase caters to these specific requirements with its flexible data structure and robust security.

Subsequently, there's the Blockchain Users entity. Their principal responsibility lies in verifying the authenticity of the certificates. They interact with the blockchain network, querying and validating the certificates' data to ensure that they have not been tampered with, maintaining the system's credibility.

Lastly, at the base level are the Graduates, who are the primary beneficiaries of the system. Graduates request and subsequently receive verifiable digital certificates from the university. These digital certificates bear a distinct advantage over traditional paper-based certificates, in that they are difficult to forge and can be instantly validated. The University entity wields a significant role, per-forming key operations such as managing the Algorand Blockchain network and creating digital certificates for graduates. By doing so, the university not only modernizes its certificate issuance process but also imparts an additional layer of trust and credibility to its graduates' credentials.

Fig. 2. System Architecture for UTM-BADVES

4.3 User Interface Design

This subsection delves into the crucial aspect of User Interface (UI) design in the UTM-BADVES system. A well-designed user interface is fundamental to any software application's success, shaping how users interact with the system and, in turn, their overall experience. The quality of the user interface directly impacts user satisfaction, system usability, and the efficiency of performing tasks.

In the context of the UTM-BADVES system, the interface design was crafted with special attention to ease of use, intuitiveness, and the specific requirements of the various users: the Admin, Graduates, and Blockchain Users. The design had to accommodate the diverse set of functions these users needed to perform, from connecting to blockchain wallets to verifying academic credentials.

This subsection will walk you through the design considerations and choices that led to the final UI design, offering a comprehensive insight into the layout, navigation, color schemes, button placements, and more. Additionally, it will highlight the design strategies employed to maintain consistency throughout the system and ensure that users have a smooth, seamless, and engaging experience. Figure 3 presents a selectively focused view of the 'Home' page of our web application, providing a detailed snapshot of key system information. This figure offers an insight into the initial user experience, capturing the design elements and the crucial introductory information presented to the user upon their first interaction with the UTM-BADVES system.

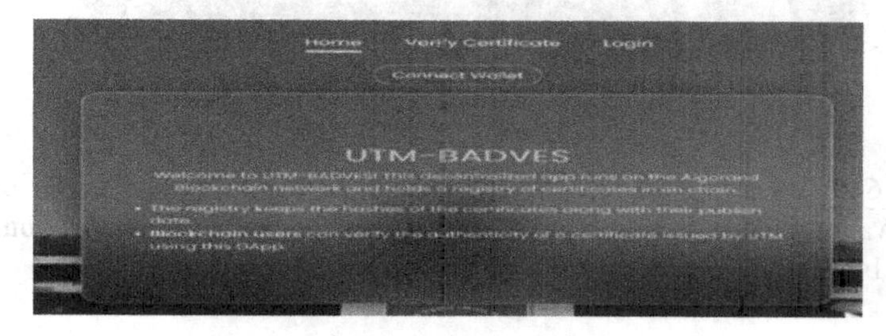

Fig. 3. 'Home' page of the web application

Figure 4 demonstrates the process for Blockchain users to validate certificate authenticity, underscoring the system's user-friendly nature. It illustrates the easy two-step verification procedure: first, users securely connect to their MyAlgo wallet, and then, by selecting the 'Verify Certificate' tab, they can confirm the certificate's validity. This figure showcases the seamless integration of blockchain technology within the interface, enabling users to perform complex operations with just a few clicks.

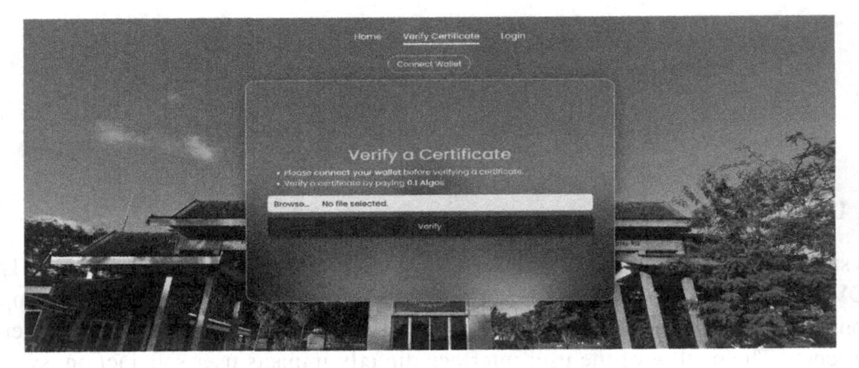

Fig. 4. Verify certificate page

Figure 5 provides an illustrative representation of the pages accessible to Admins, specifically for managing academic certificates. This includes the 'Submit Certificate' page, specifically designed for administrators, showcasing the interface that facilitates the seamless submission and management of digital certificates.

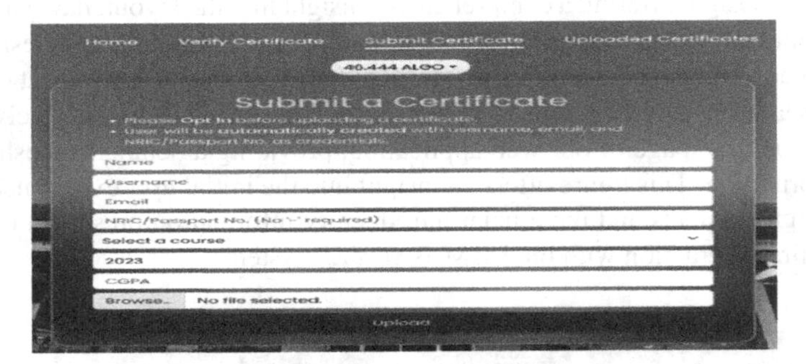

Fig. 5. Submit Certificate Page for Admins

Figure 6 shows a pop-up of a MyAlgo Connect window after a user clicks the 'Connect Wallet' button. In this example, the address of 'PSM2_First Account' acts as the normal blockchain user that wants to perform a certificate verification.

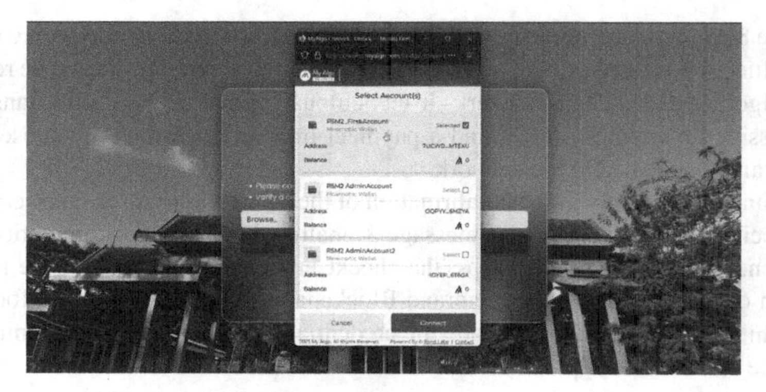

Fig. 6. A pop-up window of 'My Algo Connect'

5 Results, Testing, and Discussion

This chapter delves into the intricacies of the project's implementation and testing stages. It lays out an overview of the system's development process, illuminating the vital code snippets, the testing procedures followed, and the significance of the system's core functionalities. To ascertain the reliability and robustness of the system, a variety of testing methodologies were employed, including white box testing, black box testing, and user testing, providing a comprehensive analysis of the system's performance under diverse conditions.

Figure 7 specifically illustrates the operation of the 'check_cert' function. This function plays a pivotal role in verifying the authenticity of an academic certificate by comparing its hash value with the one stored on the blockchain. It confirms the compliance of group transaction conditions, including payment verification, and retrieves the corresponding value from the global state. Depending on whether the required value is identified, the transaction either receives approval or faces rejection. Developed in PyTeal, this 'check_cert' function operates as a dedicated handler for the certificate verification process, underscoring the intricate interplay of blockchain technology in assuring the validity of academic credentials in the UTM-BADVES system.

```
def check_cert(self):
    cert_key = ScratchVar(TealType.bytes)
    get_cert_value = App.globalGetEx(Txn.applications[0], cert_key.load())
    return Seq([
        Assert(
            And(
                # The number of transactions within the group transaction must be exactly 2.
                Global.group_size() == Int(2),
                # Check that this call is first in group
                Txn.group_index() == Int(0),

                # The number of arguments attached to the transaction should be exactly 2.
                Txn.application_args.length() == Int(2),

                # Checks for payment transaction.
                Gtxn[1].type_enum() == TxnType.Payment,
                Gtxn[1].receiver() == Global.creator_address(),
                Gtxn[1].amount() == Int(100000),
                Gtxn[1].sender() == Gtxn[0].sender(),
            ),
        ),
        # rehash the certificates hash.
        cert_key.store(Keccak256(Txn.application_args[1])),

        # check global state to see if key exists.
        get_cert_value,

        If(get_cert_value.hasValue())
        .Then(
            Approve()
```

Fig. 7. 'Check_cert' function

Figure 8 spotlight the implementation of the checkCert function. Integral to the system, this function is responsible for ascertaining the authenticity of a certificate recorded on the Algorand Blockchain network. It meticulously constructs a group transaction, encompassing an application call and a payment transaction, which are subsequently endorsed and dispatched to the network.

The function patiently awaits confirmation of the group transaction. Once confirmation is received, it logs a conclusive message, signaling the successful completion of the transaction. By executing these steps, the checkCert function guarantees the rigorous validation of a certificate on the Algorand Blockchain. In doing so, it helps fortify the trustworthiness and reliability of academic credentials, enhancing their recognition and acceptance across various platforms.

```
// CHECK CERTIFICATE: Group transaction consisting of ApplicationCallTxn and PaymentTxn
export const checkCert = async (senderAddress: string, cert: Cert, contract: Contract) => {
    console.log("Checking certificate...");

    if (algo.appId === Number(0)) return

    let params = await algo.algodClient.getTransactionParams().do();

    // Build required app args as Uint8Array
    let addArg = new TextEncoder().encode("check");
    let certHashArg = new TextEncoder().encode(cert.hash);
    let appArgs = [addArg, certHashArg];

    // Create ApplicationCallTxn
    let appCallTxn = algosdk.makeApplicationCallTxnFromObject({
        from: senderAddress,
        appIndex: algo.appId,
        onComplete: algosdk.OnApplicationComplete.NoOpOC,
        suggestedParams: params,
        appArgs: appArgs,
    });

    // Create PaymentTxn
    let paymentTxn = algosdk.makePaymentTxnWithSuggestedParamsFromObject({
        from: senderAddress,
        to: contract.creatorAddress,
        amount: 100000,
```

Fig. 8. First Half of 'checkcert' function

5.1 Algorand Blockchain and Network Implementation

Our system employs the Algorand Blockchain, a decentralized and secure platform supporting the creation and deployment of smart contracts and applications. It leverages a proof-of-stake consensus algorithm to efficiently validate transactions and create blocks. Custom functions integrated into the system, such as 'optIn', 'addCert', and 'checkCert', facilitate interaction with a smart contract on the Algorand Blockchain. The 'optIn' function empowers users to grant explicit permissions to the smart contract for accessing their account and executing specific operations. The 'addCert' function embeds a certificate on the Blockchain, thereby bolstering its immutability and transparency. Conversely, the 'checkCert' function corroborates a certificate's authenticity by consulting the smart contract. The strategic implementation of the Algorand Blockchain and these specific functions enable secure storage and validation of certificates, thus heightening trust and reliability in the system.

Furthermore, the system integrates Firebase Firestore, a cloud-based NoSQL database, offering a flexible and scalable solution for data storage and synchronization across various platforms. It employs a document-oriented model to structure data into collections and documents. Our system incorporates Firestore SDK functions like 'collection', 'doc', 'getDocs', and 'setDoc', which equip developers with the tools to interact with collections, documents, and the Firestore database. Specifically, the 'getDocs' function fetches data from one or multiple documents in a collection based on defined conditions or retrieves all documents. Concurrently, the 'setDoc' function allows data writing or updating in a document, superseding its content if it preexists, or generating a new document otherwise. In sum, Firebase Firestore provides sophisticated querying capabilities, real-time updates, and offline support, thereby solidifying its position as a potent instrument for data storage in our system.

5.2 Security Implementation

While the aforementioned implementations enhance code security and privacy, it is important to note that establishing comprehensive system security requires a robust strategy. This strategy must incorporate several considerations such as secure Firebase service configurations, adherence to secure coding practices, routine security evaluations, and compliance with recognized security guidelines and standards.

Within our system, hashing, a cryptographic technique, is used to generate a distinct, fixed-length string of characters from any given input data. As illustrated in Figure 5.47, the system utilizes the 'sha3_256' function from the 'js-sha3' library to compute the hash of file contents. The computed hash functions as a unique identifier symbolizing the file content and can be stored for future verification. The use of hashing safeguards data integrity since any minor alteration in the file will yield a drastically different hash value. Consequently, by comparing the newly calculated hash with the stored hash, any instance of file tampering can be swiftly detected.

For secure certificate storage, our system incorporates Firebase Storage. Access to the uploaded certificate files is narrowly confined to authorized users possessing valid addresses and administrative privileges, a restriction enforced through the 'Route' element. This selective access aids in preserving the confidentiality and integrity of the files. Firebase provides a suite of security tools and access controls, inclusive of authentication and Firebase Security Rules, ensuring that only authenticated and duly authorized users can access the stored files. By integrating these access controls, the system significantly curbs the risk of unauthorized exposure or alteration of the certificate files, thereby bolstering their security and reliability.

5.3 Testing

Testing constitutes a critical pillar in the realm of software development and implementation, playing a significant role in affirming the quality, reliability, and robustness of a system. It meticulously scrutinizes the system or its components in a systematic manner, unearthing any flaws, faults, or discrepancies from anticipated behavior. This rigorous inspection process aids in pinpointing any latent errors or vulnerabilities that could potentially undermine the system's performance.

Diverse testing methods, including black-box, white-box, and user acceptability testing, equip developers and stakeholders with invaluable insights into the system's functionality. This facilitates a robust understanding of the system's alignment with the set requirements, helping identify potential challenges or bottlenecks before deployment. This rigorous testing protocol is instrumental in delivering software solutions of superior quality and dependability.

By enhancing the overall user experience, mitigating associated risks, and boosting system stability, testing paves the way for a more seamless and secure interaction with the software. It is through such dedicated testing efforts that we ensure the system's optimal performance and reliability, offering the best possible experience to the end users.

The results demonstrate the successful development and implementation of the system, which capably provides users with verification services and comprehensive information regarding graduates from their certifications. As the Fig. 9 illustrate, upon successful verification, the system generates a detailed table presenting the graduate's information. This structured, post-verification display of graduate data is a testament to the system's effectiveness and its utility for end-users.

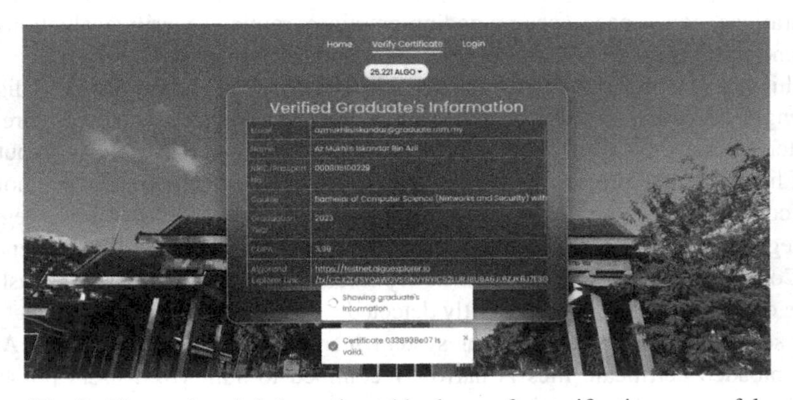

Fig. 9. The graduate's information table shown after verification successful.

The users who seek to verify certificates are equipped with a QR code and an explorer link, facilitating seamless access to the transaction information stored on the Blockchain ledger. As depicted in Fig. 10, both the QR code and the explorer link serve as direct gateways to the Blockchain explorer, enabling users to review comprehensive transaction details.

Moreover, Fig. 11 provides an insightful look into the transaction specifics, revealing important elements such as the sender's details (corresponding to the admin wallet that initiated the certificate submission into the Blockchain), the timestamp of the transaction, the unique transaction ID, and most critically, the information pertaining to the certificate owner. This detailed transaction summary not only substantiates the system's transparency but also contributes to its reliability and credibility. In terms of security, once the information is deployed to the Blockchain, it cannot be changed and tampered.

Fig. 10. Verified Graduate Information with the Blockchain ledger link information.

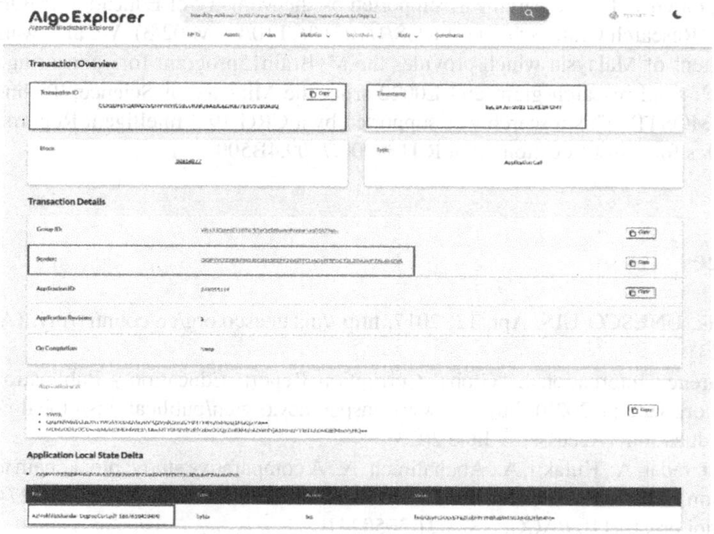

Fig. 11. The explorer which confirms the details verified and cannot be tampered.

6 Conclusion

This paper can be concluded as serves a comprehensive overview of the project, critically assessing its success in realizing its outlined objectives. It simultaneously casts a forward-looking eye towards potential enhancements, with due consideration given to the system's long-term functionality and the requisite modifications for its future deployment.

A pivotal recommendation for advancing the UTM-BADVES system would be the incorporation of an accreditation module. This could significantly augment the system's capabilities by enabling validation of the accreditation status of educational institutions or programs. Further, to address considerations of scalability and performance, it is advisable to refine the system's architecture and harness the power of distributed ledger technology for optimal performance and growth capability.

The user experience, an indispensable aspect of any system, could be substantially enhanced by integrating user feedback into the development process. By conducting rigorous usability testing and simplifying the user interface, the sys-tem can be made more intuitive and user-friendly. Additionally, the automation of the certificate submission process could streamline operations and expedite certificate issuance and validation, resulting in significant time savings for users.

In the ever-evolving landscape of technology, continuous improvement and diligent maintenance are critical to preserve the system's effectiveness and usability. These practices will ensure that the UTM-BADVES system continues to provide significant value and utility to the University of Technology Malaysia, reinforcing its position as a pioneer in leveraging Blockchain technology for academic credential verification.

Acknowledgement. This research was supported by the Ministry of Education (MOE) through Fundamental Research Grant Scheme (FRGS/1/2021/ICT10/UTM/02/3). We also want to thank the Government of Malaysia which provides the-MyBrain15program for sponsoring this work under the self-fund research grant and L0022 from the Ministry of Science, Technology and Innovation (MOSTI). This research was supported by a CRG 19.3 Intelligent Reports Based on Data Analytics for Asset Location grant R.J130000.7309.4B500.

References

1. Malaysia, UNESCO UIS, Apr. 12, 2017. http://uis.unesco.org/en/country/my, (Accessed 7 Jun 2023)
2. Transparency International, "Global Corruption Report: Education - Publications, Transparency.org (6 Apr 2020). https://www.transparency.org/en/publications/global-corruption-report-education, (Accessed 7 June 2023)
3. Ali, O., Jaradat, A., Kulakli, A., Abuhalimeh, A.: A comparative study: blockchain technology utilization benefits, challenges and functionalities. IEEE Access **9**, 12730–12749 (2021). https://doi.org/10.1109/ACCESS.2021.3050241
4. du Plessis, L., Vermeulen, N., van der Walt, J., Maekela, L.: Verification of Qualifications in Africa. Research report on the Verification of Qualifications in Africa, Northwest University, South Africa supported by the South African Qualifications Authority (2015)
5. Saunders, J.: 5 issues with paper-based management (and 3 reasons to switch). MyMobileWorkers. https://www.mymobileworkers.com/blog/the-problems-with-paper-based-management, (Accessed 7 June 2023)
6. Saberi, S., Kouhizadeh, M., Sarkis, J., Shen, L.: Blockchain technology and its relationships to sustainable supply chain management. Inter. J. Product Res. 57, 1–19, 10/17 (2018). https://doi.org/10.1080/00207543.2018.1533261
7. Ismail, R.A., et al.: Dental service system into blockchain environment . International J. Innovative Comput. **13**(1), 47–58 (2023) https://doi.org/10.11113/ijic.v13n1.394
8. Zheng, Z., Xie, S., Dai, H.-N., Chen, X., Wang, H.: An Overview of Blockchain Technology: Architecture, Consensus, and Future Trends (2017)
9. Adamopoulos, A., Davey, B., Bruno, V., Dick, M.: Blockchain collaborative network development framework. Available at SSRN 3677671 (2019)

10. De Filippi, P., Mannan, M., Reijers, W.: Blockchain as a confidence machine: the problem of trust & challenges of governance. Technol. Soc. **62**, 101284 (2020). https://doi.org/10.1016/j.techsoc.2020.101284 (8 Jan 2020)
11. Awadallah, R., Samsudin, A.: Using Blockchain in cloud computing to enhance relational database security. IEEE Access **9**, 137353–137366 (2021)
12. Harian, S.: E-Scroll dapat atasi masalah ijazah palsu (18 Feb 2019). https://www.sinarharian.com.my/article/13623/BERITA/Nasional/E-Scroll-dapat-atasi-masalah-ijazah-palsu, (Accessed 7 June 2023)
13. Emem, M.: Malaysia Launches Blockchain Certificate Verification to Combat Degree Fraud. CCN.Com. (4 March 2021). https://www.ccn.com/malaysialaunches-Blockchain-certificate-verification-to-combat-degree-fraud/?cv=1, (Accessed 7 June 2023)
14. Lim, J.: Blockchain: fighting counterfeits through digital means. Edge Markets. J. (2 May 2020). https://www.theedgemarkets.com/article/Blockchain-fighting-counterfeits-through-digital-means, (Accessed 7 June 2023)
15. Radha, S. K., Taylor, I., Nabrzyski, J., Barclay, I.: Verifiable badging system for sci-entific data reproducibility. Blockchain: Res. Appli. **2**(2), 100015 (2021). https://doi.org/10.1016/j.bcra.2021.100015
16. Rensaa, J.-A. H.: VerifyMed-Application of Blockchain technology to improve trust in virtualized healthcare services, NTNU (2020)
17. Harithuddin, A.S.M., Ninggal, M.I.H., Samian, N.: Blockchain for higher education in malaysia: a look into digital. In: Recent Advances and Applications in Blockchain Technology, p. 13. UTeM Press (2021)
18. Kamil, M., Sunarya, P.A., Muhtadi, Y., Adianita, I.R., Anggraeni, M.: BlockCert higher education with public key infrastructure in indonesia. In: 2021 9th International Conference on Cyber and IT Service Management (CITSM), pp. 1–6. IEEE (2021)
19. Aini, N., Wicaksono, S.A., Arwani, I.: Pembangunan Sistem Informasi Per-pustakaan Berbasis Web menggunakan Metode Rapid Application Development (RAD)(Studi pada: SMK Negeri 11 Malang). Jurnal Pengembangan Teknologi Infor-masi dan Ilmu Komputer **3**(9), 8647–8655 (2019)
20. Nalendra, A.: Rapid Application Development (RAD) model method for creating an agricultural irrigation system based on internet of things. IOP Conf. Ser. Mater. Sci. Eng. **1098**(2), 022103 (2021)

Factors Influencing the Digital Skills of Technical Education Students

Mohammad Fitri Hakimi Mat Hussin[1], Hanifah Jambari[1](✉), Ishak Taman[2], Muhamad Fathullah Hamzah[1], and Umi Salmah Mihad[3]

[1] School of Education, Faculty of Social Sciences and Humanities, Universiti Teknologi Malaysia, Johor, Malaysia
hanifahjambari@gmail.com
[2] Politeknik Ibrahim Sultan, Pasir Gudang, Johor, Malaysia
[3] School of Professional and Continuing Education, Universiti Teknologi Malayisa, Kuala Lumpur, Malaysia

Abstract. Digital skills are crucial technical competencies for students in higher education institutions. Weaknesses among students in utilizing digital technology can lead to moderate levels of learning achievement, consequently impacting the future graduates' marketability in the job sector. Hence, the purpose of this study is to identify the factors influencing the acquisition of digital skills among Technology with Education (TE) students at Universiti Teknologi Malaysia (UTM). This quantitative survey involved a total of 87 final-year TE students from four programmes: Bachelor of Technology with Education; Electric and Electronics, Mechanical Engineering, Building Construction and Living Skills at UTM. The study utilized a questionnaire with 21 items, encompassing three key factors: facilities, quality of teaching by lecturers, and study environment. Data analysis was conducted using Statistical Package for the Social Sciences (SPSS) software version 26.0. The findings revealed that the mean scores for all three factors were at a moderate level. Consequently, enhancing TE students' digital learning experiences should concentrate on these three factors to augment their academic achievements in today's digital era.

Keywords: Digital Skills · Factors · Technical Education

1 Introduction

Digital technology skills are essential for students in the 21st century. Anealka Aziz (2018) emphasises that in the new vision of 21st-century learning environment, students can identify the source of learning skills and knowledge. The new education ecosystem requires Gen Z and Alpha students to undergo smarter and more integrated education through technological literacy in addition to data and human literacy (Syamsul, 2018). Therefore, digital learning is very important in line with the current aspirations of the government and Industrial Revolution 4.0 (IR4.0). Almost all levels of learning, from elementary to higher education, digital learning media are used in the knowledge acquisition process. This coincides with the concept of digital learning, which is the transformation

© The Author(s), under exclusive license to Springer Nature Singapore Pte Ltd. 2024
F. Hassan et al. (Eds.): AsiaSim 2023, CCIS 1911, pp. 96–104, 2024.
https://doi.org/10.1007/978-981-99-7240-1_8

of the Malaysian education system in positioning the nation as a global education hub (Sintian et al., 2021).

2 Problem Background.

The usage level of information technology (IT) is influenced by the level of facilities available in an institution. This means that if a location has very good IT facilities, then the use of IT will also be high (Halim and Manis, 2021). This is supported by Abdul Rahman (2017) who states that there are several challenges in ensuring that school IT facilities are in good condition, such as inefficient maintenance of school computers, and difficulties in obtaining technical assistance to repair damaged computers. Although students have interest and willingness to use IT in their studies, a lack of resources and difficult access to the Internet may be the main reason for not using IT in school. The findings of Hong's (2014) study clearly show that adequate information technology facilities are crucial to carrying out digital learning smoothly and effectively.

Fadzil (2015) states that a small number of lecturers have low skills in teaching digital drawing, colouring, and animation. This is because some lecturers do not have specific qualifications such as computer or information technology degrees. Nowadays, traditional learning has become irrelevant to students because almost everyone in the entire society uses technology in their daily lives. Therefore, teachers and lecturers play an important role in mastering the use of software-related digital skills, which can have a good impact on the students' learning appetite (Siti Hajar et al., 2019). In addition, the environmental or climate factors of a school are equally important in influencing students' achievement. Rahman et al. (2021) believes that the physical environment of the place of learning is an important factor to improve the teaching and learning outcomes. The physical aspects of a learning site include the physical atmosphere of the learning site and the surrounding area.

According to the study of Halim and Manis (2021) on the integration of students with digital technology, students who believe that subjects using digital skills can provide constructive knowledge, see technology as a learning tool. It helps them to discover new perspectives of learning to solve problems and further develop the students' conceptual understanding of a subject. Therefore, this study aims to identify the factors that influence TE students' digital learning.

3 Objectives

I. Identify the facilities factor in influencing the digital skills of TE students.
II. Identify the lecturer's teaching factor in influencing the digital skills of TE students.
III. Identify the environmental factors in influencing the digital skills of TE students

4 Methodology

The research method is quantitative, i.e., a questionnaire was used as the research instrument. The respondents are final year students of UTM's Technology with Education programme. The population for this study is 116 students from four programmes. Based on the Krejci and Morgan's (1970) sample size determination table, the required study sample size is in the range of 86 to 92 respondents. Therefore, the researcher used a stratified random sampling method; each of the respondents was given a questionnaire administered through Google Forms. Eighty-seven respondents who returned the questionnaire were analysed using Statistical Package for Social Science (SPSS) software version 26.0 and the data was analysed to obtain the percentage, average mean and standard deviation. In addition, a pilot study was conducted randomly among 25 students from programs related to Technology with Education. The resulting data from this preliminary study was also analysed using SPSS software version 26.0, aiming to ascertain the questionnaire's reliability. The overall Cronbach's Alpha value obtained was 0.81, indicating a favourable level of reliability. This value underscores that the questionnaire is sufficiently reliable for use in the research.

5 Data Analysis

According to the research data analyses as contained in Tables 1, 2 and 3, the average mean scores for all three factors are at a moderate level: facilities factor, 3.40; lecturers' teaching, 3.49; and study environment factor, 3.52.

Referring to Table 1, question 5 exhibits the highest average mean of 3.57. A considerable 53 students (60.9%) strongly agreed, while 31 students (35.6%) agreed that the university's lecture halls are equipped with projectors and LCD monitors. The second-highest average mean is attributed to question 7, with mean 3.49. A notable 47 students (54%) strongly agreed that the institution provides students with free software applications. However, a minor faction of 4 students (4.6%) expressed strongly disagreed with this statement. Moving on, question 3 secured the third highest average mean at 3.40. While 43 students (49.4%) were in strongly agreed about the institution's provision of digital learning facilities, 8 students (9.2%) held an opposing view.

Among the questions, three items yielded the lowest mean values, all nearly identical: question 2, question 6, and question 1, with mean 3.39, 3.38, and 3.34, respectively. For question 2, a considerable 43 students (49.4%) acknowledged that the equipment and practical requirements at the institution functioned well, whereas only 5 students (5.7%) were strongly opposed to the statement. Concerning question 6, a significant 44 students (50.6%) strongly agreed that the laboratory provided a conducive environment with digital facilities, whereas 5 students (5.7%) held an opposing view. In the case of question 1, 1 student (1.1%) and 11 students (12.6%) disagreed that the institution's equipment and practical provisions were sufficient.

Question 4 garnered the lowest mean value, with a total of 3 students (3.4%) and 11 students (12.6%) expressing disagreed about the institution's internet access being fast and reliable. Ultimately, the mean value for the overall facility convenience factor settled at 3.40, signifying a moderate level.

Table 2 presents the findings of the conducted study and analysis regarding the impact of lecturers' teaching on the digital skills of TE students. The highest mean value, at 3.57, is attributed to question 1. A substantial 55 individuals (63.2%) strongly agreed with the assertion that lecturers conduct online classes, while 1.1% and 3.4% of students respectively expressed disagreed and strongly disagreed with the statement. Moving

Table 1. Mean Scores for Facilities Factor

No	Item	Frequency/Percentage				Mean	Standard Deviation
		SDA	DA	AG	SA		
1	Equipment and practical needs at the institution are sufficient	1 (1.2)	11 (12.6)	32 (36.8)	43 (49.4)	3.34	0.744
2	The equipment and practical needs at the institution are working well	–	5 (5.7)	43 (49.4)	39 (44.8)	3.39	0.598
3	The institution provides facilities in digital learning	–	8 (9.2)	36 (41.4)	43 (49.4)	3.40	0.655
4	Internet access in the institution is fast	3 (3.4)	11 (12.6)	36 (41.4)	37 (42.5)	3.23	0.803
5	The lecture hall is equipped with a projector and LCD	–	3 (3.5)	31 (35.6)	53 (60.9)	3.57	0.563
6	The laboratory provided is conducive and equipped with digital facilities	–	5 (5.7)	44 (50.6)	38 (43.7)	3.38	0.595
7	The institution provides free software applications to students	–	4 (4.6)	36 (41.4)	47 (54.0)	3.49	0.588
Overall Average						3.40	0.649

forward, both question 4 and question 6 yielded the second highest mean value, both at 3.55. For question 4, a total of 52 students (59.8%) strongly agreed that their lecturers offer online tests and quizzes, while for question 6, 51 students (58.6%) strongly agreed that their lecturers exhibit proficiency in utilizing digital tools relevant to the course topic.

The third highest mean value, 3.49, pertains to question 2, where 46 students (52.9%) strongly agreed that lecturers incorporate digital resources in their teaching methods, while 3 students (3.4%) held a differing viewpoint. Moving on to question 5 and question

Table 2. Mean Scores for Lecturers' Factor

No	Item	Frequency/Percentage				Mean	Standard Deviation
		SDA	DA	AG	SA		
1	Lecturer conducts online classes	1 (1.2)	3 (3.4)	28 (32.2)	55 (63.2)	3.57	0.622
2	Lecturer's teaching uses a lot of digital facilities	–	3 (3.4)	38 (43.7)	46 (52.9)	3.49	0.568
3	Lecturer masters the use of software well	–	4 (4.6)	45 (51.7)	38 (43.7)	3.39	0.578
4	Lecturer provides online tests and quizzes	–	4 (4.6)	31 (35.6)	52 (59.8)	3.55	0.586
5	Lecturer explains the guidelines for the computer applications used	–	3 (3.5)	41 (47.1)	43 (49.4)	3.46	0.567
6	Lecturer masters the digital use related to a topic well	–	3 (3.4)	33 (37.9)	51 (58.6)	3.55	0.566
7	Lecturer masters the use of digital tools well	–	3 (3.4)	42 (48.3)	42 (48.3)	3.45	0.566
Overall Average						3.49	0.579

7, their mean values stand at 3.46 and 3.45, respectively. Concerning question 5, 43 students (49.4%) strongly agreed that lecturers elucidate guidelines for utilizing computer applications, while 3 students (3.4%) held an opposing perspective. As for question 7, 42 students (48.3%) agreed that lecturers possess adeptness in using digital tools.

Furthermore, question 3 received the lowest mean value of 3.39, with 4 students (4.6%) expressing disagreed regarding lecturers' mastery of software usage, while 45 students (51.7%) endorsed the statement. Ultimately, the overall mean value for the lecturer's teaching factor stands at 3.51, signifying a moderate level of influence.

Table 3. Mean Scores for Study Environment Factor

No	Item	Frequency/Percentage				Mean	Standard Deviation
		SDA	DA	AG	SA		
1	Computer layout in the AutoCAD laboratory is neat and orderly	–	4 (4.6)	32 (36.8)	51 (58.6)	3.54	0.587
2	Computer usage guidelines are provided	1 (1.1)	2 (2.3)	38 (43.7)	46 (52.9)	3.48	0.607
3	The table in the AutoCAD lab is arranged facing the projector and LCD	2 (2.3)	3 (3.4)	34 (39.1)	48 (55.2)	3.47	0.679
4	The laboratory is equipped with digital facilities	–	2 (2.3)	41 (47.1)	44 (50.6)	3.48	0.547
5	There is air conditioning in the laboratory for the comfort of students conducting simulations	–	2 (2.3)	32 (36.8)	53 (60.9)	3.59	0.540
6	The lighting in the workshop/lab while doing the simulation is good	–	3 (3.4)	35 (40.2)	49 (56.3)	3.53	0.567
7	The layout of the workshop/laboratory facilitates student movement during the simulation	–	3 (3.4)	31 (35.6)	53 (60.9)	3.57	0.653
Overall Average						3.52	0.584

Referring to Table 3, question 5 acquired the highest mean value of 3.59. A notable 53 students (60.9%) and 32 students (36.8%) concurred that the laboratory is equipped with air conditioning to ensure the comfort of students engaged in simulations. A mere 2 individuals (2.3%) expressed disagreed with this statement. Moving on, the second-highest mean value, at 3.57, corresponds to question 7. A significant 53 students (60.9%) strongly agreed that the workshop/laboratory layout promotes smooth student movement during simulations, whereas 3 students (3.4%) held an opposing view.

The third-highest mean value, 3.54, is attributed to question 1. Among them, 51 students (58.6%) perceived the computer layout in the AutoCAD laboratory to be well-organized and tidy, while 4 students (4.6%) held differing opinions on this matter.

Turning to question 2 and question 4, they both garnered an identical mean value of 3.48. For question 2, 46 students (52.9%) confirmed the availability of computer usage guidelines, and for question 4, 44 students (50.6%) agreed that the laboratory is furnished with digital facilities. The lowest mean value recorded, 3.47, corresponds to question 3. In this context, 3 students (3.4%) and 2 students (2.3%) noted that the tables in the AutoCAD laboratory were not positioned facing the projector and LCD. However, a more substantial 48 students (55.2%) and 34 students (39.1%) strongly agreed that the laboratory tables are indeed arranged to face the projector and LCD. Ultimately, the overall mean value attained for the environmental factor is 3.52, indicating a moderate level.

6 Discussion

The results of the convenience factor study reveal that the information technology facilities provided in workshops and laboratories for TE students are at a moderate level. This is primarily due to the need for improved internet access in these spaces, which makes digital learning challenging for students. Additionally, the availability of computer or laptop units in the workshops or laboratories is limited; only specific laboratories and workshops are equipped with such devices for teaching and learning certain subjects. Awang (2007) pointed out that the utilization of information technology is influenced by the availability of facilities in educational institutions. Consequently, to ensure a continued learning experience with digital technology, colleges must guarantee that their information technology facilities and infrastructure are sufficient to meet teaching and learning needs.

Regarding the lecturer's teaching factor, the obtained results also fall within the moderate range. The majority of respondents agree that lecturers should possess competent digital skills in order to enhance student achievement by implementing diverse teaching strategies and methods involving digital technology. In accordance with the latest advancements in digital technology, learning should be more electronically oriented. For instance, when designing formal assessments like quizzes, tests, and exams, lecturers should opt for digital formats. Lecturers need to embrace digital tools and online software to improve students' proficiency in utilizing such resources.The findings also demonstrate that the lecturer's competence in terms of digital skills related to a specific topic impacts students' digital skill. Based on the data analysis, a total of seven respondents disagreed with the notion that lecturers are well-versed in software or digital technology. This could be attributed to various factors, including the lack of suitable lecturers for particular subjects. Consequently, due to the allocation of subjects to less qualified lecturers by institutional management, these lecturers encounter challenges in effectively conveying learning objectives to students (Rahman et al., 2021).

Although the study environment factor may not significantly influence TE students' digital skills, it does affect their overall learning. This is because a conducive environment for knowledge development and practical training facilitates skill acquisition. The results underscore the role of a favourable environment in promoting better subject comprehension. According to Zakaria et al. (2008), the learning environment plays a vital role in holistic and integrated individual development. In this study, the environment encompasses the learning atmosphere and ambiance within laboratories and workshops, with a focus on the layout of digital technology facilities provided by the management. Factors such as appropriate lighting and ventilation during simulation work or practice should also be considered, as they contribute to an environment that supports effective teaching and learning sessions.

7 Conclusion

In conclusion, according to the findings, the impact of the three factors influencing the digital skills of TE students is at a moderate level. Hence, all stakeholders, including students, lecturers, and institutions, must fulfil their respective responsibilities in enhancing the digital learning ecosystem. The overarching goal is to bolster the employability of graduates, particularly those in technical fields, thereby facilitating their access to favourable job opportunities and fostering a promising career path for their future.

Acknowledgment. The authors would like to acknowledge the financial support from Universiti Teknologi Malaysia for the funding under UTM Encouragement Research (UTMER) with vote number Q. J130000.3853.20J05. And the Ministry of Higher Education under Fundamental Research Grant Scheme (FRGS) with vote number, FRGS/1/2020/SS0/UTM/02/6.

References

Ahmad, A., Jinggan, N.: Pengaruh kompetensi kemahiran guru dalam pengajaran terhadap pencapaian akademik pelajar dalam mata pelajaran Sejarah. Jurnal Kurikulum & Pengajaran Asia Pasifik, Bil **3**(2), 1–11 (2015)

Hussin, A.A.: Education 4.0 made simple: Ideas for teaching. International Journal of Educ. Literacy Stud. 6(3), 92–98 (2018).

Ujai, D.S., Ruzanna, W.M., Mohamad, W.: Pengaruh Faktor Sosial Dalam Pembelajaran Bahasa Melayu Dalam Kalangan Murid Iban (Social Factors Influence in Malay Language Teaching among Iban Students). Malay Lang. Educ. J. – MyLEJ 7(1), 74–84 (2017). http://journalarticle.ukm.my/10415/1/135-260-1-SM.pdf

Fadzil, A.: Penggunaan Aplikasi Multimedia Interaktif Dalam Kemahiran Melukis. Mewarna dan Menganimasi Secara Digital oleh Azman bin Fadzil Tesis Yang Diserahkan Untuk Memenuhi Keperluan bagi Ijazah Doktor Falsafah FEBRUARI 2015, vol. 30 (2015)

Halim, F. A., Manis, A.A.: Faktor-Faktor yang Mempengaruhi Penerimaan Pembelajaran Berbantukan Permainan Digital Pelajar Kolej Vokasional Factors That Influenced the Acceptance of Digital Game Based Learning among Vocational College's Students. Online J. TE Practitioners **6**(2), 123–133 (2021). https://doi.org/10.30880/ojtp.2021.06.02.013

Husin, M. R., et al.: Perspektif Guru Terhadap Pembelajaran Pelajar Remaja. J. Human. Soc. Sci. **3**(1), 40–49 (2021). https://doi.org/10.36079/lamintang.jhass-0301.211

Nur Ali Ramadhanm, M.: Kesediaan Pelajar Memiliki Kemahiran Teknikal Tambahan: Satu Kajian Di Uthm X, pp. 1–21 (2013)

Putrawangsa, S., Hasanah, U.: Integrasi Teknologi Digital Dalam Pembelajaran di Era Industri 4.0. Jurnal Tatsqif **16**(1), 42–54 (2018). https://doi.org/10.20414/jtq.v16i1.203

Rahman, A.F.A., Taat, M.S., Hongkong, J.: Hubungan Faktor Guru dan Persekitaran dengan Pembelajaran Pendidikan Jasmani : Satu Tinjauan di Sekolah Menengah Pendalaman Sabah. Malaysian J. Soc. Sci. Human. (MJSSH), **6**(9), 143–153 (2021). https://doi.org/10.47405/mjssh.v6i9.984

Rahman, A., Sukri, M.: No - Pembangunan Instrumen Karakter Kreatif Pelajar Pendidikan Teknikal Dan Latihan Vokasional (TE). Anp J. Soc. Sci. Human. **2**(2), 112–122 (2021). https://doi.org/10.53797/anp.jssh.v2i2.16.2021

Nik, R.A.: Penggunaan teknologi maklumat dan komunikasi (ICT) di kalangan guru pelatih UTM semasa latihan mengajar. Universiti Teknologi Malaysia (2007)

Sintian, M., Kiting, R., Wilson.: Sikap Murid Terhadap Kemahiran Literasi Digital Dalam Pembelajaran Bahasa Kadazandusun Di Sekolah Menengah Sabah, Malaysia [Students Attitude Towards Digital Literacy Skill in Learning Kadazandusun Language At Secondary School, Sabah, Malaysia]. Muallim J. Soc. Sci. Human. **5**(1), 19–27 (2021). https://doi.org/10.33306/mjssh/108

Taib, S.H., Ismail, M.A., Lubis, M.A.L.A.: Inovasi kesepaduan dan strategi pengajaran dan pembelajaran di era Revolusi Industri 4.0. ASEAN Comp. Educ. Res. J. Islam Civiliz. **3**(2), 38–54 (2019)

Arifin, S.: Innovation learning in industry 4.0 era. Diakses daripada (2018). http://www.kopertis3.or.id/v6/wp-content/uploads/2019/05/M2-Inovation-PembelajaranIdustri-420-12-2018.pdf

Baki, I.A.: Panduan pelaksanaan pendidikan abad ke-21. In: Baki.Dan, V., Khas, P. (eds.) Nilai: Institut Aminuddin (2012). Falsafah Pendidikan Kebangsaan Memperkasakan Peranan Pendidikan Teknik Vokasional Dan Pendidikan Khas (2017)

Zakaria, H., Arifin, K., Ahmad, S., Aiyub, K.: Pengurusan Fasiliti Dalam Penyelenggaraan Bangunan : Amalan Kualiti , Keselamatan dan Kesihatan Universiti Kebangsaan Malaysia. J. Techno-Soc. 2015, 23–36 (2008)

Ballet Gesture Recognition and Evaluation System (Posé Ballet): System Thinking, Design Thinking, and Dynamic Improvement in Three Versions from Laboratory to Art Gallery

Apirath Limmanee(✉) and Peeraphon Sayyot(✉)

King Mongkut's Institute of Technology Ladkrabang, Bangkok 10520, Thailand
apirath.li@kmitl.ac.th, Peeraphonza2143@gmail.com

Abstract. This paper elaborates the design and implementation of "Posé Ballet," the ballet gesture recognition and evaluation system using MS Kinect camera. "Posé Ballet" has gone through continual development which can be divided into three phases according to requirements from three different user groups. These three groups are ballet students, ballet teachers and dancers, as well as beginners and laypersons. After the technical side of the system is explained, we describe our "design thinking" concepts. With these concepts, our new and improved third-version software as well as GUI are developed to serve the visitors at Bangkok Art and Culture Centre (BACC). At the end, some results and feedback are collected. These give us insight on how to improve our system further toward the goal of being standardized or commercialized.

Keywords: MS Kinect · human computer interaction · gesture recognition · ballet · education technology · serious games

1 Introduction, Motivation, and Previous Work

Have you ever heard a piece of music and simply could not get it out of your head? Inside their heads are where ballet students sometimes play the music and perform the associated dance moves. Maybe they are doing homework, trying to sleep, or brushing their teeth, but their minds and feet keep drifting back and rehearsing their performances. That is the sign that their ballet teachers have leaped out of the studios and into students' hearts. These are teachers who connect with their students and lingers in their mind even after the classes' final curtsies. This is possible for effective real-life teachers, but how can we craft this kind of effectiveness for virtual classes using information technology?

At the start of our research project, we employed Kinect sensors for implementing two versions of ballet gesture recognition and evaluation system. They are designed to suit the perception and requirements of two specific user groups. One of them is ballet students preparing for ballet exams. Another group represents professional dancers and ballet teachers. Indeed, one author of this paper has been studying ballet for more than 10 years and certified by passing some RAD (Royal Academy of Dance) exams. That

© The Author(s), under exclusive license to Springer Nature Singapore Pte Ltd. 2024
F. Hassan et al. (Eds.): AsiaSim 2023, CCIS 1911, pp. 105–124, 2024.
https://doi.org/10.1007/978-981-99-7240-1_9

is why the first version of our software is focused on practicing and evaluating dance moves used in the exam. Then, the second version is developed to satisfy the quest for beauty and perfection as required by professional dancers and ballet teachers.

However, in order to expand our research work, we constantly look for ballet teaching experts whom we can learn from such that our system can be improved. We are grateful to find two great ballet teachers. Vararom Pachimsawat, an artistic director who taught ballet at Dance Centre School of Performing Arts in Bangkok, Thailand, for more than 30 years. Another one is Nutnaree Pipit-Suksun, a ballerina who was the youngest person to win a gold medal at the Adeline Genée International Competition in 2001 [10]. We deeply appreciate her for being our model in this research project, in which her photos are displayed in our software GUI, so that users can imitate her six ballet poses. She choreographed these poses herself, giving a sense of beauty and continuity from one pose to another.

Fig. 1. "Posé Ballet" on Display at BACC.

One main contribution of this paper is to discuss three different versions of our software, as tailor made to three specific user groups. The emphasis is on the most recent version designed for beginners and laypersons. This version was displayed publicly in The International Dance Festival, organized by Friends-of-the-Arts Foundation at Bangkok Art and Culture Centre (BACC). The three software versions are developed according to the requirements discussed in Sect. 3. Prior to that, we review related research work in Sect. 2.

Although we still have not really sold our "Posé Ballet," we want it to be sellable, so this paper will discuss sellable ideas. This requires the "design thinking" part, as discussed in Sect. 5, in addition to the "system thinking" part, as discussed earlier in Sect. 4. The "design thinking" strategies discussed in Sect. 5 apply primarily to our third-version software, which was on display at BACC, serving the user group of beginners and laypersons. However, they can also later be applied to dynamically improve the first and second versions serving other two user groups consisting of ballet students, as well as dance teachers and professional dancers.

Note that, technically, the "design thinking" part deals primarily with the interaction between human users and the system. Therefore, the main discussion in Section V will be about the GUI and use cases, whereas the discussion in Sect. 4 focuses on the common fundamental components in hardware, software, as well as the recognition and evaluation framework.

2 State of the Arts

A significant portion of IT research in the context of ballet heavily relies on the utilization of Kinect cameras. These cameras are frequently referred to as sensors due to their capability to instantaneously convert captured images into numerical representations of body joint positions in three-dimensional Cartesian coordinates. This facilitates researchers in the manipulation of these values for many purposes, including dance recording, learning, practice, and the evaluation of ballet performances.

Like this paper, several research endeavors use Kinect cameras to design ballet learning and training systems. Illustrative examples encompass [1–9]. These works exhibit divergent focuses and techniques. The implemented dance environments also span a spectrum, ranging from an intelligent mirror (referred to as "Super Mirror" [7]) to entirely immersive virtual reality (VR) setups like the CAVE [2, 3]. Within this domain, [5] concentrates on aspects related to remote learning and the feedback provided to students. Several papers delve into the incorporation of artificial intelligence (AI) to analyze ballet performances and furnish feedback to users. Some rely on conventional rules for dance evaluation, involving calculations of parameters such as torso misalignment and pelvis displacement based on joint coordinates captured by the Kinect. Notably, [2, 3] propose sophisticated gesture recognition algorithms that yield highly precise outcomes.

Our work can be considered in a larger context of serious games as applied to dance teaching. Examples include [13], where games are used to teach dance in primary schools, and [14], where folklore dance is considered.

Alternatively, this work can also be included in the context of exergames since it can be considered as a computer game which combines entertainment and body movement. Gao et al. [17] shows that dance games can improve self-efficacy, social support, and physical activity level in fourth-grade and fifth-grade students. Research in exergaming also include older population. For example, Liao shows in [18] that a Kinect-based exergame helps improve cognitive function in frail older adults. Note, however, that the aforementioned works focus on psychological, physical, and cognitive benefits, while our work focuses more on the aesthetic aspects of dance.

In this work, feedback from users is collected by means of questionnaires. The questions are separated into four groups of questions. The four groups have specific aims of measuring four psychologically related concepts, which are situational interest, competition anxiety, exercise health belief, and dance health belief. These concepts are frequently discussed, not only in exergame research, but also in other related fields, such as physical education or sport psychology. We review some interesting works that study these concepts as follows.

Situational interest refers to an influence exerted by contextual stimuli related to the external environment [11]. In this case, the stimuli would be the lesson learned from "Posé Ballet." Situational interest can be elicited from the learners in a short period of time by changing their affect and cognition in a certain context, combined with emotional experiences or the external physical environment [12]. It is shown in [19] that situational interest in exergames increase the levels of physical activity metrics.

Competition is a common learning strategy. Through the experience of competing with oneself or others, learners are motivated to improve their performance and gain social recognition [15]. According to a study in sport psychology, competitions with rewards can facilitate performance and increased effort [20]. However, it is prudent to keep the level of competition anxiety in check. A study shows that if the level of anxiety is too high that the learners feel threatened, it can lead to a lower self-esteem [16].

3 Stakeholders' Requirements

In a dance studio, the teacher plays the music, shows students how to perform dance moves, explain the best way to execute each move, and correct students' missteps. We would like to develop a system for ballet students that best replicates this dancing environment. The first version of our software is thus designed to serve requirements of ballet students.

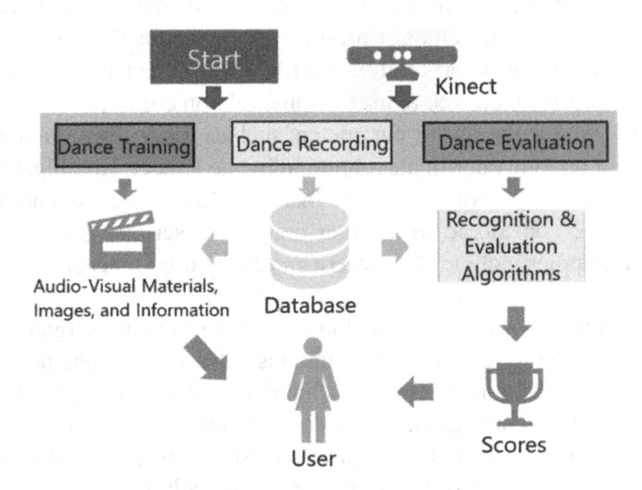

Fig. 2. Software Components

Through conversations held with various stakeholders, it becomes apparent that distinct individuals perceive ballet dance in varying ways. To develop our system with an approach centered on perception, we undertake an examination of user needs rooted in three distinct categories of ballet dance perception. These categories encompass the viewpoint of ballet students preparing for examinations, the perspective of professional dancers and dance instructors, and the outlook of novices and non-experts. The diverse perceptual categories consequently impact the configuration of the ballet learning and training system, influencing both the emphasized system components and the graphical user interface (GUI) design.

Illustrated in Fig. 2 are the three constituents comprising our software: dance training, dance recording, and dance evaluation. The interplay between user perception and these components is explicated in Tables 1 through 3.

Table 1. Software Requirements for Ballet Students Preparing for Examinations

Software Components	Requirements
Dance Recording	Dance and Songs Used in Exams (Audio-Visual and Images)
Dance Training	Mimicking Dance Moves on the Screen or Dance from Students' Own Memory with Music
Dance Evaluation	Verbal, Graphical, Audio, or Numerical

Table 2. Software Requirements for Professional Dancers and Dance Teachers

Software Components	Requirements
Dance Recording	Focus on Accuracy of Recorded Images
Dance Training	Not Needed
Dance Evaluation	Technical and Objective Feedbacks, e.g., Angles Made by Joint Coordinates

Table 3. Software Requirements for Beginners and Laypersons

Software Components	Requirements
Dance Recording	Simple and Fun Dance Moves (Audio-Visual)
Dance Training	Mimicking Dance Moves on the Screen with Music
Dance Evaluation	Short and Appealing Feedback: Verbal ("Perfect," "Very Good," "Good") or Graphical (Emoji, Stars, etc.)

4 System-Thinking: Hardware, Software, and Gesture Recognition and Evaluation Framework

This section describes three aspects of our system, which are hardware, software, as well as gesture recognition and evaluation framework.

4.1 Hardware Prototype

A hardware prototype has been meticulously designed, developed, and positioned within the dance studio of the "Dance Centre School of Performing Arts" situated in Bangkok, Thailand. The Kinect camera's elevation and the monitor's positioning are adaptable to accommodate users' heights. Users can install our software onto their laptop computers and seamlessly connect it to the hardware setup using a plug-and-play approach. Safety measures such as batteries, circuits, and an integrated loudspeaker are situated at the base, along with wheels beneath, rendering it effortlessly mobile. The visual depiction of the prototype can be observed in Fig. 3. To ensure convenient portability, only the upper portion of the prototype is transported for display at the BACC, as illustrated in Fig. 1.

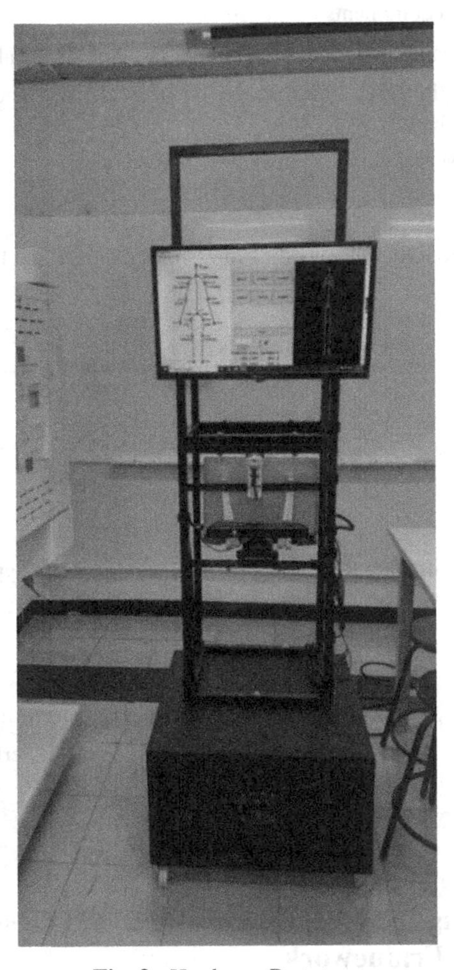

Fig. 3. Hardware Prototype

4.2 Software Structure

All versions of our software contain three main components, which are dance recording, dance training, and dance evaluation.

In general, the database and the three components of both programs can be described through block diagrams depicted in Figs. 4, 5, 6 and 7. The ways in which the three software versions are implemented certainly vary based on the specifications outlined in the preceding section (Tables 1, 2 and 3). To illustrate, the assessment algorithms for baller students are notably more intricate, given that the program needs to identify a greater number of dance flaws and compute comprehensive scores. In contrast, proficient dancers only require raw angle data generated from joint coordinates and the divergence from the desired angles. A comprehensive breakdown of the evaluation approach will be provided in Subsect. 4.3.

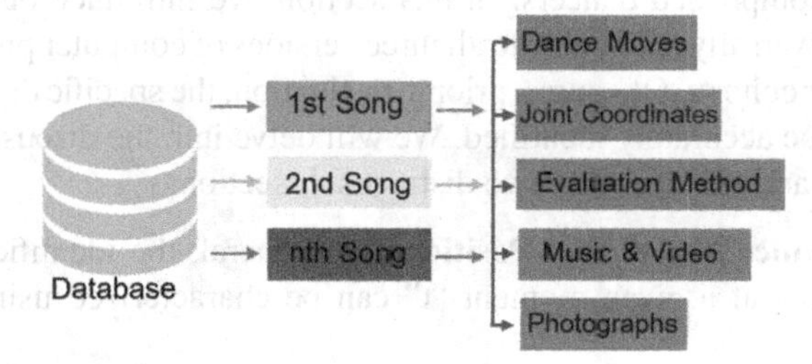

Fig. 4. Database Components

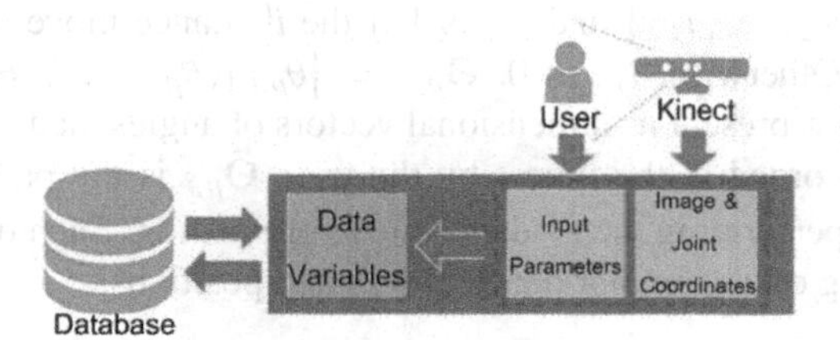

Fig. 5. Dance Recording Part

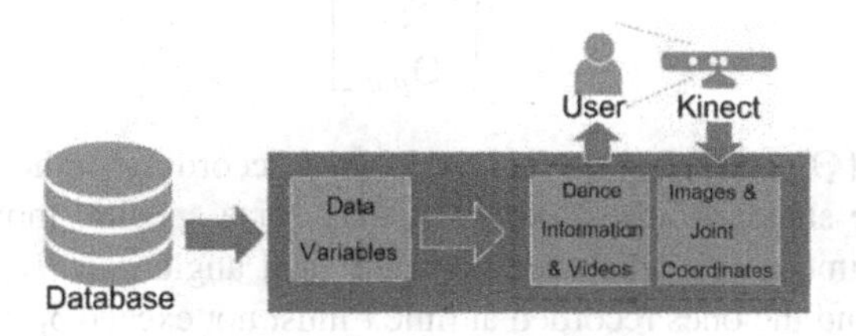

Fig. 6. Dance Training Part

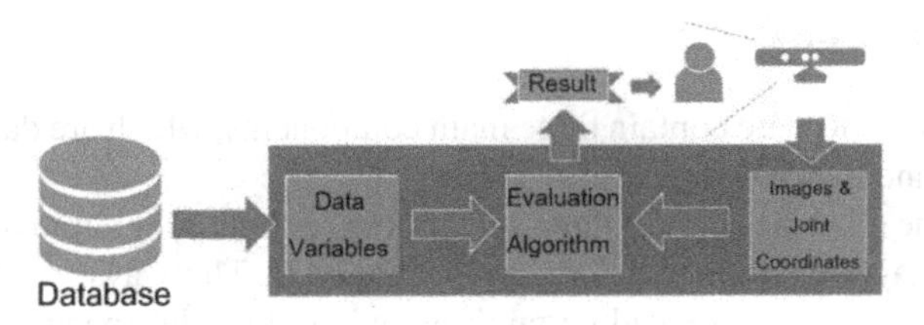

Fig. 7. Dance Evaluation Part

4.3 Gesture Recognition and Evaluation Framework

To facilitate dance recognition and assessment, the Kinect captures the user's skeletal joint coordinates during dance sequences at various time points, denoted as "t". These recorded coordinate instances are subsequently compared with predefined coordinates acquired from accomplished dancers. In this section, we introduce our mathematical structure that is universally applicable to all three versions of computer programs tailored for the distinct user cohorts. Of course, prior to evaluation, the specific dance movements or positions must be accurately identified. We will delve into the discussions regarding dance recognition and evaluation in two distinct sub-sections.

Recognition of Dance Moves and Positions. In general, the identification of dance motions and stances at a given moment "t" can be characterized using a functional vector

$$\boldsymbol{R_t} = f\left(\boldsymbol{\Theta}_{p,i}, \boldsymbol{\Theta}_t\right) \tag{1}$$

where $\boldsymbol{R_t} = \left\{r_{t,1}, r_{t,2}, \ldots, r_{t,n}\right\}$ and $r_{t,i}$ is 1 if the ith dance move or position is recognized at time t. Otherwise, $r_{t,i}$ is 0. $\boldsymbol{\Theta}_{p,i} = \left\{\theta_{p,i,1}, \theta_{p,i,2}, \ldots, \theta_{p,i,m}\right\}$ and $\boldsymbol{\Theta}_t = \left\{\theta_{t,1}, \theta_{t,2}, \ldots, \theta_{t,m}\right\}$ represent m-dimensional vectors of angles made by skeletal joints. $\boldsymbol{\Theta}_t$ is the vector recorded at the time t by the user. $\boldsymbol{\Theta}_{p,i}$ is the one pre-recorded by exemplary dancers performing the ith dance move or position. Each of $\boldsymbol{\Theta}_{p,i}$ is a part of the matrix consisting of q exemplar dance moves and positions.

$$\boldsymbol{T} = \begin{bmatrix} \boldsymbol{\Theta}_{p,1} \\ \boldsymbol{\Theta}_{p,2} \\ \vdots \\ \boldsymbol{\Theta}_{p,q} \end{bmatrix} \tag{2}$$

The function $f\left(\boldsymbol{\Theta}_{p,i}, \boldsymbol{\Theta}_t\right)$ in (1) can be designed according to the developer's standard. One possible standard in our design is that, for a specific move or position to be recognized at time t, the difference between each angle pair $\theta_{p,i,j}$ and $\theta_{t,j}$ of the prerecorded ones and the ones recorded at time t must not exceed δ_j, i.e.,

$$\left|\theta_{p,i,j} - \theta_{t,j}\right| \leq \delta_j \tag{3}$$

for all available value of j. By counting all possible independent angles from the skeletal joints in Fig. 8, we know that $1 \leq j \leq 19$ in our case. When all j are considered together,

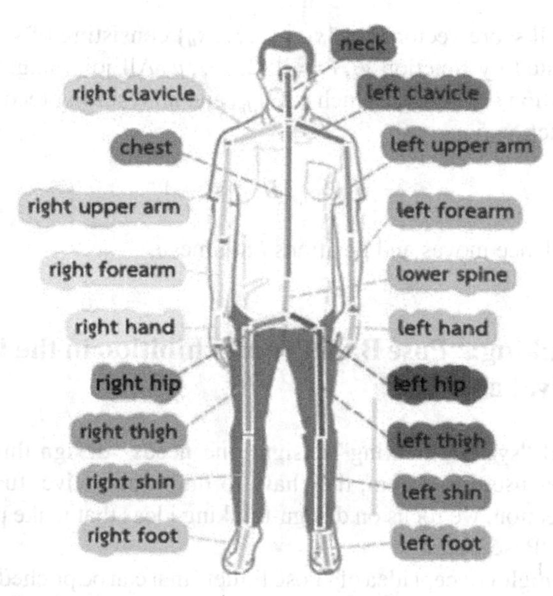

Fig. 8. Body Parts Made By Connecting Skeletal Joints

we can derive elements in $\boldsymbol{R}_t = \{r_{t,1}, r_{t,2}, \ldots, r_{t,n}\}$ based on our recognition standard as

$$r_{t,i} = \prod_{j=1}^{19} u\left(\delta_j - \left|\theta_{p,i,j} - \boldsymbol{\theta}_{t,j}\right|\right) \tag{4}$$

where

$$u(x) = \begin{cases} 1, x \geq 0 \\ 0, \textit{otherwise}. \end{cases} \tag{5}$$

Evaluation of Dance Moves and Positions. Regarding the assessment of dance maneuvers and postures, it is imperative to incorporate the angles formed by the skeletal joints as input. Differing from the dance recognition phase, the outcome of the evaluation process should indicate the quality of a recognized dance move or position, rather than merely confirming its recognition status. Furthermore, these results should elicit constructive feedback for the user's benefit. Each feedback instance is generated through a comparison of the input skeletal joint data against a pre-established criterion. In this context, we will outline the evaluation framework before subsequently implementing it within our computer programs.

The framework consists of the vector of m skeletal joint angles at time t_r, $\boldsymbol{\Theta}_{i,t_r} = \{\theta_{i,t_r,1}, \theta_{i,t_r,2}, \ldots, \theta_{i,t_r,m}\}$. We denote time by t_r with r as subscript to emphasize that only the vectors recognized as a dance move or position i from the previous step are considered as input. This input will be processed by h_i criterion functions $c_{i,k}, k = 1, 2, \ldots, h_i$, resulting in a feedback vector

$$\boldsymbol{B}_{i,t_r} = \{b_{i,t_r,1}, b_{i,t_r,2}, \ldots, b_{i,t_r,h_i}\} = \{c_{i,1}(\boldsymbol{\Theta}_{i,t_r}), c_{i,2}(\boldsymbol{\Theta}_{i,t_r}), \ldots, c_{i,h_i}(\boldsymbol{\Theta}_{i,t_r})\} \tag{6}$$

Then, the overall score vector $\boldsymbol{S} = \{s_1, s_2, \ldots, s_u\}$ consisting of scores in different u aspects is computed by function $g_l, l = 1, 2, \ldots, u$. All joint angles recognized as dance move or position i at time t_r, which is $\boldsymbol{\Theta}_{i,t_r}$, and all resulting feedbacks $\boldsymbol{B}_{i,t_r}$ serve as inputs of the function, i.e.,

$$s_l = g_l\left(\boldsymbol{\Theta}_{i,t_r}, \boldsymbol{B}_{i,t_r} | \forall i, t_r\right), \quad (7)$$

for all recognized dance moves and positions i at times t_r.

5 Design-Thinking: "Posé Ballet" on Exhibition in the International Dance Festival at BACC

Apart from a good "system-thinking" design, one needs "design-thinking" because, before the attendees use our system, they have to find it attractive, funny, and easy to play with. In this section, we focus on design-thinking ideas that make products sellable and apply them to "Posé Ballet."

We start from a high-concept idea of "Posé Ballet" that can be pitched in one sentence. After all those system components are rigorously developed in the previous section, we still need one-sentence description to get the crowd's attention at BACC. We came up with a short sentence: "Let's discover a ballerina in you" as our high concept idea. We put the sentence (in Thai translation) on our flyer, as shown in Fig. 9. It turns out that lots of attendees are interested and play with "Posé Ballet" in order to "discover a ballerina in them."

Art gallery observers are, by definition, good at observing art. They would like to play an active role in discovering artistic concepts. Therefore, we will show, in this section, how we design the GUI to engage audience and give them the joy of discovery they look for.

One of the key differences between Posé Ballet and other paintings at the art gallery is that the paintings show almost everything the audience needs to know. Though the audience might create something in their imagination while watching paintings, they still physically play a more passive role. As for our system, however, the audience cannot just passively watch. We also would not tell them beforehand what kind of dance moves they will be asked to perform. Otherwise, it would have taken a lot of fun out of it. We certainly want to show the correct way of specific dance moves by displaying our model on the screen. At the same time, we also want to subtly tell the audience to imitate the dance moves without verbally doing so. This is done by a series of interactions between the system and the audience, as depicted in the Use Case Text in Fig. 10.

Based on the use case, the software and GUI is designed based on the features and conceptual ingredients discussed in Subsects. 5.1 and 5.2.

Fig. 9. Flyer of "Posé Ballet" on Display at BACC

5.1 Features

The special features of "Posé Ballet" is summarized as follows.

1) The default recorded ballet steps provide continuity from one steps to another.
2) There is a time limit to perform each step.
3) A user can record new dance positions easily without the help of another user. (See Fig. 11)
4) All positional coordinates of all users' skeletal joints are recorded
5) All angles of users' skeletal joints are calculated from three viewpoints: front-view, side-view, and top-view. These are used for recognition and evaluation algorithms.

5.2 Conceptual Ingredients for Designing GUI

In this subsection, we will break down high-concept ideas regarding GUI design into essential ingredients. We will also relate each ingredient to the GUI of our Posé Ballet.

The Hook or Aspirational Elements. The "Hook" is a great analogy derived from fishing where one needs a shiny lure as bait to attract the fish. In our case, the bait we used is the moving skeletal image shown on the right of the invitation screen, as depicted in in Fig. 12. This image adds the coolness factor, when combining with an invitation sentence, makes even the complete strangers, with no knowledge of dance whatsoever, stop and play with Posé Ballet. From our experience, there were times when the system did not work properly. However, if the moving skeletal image still works, passer-by always stops and plays with it.

Use Case Text: Playing

- 1. System shows an invitation screen.
- 2. User gestures to starts playing.
- 3. System plays the song.
- 4. System shows the next dance screen (Explanation, model picture, skeleton).
- 5. User mimics the dance position
- 6. User gets feedback (stars, compliments, scores).
- 7. The system records the joint coordinates of user.
- 8. If there is the next dance position, return to step 4. otherwise, go to step 9.
- 9. System shows the summary page.

Fig. 10. Use Case Text: Playing "Posé Ballet"

The Larger-than-Life Elements. We know ballet is interesting, but ballet training systems using information technology do not always give best dance lessons. In dance studio, we sometimes meet great dance teachers whose dance moves leave us in awe and hence motivate us subtly. Thus, a ballet training system should be an exaggeration of real dance studios in order to compensate for the lack of personal encounter. In our case, a real-life ballet teacher is substituted by pictures of our model teacher and famous ballerina, Nutnaree Pipit-Suksun, as shown in Fig. 13. Also, the reflective mirror in the studio is replaced by the screen capable of showing users' skeletons. Thus, users will find the perception of their bodies is enhanced and can vividly compare their poses to one of the best ballet models on the screen.

The Fish-Out-Of-Water Elements. Although Posé Ballet is aimed at regular art gallery visitors who might never try ballet before, that does not mean we will make it too easy for them, otherwise, it would be boring. Simply being challenged by our GUI environment is automatically engaging and interesting. The time limit imposed by each ballet step ensures that users need to focus.

Theme and Lesson Elements. These elements offer some lessons that users can learn after they finish playing with our system. This does not mean that the lessons must be didactic or boring. It simply means that our system has to say or communicate something to the audience. In our case, we simply want to communicate that ballet can be fun and everyone can learn it. We therefore do not impose too strict criteria for passing each dance moves. After each dance move, the user will get funny feedback, either in the

Use Case Text: Recording

- 1. System shows the main menu.
- 2. User clicks "save."
- 3. System shows a recording screen.
- 4. User fills in description of the recorded position.
- 5. User presses "Capture."
- 6. System shows timer counting until time is up, while the model (user) is posing.
- 7. System saves joint coordinates and angles. (At the same time, photographer takes model's picture.)
- 8. If "OK", proceed to 9. if "Cancel", go back to 3.
- 9. If there is the next position, return to 3. Otherwise, system shows "Goodbye."

Fig. 11. Use Case Text: Recording Dance Position in "Posé Ballet"

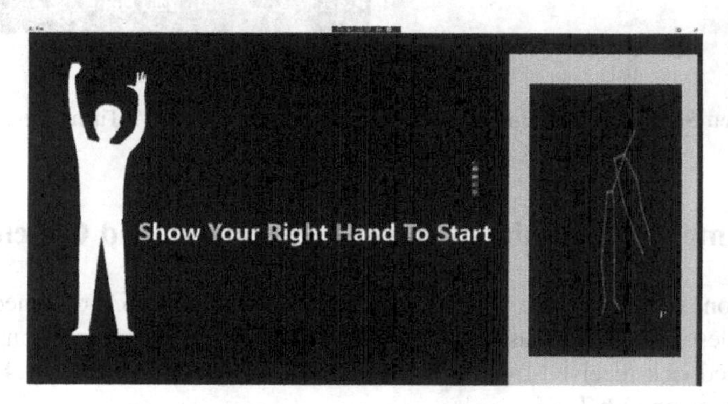

Fig. 12. Invitation Screen with the User's Skeletal Image

form of a smiling dog if they pass, or a tiring dog if they fail, as shown in Figs. 14 (a) and (b).

Fig. 13. Playing Screen Showing both the Model Ballerina and the User's Skeletal Image

(a) (b)

Fig. 14. Screen Showing (a) A Happy Dog as Positive Feedback (b) A Tiring Dog as Negative Feedback

6 Experiments, Preliminary Results, Feedback, and Conclusion

In this section, we give some preliminary results in Subsect. 6.3 obtained from the experiment described in 6.2, using the recognition and evaluation function in 6.1. We also show feedback received from users filling out our questionnaire in 6.4 – 6.6. The conclusion is given in 6.7.

6.1 Recognition and Evaluation Function Used in the Experiment

From the framework in Subsect. 4.3, we can see that the accuracy of dance move recognition and evaluation depends on how we design the functions $f\left(\boldsymbol{\Theta}_{p,i}, \boldsymbol{\Theta}_t\right)$, $c_{i,j}\left(\boldsymbol{\Theta}_{i,t_r}\right)$, and $g_l\left(\boldsymbol{\Theta}_{i,t_r}, \boldsymbol{B}_{i,t_r} | \forall i, t_r\right)$. These functions are integrated into our software and can be customized by users. We are not going to derive the optimal functions in this paper. Rather, we design some reasonable ones and show preliminary results.

In the recognition step, we use the function described by Eqs. (1) – (4) for both computer programs. The margin δ_j in (4) is tunable according to the level of difficulty needed. As for the evaluation step, we design 12 criteria $c_{i,j}(\Theta_{i,t_r}), j = 1, 2, \ldots, 12$ that return the feedbacks $b_{i,t_r,j}$ as numerical scores. After that, the overall score s_l for a complete dance song is computed as the feedbacks $b_{i,t_r,j}$ weighted by $w_{i,j}$ for all dance moves and positions i recognized at times t_r, i.e.,

$$s_l = \sum_{\forall i,t_r} \sum_{j=1}^{12} w_{i,j} b_{i,t_r,j}. \quad (8)$$

Observe that the weights assigned to distinct dance motions and positions vary. For instance, Plié places greater emphasis on feedback from knee joints, whereas Battement Tendu prioritizes feedback from foot joints.

6.2 Experimental Steps

We propose a potential experiment involving an expert, an experimenter, and a group of participants. The experiment can be conducted using the following steps:

(1) The expert utilizes heuristic guidelines to select weights in Eq. (8).
(2) Participants choose songs and execute dance moves. Simultaneously, a computer program captures and stores photographs of their performances. The program also evaluates the recognized dance moves and positions.
(3) The expert reviews the photographs from step 2. They identify and rate the accurate dance moves and positions.
(4) The experimenter contrasts the set of photos chosen by the expert in step 3 with the set recognized by the computer program in step 2. Subsequently, the experimenter generates experimental outcomes by calculating numerical values that quantify the extent of disparity between the two sets.
(5) Additionally, the expert's ratings are juxtaposed with the overall scores generated by the program. A positive outcome would involve dance moves and positions that receive higher expert ratings also obtaining elevated overall scores.

6.3 Preliminary Results

To ensure the reliability of our experimental findings, it is imperative that the participant pool is sufficiently large and encompasses varying levels of expertise. Incorporating multiple experts could also serve the purpose of triangulation. Alternatively, the integration of artificial intelligence (AI) might facilitate the adaptation of the weight in Eq. (8). These potential avenues represent directions for future exploration within our research.

Presently, our efforts are concentrated on executing the procedures outlined in Sect. 6.2. We engage one expert and enlist two ballet beginners, each representing a different gender, as participants. The preliminary outcomes are promising, as the computer program accurately identifies dance moves such as Bras Bas, Plié, and Relevé, including instances with first-position and second-position foot placements. Notably, for the first-position feet configuration, the images rated highest by the expert align with those receiving the highest overall scores from the computer program. Concerning the second-position feet arrangement, disparities exist between the expert's preferences and the program's selections, although their visual resemblance is striking. A comprehensive summary of these findings can be found in Table 4.

Table 4. Preliminary Gesture Recognition and Evaluation Results with Different Dance Moves and Positions.

Dance Moves & Positions	Gesture Recognition	Comparison of Best-Score Picture with Expert's Top-Rate Picture
Bras Bas, 1st Feet	Correct	Same Picture
Plié, 1st Feet	Correct	Same Picture
Relevé, 1st Feet	Correct	Same Picture
Bras Bas, 1st Feet	Correct	Very Similar Picture
Plié, 1st Feet	Correct	Very Similar Picture
Relevé, 1st Feet	Correct	Very Similar Picture

Table 5. Average Score from the Questionnaire Measuring the Situational Interest of Users.

Question	Average Score
Did you find the activity enjoyable?	4.31
Did you feel a sense of challenge during the activity?	4.04
Did you feel a sense of accomplishment during the activity?	4.00
Did you feel engaged and focused on the activity?	4.23
Did you feel a sense of flow during the activity?	4.15
Did you feel a sense of excitement or anticipation during the activity?	4.27

6.4 Feedback Measuring Users' Situational Interest

We show feedback received from users filling out our questionnaire. The questions aim at measuring the level of situational interest in users. The participants of this study are 26 first-year bachelor-of-technology students at Faculty of Industrial Education, King Mongkut's Institute of Technology Ladkrabang (KMITL). A 5-point Likert scale is applied to measure the degree of agreement. 1 stands for strongly disagree and 5 stands for strongly agree (Table 5).

6.5 Feedback Measuring Users' Competition Anxiety

Competition is a common learning strategy. Through the experience of competing with oneself or others, learners are motivated to improve their performance and gain social recognition [15]. This is somehow related to the "fish-out-of-water" element discussed in the previous section. There should be a sense of uneasiness or competition to get the learners motivated. According to a study in sport psychology, competitions with rewards can facilitate performance and increased effort [18]. However, it is prudent to keep the level of competition anxiety in check. A study shows that if the level of anxiety is too high that the learners feel threatened, it can lead to a lower self-esteem [16]. From Table 6, we can see that the response is moderate, i.e., in the range from 2.5 to 3.5. In the future work, we may try to vary the level of difficulty and study the relationship between the competition anxiety and the situational interest.

Table 6. Average Score from the Questionnaire Measuring the Competition Anxiety of Users.

Question	Average Score
How anxious do you feel when you are about to use and really use Posé Ballet?	3.46
How stressed do you feel when you are about to use and really use Posé Ballet?	2.96
How much pressure you feel to perform well when you are about to use and really use Posé Ballet?	2.96
How much do you worry about making mistakes when using Posé Ballet?	3.15
How much do you feel like you are being judged by others when using Posé Ballet?	3.35
How much do you compare yourself to others when using Posé Ballet?	3.08

6.6 Feedback Measuring Users' Exercise Health Belief and Dance Health Belief

As "Posé Ballet" can be considered an exergame which mixes physical exercising and game entertainment, we also ask questions to measure the level of "Exercise Health Belief" as well as "Dance Health Belief," in order to assess participants' belief regarding relationships between health and exercising or dancing, as shown in Tables 7 and 8. In the future, when more results are collected from several groups, we can study how the exercise/dance health belief of each group relates to their situational interest of "Posé Ballet."

Table 7. Average Score from the Questionnaire Measuring the Exercise Health Belief of Users.

Question	Average Score
How much do you believe that exercising regularly can improve your physical health?	4.76
How much do you believe that exercising regularly can improve your mental health?	4.59
How much do you believe that exercising regularly can improve your overall well-being?	4.82

(continued)

Table 7. (*continued*)

Question	Average Score
Do you think that exercise can help you manage stress?	4.32
How much do you believe that exercising regularly is an effective way to prevent or manage health conditions?	4.32
Do you believe that exercise can reduce the risk of chronic diseases?	4.15
Do you believe that exercise can help you maintain a healthy weight?	4.35
How much do you feel that exercising regularly is important to you?	3.88
Do you regularly engage in exercising activity?	2.85
How likely you are to engage in regular exercise in the future?	3.85

Table 8. Average Score from the Questionnaire Measuring the Dance Health Belief of Users.

Question	Average Score
How much do you believe that dancing regularly can improve your physical health?	3.97
How much do you believe that dancing regularly can improve your mental health?	4.14
How much do you believe that dancing regularly can improve your overall well-being?	4.05
Do you think that dancing can help you manage stress?	3.68
How much do you believe that dancing regularly is an effective way to prevent or manage health conditions?	3.5
Do you believe that dancing can reduce the risk of chronic diseases?	3.35
Do you believe that dancing can help you maintain a healthy weight?	3.65
How much do you feel that dancing regularly is important to you?	2.62
Do you regularly engage in dancing activity?	1.76
How likely you are to engage in regular dancing in the future?	3.85

6.7 Discussion and Conclusion

The limitation from preliminary results in 6.3 is that the number of samples (positional coordinates of users' skeletons) are very limited. At present, we still keep collecting more samples from university students who give us permission to do so. We expect more quantitatively accurate results in future work.

Situational interest refers to an influence exerted by contextual stimuli related to the external environment [11]. In 6.3, the stimuli would be the lesson learned from "Posé Ballet." Situational interest can be elicited from the learners in a short period of time by changing their affect and cognition in a certain context, combined with emotional experiences or the external physical environment [12]. Situational interest From the obtained score, we can conclude that the users have high situational interest scores, as the average is more than 4 in every question.

Regarding competition anxiety from Table 6, we can see that the response is moderate, i.e., in the range from 2.5 to 3.5. In the future work, we may try to vary the level of difficulty and study the relationship between the competition anxiety and the situational interest.

It can be seen from Table 7 that the first seven questions receive average score of more than 4, so, in general, the participants believe that exercising can help preventing diseases and maintaining physical fitness. However, when the last three questions are directed toward the participants' regular exercising behavior, the average scores are significantly lowered. We can conclude that, while most participants believe exercise is good for health, some of them find it difficult to exercise regularly.

As can be seen in Table 8, when we replace the word "exercise" from Table 7 with "dancing," the score is significantly reduced. We can infer that some participants do not consider dancing an exercise. When comparing "dancing" with "exercise," they believe the latter provides better health benefit.

However, we can observe that one pattern in Table 7 is also shown in Table 8. The last three questions receive significantly lower score than the first seven questions. We reach the similar conclusion that, while most participants believe dancing is good for health, some, if not most, of them find it difficult to dance regularly.

In conclusion, even when most participants do not regularly dance and do not consider dancing as healthy as other forms of exercise, they still find "Posé Ballet" interesting and enjoyable. Perhaps, during a short while they play with "Posé Ballet," they can discover ballerinas inside themselves.

Acknowledgements. This work is financially supported by King Mongkut's Institute of Technology Ladkrabang (KMITL) (Grant No. 2565–02-03–009). We would like to thank Vararom Pachimsawat and Nutnaree Pipit-Suksun, two wonderful ballerinas, for their advice and effort, without which this project would not be a success. We are also grateful to Bangkok Arts and Culture Centre (BACC) and Dance Centre School of Performing Arts for venue support.

References

1. Limmanee, A., Buranasuk, S.: Perception-driven ballet gesture recognition and evaluation system using kinect. In: 2021 IEEE International Conference on Imaging Systems and Techniques (IST), pp. 1–6, New York (2021)
2. Kyan, M., et al.: An approach to ballet dance training through MS kinect and visualization in a cave virtual reality environment. ACM Trans. Intell. Syst. Technol. (TIST) **6**(2), 1–37 (2015)
3. Muneesawang, P., et al.: A machine intelligence approach to virtual ballet training. IEEE Multimedia **22**(4), 80–92 (2015)
4. Trajkova, M., Cafaro, F.: E-ballet: designing for remote ballet learning. In: Proceedings of the 2016 ACM International Joint Conference on Pervasive and Ubiquitous Computing: Adjunct, pp. 213–216, Heidelberg (2016)

5. Knudsen, E.W., et al.:Audio-Visual Feedback for Self-Monitoring Posture in Ballet Training. In: New Interfaces for Musical Expression 2017, pp. 71–76, Copenhagen (2017)
6. Trajkova, M.: Towards AI-Enhanced Ballet Learning. In: Proceedings of the 6th International Conference on Movement and Computing, pp. 1–5 (2019)
7. Marquardt, Z., Beira, J., Em, N., Paiva, I., Kox, S.: Super Mirror: A Kinect Interface for Ballet Dancers. In: Proc. International Conference on Control, Automation and Systems, pp. 1619–1624 (2012)
8. Trajkova, M.: Designing AI-Based Feedback for Ballet Learning. In: Extended Abstracts of the 2020 CHI Conference on Human Factors in Computing Systems, pp. 1–9 (2020)
9. Limmanee, A., Poemyanwantana, P.: Computer Software for Practice and Evaluation of Ballet Lessons by Gesture Recognition Using Kinect. In: ECTI-CARD 2020, pp. 417 – 421 (2020)
10. Wikipedia, https://en.wikipedia.org/wiki/Ommi_Pipit-Suksun, Accessed 20 Aug 2023
11. Renninger, K.A., Hidi, S.E.: Interest development and learning. In Renninger, K. A., Hidi, S. E. (eds.) The Cambridge handbook of motivation and learning, pp. 265–290. Cambridge University Press (2019)
12. Hidi, S., Renninger, K.A.: The four-phase model of interest development. Educ. Psychol. **41**(2), 111–127 (2006)
13. Wang, Y., Liu, Q.: Effects of game-based teaching on primary students' dance learning: the application of the personal active choreographer. Int. J. Game-Based Learn. (IJGBL) **10**(1), 19–36 (2020)
14. Bakalos, N., Rallis, I., Doulamis, N., Doulamis, A., Voulodimos, A., Vescoukis, V.: Motion primitives classification using deep learning models for serious game platforms. IEEE Comput. Graph. Appl. **40**(4), 26–38 (2020)
15. Seaborn, K., Fels, D.I.: Gamification in theory and action: a survey. Int. J. Hum. Comput. Stud. **74**, 14–31 (2015)
16. Yang, J.C., Lin, M.Y.D., Chen, S.Y.: Effects of anxiety levels on learning performance and gaming performance in digital game-based learning. J. Comput. Assist. Learn. **34**(3), 324–334 (2018)
17. Gao, Z., Huang, C., Liu, T., Xiong, W.: Impact of interactive dance games on urban children's physical activity correlates and behavior. J. Exerc. Sci. Fit. **10**(2), 107–112 (2012)
18. Liao, Y.Y., Chen, I.H., Hsu, W.C., Tseng, H.Y., Wang, R.Y.: Effect of exergaming versus combined exercise on cognitive function and brain activation in frail older adults: a randomised controlled trial. Ann. Phys. Rehabil. Med. **64**(5), 101492 (2021)
19. Pasco, D., Roure, C.: Situational interest impacts college students' physical activity in a design-based bike exergame. J. Sport Health Sci. **11**(2), 172–178 (2022)
20. Shi, X., Kavussanu, M., Cooke, A., McIntyre, D., Ring, C.: I'm worth more than you! effects of reward interdependence on performance, cohesion, emotion and effort during team competition. Psychol. Sport Exerc. **55**, 101953 (2021)

Structured Teaching Using Drone Simulators for Students' Confidence in Real Flight

Ahmad Nabil Md Nasir[1(✉)], Mahyuddin Arsat[1], Muhammad Khair Noordin[1], Mohd Akhmal Muhamad Sidek[2], and Mohammad Zakri Tarmidi[3]

[1] Fakulti Sains Sosial Dan Kemanusiaan, Universiti Teknologi Malaysia, 81310 Johor Bahru , Johor, Malaysia
ahmadnabil@utm.my

[2] Fakulti Kejuruteraan Kimia Dan Kejuruteraan Tenaga, Universiti Teknologi Malaysia, 81310 Johor Bahru, Johor, Malaysia

[3] Fakulti Alam Bina Dan Ukur, Universiti Teknologi Malaysia, 81310 Johor Bahru , Johor, Malaysia

Abstract. A structured learning approach through the constructivism approach is a method that helps students build understanding in phases, from easy to difficult. The use of drone simulators in the subject of drone handling is a practical way to build understanding for students in phases. Good student understanding gives them the confidence to operate drones from virtual simulators to real through real drone operators. The study wants to see the relationship between the use of drone simulators and students' confidence in operating real drones. A questionnaire served as the main research instrument. 35 students participated as purposive samples. This investigation can be divided into three sections; initial knowledge data before entering class, drone simulator data as simulator data, and real flight data collected afterward as final data - using both descriptive analyses such as mean, frequency analysis as well as inferential paired-sample T-Test to find relationships among them all. The findings of the study show that there is a significant difference between the students' confidence before and after participating in the class on the actual handling of drones. This shows that the handling of dangerous drones in terms of safety needs to start with structured drone simulators to ensure that students can master the knowledge, technical and practical aspects of drones. It is hoped that this structured and phased teaching and management process can be done and developed for other courses as well. The findings of the study indicate that using structured drone simulators positively impacts students' confidence in operating real drones. The results highlight the importance of starting with simulators to ensure that students can master the necessary knowledge, technical skills, and practical aspects of drone handling. This approach is seen as a safer way to introduce students to handling potentially dangerous drones while building their competence and self-assurance.

Keywords: Drone Simulator · Structured Teaching Method · Student's Confidence

© The Author(s), under exclusive license to Springer Nature Singapore Pte Ltd. 2024
F. Hassan et al. (Eds.): AsiaSim 2023, CCIS 1911, pp. 125–136, 2024.
https://doi.org/10.1007/978-981-99-7240-1_10

1 Introduction

Drone simulation in education has gained widespread acclaim as an effective and engaging means for teaching various topics. According to [1], drone simulators allow students to gain hands-on experience without risky physical equipment or risks associated with flying real drones, offering real world advantages at no additional expense or risks to themselves or to society in general. In education, drone simulators have a very quick application to make students more educated and realize in adopt the technology in learning.

Drone simulators have quickly become an indispensable way for teaching STEM concepts [2, 3] like science, technology, engineering, and math (STEM). Students can study aerodynamics, physics, and geometry by playing virtual reality drone games; playing with drones allows students to witness first-hand how different factors influence flight. While in teaching programming and coding, drones typically require programming skills for autonomous operation. Simulation environments offer students a safe place to explore programming languages like Python and JavaScript in an autonomous drone control setting - giving them time to study these programming languages safely before creating programs, programming them and then testing their programs against their virtual drone [4].

In specific courses such as in Geographic Information Systems (GIS), drones equipped with sensors and cameras are increasingly being utilized as platforms for creating GIS apps, and simulations may assist students in comprehending how drones take aerial photographs to gather data for analysis and mapping purposes [5, 6]. Furthermore, students can test their knowledge with GIS software by processing drone data to create 3D maps or models of these. In extend of it, drone simulations offer students a great way to learn mission planning and navigation techniques. Students can create flight strategies, identify waypoints, and even simulate complex missions such as rescue/search missions/environmental monitors with these simulators, which also allows for evaluations to evaluate effectiveness as well as adjustments depending on requirements.

There also many research in education, stated that drone simulators as a virtual laboratory can be an offer tools for hosting online challenges and contests that foster collaboration, problem-solving and the creative thinking skills in students [7]. Such events could involve race simulations, obstacle courses or tasks which call upon piloting skills with strategic planning abilities - something many drone simulators facilitate through virtual reality technologies such as VR.

Drone simulations in the classroom are an effective tool for education particularly in fields such as engineering computers, aviation, computer science, geography, and environmental science [8]. By using drone simulator, they will offer an interactive learning experience that is usually more enjoyable and effective than conventional lectures-based learning [4]. Simulations allow students to pilot drones in a range of situations and conditions without the risk of the real-world flight. In the other part, drone simulators are ideal for practicing and improve student's drone piloting skills prior to applying for an official drone pilot's licence [9]. This practicing by using drone simulator will reduce the risk because of learning to operate a drone could be a risk especially for novices. Drones are costly and could cause injuries or damage in the event of a fall. Simulators offer a safe and secure environment where students can learn to control drones without the dangers [10, 11]. Using the drone simulator also can be seen as cost effective. Even though purchasing a drone simulator as well as the needed hardware could result in an initial cost, it is more affordable over the long term compared to the cost of fixing or replacing damaged drones during training.

Utilizing a drone simulator in students' training can have huge impact on the performance of students and this is apparent in comparing the performance of students prior to and following when the drone simulator was introduced [9]. When first learning about the simulator for drone, students might have difficulty grasping the basic concepts of drone operations [1]. Once they have experienced the simulator these abstract concepts will be more tangible and lead to more understanding and better application [11].

2 Drone Technology Subject

Drone technology is a subject that offered through Universiti Teknologi Malaysia's Institute for Life-Ready Graduate (UTM-Ileague). This course forms part of the University's general program designed to recognize student participation in extracurricular activities and understand how their participation enhances learning processes. The ultimate objective is for participants to comprehend the advantages offered by extracurricular involvement for academic performance and development. Utilizing experiential learning reflection, self-reflection, and case studies as tools for students' knowledge acquisition.

As we navigate towards an innovative technological revolution 4.0 and its associated new challenges, this course strives to give UTM students an in-depth knowledge and understanding of drone technology and how best to apply it in various scenarios. Students will develop an overall knowledge of drones as well as their applications. The main goals for this class include improving cooperation and leadership capabilities as well as adaptability skills that promote ethical conduct.

Learnings skills are acquired via assignments included within a course as well as assignments completed outside it. Skills of leadership, teamwork and academic performance can be tested using simulators as well as realistic flight simulator tasks. Students design and construct drones which operate properly and perform as promised, in conjunction with using drone simulator software as well as commercially available models which have proven extremely helpful for open spaces. This course allows students the chance to gain knowledge by serving others. Students who are enrolled in courses that improve their knowledge through service learning should share their experiences with the rest of society in compliance with regulations and ethical guidelines when using drones to instruct others and to ensure that both are secure.

This 14-week class consists of four major parts. These are: theoretical concepts, using drone simulators, real fly on commercial drones; explanation of service-learning principles and its implementation; as well as demonstration. As part of their theory portion, students will gain exposure to aviation fundamentals including knowledge from International Civil Aviation Organization (ICAO), unmanned aerial system regulations imposed by Civil Aviation Authority of Malaysia (CAAM), as well as drone fundamental operations. Students in this course will use DJI Flight Simulator as the official simulator to teach participants to operate drones safely and responsibly. The lecturer will use DJI Flight Simulator during class time for instruction purposes. Drone simulators help students use the correct control panel during deployment to ensure a seamless process and improve confidence levels in students using drones. Drone simulation can give an in-depth knowledge of operations while building confidence among users.

Next step in teaching drone classes will involve piloting real drones outdoors using commercial models. They will only gain confidence enough to fly the drone freely once reaching certain thresholds in terms of confidence in using simulator. As they fly commercial drones with confidence. Drone flights take place in open areas at low, medium, or high task order. Finally, this course emphasizes explaining the service of learning to students as a service to society - sharing concepts covered during class about drone research with everyone involved in class.

3 Constructivism Learning Approach

Constructivist learning entails a series of steps designed to foster understanding and knowledge creation [12, 13], although its exact order depends on circumstances or topics specific to an individual learner. Together these steps constitute an overall framework to implement constructionist learning methods successfully. This approach will start by tapping into existing knowledge. This style to activate students' prior experiences and knowledge related to an issue, whether through discussions, brainstorming sessions, or activities prior to assessments, tapping into existing knowledge can establish an essential basis from which new concepts may emerge. There also will create an engaging learning experience through challenging or provocative issues or problems by providing questions or challenges which spur curiosity and require analysis, investigation and problem-solving capabilities - this forms the cornerstone for inquiry-based and active learning methods.

By using this approach, the student will have an opportunity to explore and investigate. This method will encourage students to conduct independent or group research of any topic of their choosing by giving them sources such as videos, books, articles, or online material that allows them to collect data from multiple angles and form opinions on various issues [14, 15]. Encourage experiments as well as engaging them in hands-on activities which build knowledge or any studies which expand upon it [16]. Engaging in discussion and collaboration toward this learning style will encourage students to pose inquiries that spark further research of a subject area. Establish an environment conducive to inquiry-driven culture by supporting and rewarding your student inquiries, helping refine questions while cultivating worthwhile investigations, while inspiring critical thinking skills through open-ended questioning or suggestions. It also will engage students in discussions and collaborative activities by inspiring their participation, communicating their views and experiences to other students in a safe learning environment that welcomes and respects all views - regardless of origin - while creating welcoming spaces where students can exchange views about experiences shared between classes; challenge assumptions held by others, find meaning through these experiences, or find themselves.

The main purpose of this approach in teaching technology is for the student will have conceptualizing concepts and synthesizing them the knowledge by connecting new data with previous information as well as establishing it into their existing understanding

[17]. Provide assistance when organizing ideas; assist with drawing concept maps or using visual representations to explain what students think about Providing students with opportunities to use what they have learned in real-world settings, either to solve real world issues or create connections across disciplines. Make sure they grasp both its significance and value for learning. Continuous improvement is the key to succeeding in the constructivism approach. Encouraging the students to further their education by reviewing knowledge, reflecting upon feedback, and refining thought processes is vital in creating lasting knowledge gains for our society and economy. Teaching should always remain ongoing, so students never stop growing through experience.

4 Methodology

This research employed quantitative methods. A questionnaire served as the main research instrument. 35 students participated as purposive samples because this investigation focused on those taking Drone Technology courses for semester 2 2022/2023. This investigation can be divided into three sections; initial knowledge data before entering class, drone simulator data as simulator data, and real flight data collected afterward as final data - using both descriptive analyses such as mean, frequency analysis as well as inferential paired-sample T-Test to find relationships among them all.

5 Result

Based on the findings of the study, the researcher detailed the findings through 3 levels, namely before, during and after the class was conducted throughout the 14 weeks.

5.1 Preliminary Data Before Entering the Course

To gather these preliminary data, the researcher needs to collect demographic information about students as well as other information to examine the students' initial knowledge of whether they've flown a drone before, the utilization of a drone simulator and their level of confidence they have to fly a drone.

Table 1. below shows the gender of students.

Gender	Frequency	Percentage (%)
Male	21	60
Female	14	40
Total	35	100

Table 2. shows the frequency and percentage of experience piloting real drone before entering the course.

Do you have any experience piloting real drones?	Frequency	Percentage (%)
Yes	15	42.9
No	20	57.1
Total	35	100

Table 3. shows the frequency and percentage of e overall knowledge about drone before entering the course.

How would you rate your overall knowledge about drone? 1 to 10 (1 - less, 10 most knowledgeable) before entering this course	Frequency	Percentage (%)
1	11	31.4
2	9	25.7
3	4	11.4
4	4	11.4
5	6	17.1
6	1	2.9
7	0	0.0
8	0	0.0
9	0	0.0
10	0	0.0
Total	35	100

Table 4. shows the frequency and percentage of level of confidence in flying drone before entering this course.

Level of confidence in flying drone before entering this course? 1 to 10 (1 - less, 10 most confident)	Frequency	Percentage (%)
1	15	42.9
2	20	57.1
Total	35	100

5.2 Drone Simulator Data

Information gathered for step two includes details regarding drone simulator usage as it relates to control panel usage, such as level of difficulty operating drone simulator and importance it plays when beginning flights; proficiency gained while using drone simulator can then allow one to fly their drone independently after mastery has been obtained through continuous practice using it.

Table 5. below shows the results of understanding the simulator control for drone simulator in class.

How easy was it to understand the simulator controls?	Frequency	Percentage (%)
Easy	20	57.1
Moderate	15	42.9
Hard	0	0.0
Total	35	100

Table 6. below shows the results of drone simulator accurately represent the physical sensations and controls of a real drone.

Does the drone simulator accurately represent the physical sensations and controls of a real drone?	Frequency	Percentage (%)
Strongly Agree	13	37.1
Agree	19	54.3
Disagree	3	8.6
Strongly Disagree	0	0
Total	35	100

Table 7. below shows the results of does simulator improved your understanding of drone piloting.

Has using the simulator improved your understanding of drone piloting?	Frequency	Percentage (%)
Strongly Agree	15	42.9
Agree	20	57.1
Disagree	0	0.0
Strongly Disagree	0	0.0
Total	35	100

5.3 Real Flight Data

Researchers obtain this final data as part of their analysis to analyze its relevance for actual flights using simulators; such data includes remote control panel operation difficulties and pilot confidence when flying an actual drone.

Table 8. below shows the results the student enjoys piloting real drone.

Are you enjoying piloting real drone?	Frequency	Percentage (%)
Yes	35	100.0
No	0	0.0
Total	35	100

Table 9. below shows the results of does simulator make your real flying easier.

Does the drone simulator make your real flying easier?	Frequency	Percentage (%)
Strongly Agree	18	51.4
Agree	17	48.6
Disagree	0	0.0
Strongly Disagree	0	0.0
Total	35	100

Table 10. shows the frequency and percentage of level of confidence in flying drone after entering this course.

Level of confidence in flying drone after entering this course? 1 to 10 (1 - less, 10 most confident)	Frequency	Percentage (%)
7	2	5.7
8	5	14.3
9	11	31.4
10	17	48.6
Total	35	100

Based on the findings of the study, it shows that there is a change in confidence in drone flight before and after the teaching session. This is indicated by the significant value indicating values below 0.05. This finding also shows that after going through a teaching session that includes a drone simulation session, students are seen to be able to master and be confident to fly a real drone.

Table 11. shows the comparison level of confidence in flying drone before and after entering this course.

Paired Samples Statistics

		Mean	N	Std. Deviation	Std Error Mean
Pair 1	confident_pre	15.71	35	5.021	0.849
	confident_post	92.29	35	9.103	1.539

Paired Samples Correlations

		N	Correlation	Sig
Pair 1	confident_pre & confident_post	35	0.800	0.000

Paired Samples Test

		Mean	Std. Deviation	Std Error Mean	95% Confidence Interval of the Difference		t	df	Sig. (2-tailed)
					Lower	Upper			
Pair 1	confident_pre - confident_post	-76.571	5.913	0.999	-78.602	-74.54	-76.617	34	0.000

6 Discussion

Based on the findings of the study, the researcher detailed the findings through 3 levels, namely before, during and after the class was conducted throughout the 14 weeks.

6.1 Before Entering the Course

Based on the findings of the study, it clearly shows that most students have never flown a drone. Although there are students who have flown a drone before the class, most of them show that they are in the low of drone knowledge. This is also seen in line with their low enough confidence to fly a real drone before the class starts.

Unstructured knowledge through unguided experimentation, only through trial and error, unguided self-reading makes them only know but not knowledgeable. It is in line with the findings [9] which shows that the initial knowledge is not focused and structured, making students only know randomly without the main guidance.

6.2 During the Course

The structured class approach, starting with low level, medium level, and high-level aspects of the use of real drone simulation makes the students more directed and guided. Based on the findings, the use of drone control in a simulator is seen to be easy to use, helping to provide an understanding of drone flight as well as increasing students' confidence in understanding the knowledge and skills in operating a real drone through simulation. The study, in line with data from [1], that stated the use of drone simulators not only provides students with real situations without using real equipment, but also reduces the risk of accidents and costs involved. It not only has a positive impact on students and society in general.

Most students show satisfaction, and they think that the use of simulation in drone teaching improves their understanding well. A structured approach through hands-on teaching based on the theory of constructivism gives students the advantage to learn in a more structured way.

6.3 Real Flight Data

The real flight process is a new moment for students. Armed with the knowledge of using a drone simulator, they experience more challenging real flights. After several flight sessions using real drones, the students were seen having a lot of fun and were seen to be confident in their handling. This is clearly obtained through the findings that show they are quite confident and explain that the use of the drone simulator before the actual flight is quite helpful to them. Their real confidence level increased quite well, through real flight which started with the input of knowledge and skills using the drone simulator. Utilizing a drone simulator in students' training can have huge impact on the performance of students and this is apparent in comparing the performance of students prior to and following when the drone simulator was introduced [9].

Based on the findings of the study, it clearly shows that the confidence level of students increases well before and after the drone class takes place. The approach during the class with a drone simulator and ending with an actual flight makes students more prepared and confident. It clearly shows that the structured drone teaching approach through the constructivist teaching approach successfully achieves the real objective of the class. The basic concept of drones needs to be well understood by students at the beginning of the class, it is very important to ensure that students can interact well when operating a real drone [1]. Abstract concepts that are difficult to understand will be easily acquired if basic knowledge can be understood well. Practical along with theoretical and technical understanding is quite important in ensuring students can fly a real drone [11].

By using drone simulator, they will offer an interactive learning experience that is usually more enjoyable and effective than conventional lectures-based learning [4]. Simulations allow students to pilot drones in a range of situations and conditions without the risk of a real-world flight. In the other part, drone simulators are ideal for practicing and improving student's drone piloting skills prior to applying for an official drone pilot's license [9]. Practicing by using drone simulator will reduce the risk because learning to operate a drone could be a risk especially for novices. Drones are costly and could cause injuries or damage in the event of a fall. Simulators offer a safe and secure environment where students can learn to control drones without the dangers [10, 11]. Using the drone simulator also can be seen as cost effective. Even though purchasing a drone simulator as well as the needed hardware could result in an initial cost, it is more affordable over the long term compared to the cost of fixing or replacing damaged drones during training.

7 Conclusion

Increasing the level of student learning needs to be emphasized. The beginning of a basic level to a higher level needs to be well structured. The use of simulators for the subject of drones has enough impact on students' ability in terms of knowledge, techniques, and skills in operating the actual drones' flight. Therefore, the structured approach for this subject which started with drone simulator learning is good enough and should be continued. The students' confidence can be improved, especially towards the handling of drones which requires a high level of concentration and safety.

References

1. Obaid, M., Mebayet, S.: Drone controlled real live flight simulator. J. Phys.: Conf. Series. **1818**(1), 012104 https://doi.org/10.1088/1742-6596/1818/1/012104 (2021)
2. Tezza, D., Garcia, S., Andujar, M.: Let's Learn! An Initial Guide on Using Drones to Teach STEM for Children. In: Zaphiris, P., Ioannou, A. (eds.) Learning and Collaboration Technologies. Human and Technology Ecosystems: 7th International Conference, LCT 2020, Held as Part of the 22nd HCI International Conference, HCII 2020, Copenhagen, Denmark, July 19–24, 2020, Proceedings, Part II, pp. 530–543. Springer International Publishing, Cham (2020). https://doi.org/10.1007/978-3-030-50506-6_36
3. Jemali, N.J.N., Rahim, A.A., Rosly, M.R.M., Susanti, S., Daliman, S., Muhamamad, M., Karim, M.F.A.: Adopting drone technology in STEM education for rural communities. IOP Conf. Series: Earth Environ. Sci. **1064**(1), 012017 (2022). https://doi.org/10.1088/1755-1315/1064/1/012017
4. Chou, P.-N.: Smart technology for sustainable curriculum: using drone to support young students' learning. Sustainability **10**(10), 3819 (2018). https://doi.org/10.3390/su10103819
5. Samad, A., Kamarulzaman, N., Hamdani, M., Mastor, T., Hashim, K.: The potential of Unmanned Aerial Vehicle (UAV) for civilian and mapping application. In: 2013 IEEE 3rd International Conference on System Engineering and Technology. https://doi.org/10.1109/ICSENGT.2013.6650191 (2013)
6. Unger, D.R., Kulhavy, D.L., I-Kuai Hung, Yanli Zhang, Pat Stephens Williams,: Integrating drones into a natural-resource curriculum at Stephen F. Austin State University. J. Forestry **117**(4), 398–405 (2019). https://doi.org/10.1093/jofore/fvz031
7. Irwanto, I.: using virtual labs to enhance students' thinking abilities, skills, and scientific attitudes. https://doi.org/10.31227/osf.io/vqnkz (2018)
8. Walters, B., Potetz, J., Fedesco, H.N.: Simulations in the classroom: an innovative active learning experience. Clin. Simul. Nurs. **13**(12), 609–615 (2017). https://doi.org/10.1016/j.ecns.2017.07.009
9. Golmohammadi, B., Kashani, V., Mohsenabad, D.: Comparing the effect of using simulators and mental imagery of piloting techniques on the performance of UAV athlete pilots. J. Military Med. **19**(5), 440–450 (2018)
10. Postal, G., Pavan, W., Rieder, R.A.: Virtual environment for drone pilot training using VR devices. In: 2016 XVIII Symposium on Virtual and Augmented Reality (SVR). https://doi.org/10.1109/SVR.2016.39 (2016)
11. Bin, Hu., Wang, J.: Deep learning based hand gesture recognition and UAV flight controls. Int. J. Autom. Comput. **17**(1), 17–29 (2019). https://doi.org/10.1007/s11633-019-1194-7
12. Anderson, D., Lucas, K.B., Ginns, I.S.: Theoretical perspectives on learning in an informal setting. J. Res. Sci. Teach. **40**(2), 177–199 (2003). https://doi.org/10.1002/tea.10071

13. Fox-Turnbull, W., Snape, P.: Technology teacher education through a constructivist approach. Design Technol. Educ.: an Int. J. **16**(2), 45–56 (2011)
14. Liu, S.: Factors related to pedagogical beliefs of teachers and technology integration. Comput. Educ. **56**(4), 1012–1022 (2011). https://doi.org/10.1016/j.compedu.2010.12.001
15. Nasir, A.N.M., Ahmad, A., Udin, A., Abd Wahid, N. H., Suhairom, N.: Vocational college's students preferences on practical teaching methods for electronic subject. J. Tech. Educ. Train. **12**(3), 180–188. Retrieved from https://publisher.uthm.edu.my/ojs/index.php/JTET/article/view/3961. (2020)
16. Qasem, M.: Constructivist learning theory in physiotherapy education: a critical evaluation of research. J. Novel Physiotherapies. **5**, 253 (2015). https://doi.org/10.4172/2165-7025.1000253
17. Divaharan, S., Lim, W., Tan, S.: Walk the talk: immersing pre-service teachers in the learning of ICT tools for knowledge creation. Australasian J. Educ. Technol. **27**, 1304-1318 (2011). https://doi.org/10.14742/ajet.v27i8.895

Use of Drone Flight Simulator for Bridging Theories of UAV Systems into Practice: A Pilot Study

Mahyuddin Arsat[1](✉), Ahmad Nabil Md Nasir[1], Lukman Hakim Ismail[2], Muhammad Khair Noordin[1], Adibah Abdul Latif[1], Zainal Abidin Arsat[3], and Khairul Mirza Rosli[1]

[1] Fakulti Sains Sosial dan Kemanusiaan, Universiti Teknologi Malaysia, 81310 Johor Bahru, Malaysia
mahyuddin@utm.my

[2] Fakulti Kejuruteraan Elektrik, Universiti Teknologi Malaysia, 81310 Johor Bahru, Malaysia

[3] Fakulti Kejuruteraan dan Teknologi Mekanikal, Universiti Malaysia Perlis, 02600 Arau, Perlis, Malaysia

Abstract. Unmanned aerial vehicles (UAVs) are used in a variety of industries for a wide range of purposes. The UAV industry in Malaysia is experiencing an increase in demand for qualified workers, which has prompted Technical and Vocational Education and Training (TVET) institutions to introduce certification programs. These programs often rely on actual UAVs for training, which can be costly, restrict possibilities for hands-on practice, and present safety issues. Therefore, drone simulators have emerged as a learning tool to address the issue, recreating real-world operational environments for virtual pilot training and interaction. In order to comprehend participants' understanding with learning through drone simulators, a qualitative research approach was adopted in this study. The researchers used an Unmanned Aerial Vehicle Technology Laboratory course as a pilot study to evaluate how students developed their understanding of flying principles while using UAV simulators. Semi-structured interviews were selected. A purposive sampling strategy was used to identify individuals who had completed the drone simulator sessions. Individual interviews were carried out to delve deeper into the experiences of the individuals. The findings of the study reveal three main categories: self-awareness, contextual understanding, and evaluating the effectiveness of practice. Self-awareness refers to students' ability to reflect on their learning process and recognize how their theoretical understanding informs their practical application. Contextual understanding relates to students' capacity to apply theoretical concepts within specific contexts. Evaluating the effectiveness of practice involves students critically assessing their own practice and identifying areas for improvement.

Keywords: Drone Simulator · Students Learning Outcomes · Simulation-based Learning

© The Author(s), under exclusive license to Springer Nature Singapore Pte Ltd. 2024
F. Hassan et al. (Eds.): AsiaSim 2023, CCIS 1911, pp. 137–145, 2024.
https://doi.org/10.1007/978-981-99-7240-1_11

1 Introduction

Drones, also known as unmanned aerial vehicles (UAVs), have quickly developed into adaptable platforms with uses in a wide range of industries, including surveillance, aerial photography, package delivery, agriculture, and disaster management [1, 2]. In Malaysia, the field of UAVs or drones has witnessed significant growth [3], leading to an increasing need for skilled professionals in this sector. To address this demand, several TVET institutions in Malaysia offer certification programs focused on UAV technology. Traditionally, TVET institutions have employed a combination of theoretical instruction and practical hands-on training to impart UAV-related knowledge and skills. This includes classroom lectures, practical exercises, and field training to familiarize students with UAV operations, flight dynamics, maintenance, and safety protocols. While these approaches are valuable, they often rely on costly physical UAVs for training purposes, limiting the availability of hands-on practice opportunities and potentially posing safety concerns.

To address the limitations of traditional teaching practices in TVET institutions, the integration of simulation-based learning for drone training emerges as a promising solution. Simulation-based learning is a valuable learning experience and learning tool for bridging the gap between theoretical concepts and practical skills [4]. Various types of scaffolding can support simulation-based learning at various stages of knowledge and skill development [5]. A drone simulator replicates the real-world operational environment and allows students to virtually pilot and interact with a drone using realistic flight controls. This software-based platform emulates the flight dynamics, controls, and ambient conditions that a UAV would experience in a virtual environment. Drone simulators provide a secure, economical, and effective way to educate operators and test UAV systems without the need for actual hardware by utilising realistic physics engines, accurate 3D models, and simulated scenarios.

It is evident the usage of drone simulators to translate theoretical understanding into real-world UAV system implementations. Several studies have found that drone simulators can help improve training and operational capabilities in UAV systems [6, 7] for example, conducted simulator-based training and discovered that simulators provided a safe and cost-effective environment for beginner pilots to learn critical skills and confidence in flying UAVs. Similarly [8], used a simulator and found that as the number of trials increased, participants became accustomed to the cognitive load of visual/auditory tasks, and the number of repetitions lowered stress and anxiety levels, increased attention, and enhanced game performance. According to [9] a study on multimodal augmented reality applications for training aviation traffic operations, the study shows that the augmented reality has the potential to complement conventional flight instruction by bridging the gap between theory and practice.

2 Simulation-Based Learning

Teaching with virtual simulators has shown significant contributions to the development of students' skills compared to traditional video-based training methods [10]. The use of virtual simulators provides a more immersive and realistic learning experience, allowing students to interact with the simulated environment and practice their skills in a controlled setting. This interactive approach enhances the students' ability to apply theoretical knowledge to real-life scenarios and make real-time decisions in the context of UAV operations [11]. By replicating the feeling of operating a real aircraft, the simulator enhances the trainees' capabilities and improves their decision-making skills in complex flying environments. While simulation-based experiential learning offers significant safety and risk-reduction benefits, it is important to consider the practical implications and outcomes of integrating such learning methods [12]. It is crucial to recognize that positive outcomes in terms of learning efficiency and cost-effectiveness are not always guaranteed. Therefore, a comprehensive assessment of the benefits and limitations of simulation-based learning is necessary to ensure its effectiveness and suitability in the specific context of UAV skill development.

Despite the potential challenges, simulation-based learning offer an active learning environment that supports the transfer of skills from training to real work settings, particularly for controllers and operators [13]. This approach promotes adaptation to changing training content and fosters an innovative approach to flight safety. By engaging in simulated scenarios, trainees can practice critical tasks, develop situational awareness, and acquire the necessary skills to effectively respond to challenging situations in real-world UAV operations. This article aims to provide a knowledge of understanding of drone simulators in bridging the gap between theoretical of flight principles in UAV system and real-world applications. A pilot study has been conducted to examine the students understanding on the four key principles: i) lift, thrust, drag, and weight forces. The utilization of drone simulators allows students to apply and observe these principles in a virtual environment, facilitating the transfer of theoretical knowledge into practical skills.

3 Research Methodology and Pilot Study

To fully grasp the understanding with their learning process with drone simulators, a qualitative research methodology has been employed. A course entitled Unmanned Aerial Vehicle Technology Laboratory were selected as a pilot study. This study main goal was to investigate the abilities students developed their understanding on the principles of flight while operating UAV simulators. The selection of semi-structured interviews as the primary data collection method was based on the flexibility it offers. Semi-structured interviews gave researchers the chance to collect in-depth, complex, and nuanced information by letting participants express their ideas, experiences, and insights in their own words [14, 15].

A purposive sampling technique was utilized to select the participants who had completed the drone simulator sessions. The number of participants (seven) for pilot study was considered sufficient to gather a range of perspectives and experiences while allowing for in-depth analysis of the data. The semi-structured interviews were performed individually. The interviewer used an interview protocol (the research instrument) that allowed for probing and follow-up questions to probe further into the participants' experiences during the interview process. The data from the interviews were analysed using thematic analysis, which concentrated on identifying common themes, patterns, and insights that appeared in the participants' narratives. Through iterative approach, the participants' experiences were thoroughly examined, and similarities and differences between their responses were found (Table 1).

Table 1. Teaching approaches versus learning activities

Duration (week)	Approach	Learning activities
1 – 6	Theoretical	Theoretical lessons on UAV technology, principles of flight, aerodynamics, sensor technology and its application in various industries
7	Simulation	Controlling flight: i) Student manipulate the flight controls and observe the drone's behaviour ii) Execute staged-tasks such as flying in a straight line, flying in multiple altitudes, and performing turns iii) Performing combinations of controls; pitch, roll and yaw, create a unique drone's movements
8 – 9		Flying and manoeuvres: i) Execute flying challenges thru time limit ii) Performing tasks in various obstacle course and indoor areas. It requires to perform specific manoeuvres e.g. flipping, coordinated flight and vertical ascents or descents
10 – 14	Practical	Complete flight training course in stages Students are require to perform flying in open area or dedicated areas with specific flight paths and tasks

Unmanned Aerial Vehicle Technology Laboratory is a comprehensive course to facilitate undergraduate students in TVET for theoretical knowledge and practical understanding of UAV technology. This course is chosen and signed as case for a pilot study. The course duration spans 14 weeks, offering a well-structured and progressive learning experience for undergraduate TVET students. Throughout the course, students engage with specific modules designed to foster a deeper understanding of drone technology. Central to the course structure is the incorporation of the simulator as a key teaching tool. The simulator facilitates the translation of theoretical concepts, such as the principles of flight, into hands-on practice, enabling students to grasp the practical aspects of UAV operations.

4 Research Findings

Thematic analysis of qualitative data in Table 2 reveals three categories that characterize the process of bridging theory into practice: self-awareness, contextual understanding, and evaluating effectiveness. Self-awareness has been identified through a careful analysis of primary codes and their related sub-codes, it suggest students' capacity for reflection and their ability to apply theoretical knowledge to real-world situations. Contextual

Table 2. Finding of thematic analysis

Themes		Category (Primary Codes)	Key words (Sub-codes)
Self-awareness	Bridging theory into practice	Change Perception	a) Importance of learning through simulator
			b) Skill set
		Knowing the knowledge gaps	a) Identify specific misunderstanding
			b) Recognise the unclear concepts
			c) Identify challenging aspects
		Developing understanding	a) Able to explain
			b) Apply theory into simulation
Contextual understanding		Application of flight principles in different scenarios	a) Demonstrate of skills in different aerial manoeuvres
			b) Demonstrate of skills in emergency conditions
		Connecting theory to real-life situations	a) Apply safety measures
			b) Identify the similarity between simulation and actual flight
Evaluating the effectiveness of practice		Self-assessment to improve	a) Identify the strengths and weaknesses
			b) Monitor performance
			c) Utilise feedback
		Comparing practice thru simulator and traditional learning	a) Effectiveness of hands-on experience
			b) Peer teaching (Role as co-pilot)
			c) Skill retention

understanding encapsulates a pivotal aspect of students' engagement with drone simulators, explaining their adeptness at situating theoretical constructs within real-world contexts. Evaluating effectiveness entails reflecting on outcomes to refine practices.

4.1 Self-awareness

Within the theme of "Self-awareness," several salient primary codes have emerged, each shedding light on distinct dimensions of students' reflective capabilities. The first primary code, "Change Perception," encapsulates sub-codes that unveil how students perceive and evaluate their learning experiences through the lens of simulator-based instruction. Sub-code (a), "Importance of learning through simulator," delves into students' recognition of the unique advantages offered by simulator-driven learning, while sub-code (b), "Skill set," delves into the perceptible development of skills fostered by their interactions with the simulator environment.

The second primary code, "Knowing the knowledge gaps," delves deeper into the self-awareness of students by uncovering their aptitude for identifying gaps in their comprehension. This code is expounded through sub-codes (a), (b), and (c), which respectively explore their ability to pinpoint specific misunderstandings, recognize unclear concepts, and acknowledge challenging aspects that require further exploration. Lastly, the third primary code, "Developing understanding," encapsulates sub-codes that elucidate the transformative nature of students' learning journeys. Sub-code (a), "Able to explain," delves into their articulation of acquired knowledge,

Excerpt 1 (Student 4):

"When I combined the remote controls, the pitch, roll, and yaw. It created smooth flight manoeuvres."

Student 4 displays ability to explain the relationship between various control inputs (pitch, roll, and yaw) and the combined impacts on the drone's flight manoeuvres. The student's description of performing a coordinated turn while maintaining altitude by applying forward pitch and rolling to the right demonstrates their awareness of their own actions as well as their comprehension of how those actions effect the drone's behaviour. While sub-code (b), "Apply theory into simulation," delves into their adeptness at translating theoretical constructs into practical scenarios facilitated by the simulator platform. Student 1 and student 7 demonstrates ability to apply theoretical of pitch control adjustment and the impact on the drone's height (application in simulator).

Excerpt 2 (Student 1):

"I adjust the remote control to the left and right. It moves the drone roll clockwise and anti-clockwise"

Excerpt 3 (Student 7):

"I adjusted the pitch control up and down and observe the drone. The drone tilted, nose up and down causing drone fly up and down. When I push harder the pitch, the drone flying upward, and when I pull the pitch, the drone flying downward"

4.2 Contextual Understanding

This theme is rooted in a comprehensive scrutiny of primary codes and its corresponding sub-codes, which collectively unveil the interplay between theoretical comprehension and its practical application. The first primary code, "Application of flight principles in different scenarios," elucidates their capability to extend their grasp of flight theory beyond mere theory, culminating in the execution of intricate aerial manoeuvres. Sub-code (a), "Demonstration of skills in different aerial manoeuvres," highlights on the student adeptness at translating theoretical knowledge into aerial performances, thereby accentuating their practical skill.

Excerpt 1 (Student 3):

"I need to know how gives the input on the remote control so my drone can accomplish specific tasks and flight paths. I need to control the yaw, so I could spin my drone without changing its altitude."

Student 3 exhibits contextual understanding by explaining the implications of yaw control, which entails rotating the drone vertically. The student's understanding of the precise function and application of yaw control in adjusting the drone's orientation during flight is demonstrated by their recognition that yaw input allows them to change the drone's heading without causing altitude or tilting. Sub-code (b), "Demonstration of skills in emergency conditions," delves deeper into their capacity to wield theoretical underpinnings to navigate basic flight scenarios.

The second primary code, "Connecting theory to real-life situations," explores the subtle ways in which students bridge the gap between abstract theoretical ideas and concrete real-world contexts. They systematically follow safety procedures and security measures, as detailed in Sub-Code (a), "Apply safety measures," putting theory into practise with great care to guarantee safe aircraft operations. Their ability to recognise similarities between simulated scenarios and real-world flying conditions is highlighted by Sub-code (b), "Identify the Similarity between Simulation and Actual Flight," demonstrating their discernment and acute contextual awareness.

> *Excerpt 2 (Student 6):*
> *"I could see that by adjusting the roll control, which requires tilting the drone left or right, has an impact on the tilting behaviour of the drone. When I rolled the drone to the right, it leaned gently in that direction, making it easier to manage turns. Rolling to the left produced a tilting motion."*

Student 6 demonstrates certain of understanding the similarity of flight control in simulation and actual flight by explaining how roll control significantly affects the drone's tilting behaviour. Students' grasp of the function of roll control in guiding the drone past obstacles is demonstrated by their observation of how rolling the drone to the right or left causes the same tilting motions.

4.3 Evaluating the Effectiveness of Practice

The comprehensive study of the gathered data produced a clear thematic pattern that has been named "Contextual Understanding." This overarching theme elaborates on a crucial aspect of students' interaction with drone simulators, illuminating their skill in placing theoretical knowledge within practical contexts. Several primary codes have emerged within the theme of "Contextual Understanding," each providing unique insights into how students learn to situate theoretical ideas within actual contexts. The first primary code, "Self-assessment to improve," explores students' tendency for self-reflection, and sub-codes (a), (b), and (c) highlight how well-versed they are in recognising their strengths and weaknesses, methodically monitoring their progress, and effectively utilising useful feedback to improve their flight skills.

The ability of students to analyse the effectiveness of their application of theory and to critically evaluate their own practise is relevant to this theme. Students who evaluate the effectiveness of their practise take the time to consider their own strengths, weaknesses, and difficulties.

Excerpt 1 (Student 5):

"I found that the impact of abrupt control inputs was an interesting flight characteristic. Pitch, roll, and yaw alterations that I made abruptly and aggressively caused the drone to move more quickly and exaggeratedly. In order to achieve stable flight and prevent abrupt or unpredictable behaviour, I need to give smooth and progressive inputs. Learning this made me think about how skilful and precise drone piloting is."

Excerpt 2 (Student 7):

"I could make a precise turn while retaining altitude by simultaneously applying forward pitch and rolling to the right. As I coordinated the drone's movements to achieve particular flight patterns or follow a specified flight path."

Student 5 evaluates the effectiveness of their practise by considering how abrupt control inputs affect the drone's behaviour. The student's understanding that abrupt or aggressive adjustments to the controls cause quick and rapid movements; become evidence of their evaluation that smooth, progressive inputs are necessary for obtaining steady flight. Understanding the value of accuracy and dexterity when operating a drone shows that they have evaluated their own methods. The student 7's description of controlling the drone's motions to produce particular flight patterns or follow a specified flight path indicates their assessment of the effectiveness of their practice and the satisfaction they gain from doing so.

The following primary code, "Comparing practise through simulator and traditional learning," delves into the subtle region of students' comparative discernment of simulator-based learning against traditional educational techniques. Sub-code (a), "Effectiveness of hands-on experience," dives into the students' assessment of the efficacy gained from hands-on engagement in experiential learning. Meanwhile, sub-code (b), "Peer teaching (Role as co-pilot)," captures the distinct component of collaborative learning in which students serve as co-pilots in each other's learning experiences, therefore improving their practical understanding of flight principles. Furthermore, sub-code (d), "Skill retention," chronicles their observations into the long-term sustainability of acquired skills, distinguishing the different pathways of knowledge assimilation facilitated by the two diverse educational approaches.

5 Conclusion

These findings provide understanding on the use of drone flight simulator as a learning tool to bridge theory into practice and identify areas for development in students' application of theory in real-world scenarios. It is evidence that simulation-based learning activities i) flight control manipulation and ii) stability and manoeuvrability exploration were able to accommodate students understanding on the principle of flight; i.e. lift, thrust, drag, and weight forces. Educators or teachers can facilitate students in effectively applying theoretical knowledge to actual settings (during simulation-based learning activities) by increasing the student's self-awareness, contextual understanding, and evaluating their effectiveness of practice.

References

1. Ab Rahman, A.A., et al.: Applications of drones in emerging economies: a case study of Malaysia. In: 2019 6th International Conference on Space Science and Communication (IconSpace), pp. 35–40. Johor Bahru, Malaysia (2019). https://doi.org/10.1109/IconSpace.2019.8905962
2. Gohari, A., Ahmad, A.B., Oloruntobi, O.O.: Recent drone applications in Malaysia: an overview. Int. Arch. Photogrammetry Remote Sens. Spat. Inf. Sci. **XLVIII-4/W6-2022**, 131–137 (2023). https://doi.org/10.5194/isprs-archives-XLVIII-4-W6-2022-131-2023
3. Mohd, I.D., Aripin, E.M., Zawawi, A., Ismail, Z.: Factors influencing the implementation of technologies behind industry 4.0 in the Malaysian construction industry. MATEC Web of Conf. **266**, 01006 (2019). https://doi.org/10.1051/matecconf/201926601006
4. Morgan, P., Cleave-Hogg, D., DeSousa, S., Lam-mcculloch, J.: Applying theory to practice in undergraduate education using high fidelity simulation. Med. Teach. **28**, e5–e10 (2006). https://doi.org/10.1080/01421590600568488
5. Chernikova, O., Heitzmann, N., Stadler, M., Holzberger, D., Seidel, T., Fischer, F.: Simulation-based learning in higher education: a meta-analysis. Rev. Educ. Res. **90**(4), 499–541 (2020). https://doi.org/10.3102/0034654320933544
6. He, Y.R., Wang, X.R., Chen, Q.J., Leng, P.: Design and implementation of virtual simulation teaching system for UAV based on WEBGL. Int. Arch. Photogrammetry Remote Sens. Spat. Inf. Sci. **XLII-3/W10**, 1239–1246 (2020). https://doi.org/10.5194/isprs-archives-XLII-3-W10-1239-2020
7. Ribeiro, R., et al.: Web AR solution for UAV pilot training and usability testing. Sensors **21**, 1456 (2021). https://doi.org/10.3390/s21041456
8. Koç, D., Seçkin, A.Ç., Satı, Z.E.: Evaluation of participant success in gamified drone training simulator using brain signals and key logs. Brain Sci. **11**(8), 1024 (2021). https://doi.org/10.3390/brainsci11081024.PMID:34439643;PMCID:PMC8392183
9. Moesl, B., Schaffernak, H., Vorraber, W., Braunstingl, R., Koglbauer, I.V.: Multimodal augmented reality applications for training of traffic procedures in aviation. Multimodal Technol. Interact. **7**(1), 3 (2022). https://doi.org/10.3390/mti7010003
10. Ismailoğlu, E., Orkun, N., Eser, I., Zaybak, A.: Comparison of the effectiveness of the virtual simulator and video-assisted teaching on intravenous catheter insertion skills and self-confidence: A quasi-experimental study. Nurse Educ. Today **95**, 104596 (2020). https://doi.org/10.1016/j.nedt.2020.104596
11. Obaid, M., Mebayet, S.: Drone controlled real live flight simulator. J. Phys. Conf. Ser. **1818**, 012104 (2021). https://doi.org/10.1088/1742-6596/1818/1/012104
12. McLean, G., Lambeth, S., Mavin, T.: The use of simulation in Ab initio pilot training. Int. J. Aviat. Psychol. **26**, 36–45 (2016). https://doi.org/10.1080/10508414.2016.1235364
13. Knecht, C.P., Muehlethaler, C.M., Elfering, A.: Nontechnical skills training in air traffic management including computer-based simulation methods: from scientific analyses to prototype training. Aviat. Psychol. Appl. Hum. Factors **6**(2), 91–100 (2016). https://doi.org/10.1027/2192-0923/a000103
14. Patton, M.Q.: Qualitative Research and Evaluation Methods (3rd ed.). SAGE Publications (2002)
15. Wilson, C.: Semi-structured interviews. In: Interview Techniques for Ux Practitioners, pp. 23–41 (2014). https://doi.org/10.1016/B978-0-12-410393-1.00002-8

Enhancement of System Network Based on Voltage Stability Indices Using FACTS Controllers

Nur Izzati Aslan[1], Norazliani Md. Sapari[1](✉), Muhammad Syafiq Bin Md Yusoff[1], Khairul Huda Yusof[2], and Mohd Rohaimi Mohd Dahalan[3]

[1] Faculty of Electrical Engineering, Universiti Teknologi Malaysia, Johor, Malaysia
norazliani.ms@utm.my

[2] Faculty of Information Science and Engineering, Management and Science University, Shah Alam, Selangor, Malaysia

[3] School of Electrical Engineering, College of Engineering, Universiti Teknologi MARA, 40450 Shah Alam, Selangor, Malaysia

Abstract. As a result of rising load demand (particularly reactive load) and a lack of available generation sources, the power system is being operated close to its voltage instability point which may lead to voltage collapse. Therefore, to prevent voltage collapse, it is imperative to constantly monitor the power system's voltage stability. This study uses Flexible AC Transmission System (FACTS) controllers to improve voltage stability in power systems. The goal is to assess how well voltage stability is improved by FACTS controllers. The MATPOWER toolbox in MATLAB is used to conduct load flow analysis on the IEEE 30 bus test system as a case study. Two voltage stability indices, the Fast Voltage Stability Index (FVSI) and the Novel Line Stability Index (NLSI) were used in the study to evaluate the system's voltage stability. These indicators offer important details about the system's capacity to sustain acceptable voltage levels under various operating circumstances. The simulation results reveal that the SVC-enabled shunt compensation method is the most successful at improving voltage stability. It is followed by shunt-series compensation, which uses the UPFC and exhibits notable improvements. Comparatively less successful at improving voltage stability is series correction, as represented by the TCSC. The result of this study provides insight into how FACTS controllers should be used to improve voltage stability. According to the findings, shunt compensation-based FACTS controllers perform better than series compensation-based controllers. This paper highlights the contribution to improving system stability by using FACTS devices.

Keywords: Voltage Stability · FACTS Controllers · SVC · TCSC · UPFC · Fast Voltage Stability Index (FVSI) · Novel Line Stability Index (NLSI) · Load Flow Analysis

© The Author(s), under exclusive license to Springer Nature Singapore Pte Ltd. 2024
F. Hassan et al. (Eds.): AsiaSim 2023, CCIS 1911, pp. 146–160, 2024.
https://doi.org/10.1007/978-981-99-7240-1_12

1 Introduction

Voltage stability is a vital component of power system performance and is essential for providing a consistent and effective supply of electricity [1]. Maintaining voltage stability gets harder as there is a greater demand for electricity and renewable energy sources are integrated. Voltage instability initiated the serious stability condition and network security by causing voltage collapse, voltage swings, and even cascading failures.

The FACTS controllers have emerged as an efficient instrument for improving power system voltage stability in response to these challenges. The FACTS controllers enable operators to control voltage levels and improve the system's stability margins by providing accurate monitoring and compensation capabilities. SVC, TCSC, and UPFC have attracted significant attention among the different FACTS controllers due to their efficiency and flexibility in improving voltage stability. In a real-world power system situation, this research focuses on examining the potential of SVCs, TCSCs, and UPFCs to enhance voltage stability. The performance of these FACTS controllers is assessed using a case study involving the IEEE 30 bus test system, which is frequently used as a benchmark system in power system research. MATLAB's MATPOWER toolbox is used to do load flow simulations as part of the analysis, which accurately simulates the behaviours of the system under various operating conditions.

Two voltage stability indices are used to assess the reliability of this system, namely: Fast Voltage Stability Index (FVSI) and Novel Line Stability Index (NLSI). The indices offer valuable information on the system's ability to maintain acceptable voltage levels, as well as identifying critical elements that may have an impact on overall stability. Furthermore, the accuracy of various FACTS controllers in terms of voltage stability enhancement can be assessed and compared using these indices. The study aims at guiding how to select and install FACTS control units for improving voltage stability in power systems. Valuable information on the optimum use of these controllers to improve voltage stability can be gained through analysis of simulation results and comparison between SVCs, TCSCs, and UPFCs.

2 Background

Numerous studies have been carried out to address this problem and improve the stability of electrical networks. The use of FACTS controllers has emerged as a viable solution as researchers have investigated numerous methods and techniques to evaluate and enhance voltage stability over time. The existing research on improving voltage stability with FACTS controllers is thoroughly reviewed in this section.

2.1 Voltage Stability Assessment

Voltage stability value is known as an indicator of the stability of the system as mentioned in previous research [5–8]. These indices provide a quantitative assessment of the power system's ability to maintain a stable voltage profile under normal and abnormal conditions. Traditional approaches, including load flow analysis and eigenvalue analysis, have limits in capturing transient and dynamic behaviours, but they offer useful insights into system stability. Voltage stability indices, such as the Fast Voltage Stability Index (FVSI) and the Novel Line Stability Index (NLSI), have been developed recently. These indices combine dynamic properties and offer a more precise assessment of voltage stability.

2.2 FACTS Controllers for Voltage Stability Enhancement

Recently, FACTS devices have become increasingly appealing for long-distance power transmission due to the rapid improvement in power electronics. Power systems can function more flexibly, safely, and economically with the help of FACTS devices. The main challenge to extending the capacity of the transmission line is the limitation of the FACTS capacity device itself to operate at the lowest cost while considering the security of transmission limitation [7, 8]. Widely utilized FACTS components including SVCs, TCSCs, and UPFCs provide special benefits for controlling voltage levels and enhancing system stability. Shunt compensation is offered by SVCs, aiding in the regulation of reactive power and voltage profiles. Series compensation provided by TCSCs enables better control of power flow and voltage stability. Shunt and series compensation are combined by UPFCs to provide a thorough improvement in voltage stability.

To assess the effectiveness of SVCs, TCSCs, and UPFCs under various operating scenarios and system configurations, comparative assessment was carried out. The benefits of FACTS controllers on voltage stability have been established in simulation studies utilizing a variety of test systems, including the IEEE 30 bus test system, by improving voltage profiles, lowering line losses, and boosting system security. Studies have demonstrated that careful FACTS controller placement and number selection can greatly increase voltage stability while lowering associated expenses.

3 Methodology

Figure 1 shows a standard IEEE 30 bus test system was used which consists of 5 generator buses, 24 load buses, and 41 interconnected lines. And used in a per unit system, with a base 100MVA.

The MATPOWER toolbox in MATLAB is used to perform load flow analysis. This analysis is intended to determine the stable operating conditions of the IEEE 30 bus test system that includes bus voltage degrees and angles, line flow characteristics, downstream power generation and consumption. Newton-Raphson used to solve the system of non-linear equations that describe the power balance at each bus in the system. The results of the power flow analysis will be used to visualize the performance of the system, such as the voltage and current values at each bus, and the real and reactive power flows in the transmission lines.

3.1 Voltage Stability Indices

In contemporary power systems, the voltage stability indices can be used to determine the state of a power system's voltage stability. The index for each bus in the IEEE 30 bus test system will be determined using the power flow solution. The buses will be ranked according to their index values once the values for each bus have been determined. Buses

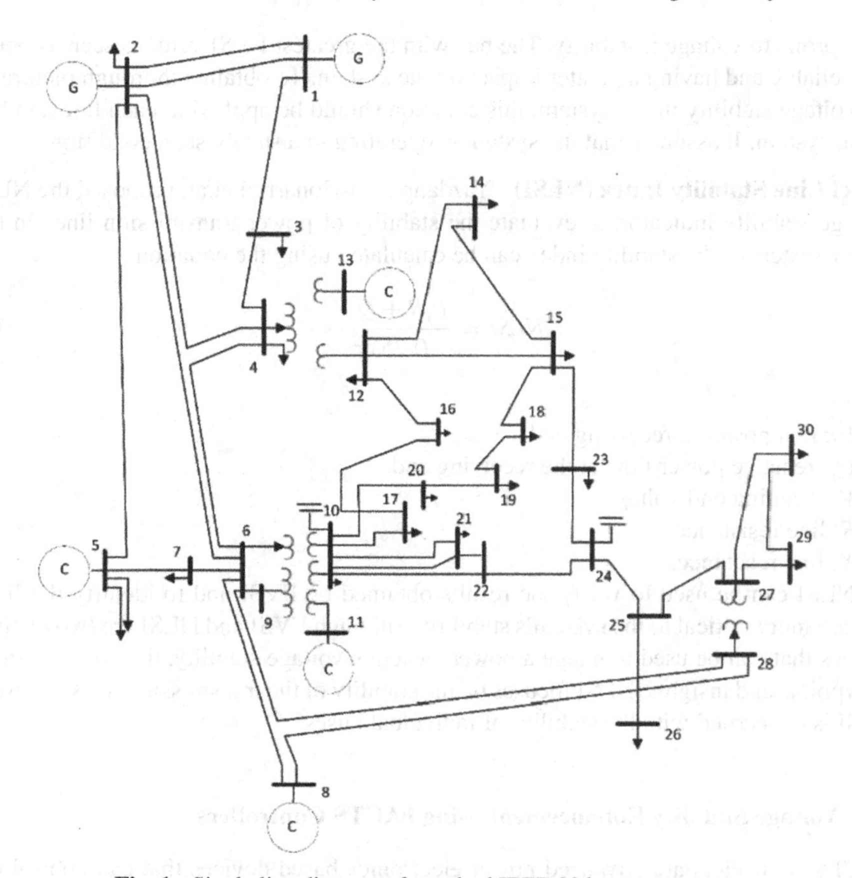

Fig. 1. Single line diagram of standard IEEE 30 bus test system.

with higher index values (less stable) were ranked higher, and this ranking will be used to determine which buses are nearly to collapse. The FVSI and NLSI forms of voltage stability index will be employed in this study to determine the bus index.

Fast Voltage Stability Index. The author [13] has proposed FVSI to evaluate the stability value of the system. Based on this formula, the unstable point is if the value is approaching unity [14]. The stability index is calculated by:

$$FVSI_{ij} = \frac{4Z^2 Q_j}{V_i^2 X} \tag{1}$$

where:

Z: line impedance.

Q_j: reactive power flow at the receiving end.

V_i: sending end voltage.

X: reactance of the line connecting the bus i and j.

Equation (1) is based on the idea that the only factor influencing the voltage change is the variation in reactive power. This equation can be used to identify the bus that is most prone to voltage instability. The bus with the greatest FVSI value is seen as being less reliable and having a greater impact on the system. To obtain a thorough picture of the voltage stability in the system, this equation should be applied to each line and bus in the system. It assumes that the system is operating in a steady state condition.

Novel Line Stability Index (NLSI). Yazdanpanah-Goharrizi et al. proposed the NLSI voltage stability indicator, to evaluate the stability of power transmission lines in the power systems. The stability index can be calculated using the equation:

$$NLSI = \frac{P_j R + Q_j X}{0.25 V_i^2} \tag{2}$$

where:

P_j: real power at receiving end.
Q_j: reactive power flow at the receiving end.
V_i: sending end voltage.
R: line resistance.
X: line reactance.

NLSI can be used to verify the results obtained by FVSI and to identify the lines that are more critical to the system's stability. Although FVSI and NLSI are two distinct indices that can be used to gauge a power system's voltage stability, they offer various viewpoints and insights. NLSI focuses on the stability of the transmission lines, whereas FVSI is concerned with the stability of individual buses.

3.2 Voltage Stability Enhancement using FACTS Controllers

FACTS controllers are advanced power electronics-based devices that can control the flow of power in a power system. By using these controllers, the power system's voltage stability can be improved, leading to a more reliable and stable power supply. This is achieved by controlling the reactive power flow and improving the system's impedance characteristics, which in turn helps to prevent voltage collapse.

Modelling of Static Var Compensator (SVC). The SVC model diagram is shown in Fig. 2. A limited firing angle or susceptance limit can be used to model an SVC as a varied reactance. The reactive power restriction is considered in the SVC modelling and acts as a variable susceptance connected in parallel to the network. It may be inductive or capacitive. The SVC injects Q_k, the reactive power it draws, into bus k as follows:

$$Q_k = -V_k^2 B_{SVC} \tag{3}$$

where:

V_k: terminal voltage at bus k.
B_{SVC}: susceptance of SVC.

The reactive power for SVC is set within −100MVAR and 100MVAR, to get the improvement of system voltage.

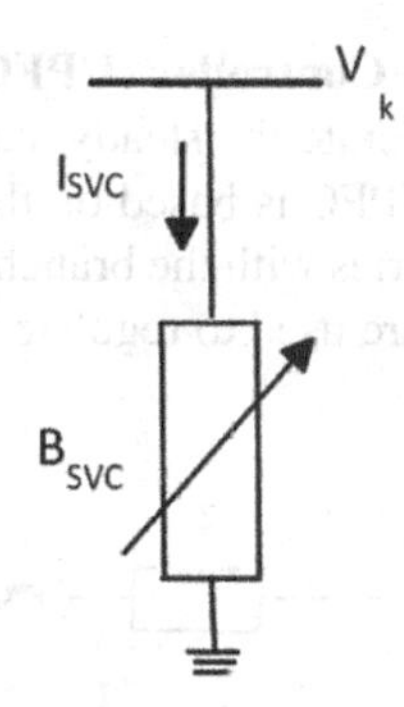

Fig. 2. SVC model

Modelling of Thyristor Controlled Series Compensator (TCSC). The series impedance of the TCSC, its control strategies, and its interactions with the rest of the power system are all considered by the TCSC model in Fig. 3. In this study, two models are utilized to simulate the effects of TCSC for both dynamic and steady-state applications on the power network. These models are the reactance firing angle model and the adjustable series reactance model. Utilizing the less complex variable series reactance model, the transmission line's impedance is altered by altering its reactance. The power To limit the power flow across a branch to a certain amount, the TCSC can function in an inductive or a capacitive zone. The following equation can be used to express the effective impedance of a transmission line using TCSC:

$$X_{ij} = X_{line} + X_{TCSC} \tag{4}$$

$$X_{TCSC} = k * X_{line} \tag{5}$$

Where:

X_{ij}: effective reactance of the line.
X_{line}: original reactance of the line.
X_{TCSC}: reactance of TCSC.
k: degree of compensation.
The value of k varies in the range of $-0.2 \leq k \leq 0.7$.

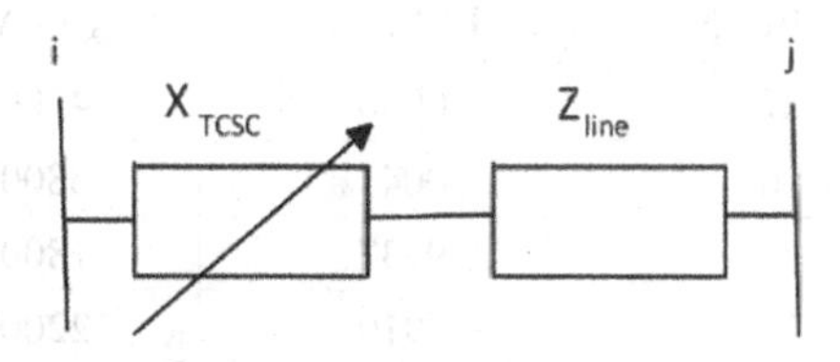

Fig. 3. TCSC model

To prevent overcompensation, the compensation level of the TCSC is changed in this project between 20% inductive and 70% capacitive. Up until the ideal reactance value is attained, the original reactance of line for this model varies between −70% and 20% of the reactance of line.

Modelling of Unified Power Flow Controller (UPFC). In this project, the SVC and TCSC models are combined to generate the steady-state model of UPFC, as shown in Fig. 4. In this work, the limit for UPFC is based on the selection for SVC and TCSC. The TCSC model is connected in series with the branch between bus i and bus j, making up the UPFC. The chosen devices are used to regulate the UPFC model.

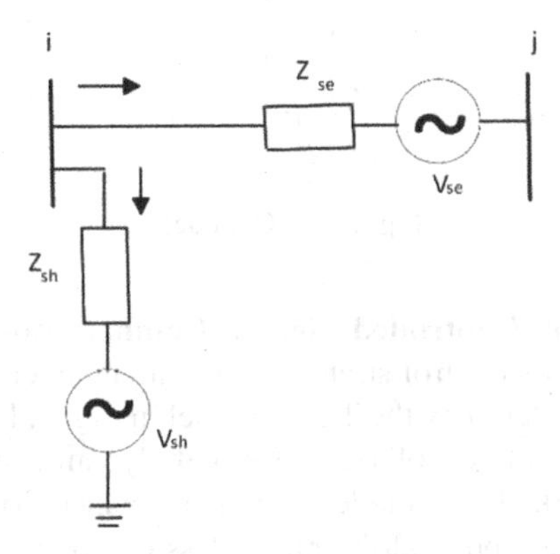

Fig. 4. UPFC model

4 Results and Analysis

4.1 Analysis for Stability Index Ranked

Tables 1 and 2 presented the bus ranking based on the decreasing order of the reactive power value for FVSI and NLSI respectively. For FVSI, Bus 27 is the weakest bus due to the least reactive power value, 0.045MVAR. While for NLSI, branch 3 (bus 27 – 30) indicated the weakest bus.

Table 1. FVSI Bus ranking (from weakest to strongest) based on maximum load ability

Rank	Bus No.	FVSI	Q_{max} (VAR)
1	27	1.0030	45000
2	6	1.0047	68000
3	29	0.9937	118000
4	25	0.9910	122000
5	28	0.9920	155000
6	12	0.9946	169000
7	24	0.9912	216000

(continued)

Table 1. (*continued*)

Rank	Bus No.	FVSI	Q_{max} (VAR)
8	2	0.9936	244000
9	15	0.9914	272000
10	10	0.9930	313000
11	23	0.9908	331000
12	4	0.9910	358000
13	14	0.9921	395000
14	9	0.9914	441000
15	8	0.9912	617000
16	1	0.9904	622000
17	16	0.9914	676000
18	22	0.9913	699000
19	18	0.9904	1433000
20	5	0.9903	1694000
21	19	0.9900	4968000
22	3	0.9900	15439000
23	21	0.9900	60028000
24	7	0.0000	100001000
25	11	0.0000	100001000
26	13	0.0000	100001000
27	17	0.0000	100001000
28	20	0.0000	100001000
29	26	0.0000	100001000
30	30	0.0000	100001000

Table 2. NLSI Bus ranking (from weakest to strongest) based on maximum load ability

Rank	Branch	(From Bus – To Bus)	NLSI	Q_{max} (VAR)
1	38	27–30	0.9921	4130
2	12	6–10	0.9919	4430
3	39	29–30	0.9907	5500
4	37	27–29	0.9913	5900
5	36	28–27	0.9904	6190

(*continued*)

Table 2. (*continued*)

Rank	Branch	(From Bus – To Bus)	NLSI	Q_{max} (VAR)
6	34	25–26	0.9910	6520
7	33	24–25	0.9913	7510
8	32	23–24	0.9901	9170
9	17	12–14	0.9902	9520
10	15	4–12	0.9903	9530
11	22	15–18	0.9901	11250
12	11	6–9	0.9904	11790
13	13	9–11	0.9904	11790
14	25	10–20	0.9904	11790
15	35	25–27	0.9904	11790
16	20	14–15	0.9901	12370
17	5	2–5	0.9904	12380
18	19	12–16	0.9904	12380
19	30	15–23	0.9904	12380
20	40	8–28	0.9904	12380
21	2	1–3	0.9902	13030
22	21	16–17	0.9901	13030
23	6	2–6	0.9907	13760
24	31	22–24	0.9907	13760
25	3	2–4	0.9902	14560
26	28	10–22	0.9906	16510
27	16	12–13	0.9901	17680
28	18	12–15	0.9902	19040
29	23	18–19	0.9901	19040
30	8	5–7	0.9902	20630
31	14	9–10	0.9904	22510
32	9	6–7	0.9900	30940
33	26	10–17	0.9900	30940
34	24	19–20	0.9901	35360
35	27	10–21	0.9903	35370
36	1	1–2	0.9901	41260
37	41	6–28	0.9902	41260
38	4	3–4	0.9901	61880
39	7	4–6	0.9901	61880
40	10	6–8	0.9900	61900
41	29	21–22	0.9900	123750

4.2 Voltage Profile

Figure 5 shows the performance of system voltage due to FACTS installation in the weakest bus (Bus 27). The SVC was connected in parallel with the weakest bus in the system.

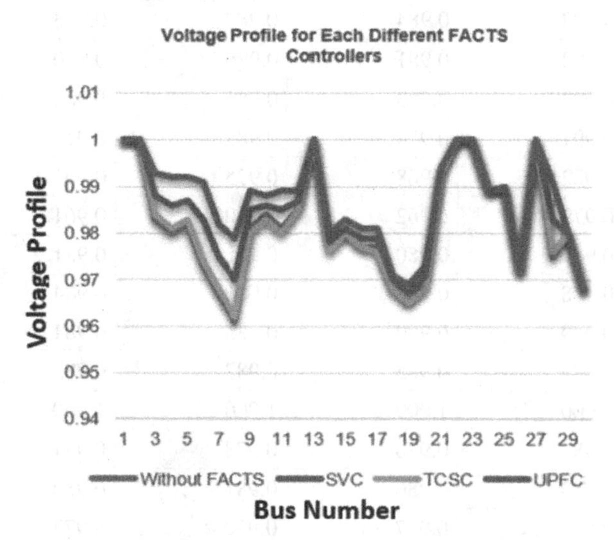

Fig. 5. Voltage performance based on different FACTS controller

As presented in Table 3, the voltage performance of the system's bus in general has greatly improved with FACTS devices installed in the system as compared to the performance of the system without FACTS devices. Installing the system with the TCSC controller only slightly improves bus voltage as compared to the voltage performance under the base scenario. However, when SVC and UPFC controllers are fitted, the bus voltage significantly improves.

The IEEE 30 bus test system's bus numbers 27 and 6 are shunt connected to the SVC model since it was determined that these buses are the optimum places to install SVC (unstable bus). The reactive power is simulated within the range of −100 MVAR to 100 MVAR. The optimal reactive power for SVC was discovered to be −1 MVAR at bus 27 and −40 MVAR at bus 6, where there is a significant improvement in the overall voltage profile and a large rise in voltage magnitude.

In contrast, a series controller will have its controller put in series with the lines. Since the branch was determined to be the most unstable based on computed NLSI, TCSC was installed on the line that connects buses 27 and 30. Branch 38 has the lowest line losses and the best overall voltage profile, and its reactance value is −0.1 MVAR of the line's initial reactance. Table 3 shows the comparison of SVC correction based on the voltage magnitude of each bus in the system.

Branch 38 (Bus 27–30) is where the UPFC model is attached. Bus 27, a bus in branch data, is connected to the UPFC's shunt portion. Since UPFC is a combination of SVC and TCSC models, the choice of SVC and TCSC models also affects the operating limit.

Table 3. Voltage magnitude analysis based on different types of SVC

Bus	SVC	TCSC	UPFC	Without FACTS
1	1.000	1.000	1.000	1.000
2	1.000	1.000	1.000	1.000
3	0.993	0.984	0.988	0.983
4	0.992	0.981	0.986	0.980
5	0.992	0.983	0.987	0.982
6	0.991	0.974	0.983	0.973
7	0.982	0.968	0.975	0.967
8	0.979	0.962	0.970	0.961
9	0.989	0.980	0.985	0.981
10	0.988	0.983	0.986	0.984
11	0.989	0.980	0.985	0.981
12	0.989	0.985	0.987	0.985
13	1.000	1.000	1.000	1.000
14	0.98	0.978	0.978	0.977
15	0.983	0.980	0.981	0.980
16	0.981	0.977	0.979	0.977
17	0.981	0.976	0.979	0.977
18	0.971	0.968	0.970	0.968
19	0.969	0.965	0.967	0.965
20	0.973	0.968	0.971	0.969
21	0.994	0.993	0.994	0.993
22	1.000	1.000	1.000	1.000
23	1.000	1.000	1.000	1.000
24	0.989	0.989	0.989	0.989
25	0.990	0.990	0.990	0.990
26	0.972	0.972	0.972	0.972
27	1.000	1.000	1.000	1.000
28	0.991	0.976	0.983	0.975
29	0.980	0.980	0.980	0.980
30	0.968	0.969	0.968	0.968

Reactive power and reactance value were determined to combine best at −10 MVAR and −0.1 MVAR of the line's initial reactance. After adding a UPFC device to the test system, it was discovered that the voltage profile had much improved and that line losses had decreased noticeably.

The overall voltage profile of the system has been improved by the addition of FACTS devices to low-voltage profile systems. Because bus 27's voltage is already at its max value of 1.1 p.u. and cannot be increased further without violating the voltage limitations, the magnitude of the voltage there is unaffected. The presence of FACTS is influenced and affected by its region, due to the increase in voltage magnitude for other buses in the system.

SVC and UPFC are found to be the best FACT types for improving voltage profiles since there is a significant rise in bus voltage magnitudes when compared to TCSC. By controlling the terminal voltage and supplying reactive power, which effectively increases the bus voltage magnitudes, the SVC provides reactive shunt compensation. The voltage stability limits are effectively increased by series capacitive compensation offered by TCSC, on the other hand. The voltage profile can be improved by controlling real and reactive power values.

Figures 6 and 7 show the value for real and reactive power line losses. Overall, real and reactive power line losses are larger without FACTS devices. The power flow in lines has effectively improved, and power losses in lines have been reduced. SVC enhances power flows in lines more than UPFC and TCSC do among all the FACTS devices employed in this project. At some branches, the UPFC also decreased line losses more effectively than SVC.

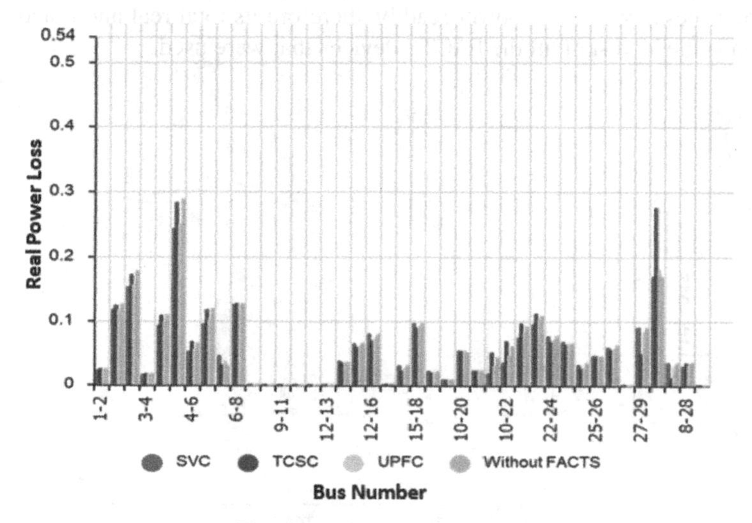

Fig. 6. Real power line losses

Figure 8 shows the total real power and reactive power line losses. Since FACTS devices can successfully be employed to boost the transfer capabilities of the available lines, the overall line losses are greatly decreased with the installation of FACTS devices. It has been found that SVC followed by UPFC improves power flow on the lines more effectively than TCSC because it has lower total line losses for both real power and reactive power loss. Unlike in Fig. 9, the total real power generation was not significantly impacted compared to the total reactive power generation because it is more crucial

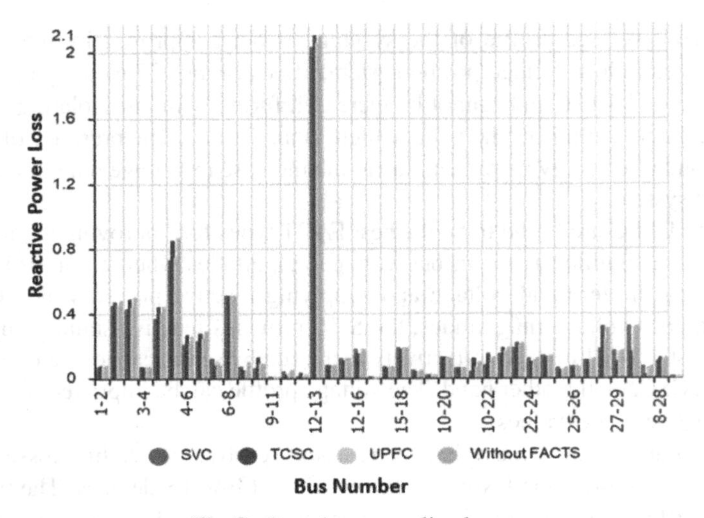

Fig. 7. Reactive power line losses

to reduce reactive power generation to obtain high-quality power supplies at reduced production costs. SVC effectively reduces reactive power flow in line because it can supply the necessary reactive power readily, therefore its total real and reactive power generation is the lowest of all the FACTS devices that were used.

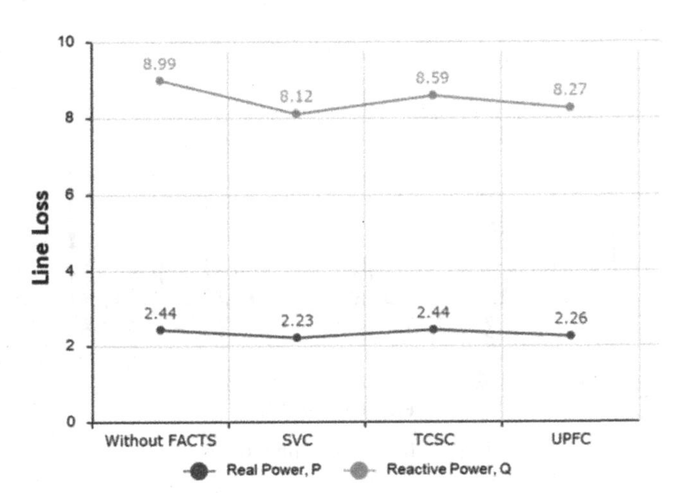

Fig. 8. Total real power and reactive power line losses

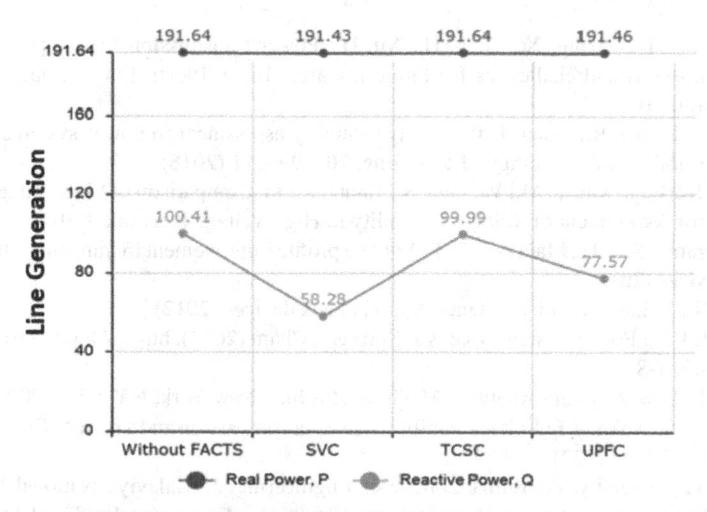

Fig. 9. Total real power and reactive power line generation

5 Conclusion and Future Studies

In conclusion, the load flow analysis was carried out perfectly to solve the power flow equation. As a result, complex voltages, and currents at each bus in the system, as well as the real and reactive power injections at each bus were obtained and analyzed. Two voltage stability indices, FVSI and NLSI are used in the study.

The simulation results reveal that the SVC-enabled shunt compensation method is the most successful at improving voltage stability. It is followed by shunt-series compensation, UPFC which exhibits notable improvements. Comparatively less successful at improving voltage stability is series correction, TCSC.

The FACTS controllers' presence and the effect they have on system performance were considered when recalculating the voltage stability indices. According to simulation results, the weakest bus and line became the strongest in terms of voltage stability when FACTS controllers were included.

Acknowledgement. The authors would like to acknowledge the financial support of this research. This work was funded by Universiti Teknologi Malaysia (UTM) under **UTM Fundamental Research** (UTMFR), Vot number: Q.J130000.3851.21H87.

References

1. Zhang, Y., Zhang, Y.: Multi-criteria decision analysis for power transmission line routing. Renew. Sustain. Energ. Revolution **81**, 1349–1357 (2018)
2. Al-Dafeeri, A.A.: Design and optimization of composite transmission lines. Int. J. Eng. Res. Technol. **33**(3), 1357–1366 (2018)
3. Al-Sarawi, M.A.: Integration of renewable energy sources into transmission systems using HVDC technology. IEEE Tran. Power Syst. **34**(2), 894–903 (2019)

4. Alhassan, A.B., Zhang, X., Shen, H., Xu, H.: Power transmission line inspection robots: a review, trends and challenges for future research. Int. J. Electr. Power Energ. Syst. **118**, 105862 (2020)
5. Ratra, S., Tiwari, R., Niazi, K.R.: Voltage stability assessment in power systems using line voltage stability index. Comput. Electr. Eng. **70**, 199–211 (2018)
6. Painuli, S., Singh Rawat, M., Vadhera, S., Tamta, R.: In: Comparison of Line Voltage Stability Indices for Assessment of Voltage Instability in High Voltage Network. (2018)
7. Gunasegaran, S. A.L., Malaysia, U.T.: Voltage profile improvement in transmission line using facts devices (2012)
8. Aziah, N.: Voltage stability enhancement via facts devices (2012)
9. Kimball, J.W.: Power System Analysis. Springer, Cham (2022). https://doi.org/10.1007/978-3-030-84767-8
10. Hill, D.J.: Power System Analysis. McGraw-Hill Inc., New York, NY, USA (2000)
11. Salama, H.S., Vokony, I.: Voltage stability indices–a comparison and a review. Comput. Electr. Eng. **98**, 107743 (2022)
12. Poornima University, Poornima College of Engineering, J. Malaviya National Institute of Technology, and Institute of Electrical and Electronics Engineers. In: Third International Conference & Workshops on Recent Advances and Innovations in Engineering (ICRAIE-2018) (2018)
13. Musiry Indices for Pin, I., Rahman, T.K.A.: Novel fast voltage stability index (FVSI) for voltage stability analysis in power transmission system. In: 2002 Student Conference on Research and Development Proceedings, Shah Alam, Malaysia (2002)
14. Kumar, S.K.N., Renuga, P.: FVSI based reactive power planning using evolutionary programming. In: ICCCCT-10 (2010)

Analysis of Multi Criteria Decision Making (MCDM) Techniques for Load Shedding in Islanded Distributed System

Muhammad Syafiq Bin Md Yusoff[1], Norazliani Md Sapari[1](✉), Nur Izzati Aslan[1], and Mohd Rohaimi Mohd Dahalan[2]

[1] Faculty of Electrical Engineering, Universiti Teknologi Malaysia, Johor, Malaysia
norazliani.ms@utm.my

[2] School of Electrical Engineering, College of Engineering, Universiti Teknologi MARA, 40450 Shah Alam, Selangor, Malaysia

Abstract. The modelling of a load shedding scheme in the islanded IEEE 33 bus system utilizing AHP and TOPSIS with voltage stability index and load value requirement is an important piece of research that attempts to improve the stability and reliability of power systems. Load shedding is an essential part of the operation of a power system, as it prevents blackouts and system failures. This research seeks to investigate and evaluate AHP and TOPSIS, two multi-criteria decision-making methodologies, for the selection of loads to be shed in a power system. In the analysis, the voltage stability index and load value requirements are also examined. The value of this study can be utilized to enhance the stability and dependability of power systems. This work can also serve as a basis for future research on power systems and multi-criteria decision-making. This study concludes with a complete examination of the AHP and TOPSIS methods for load shedding in power systems, demonstrating the potential for these methods to improve the stability and dependability of power systems.

Keywords: Load Shedding · Distribution Network · Multi-Criterion Decision Making method · Analytical Hierarchy Process · Technique for Order of Preference by Similarity to Ideal Solution · Under Frequency Load Shedding Scheme

1 Introduction

Load shedding is used in power systems where the overall demand for electrical power is far higher than the amount of power produced. The generator's overload breakers would automatically shut down the entire power plant if load shedding wasn't done, protecting the alternators from very serious damage. In addition to being very time-consuming and expensive, such damage would be difficult to restore. In general, load shedding refers to the removal of some or all the loads from a power system to maintain the operation of the system's remaining capacity. This load decrease is a reaction to system disruption and potential follow-up disruptions that lead to a generation deficiency condition. This state

© The Author(s), under exclusive license to Springer Nature Singapore Pte Ltd. 2024
F. Hassan et al. (Eds.): AsiaSim 2023, CCIS 1911, pp. 161–176, 2024.
https://doi.org/10.1007/978-981-99-7240-1_13

is frequently brought on by faults, loss of generation, switching errors, lighting strikes, and others. The balance between generations and loads could be upset by a rapid loss of generation because of exceptional circumstances such as a generator malfunction or line tripping, which would lead to a fall in system frequency. Frequency stability has grown to be a top issue for power system operators considering the recent occurrence of several outages on power systems throughout the world [1].

Frequency is a crucial indicator of how well active power generation and load are balanced. The frequency will decrease if active power generation in large-scale power networks or microgrids is insufficient owing to generation loss. System blackouts and restoration failure may result from significant frequency loss. Systems that integrate more sporadic renewable energy are more susceptible to frequency distortion. Frequency decline must be stopped to spare power systems from severe consequences. Under frequency load shedding (UFLS), a crucial approach for recovering frequency in under-frequency conditions rebalances generation and load by shedding the right number of loads [2]. Using genetic and particle swarm optimization, a technique for carrying out optimal load-shedding in contingency situations is presented. The effectiveness of various algorithms is contrasted after they are introduced. This is a useful and appropriate way for operating and securing power systems to withstand phenomena like voltage instability and line overload. The objective function of this strategy comprises, subject to operational and security constraints, minimizing the cost-of-service disruption. Outages of the line or transformer and the generator are considered contingencies. The Multiple Criterion Decision Making (MCDM) such as the Analytical Hierarchy Process (AHP) and Technique for Order of Preference by Similarity to Ideal Solution (TOPSIS) were applied in a way in the proposed method. In these cases, the IEEE 33 bus system is used to test the proposed method and present the result.

2 Islanding Operation

This section will explain the issue of islanding that will be the subject of study. Islanding is the condition in which a distributed generation (DG) system continues to generate electricity after becoming detached from the primary power grid. This may be the result of a faulty power grid or a deliberate disconnection for repair or other purposes. Islanding can present DG systems with various technical and safety challenges.

For instance, when a DG system is islanded, an uneven power flow can develop, resulting in system overload and probable damage. Furthermore, the DG system may no longer be coordinated with the power grid, resulting in unstable and unpredictable power output.

Moreover, islanding might pose safety issues for utility workers who may be unaware that the DG system is still generating electricity, as they may be operating on the grid under the false impression that it has been de-energized.

Utilities and DG system owners can include anti-islanding protection mechanisms, such as automated shut-off switches, voltage and frequency monitoring, and other control systems, to limit the dangers associated with islanding. These methods aid in detecting islanding conditions and guarantee that the DG system disconnects swiftly and safely.

Islanding is a serious problem for DG systems, and utilities and owners of DG systems must install anti-islanding measures to maintain the safety and dependability of the power grid. Therefore, in this study, the UFLS scheme using AHP and TOPSIS methods was presented to optimize the result and overcome this issue.

3 MCDM Method

This section explains about MCDM methods and algorithms used in this study which are AHP and TOPSIS methods. The act of making decisions is a human action in which the person making the decision may scarcely avoid the influence of various external factors that will ultimately shape the decision that will be successful. Multicriteria Decision-Making (MCDM), one of the most significant and rapidly developing subfields in operations research and management science, aims to make this successful decision [10]. There is various method in MCDM itself such as Elimination and Choice Expressing Reality (ELECTRE), Preference Ranking Organization Method for Enrichment Evaluation (PROMETHEE), Multi-Attribute Utility Theory (MAUT), Technique for Order of Preference by Similarity to Ideal Solution (TOPSIS), Analytical Hierarchy Process (AHP), and others. The focus of this study is to compare the result between AHP and TOPSIS methods in determining the most optimum bus rank to operate the load shedding scheme.

3.1 AHP Method

The additive notion created by prior approaches like MAUT is the foundation of the "Analytical Hierarchy Process" (AHP), which was first proposed by Saaty in 1980. The weights generated by AHP are used to calculate the weights for the alternatives and the criteria. The "eigenvalue" method, based on pairwise comparisons between criteria and DM preferences, is used for weight extraction. "Eigen" is German for "particular". AHP and other methods use fuzzy logic, although Saaty and most scholars do not support this practice because AHP is already fuzzy [11]. When a decision maker must weigh numerous factors, the Analytical Hierarchy Process (AHP) is a method for ranking decision options and choosing the optimal one. It responds to the query "Which one?" When using AHP, the decision maker chooses the alternative that best satisfies his or her choice criteria and creates a numerical score to rank each alternative decision based on how well each alternative satisfies those requirements [3]. Figure 1 shows the AHP scheme.

Some researchers recommend comparing the outcomes of an issue handled using approach A with the same problem solved using procedures B and C. It seems like a wonderful idea, but it would be challenging because the three methods are likely to produce diverse outcomes. There is a strong likelihood that the three methods' results are "right" if they all agree. However, in this situation, one must exercise caution when concluding the outcome's comparison.

The three answers may be consistent if, for example, the same problem is addressed using AHP, which provides criteria weights, and then those weights are used for other methods, such as TOPSIS ("hybrid TOPSIS"). However, because the last procedures use partial values generated from the first, one must be aware that in this instance, the findings may be deceptive. Additionally, because AHP weights are derived from DM's preferences, they are subjective, and the last two approaches pick up on this subjectivity.

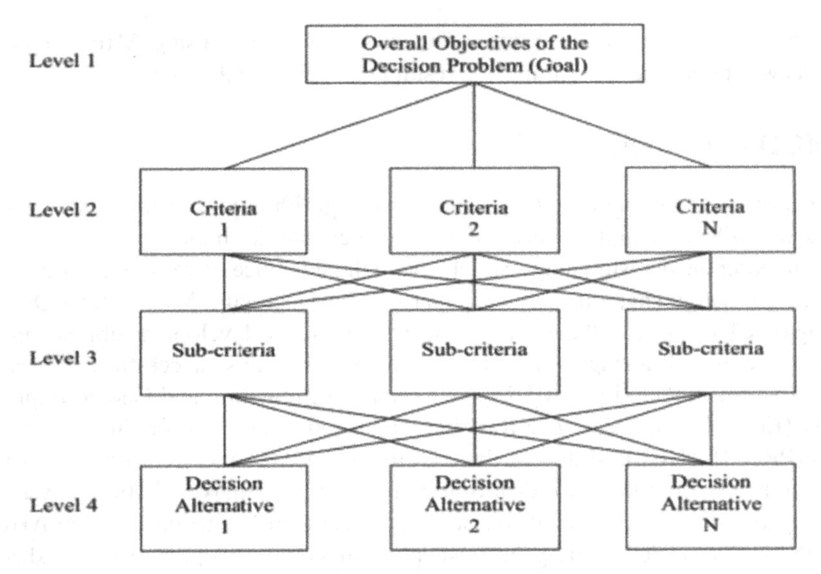

Fig. 1. AHP Scheme level

AHP Algorithms. Firstly, develop a hierarchy structure: Come out with the necessary criterion that needs to be included in this case, load category (LV) and Stability Indices (SI) for the upper level. Then, list down the alternative of each criterion which is the number of buses for the lower level. For example, in Fig. 2. Equation (1) until (17), briefly explained the parameters required in the AHP process proposed in this work.

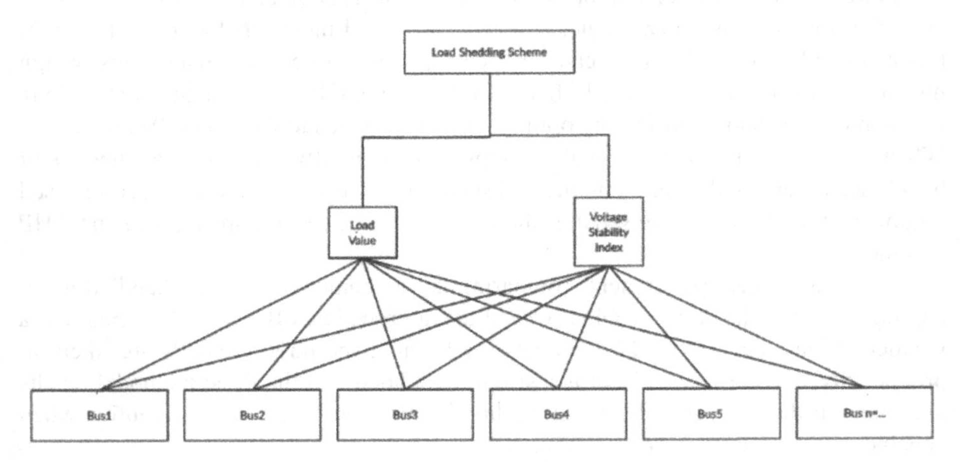

Fig. 2. Hierarchy for this study

Step 1. Develop matrix for each criterion.

$$C_{LoadValue} = \begin{bmatrix} C_{LV\_Bus1} \\ C_{LV\_Bus2} \\ \cdots \\ C_{LV\_BusN} \end{bmatrix} = \begin{bmatrix} LoadValueBus1 \\ LoadValueBus2 \\ \cdots \\ LoadValueBusN \end{bmatrix} \quad (1)$$

$$C_{StabiltyIndices} = \begin{bmatrix} C_{SI\_Bus1} \\ C_{SI\_Bus2} \\ \cdots \\ C_{SI\_BusN} \end{bmatrix} = \begin{bmatrix} SIBus1 \\ SIBus2 \\ \cdots \\ SIBusN \end{bmatrix} \quad (2)$$

$$\sum C_{LV} = \sum_{i=1}^{N} C_{LV\_BusN} \quad (3)$$

$$\sum C_{SI} = \sum_{i=1}^{N} C_{SI\_BusN} \quad (4)$$

Step 2. Pairwise comparison for criterion– To determine normalized criterion weight.

$$Pairwise_{criterion} = \begin{matrix} \\ LV \\ SI \end{matrix} \begin{matrix} LV & SI \\ \left[\begin{matrix} 1 \\ \frac{\sum C_{SI}}{\sum C_{LV}} \end{matrix} \right. & \left. \begin{matrix} \frac{\sum C_{LV}}{\sum C_{SI}} \\ 1 \end{matrix} \right] \end{matrix} \quad (5)$$

$$Pairwise_{criterion}{}^{2} = \begin{bmatrix} 1 & \frac{\sum C_{LV}}{\sum C_{SI}} \\ \frac{\sum C_{SI}}{\sum C_{LV}} & 1 \end{bmatrix} \begin{bmatrix} 1 & \frac{\sum C_{LV}}{\sum C_{SI}} \\ \frac{\sum C_{SI}}{\sum C_{LV}} & 1 \end{bmatrix} \quad (6)$$

$$Criterion = \begin{bmatrix} \sum Row1 \\ \sum Row2 \end{bmatrix} \quad (7)$$

$$CSumCriterion = \sum Row1 + \sum Row2 \quad (8)$$

$$WeightCriterion = \begin{bmatrix} \frac{\sum Row1}{SumCriterion} \\ \frac{\sum Row2}{SumCriterion} \end{bmatrix} = \begin{bmatrix} Weight_{LV} \\ Weight_{VSI} \end{bmatrix} \quad (9)$$

Step 3. pairwise comparison alternative which is lower level. To determine the alternative weight

$$WPairwise_{alternative} = \begin{matrix} \\ C_{LV1} \\ C_{LV2} \\ C_{LVN} \end{matrix} \begin{matrix} C_{LV1} \; C_{LV2} \; C_{LVN} \\ \begin{bmatrix} 1 & P_{12} & P_{1N} \\ P_{21} & 1 & P_{2N} \\ P_{N1} & P_{N2} & 1 \end{bmatrix} \end{matrix} \quad (10)$$

$$Pairwise_{alternative}{}^{2} = \begin{bmatrix} 1 & P_{12} & P_{1N} \\ P_{21} & 1 & P_{2N} \\ P_{N1} & P_{N2} & 1 \end{bmatrix} \begin{bmatrix} 1 & P_{12} & P_{1N} \\ P_{21} & 1 & P_{2N} \\ P_{N1} & P_{N2} & 1 \end{bmatrix} \quad (11)$$

$$\text{AlternativeLoadValue}=\begin{bmatrix}\sum \text{Row1}\\ \sum \text{Row2}\\ \sum \text{RowN}\end{bmatrix} \tag{12}$$

$$\text{SumAlternativeLoadValue}=\sum \text{Row1}+\sum \text{Row2}+\sum \text{RowN} \tag{13}$$

$$\text{WeightAlternativeLoadValue}=\begin{bmatrix}\frac{\sum \text{Row1}}{\text{SumAlternativeLoadValue}}\\ \frac{\sum \text{Row2}}{\text{SumAlternativeLoadValue}}\\ \frac{\sum \text{RowN}}{\text{SumAlternativeLoadValue}}\end{bmatrix} \tag{14}$$

$$\text{WeightAlternativeLoadValue}=\begin{bmatrix}\text{Weight}_{\text{LV\_ Bus1}}\\ \text{Weight}_{\text{LV\_ Bus2}}\\ \text{Weight}_{\text{LV\_ BusN}}\end{bmatrix} \tag{15}$$

And repeat the step for alternative Stability Indices.

Step 4. Multiply the value of the weight alternative and weight criterion to obtain the bus score.

$$\text{Score}_{\text{Bus}}=\begin{bmatrix}\text{Weight}_{\text{LV\_ Bus1}} & \text{Weight}_{\text{SI\_ bus1}}\\ \text{Weight}_{\text{LV\_ Bus2}} & \text{Weight}_{\text{SI\_ bus2}}\\ \text{Weight}_{\text{LV\_ BusN}} & \text{Weight}_{\text{SI\_ busN}}\end{bmatrix}\begin{bmatrix}\text{Weight}_{\text{LV}}\\ \text{Weight}_{\text{SI}}\end{bmatrix} \tag{16}$$

Step 5. The list of bus ranks produced based on the score obtained. (The highest score will be shed first)

$$\text{Bus ranking}=\begin{bmatrix}\text{BR}_1\\ \text{BR}_2\\ \text{BR}_3\end{bmatrix} \tag{17}$$

3.2 TOPSIS Method

The most well-known method for resolving MCDM issues is called Technique for Order Performance by Similarity to Ideal Solution (TOPSIS), which was put forth by Hwang & Yoon in 1981. This approach is predicated on the idea that the selected alternative should be closest to the Positive Ideal Solution (PIS), or the solution that minimizes costs and maximizes benefits, and the furthest from the Negative Ideal Solution (NIS) as shown in Fig. 3 [12]. Equation (18) until (23) formulated the step for TOPSIS algorithms proposed in this work.

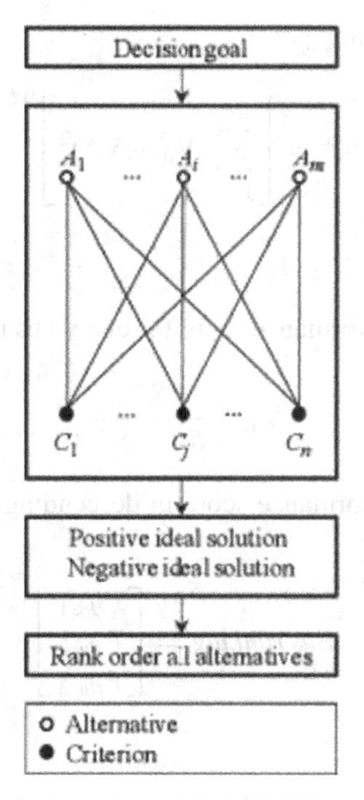

Fig. 3. Overview of the TOPSIS method

TOPSIS Algorithms.
Step 1. develop a decision matrix.

$$C_{\text{LoadValue}} = \begin{bmatrix} C_{\text{LV\_Bus1}} \\ C_{\text{LV\_Bus2}} \\ \cdots \\ C_{\text{LV\_BusN}} \end{bmatrix} = \begin{bmatrix} \text{LoadValueBus1} \\ \text{LoadValueBus2} \\ \cdots \\ \text{LoadValueBusN} \end{bmatrix} \tag{18}$$

$$C_{StabiltyIndices} = \begin{bmatrix} C_{SI\_Bus1} \\ C_{SI\_Bus2} \\ \cdots \\ C_{SI\_BusN} \end{bmatrix} = \begin{bmatrix} SIBus1 \\ SIBus2 \\ \cdots \\ SIBusN \end{bmatrix} \tag{19}$$

Step 2. Determine the weightage for each criterion (0.5) and calculate the weight of the normalised decision matrix using the formula:

$$r_{ij} = \frac{0.5 \times X_{ij}}{\sqrt{\sum_{j=1}^{n} X_{ij}^2}} \tag{20}$$

Step 3. Determine the positive value and negative value based on the criterion. This value will be the indicator to calculate Positive Solution Ideal (PSI) and Negative Solution

Ideal (NIS) by using the formula:

$$S_i^{\pm} = \left[\sum_{j=1}^{m} (V_{ij} - V_j^{\pm})^2\right]^{0.5} \tag{21}$$

$V_j^+ = \text{Min (LV \& VSI)}$

$V_j^- = \text{Max (LV\&VSI)}$

Step 4. Calculate the performance score for every bus using the formula:

$$P_i = \frac{S_i^-}{S_i^+ + S_i^-} \tag{22}$$

Step 5. Arrange the performance score in descending order and the highest score will be shed first.

$$Bus\ ranking = \begin{bmatrix} P_{Bus1} \\ P_{Bus2} \\ P_{Bus3} \end{bmatrix} \tag{23}$$

4 Proposed Method

This section will go through the process of the simulation of islanded IEEE 33 bus system via MATLAB Simulink and the calculation of stability indices.

4.1 Test System

This section briefly described the test system used in this work. The parameters for the standard test system and DG are presented in Tables 1 and 2 respectively.

Table 1. Parameter of IEEE bus system

To	From/Load	P (MW)	Q (Mvar)	S (MVA)	R (Ω/km)	X (Ω/km)	Z (Ω/km)	V
1	2	0.100	0.060	0.11662	0.0922	0.0470	0.1034	12660
2	3	0.090	0.040	0.09849	0.4930	0.2511	0.5532	12660
3	4	0.120	0.080	0.14422	0.3660	0.1864	0.4107	12660
4	5	0.060	0.030	0.06708	0.3811	0.1941	0.4276	12660
5	6	0.060	0.020	0.06325	0.8190	0.7070	1.0819	12660
6	7	0.200	0.100	0.22361	0.1872	0.6188	0.6465	12660

(continued)

Table 1. (*continued*)

To	From/Load	P (MW)	Q (Mvar)	S (MVA)	R (Ω/km)	X (Ω/km)	Z (Ω/km)	V
7	8	0.200	0.100	0.22361	0.7114	0.2351	0.7492	12660
8	9	0.060	0.020	0.06325	1.0300	0.7400	1.2682	12660
9	10	0.060	0.020	0.06325	1.0440	0.7400	1.2796	12660
10	11	0.045	0.030	0.05408	0.1966	0.0650	0.2070	12660
11	12	0.060	0.035	0.06946	0.3744	0.1238	0.3943	12660
12	13	0.060	0.035	0.06946	1.4680	1.1550	1.8679	12660
13	14	0.120	0.080	0.14422	0.5416	0.7129	0.8953	12660
14	15	0.060	0.010	0.06083	0.5910	0.5260	0.7911	12660
15	16	0.060	0.020	0.06325	0.7463	0.5450	0.9241	12660
16	17	0.060	0.020	0.06325	1.2890	1.7210	2.1502	12660
17	18	0.090	0.040	0.09849	0.7320	0.5740	0.9302	12660
2	19	0.090	0.040	0.09849	0.1640	0.1565	0.2266	12660
19	20	0.090	0.040	0.09849	1.5042	1.3554	2.0247	12660
20	21	0.090	0.040	0.09849	0.4095	0.4784	0.6297	12660
21	22	0.090	0.040	0.09849	0.7089	0.9373	1.1751	12660
3	23	0.090	0.050	0.10296	0.4512	0.3083	0.5464	12660
23	24	0.420	0.200	0.46519	0.8980	0.7091	1.14421	12660
24	25	0.420	0.200	0.46519	0.8960	0.7011	1.1377	12660
6	26	0.060	0.025	0.06500	0.2030	0.1034	0.2278	12660
26	27	0.060	0.025	0.06500	0.2842	0.1447	0.3189	12660
27	28	0.060	0.020	0.06325	1.0590	0.9337	1.4118	12660
28	29	0.120	0.070	0.13892	0.8042	0.7006	1.0665	12660
29	30	0.200	0.600	0.63246	0.5075	0.2585	0.5695	12660
30	31	0.150	0.070	0.16553	0.9744	0.9630	1.3699	12660
31	32	0.210	0.100	0.23259	0.3105	0.3619	0.4768	12660
32	33	0.060	0.040	0.07211	0.3410	0.5302	0.6303	12660
		3.715	**2.300**	**4.54900**	**20.5784**	**17.7843**	**27.6378**	

4.2 Voltage Stability Index Calculation

The voltage stability index in distribution networks is crucial for both effective energy management and dependable electricity distribution. The voltage stability index can be a criterion for the quality of the power because it represents how dependable the distribution network is in terms of supply and demand. For distribution networks, the quadratic equation of sending end voltages can be used to derive the stability index in

Table 2. Parameter of Distributed Generation (DG)

Generator	Rated Power, MW	Rated Frequency, Hz	Rated Voltage, VLL-rms	Location
1	2.00	50	12660	Bus 6
2	1.00	50	12660	Bus 18
3	1.65	50	12660	Bus 33

Eq. (24), with 0 as the critical value and 1 as the stable value [14].

$$SI = 2V_s^2V_r^2 - V_r^4 - 2V_r^2(PR + QX) - |Z|^2\left(P^2 + Q^2\right) \tag{24}$$

Figure 4 shows the illustration of the distribution line network of sending end voltage that is used in the calculation of the voltage stability index. The calculation involves the per unit value of the parameter of the distribution network. Assuming the base voltage and base power are equal to 12.66 kV and 100 MVA and using the value of the power and voltage of the test system obtained from the simulation to calculate the voltage stability index for AHP and TOPSIS criterion.

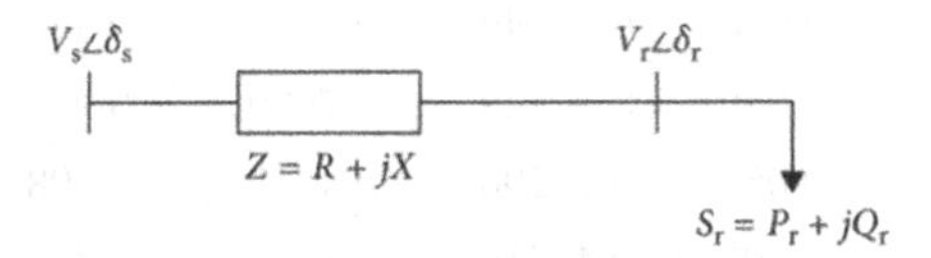

Fig. 4. Line diagram of two buses in the distribution network

4.3 Islanding Operation

By using a step signal to send the signal to the national grid's circuit breaker to perform the islanding operation in the test system. The circuit breaker of bus 1 will disconnect the national grid from the test system causing the system to be fully independent and operate using DG which expected outcome made the voltage stability of the system decreased due to power losses from the national grid. Therefore, the first load shedding scheme occurs to increase the voltage stability index near to stable value. The bus ranks are obtained from the AHP and TOPSIS method and are used to perform a load-shedding scheme.

4.4 Under Frequency Load Shedding (UFLS)

UFLS is performed at 0.99 pu of the frequency of rotational machinery to avoid the machine from damage. When the machines operate under 0.99 pu for a long period the machine will break and cause a blackout. To avoid these issues occurring, the UFLS must be operated when the frequency of the DG reaches 0.99 (49.5 Hz for a rated frequency 50 Hz) to ensure the generator can operate safely and well [1].

5 Result and Analysis

Table 3 presents the value of VSI for each bus in the system. Based on the result obtained, line 24 to 25 has the lowest voltage stability index which is the most critical line. In this work, the load value and voltage stability index were used as criteria for AHP and TOPSIS methods as indicated in Table 4.

Table 3. Voltage stability indices for every branch in standard condition

To	From/ Load	P, p.u	Q, p.u	S, p.u	R, p.u	X, p.u	Z, p.u	V, p.u	VSI
1	2	0.001002	0.000601	0.001168	0.05753	0.02932	0.06457	0.9980	0.9987
2	3	0.000914	0.000406	0.001000	0.30760	0.15667	0.34519	0.9903	0.9910
3	4	0.001231	0.000820	0.001479	0.22836	0.11630	0.25627	0.9879	0.9611
4	5	0.000622	0.000311	0.000696	0.23778	0.12110	0.26684	0.9859	0.9520
5	6	0.000653	0.000218	0.000689	0.51099	0.44112	0.67505	0.9910	0.9439
6	7	0.002194	0.001097	0.002453	0.11680	0.38608	0.40337	0.9910	0.9633
7	8	0.002196	0.001098	0.002456	0.44386	0.14668	0.46747	0.9866	0.9622
8	9	0.000664	0.000221	0.000700	0.64264	0.46170	0.79130	0.9836	0.9463
9	10	0.000669	0.000223	0.000705	0.65138	0.46170	0.79841	0.9812	0.9350
10	11	0.000502	0.000335	0.000603	0.12266	0.04056	0.12919	0.9805	0.9266
11	12	0.000670	0.000391	0.000775	0.23360	0.07724	0.24604	0.9793	0.9237
12	13	0.000678	0.000396	0.000785	0.91592	0.72063	1.16543	0.9790	0.9179
13	14	0.001368	0.000912	0.001644	0.33792	0.44480	0.55860	0.9808	0.9168
14	15	0.000688	0.000115	0.000698	0.36874	0.32818	0.49363	0.9822	0.9246
15	16	0.000692	0.000231	0.000730	0.46564	0.34004	0.57658	0.9838	0.9299
16	17	0.000707	0.000236	0.000745	0.80424	1.07378	1.34156	0.9920	0.9347
17	18	0.001068	0.000475	0.001168	0.45671	0.35813	0.58038	0.9947	0.9672
2	19	0.000902	0.000401	0.000987	0.10232	0.09764	0.14144	0.9974	0.9916
19	20	0.000902	0.000401	0.000987	0.93851	0.84567	1.26331	0.9938	0.9873
20	21	0.000902	0.000401	0.000987	0.25550	0.29849	0.39290	0.9931	0.9748
21	22	0.000902	0.000401	0.000987	0.44230	0.58481	0.73323	0.9925	0.9714
3	23	0.000913	0.000507	0.001045	0.28152	0.19236	0.34096	0.9867	0.9611
23	24	0.004262	0.002030	0.004721	0.56028	0.44243	0.71390	0.9799	0.9412
24	25	0.004262	0.002029	0.004720	0.55904	0.43743	0.70984	0.9765	0.9156
6	26	0.000654	0.000273	0.000709	0.12666	0.06451	0.14214	0.9898	0.9644
26	27	0.000656	0.000273	0.000711	0.17732	0.09028	0.19898	0.9884	0.9597

(*continued*)

Table 3. (*continued*)

To	From/ Load	P, p.u	Q, p.u	S, p.u	R, p.u	X, p.u	Z, p.u	V, p.u	VSI
27	28	0.000668	0.000223	0.000704	0.66074	0.58256	0.88088	0.9852	0.9531
28	29	0.001354	0.000790	0.001568	0.50176	0.43712	0.66546	0.9832	0.9400
29	30	0.002267	0.006802	0.007170	0.31664	0.16128	0.35535	0.9818	0.9310
30	31	0.001734	0.000809	0.001914	0.60795	0.60084	0.85476	0.9870	0.9260
31	32	0.002446	0.001165	0.002709	0.19373	0.22580	0.29752	0.9897	0.9474
32	33	0.000707	0.000471	0.000849	0.21276	0.33081	0.39332	0.9949	0.9586
		0.04005	**0.02506**	**0.04926**	**12.83938**	**11.09607**	**17.24389**		

Table 4. Bus rank and score from AHP and TOPSIS

TOPSIS		AHP	
Bus ranking	Score	Bus ranking	Score
25	1.000	25	0.154
24	0.882	24	0.142
22	0.489	22	0.087
21	0.481	21	0.086
20	0.449	4	0.081
4	0.451	20	0.080
2	0.449	2	0.079
3	0.441	3	0.079
19	0.440	19	0.078
23	0.422	23	0.075
5	0.332	30	0.068
30	0.345	5	0.062
8	0.309	8	0.062
32	0.293	32	0.059
7	0.247	31	0.056
6	0.283	7	0.055
31	0.279	6	0.053
14	0.227	14	0.053
29	0.190	29	0.048

(*continued*)

Table 4. (*continued*)

TOPSIS		AHP	
Bus ranking	Score	Bus ranking	Score
13	0.187	13	0.043
12	0.185	12	0.042
11	0.172	10	0.041
10	0.162	15	0.040
15	0.180	11	0.039
9	0.144	9	0.037
16	0.136	16	0.037
17	0.097	17	0.033
18	0.084	18	0.030
28	0.071	28	0.029
27	0.064	27	0.028
26	0.059	26	0.028
33	0.028	33	0.023

Table 5 shows the comparison of the VSI index for AHP and TOPSIS to get the bus rank for AHP and TOPSIS.

Table 5. Comparative for VSI in stable, islanded and load shed conditions.

To	From/ Load	VSI value				
		Stable	Islanded	AHP	TOPSIS	Conventional
1	2	0.9987	0.8824	0.9071	0.9071	0.9070
2	3	0.9910	0.8819	0.9065	0.9065	0.9065
3	4	0.9611	0.8882	0.9130	0.9130	0.9129
4	5	0.9520	0.9044	0.9250	0.9250	0.9249
5	6	0.9439	0.9217	0.9380	0.9380	0.9379
6	7	0.9633	0.9690	0.9737	0.9737	0.9736
7	8	0.9622	0.9572	0.9626	0.9626	0.9626
8	9	0.9463	0.9521	0.9565	0.9565	0.9564
9	10	0.9350	0.9461	0.9497	0.9497	0.9496
10	11	0.9266	0.9432	0.9459	0.9459	0.9459

(*continued*)

Table 5. (*continued*)

To	From/ Load	VSI value				
		Stable	Islanded	AHP	TOPSIS	Conventional
11	12	0.9237	0.9433	0.9457	0.9457	0.9457
12	13	0.9179	0.9430	0.9448	0.9448	0.9448
13	14	0.9168	0.9461	0.9470	0.9470	0.9470
14	15	0.9246	0.9484	0.9494	0.9494	0.9494
15	16	0.9299	0.9538	0.9546	0.9546	0.9546
16	17	0.9347	0.9620	0.9620	0.9620	0.9620
17	18	0.9672	0.9794	0.9798	0.9798	0.9798
2	19	0.9916	0.8823	0.9070	0.9070	0.9069
19	20	0.9873	0.8785	0.9030	0.9030	0.9029
20	21	0.9748	0.8674	0.8916	0.8916	0.8915
21	22	0.9714	0.8644	0.8885	0.8885	0.8884
3	23	0.9611	0.8883	0.9131	0.9131	0.9130
23	24	0.9412	0.8698	0.9001	0.9001	0.9062
24	25	0.9156	0.8462	0.8939	0.8939	0.8877
6	26	0.9644	0.9701	0.9748	0.9748	0.9748
26	27	0.9597	0.9691	0.9734	0.9734	0.9734
27	28	0.9531	0.9675	0.9712	0.9712	0.9712
28	29	0.9400	0.9560	0.9587	0.9587	0.9587
29	30	0.9310	0.9487	0.9505	0.9505	0.9505
30	31	0.9260	0.9523	0.9530	0.9530	0.9530
31	32	0.9474	0.9716	0.9715	0.9715	0.9715
32	33	0.9586	0.9801	0.9799	0.9799	0.9799

Based on the result obtained from the simulation, the VSI for islanded decreased and to increase the VSI back to avoid the backlash to the generator, the load shedding scheme was used. The result above shows the comparison between the bus rank using AHP, TOPSIS and conventional methods. For AHP and TOPSIS, these methods prefer to shed load 25 while conventional methods prefer to shed load at bus 24. The greater amount of load shed the greater amount for VSI increased. But in terms of load value, bus 24 and bus 25 had the same amount of load but to determine which load to shed, a variety of criteria are used. In this study, the VSI is included to determine the score of the load and which load to shed first. In Table 5, the result shows the effectiveness of the bus rank produced by the AHP and TOPSIS method to determine which load to shed.

6 Conclusion

For the evaluation of load-shedding schemes in the IEEE 33 bus system, the analysis of multi-criteria decision-making (MCDM) techniques, particularly the AHP and TOPSIS methods, has yielded important insights for decision-makers in the power systems domain.

Like this, using AHP gave decision-makers a hierarchical framework for making decisions, allowing them to assess and provide relative weights to criteria and sub-criteria. The importance and preferences of several factors were evaluated using pairwise comparisons and the AHP algorithm, resulting in an overall rating of the alternative load-shedding schemes.

Other than that, the distribution network operators can assess various load-shedding options based on how closely resemble the ideal solution by applying TOPSIS as another option to perform a load-shedding scheme. Even though, the result for the whole bus rank is slightly different but for the highest rank of AHP and TOPSIS. These show that the AHP and TOPSIS are reliable methods to perform the load-shedding scheme. Other than that, the load-shedding scheme using AHP and TOPSIS produced a higher voltage stability index than the conventional method proof of the effectiveness of the AHP and TOPSIS method in this study.

Acknowledgement. The authors would like to acknowledge the financial support of this research. This work was funded by Universiti Teknologi Malaysia (UTM) under **UTM Fundamental Research** (UTMFR), Vot number: Q.J130000.3851.21H87.

References

1. Darbandsari, A., Amraee, T.: Under frequency load shedding for low inertia grids utilizing smart loads. Int. J. Electr. Power Energy Syst. **135** (2022)
2. Li, C., et al.: Continuous under-frequency load shedding scheme for power system adaptive frequency control. IEEE Trans. Power Syst. **35**(2), 950–961, March 2020
3. Nguyen, A.T., Le, N.T., Quyen, A.H., Phan, B.T.T., Trieu, T.P., Hua, T.D.: Application of AHP algorithm on power distribution of load shedding in island microgrid. Int. J. Elec. Comput. Eng. **11**(2), 1011–1021, April 2021
4. Eissa Mohammed, K., Osman Hassan, M.: An Intelligent Load Shedding System Application in Sudan National Grid لكهربا في شبكة لذكيا للأحمال خفض تطبيق القومية السودانية (2018)
5. Wahyudi, R., Yamashika, H., Teknik, F., Muhammadiyah Sumatera Barat, U.: Analisa Stabilitas Transien Pada Jaringan Distribusi Radial IEEE 33 Bus Terhubung Dengan Energi Terbarukan (Photovoltaic). Ensiklopedia Res. Commun. Serv. Rev. **1**(3), 176–182, August 2022
6. Saravanan, V., Manickam, R., Selvam, M., Ramachandran, M.: Optimization in remanufacturing View project A Study on Artificial intelligence with Machine learning and Deep Learning Techniques View project Electrical and Automation Engineering Interaction between Technical and Economic Benefits in Distributed Generation
7. Gouveia, J., Gouveia, C., Rodrigues, J., Carvalho, L., Moreira, C.L., Lopes, J.A.P.: Planning of distribution networks islanded operation: from simulation to live demonstration. Elec. Power Syst. Res. **189**, 106561, December 2020

8. Wang, C., et al.: Underfrequency load shedding scheme for islanded microgrids considering objective and subjective weight of loads. IEEE Trans. Smart Grid (2022)
9. Larik, R.M., Mustafa, M.W., Aman, M.N.: A critical review of the state-of-art schemes for under voltage load shedding. Int. Trans. Elec. Energy Syst. **29**(5), e2828, May 2019
10. Multi-criteria decision analysis methods for energy sector's sustainability assessment_ Robustness analysis through criteria weight change _ Elsevier Enhanced Reader
11. Papathanasiou, J., Ploskas, N.: Springer Optimization and Its Applications 136 Multiple Criteria Decision Aid Methods, Examples and Python Implementations
12. Eminoglu, U., Hocaoglu, M.H.: Ranking method of equipment failure risk in shipboard power system. A voltage stability index for Radial Distribution Networks. In: 2007 42nd International Universities Power Engineering Conference (2007)

Comparison of Diesel and Green Ship Carbon Emissions with A-Star Route Optimization

Mohammad Zakir Bin Mohd Ridwan, Malcolm Yoke Hean Low(✉), and Weidong Lin

Singapore Institute of Technology, 567739 Singapore, Singapore
malcolm.low@singaporetech.edu.sg

Abstract. Approximately 2% of the world's carbon dioxide (CO2) emissions are attributed to international shipping, with the main source of carbon emissions from ships coming from the marine engines' consumption of fossil fuels, especially heavy fuel oil. With the use of the A* algorithm for route optimization, this study compares the carbon emissions of diesel and greener ships by simulating the routing of 5000 TEU container ships through different weather conditions. In the study, data from dependable sources are gathered, carbon emissions are calculated, and routes are optimised utilising the A* algorithm. The outcomes show the discrepancies in carbon emissions between diesel and green ships, and the algorithm's success in lowering emissions through optimal route planning is demonstrated. The research highlights the significance of switching to more environmentally friendly options whilst providing useful information for the shipping industry. The study concludes with suggestions for more investigation and advancements in carbon emission reduction tactics. By utilizing the A* algorithm, this research enhances the study of ship emissions analysis while developing environmentally responsible behaviour in the marine industry.

Keywords: A* algorithm · Decarbonization · Diesel Engine · Fuel Consumption · Sustainability · Weather-routing

1 Introduction

The marine sector is essential to international trade as it links countries and makes it easier for goods to travel between continents. However, this industry also makes a sizable contribution to carbon emissions, which has prompted worries about its effects on the environment. It is crucial to evaluate the carbon emissions of various ship types and investigate measures for emission reduction as the world works to slow down climate change and move toward a more sustainable future. This study uses the A* route-optimization algorithm to compare the carbon emissions of diesel and green ships.

Approximately 2% of the world's carbon dioxide (CO2) emissions are attributed to international shipping, according to the Fourth IMO GHG Study 2020 [1], a specialized body of the United Nations. The main source of carbon emissions from ships is the marine engines' consumption of fossil fuels, especially heavy fuel oil. At the current rate

© The Author(s), under exclusive license to Springer Nature Singapore Pte Ltd. 2024
F. Hassan et al. (Eds.): AsiaSim 2023, CCIS 1911, pp. 177–193, 2024.
https://doi.org/10.1007/978-981-99-7240-1_14

of growth of the maritime industry, greenhouse gas (GHG) emissions could increase to 10% by 2050 [2]. Hence, this prompts an increased interest in investigating alternatives to conventional diesel-powered ships and implementing cleaner technology, such as green or low-carbon ships.

Understanding the carbon emissions linked to various ship types and improving shipping routes can dramatically reduce emissions. The A* algorithm, a popular pathfinding method in computer science, presents a promising approach for choosing the best routes while considering heuristic decision-making. This method was used by Zhang et al. [3] with the introduction of Bezier Curves to determine shorter paths, in a more effective manner. Ari et al. [4] utilized the A* algorithm along with another technique called 8-adjacency integer lattice discretization to find quicker routes in cold waters, increase fuel efficiency and decrease the environmental impacts caused by the ship.

The A* algorithm is a method for finding the best path between two points. It considers both the distance already covered and the anticipated distance to the destination. The algorithm begins at the starting position and scans the surrounding areas. It predicts how much more it would cost to travel to the destination from each nearby location after computing the cost of getting there. Based on these calculations, it chooses the closest place that has the most potential, then it continues the procedure until it either reaches the destination or runs out of options. The A* algorithm effectively determines the ideal path by considering both the actual distance travelled and the anticipated distance to the destination.

In this paper, we will investigate the fuel consumption and carbon emission of ships with different types of engines (green vs brown) when navigating across a route produced by A* routing algorithms while considering different windspeed scenarios.

2 Methodology

Simulations will be conducted in FlexSim software (http://www.flexsim.com). FlexSim is a powerful simulation software used to model and optimize complex systems. The goal is to calculate the carbon emissions from container vessels based on different weather scenarios. The average time taken for the ships to travel from one port to another will be collected. When faced with different wind conditions, this would result in noticeable speed loss. With extended time out at sea due to these weather conditions as compared to calm seas, while producing a steady engine power output, the speed of the vessels may vary during the voyage. Allowing an increase in engine output to compensate for the speed loss will increase fuel consumption. Hence, comparisons will be made with regard to time taken, with and without speed loss compensation, and ultimately calculating the carbon emissions for the different scenarios. This simulation will work in tandem with the built-in A* algorithm of FlexSim to ensure optimized navigation out at sea.

2.1 Vessel Particulars

5000 twenty-foot equivalent unit (TEU) container ships with the particulars listed in Table 1 will be used to run the simulations. Five similar ships have been modelled to find the average outcome of the different weather conditions.

Table 1. Vessel Particulars.

Description	Value
Length Overall, L_{OA} (m)	289.56
Length between Perpendiculars, L_{pp} (m)	280
Beam (m)	32.31
Draft (m)	12.04
Block Coefficient, C_B	0.7
Displacement (MT)	76246.43
Gross Tonnage (MT)	70927.95
Design Speed, v_{ref} (knots)	24.6
Design Speed, v_{ref} (m/s)	10.66
Optimal Speed, v (knots)	18.8
Optimal Speed, v (m/s)	9.67
Froude Number, F_n	0.18

Design speed, v_{ref}, is based on [1] as the average speed for which container ships of 5000–7999 TEU category are designed. Optimal Speed, v, is based on the same report, where the optimal power consumption and carbon emissions are taken into consideration for the size of the specific category of ships. The calculations from here on are based on optimal speed.

2.2 Route Information

The route from Devonport, Australia to Melbourne, Australia has been selected, passing through the Roaring Forties. The Roaring Forties is a term used to describe a region of strong westerly winds found in the Southern Hemisphere, specifically between the latitudes of 40 and 50°. The interaction between the rotation of the Earth, atmospheric pressure systems, and the form of the Earth's surface causes these winds, which are renowned for their power and consistency. Windspeeds in this region range from 15 to 35 knots, clocking in at 8 on the Beaufort Scale, at its peak.

2.3 Sea State

'Sea state' describes the physical state of the ocean's surface. This includes the size, frequency, and amplitude of waves produced by the wind as well as other elements such as swells and currents. It is a crucial factor in weather prediction, marine operations, and determining the safety of maritime activities.

The Beaufort Scale is a widely used method of determining and reporting the sea state. The scale ranges from 0 to 12, according to the current windspeed. Accompanying each Beaufort Number (BN) may also be a visual description of the state of the sea and the land (Table 2). For example, a BN of 8 represents a gale condition with a wind speed of 28–33 knots. For this simulation, the Beaufort scale is taken into consideration.

Table 2. Beaufort Scale.

Beaufort Number (BN)	Description	Wind Speed (knots)	Beaufort Number (BN)	Description	Wind Speed (knots)
0	Calm	<1	7	Near Gale	28–33
1	Light Air	1–3	8	Gale	34–40
2	Light Breeze	4–6	9	Strong Gale	41–47
3	Gentle Breeze	7–10	10	Storm	48–55
4	Moderate Breeze	11–16	11	Violent Storm	56–63
5	Fresh Breeze	17–21	12	Hurricane	64+
6	Strong Breeze	22–27			

2.4 Wind Speed Data Collection

The windspeed is collected from the European Centre for Medium-Range Weather Forecasts (ECMWF), an organization that specializes in weather prediction and climate research, via Windy.com. Figure 1 shows the map of the route which is divided into a 15x15 grid with cells labelled from 1–15 and A-O.

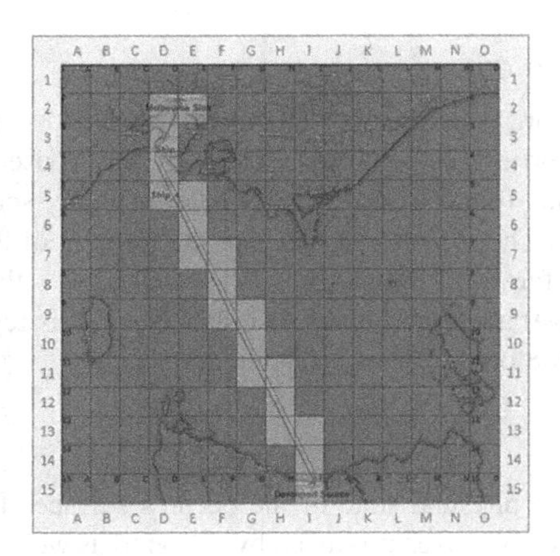

Fig. 1. FlexSim A* Simulation Grid.

The basic simulation is run to observe the route taken by the A* algorithm. Only the top windspeeds of the hour for each of the intercepted cells are recorded, for 7 July 2023 to 8 July 2023. Windspeeds are categorized into their respective BN.

2.5 Estimation of Ship Speed Loss Based on Beaufort Number

While a ship is sailing in real sea conditions, it may face irregular waves and changing windspeeds. This will add to any resistance already encountered. To forecast the unintentional speed loss of a ship brought on by erratic wind and waves, Kwon [5] suggested an approximation approach for calculating the resistance added by wind and waves. This method is easy to use, practical, highly adaptable, and quick because it does not require specific ship specifications such as ship hull lines. Due to reasons such as confidentiality, obtaining said hull lines is not easy. Kwon's methods will be employed in this study to estimate the speed loss from fast winds. The method uses the following formulas:

$$F_n = v_1/\sqrt{L_{pp} \times g} \tag{1}$$

$$\Delta v = v_1 - v_2 \tag{2}$$

$$\frac{\Delta v}{v_1}100\% = C_\beta C_U C_{Form} \tag{3}$$

Where F_n is the Froude Number; v_1 is the ship speed in still water (m/s); L_{pp} is the length between perpendiculars (m); g is the gravitational acceleration (m/s^2); Δv is the involuntary speed loss (m/s); v_2 is the actual ship speed in wind and waves (m/s); C_β is the direction reduction coefficient given in Table 3, which is related to the encounter angle (Fig. 2) and the Beaufort Number (BN) (Table 2); C_U is the speed reduction coefficient, as shown in Table 4; C_{Form} is the ship form coefficient; and ∇ is the displacement (m^3), as shown in Table 5. With the coefficients calculated, the respective decrease in ship speed is seen in Table 6 based on the optimal speed, v.

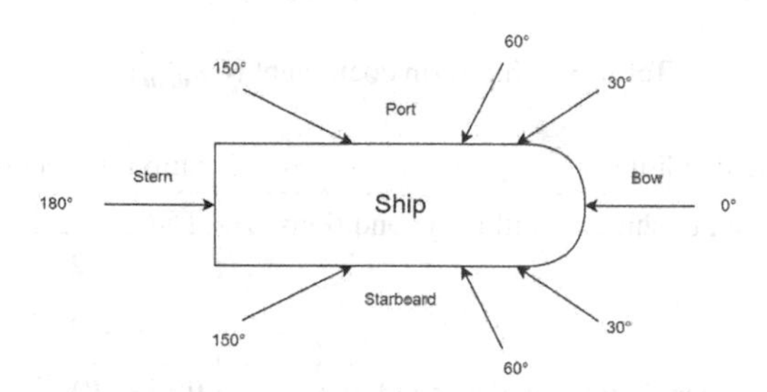

Fig. 2. Ship wind encounter angle.

2.6 Simulation Conditions

Four scenarios are run for the simulation. For each simulation, the area is divided into a 15 x 15 grid, with a total of 225 cells. However, only cells that the ships intercept while travelling from one port to another, are considered. These cells will then change colors, according to their respective windspeeds, and the Beaufort Number (BN), attached to

Table 3. Direction reduction coefficient (C_β).

Encounter Direction	Encounter angle (β)	Direction reduction coefficient (C_β)
Head sea	0 –30 °	$2C_\beta = 2$
Bow sea	30 –60 °	$2C_\beta = 1.7 - 0.03(BN - 4^2)$
Beam sea	60 –50 °	$2C_\beta = 0.9 - 0.06(BN - 6^2)$
Following sea	150–180 °	$2C_\beta = 0.4 - 0.03(BN - 8^2)$
For this simulation, all wind encountered will be assumed to be from head sea		

Table 4. Speed reduction coefficient (C_U).

Block Coefficient (C_B)	Ship Loading Conditions	Speed Reduction coefficient (C_U)
0.55	Normal	$1.7 - 1.4F_n - 7.4(F_n)^2$
0.60	Normal	$2.2 - 2.5F_n - 9.7(F_n)^2$
0.65	Normal	$2.6 - 3.7F_n - 11.6(F_n)^2$
0.70	Normal	$3.1 - 5.3F_n - 12.4(F_n)^2$
0.75	Full Load or Normal	$2.4 - 10.6F_n - 9.5(F_n)^2$
0.80	Full Load or Normal	$2.6 - 13.1F_n - 15.1(F_n)^2$
0.85	Full Load or Normal	$3.1 - 18.7F_n + 28.0(F_n)^2$
0.75	Normal	$2.6 - 12.5F_n - 13.5(F_n)^2$
0.80	Normal	$3.0 - 16.3F_n - 21.6(F_n)^2$
0.85	Normal	$3.4 - 20.9F_n + 31.8(F_n)^2$

Table 5. Ship form coefficient (C_{Form}).

Ship type and loading conditions	Ship form coefficient (C_{Form})
All ships (except container ships) in full load conditions	$\frac{0.5BN+BN^{6.5}}{2}$ $2.7\times\nabla^{3}$
All ships (except container ships) in ballast conditions	$\frac{0.7BN+BN^{6.5}}{2}$ $2.7\times\nabla^{3}$
Container ships in normal loading conditions	$\frac{0.5BN+BN^{6.5}}{2}$ $22\times\nabla^{3}$

those windspeeds. These cells will change colors for every real-time hour that passes. Figure 3 shows the FlexSim map configuration for the four scenarios.

Table 6. Involuntary speed loss and actual ship speed in respective BN, with operating speed of 18.8 knots.

Beaufort Number (BN)	Percent involuntary speed loss ($\%\Delta v$)	Actual ship speed (v_2)
0	0.00	9.67
1	0.00	9.67
2	0.00	9.67
3	0.06	9.67
4	0.36	9.64
5	1.53	9.52
6	4.99	9.19
7	13.60	8.36
8	32.39	6.54

Scenario 1 represents a calm sea scenario without any additional wind or wave resistance that the vessel may face. The ship will sail smoothly to its destination. The voyage time taken is 10 h. Scenario 2 will consider the wind effects and the vessel will slow down according to the involuntary speed loss it experiences according to the respective Beaufort force of the cell that it passes through in the simulation. Additional

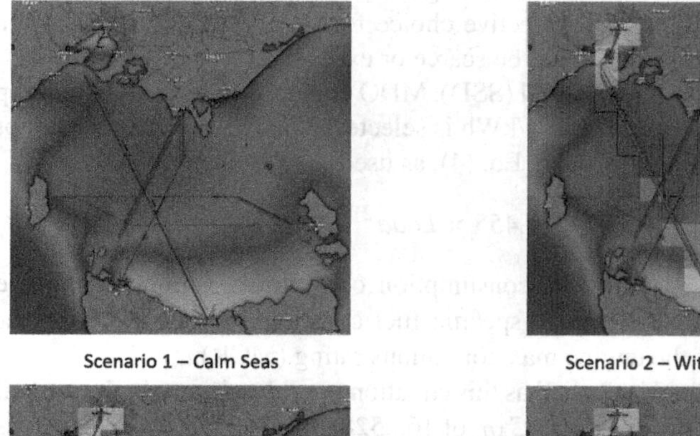

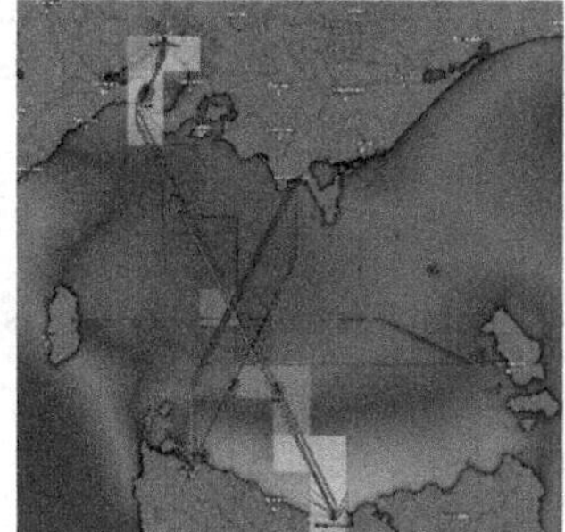

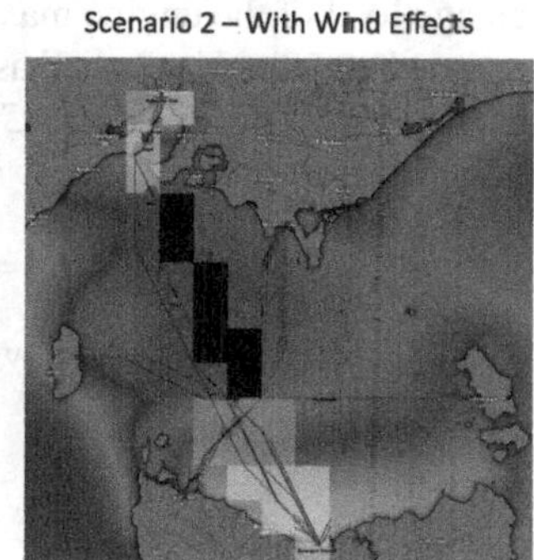

Scenario 4 - With wind effects and increased speed while avoiding BN8 cells in black

Fig. 3. Simulation Scenarios.

time will elapse without any speed increase to make up for the added resistance. In scenario 3, speed is increased to make up for the loss time to still ensure the voyage time is about 10 h. With the speed increase, the ship will no longer operate at optimum speed. The engine rating remains the same and is used for all conditions. Increasing the speed will affect the power demand, hence consuming more fuel. Scenario 4 is similar to Scenario 3, but the vessel will route to neighbouring cells that do not reach the BN8, or windspeeds above 34 knots, using A* algorithm. Only cells of up to BN7 are allowed for the ship to travel, observing the fuel consumption from added distance travelled, as opposed to increasing speed to keep up with high wind speeds.

3 Engine Fuel Consumption Calculations

Two types of engines and their emissions are to be calculated and compared. A marine diesel oil (MDO) engine and a dual-fuel engine with a 95-5 Ammonia-MDO ratio. Values used will be based on [1], where the average values for most parameters can be found.

3.1 MDO Engine

Marine vessels frequently use MDO engines as it is broadly available and affordable. MDO engines offer adaptability, dependable performance, and the capacity to manage a range of load circumstances. MDO engines can still have pollution control technology installed in them even if they are not as ecologically beneficial as alternative fuels. MDO engines are a practical and cost-effective choice for marine propulsion, especially in areas where other fuel options could be scarce or expensive.

For this study, a slow-speed diesel (SSD), MDO engine, built post-2001, and specific fuel consumption, SFC_{Base}, of 165 g/kWh is selected. The specific fuel consumption at specific loads can be found by using Eq. (4), as used in [1]:

$$SFC_{ME} = SFC_{Base} \times (0.455 \times Load^2 - 0.71 \times Load + 1.28) \tag{4}$$

where SFC_{ME} is the specific fuel consumption of the main engine at the specified load (g/kWh); SFC_{Base} is the base specific fuel consumption (g/kWh); and $Load$ is the percentage load of the engine max continuous rating (MCR).

An assumed 80% load is applied as this equation provides the engine's most efficient load at that capacity. This gives a SFC_{ME} of 165.528 g/kWh. The total MDO consumed by the main engine at optimal speed, $FC_{v,ME}$, is given as:

$$FC_{v,ME} = SFC_{ME} \times W_{ME,Ship} \tag{5}$$

Where $W_{ME,Ship}$ is the propulsive power demand of the main engine. $W_{ME,Ship}$ is given as:

$$W_{ME,Ship} = \frac{CF_{Ship} \times W_{REF} \times (\frac{t}{t_{REF}})^{0.66} \times (\frac{v}{v_{REF}})^3}{CF_{Weather} \times CF_{Fouling}} \tag{6}$$

Where W_{REF} is the rated average main engine power at MCR (kW) as per [1] Table 81 in Annex N; t, t_{REF} is the actual draught and design draught of the ship, which is assumed as 1 when there is no data of the value; v, v_{REF} is the actual speed and design speed of the ship, where this ratio is also set to 1; and CF_{Ship} is ship-specific correction factor which is based on Table 44 in [1].

$CF_{Weather}$ is the correction factor for the influence of wind and waves, which is based on Table 44 in [1]. As this study will be conducted for several scenarios, this value is set a 1 for Scenario 1 to serve as a control. This value is set at 0.867 for Scenario 2, 3 and 4 as this accounts for an average of 15% power increase for container vessels. $CF_{Fouling}$ is the correction factor for the influence of hull fouling/hull roughness as per Table 44 in [1].

3.2 Dual-Fuel 95-5 Ammonia-MDO Engine

Dual-fuel engines provide a variety of benefits, including greater efficiency, reduced emissions, compliance with laws, operational flexibility, and noise reduction. They offer cost-effectiveness and environmental advantages by allowing ships to use cleaner-burning natural gas while also having the ability to switch to diesel as needed. The dual-fuel usage is necessary for a 'green' engine as the diesel oil is still required to be used as a pilot fuel to begin combustion.

This type of dual-fuel engine is selected as studied by Zincir B. [6] to be a promising alternative fuel as ammonia has low carbon and sulphur content within its chemical structure. Ammonia has also been used by internal combustion engines and fuel cells in the past as fuel [7].

Due to its special characteristics, ammonia has drawn interest as a potential fuel for engines. First, ammonia has a high energy density, which enables it to store and release a lot of energy during combustion [8]. In engines, where energy density is essential for effective power generation or propulsion, this qualifies it for usage in certain systems. Second, burning ammonia does not release carbon dioxide (CO_2) since it has a low carbon concentration. As a result, it presents a compelling replacement for fossil fuels since it can lessen greenhouse gas emissions and slow down climate change. We can transition to cleaner, more sustainable energy and transportation systems by using ammonia as an engine fuel.

Ammonia, (NH_3), is a substance made of nitrogen and hydrogen. It is a gas that has no color and a strong stench that is frequently utilized in industrial processes for things such as fertilizers, cleaners, and refrigeration systems. There are various categories of ammonia known as blue ammonia, brown ammonia, and green ammonia. However, these names refer to the different ways to produce or use ammonia, each with its unique properties and ramifications.

To manufacture blue ammonia, a method known as carbon capture and storage (CCS) or carbon capture, utilization, and storage (CCUS) is used [9]. Through a procedure known as steam methane reforming, carbon dioxide (CO2) emissions from industrial sources, such as power plants or chemical factories, are captured and combined with hydrogen derived from natural gas or renewable sources. Blue ammonia is the name given to the resultant ammonia.

The "blue" in blue ammonia denotes the fact that the subterranean storage of the carbon emissions linked to ammonia manufacturing reduces greenhouse gas emissions.

Because it lessens CO2 emissions, blue ammonia is seen as a more environmentally friendly alternative to conventional ammonia production techniques.

On the other hand, green ammonia is created through the electrolysis process using sustainable energy sources like solar or wind energy. Using an electric current, electrolysis entails dividing water into hydrogen and oxygen. After being created, the hydrogen is coupled with nitrogen to create ammonia.

The utilization of renewable energy and the lack of carbon emissions during production are reflected in the term "green" in green ammonia. A potential sustainable energy transporter and method to store and move renewable energy, green ammonia has attracted interest. It can be used as a feedstock for making different chemicals or as a fuel for creating electricity.

Brown ammonia is a term used to denote ammonia made from coal or other high-carbon feedstocks rather than a specific type of ammonia. Brown ammonia has higher carbon emissions than blue or green ammonia due to the carbon-intensive nature of coal-based production methods. Compared to blue or green ammonia, it is thought to be less sustainable and environmentally benign.

Ammonia Fuel Consumption. With $FC_{v,ME}$ calculated in Sect. 3.1, this value can be used to find the equivalent amount of heat produced by MDO in the MDO engine, as shown below:

$$Q_{MDO} = FC_{v,ME} \times LHV_{MDO} \times 1000 \tag{7}$$

Where Q_{MDO} is the amount of heat produced by MDO in the MDO engine (MJ); and LHV_{MDO} is the lower heating value of MDO (MJ/kg), which is 42.5 MJ/kg, taken from Table 1 of a study done by Nader R. [7] to compare different alternative fuel properties.

The amount of ammonia consumed can be found by using 95% of the total heat produced and working backwards by dividing by the LHV of ammonia (MJ/kg).

$$FC_{v,ME,Ammonia} = \frac{Q_{MDO} \times 95\%}{1000 \times LHV_{Ammonia}} \tag{8}$$

Where $FC_{v,ME,Ammonia}$ is the amount of ammonia fuel consumed at the operating speed, by the main engine (MT); and $LHV_{Ammonia}$ is given as 18.5 MJ/kg from Table 1 of the same study by Nader R. [7].

MDO Fuel Consumption. The amount of MDO consumed in this engine is 5% of $FC_{v,ME}$, found in Sect. 3.1.

$$FC_{v,ME,MDO} = FC_{v,ME} \times 5\% \tag{9}$$

Where $FC_{v,ME,MDO}$ is the amount of MDO fuel consumed at the operating speed by the main engine (MT).

4 Carbon Emission Calculations

Burning fossil fuels in marine engines results in the release of carbon dioxide and other greenhouse, which causes carbon emissions in shipping. Shipping accounts for around 90% of world trade [8], making it vital but carbon intensive. Main propulsion, auxiliary power, and fuel quality are significant contributors. Improved vessel efficiency, the exploration of alternative fuels like ammonia and biofuels, the development of cutting-edge technology like electric propulsion and wind-assisted systems, and the implementation of laws by organizations like the IMO are all efforts to minimize emissions. These steps are intended to reduce climate change and attain carbon emission 0 levels. The shift of the shipping industry to greener methods is crucial for environmental sustainability.

This study will investigate the following emissions: Carbon Dioxide, (CO_2), Sulphur Oxides, (SO_x), Nitrogen Oxides, (NO_x), Methane, (CH_4), Carbon Oxide, (CO) and Nitrous Oxide, (N_2O).

As ammonia does not produce carbon and sulphur [9], CO_2, CO and SO_x will be excluded from those calculations. There is not enough evidence for methane emission factors for ammonia, hence, the calculation is omitted from the total. Methane emission is the lowest among the stated gases coming in at less than 5kg for all scenarios, as compared to several tons for other gases, and it is reasonable to omit ammonia-based methane emissions from the calculations.

4.1 Carbon Dioxide (CO_2) Emissions

CO_2 Is a fuel-based emission that is dependent on the amount of fuel consumed. The amount of CO_2 emitted can be calculated as:

$$CO_2\ \ emitted = Total\ \ Fuel\ \ Consumed \times EF_{f,CO_2} \tag{10}$$

Where $CO_2\ \ emitted$ is in tonnes; $Total\ \ Fuel\ \ Consumed$ is in tonnes; and EF_{f,CO_2} is the fuel-based emission factor for CO_2, which is 3.206 (g of CO_2/g of fuel) for MDO, based on Table 21 of [1].

4.2 Sulphur Oxides (SO_x) Emissions

SO_x Is a fuel-based emission that is dependent on the amount of fuel consumed. The amount of SO_x emitted can be calculated as:

$$SO_x\ \ emitted = Total\ \ Fuel\ \ Consumed \times EF_{f,SO_x} \tag{11}$$

Where $SO_x\ \ emitted$ is in tonnes; $Total\ \ Fuel\ \ Consumed$ is in tonnes; and EF_{f,SO_x} is the fuel-based emission factor for SO_x, which is given as:

$$EF_{f,SO_x} = 2 \times 0.97753 \times S \tag{12}$$

Where S is the fuel sulphur content fraction given as (g of SO_x/ g of fuel) and they are presented in Table 22 of [1] as percentages. The value used here will be based on the latest available year recorded, which is at 7%.

4.3 Nitrogen Oxides (NO_x) Emissions

NO_x Is an energy-based emission that is dependent on the amount of power output of the vessel. The amount of NO_x emitted can be calculated as:

$$NO_x emitted = \frac{W_{ME,Ship} \times EF_{e,NO_x}}{1000000} \tag{13}$$

Where NO_xemitted is in tonnes; $W_{ME,Ship}$ is the propulsive power demand of the main engine (kWh) as calculated from Eq. (6); and EF_{e,NO_x} is the energy-based emission factor of NO_x, which is 28.2 g/kWh for 95% ammonia engine, based on the study by Zincir B. [6]. This value is 14.4 g/kWh for the MDO engine based on [1].

4.4 Methane (CH_4) Emissions

Methane is an energy-based emission that is dependent on the amount of power output of the vessel. The amount of methane emitted can be calculated as:

$$CH_4emitted = \frac{W_{ME,Ship} \times EF_{e,CH_4}}{1000000} \tag{14}$$

Where CH_4emitted is in tonnes; and EF_{e,CH_4} is the energy-based emission factor of methane, which is 0.01 for MDO engines based on Table 55 of [1].

4.5 Carbon Oxide (CO) Emissions

CO Is an energy-based emission that is dependent on the amount of power output of the vessel. The amount of CO emitted can be calculated as:

$$COemitted = \frac{W_{ME,Ship} \times EF_{e,CO}}{1000000} \tag{15}$$

Where COemitted is in tonnes; $EF_{e,CO}$ is the energy-based emission factor of CO, which is 0.044 based on Table 57 of [1], for an MDO engine.

4.6 Nitrous Oxide (N_2O) Emissions

N_2O Is an energy-based emission that is dependent on the amount of power output of the vessel. The amount of N_2O emitted can be calculated as:

$$N_2Oemitted = \frac{W_{ME,Ship} \times EF_{e,N_2O}}{1000000} \tag{16}$$

Where N_2Oemitted is in tonnes; and EF_{e,N_2O} is the energy-based emission factor of N_2O, which is 1.95 based on [6] for a 95% ammonia engine.

5 Results and Discussions

5.1 Voyage Duration

Each scenario is run with five vessels with the same specifications. The time recorded for calculations is only the voyage time, which the point from which the vessel is loaded and sailing towards the next port. No loading, port, or bunkering times are considered. The average time taken for these five ships is taken for the calculation of the following sections: carbon emissions and fuel costs. Both metrics require the time taken in hours. Table 7 shows the time taken for each of the scenarios.

Table 7. Time taken (s) for each scenario.

	Scenario 1	Scenario 2	Scenario 3	Scenario 4
Ship 1	39247	40632	35826	36700
Ship 2	35734	41069	35389	36700
Ship 3	35206	41069	35826	36700
Ship 4	35114	41506	35826	36700
Ship 5	34699	41069	35826	37137
Average (s)	36000	41068.96	35738.73	36787.30
Average (hours)	10.00	11.41	9.93	10.22

All recorded times are obtained from FlexSim while the ship is in the loaded condition only. There are also two more voyages consisting of eight other similar vessels to simulate real-world marine traffic and obstruction. This will prompt the A* algorithm to search for other routes to prevent collision with neighbouring vessels.

The time taken for each ship may vary slightly as they face different weather conditions that reduce their speed differently. This is most obvious in scenario 1 and 2 as the introduction of weather effects significantly slow down the vessel. In scenario 3, speed is increased to make up for the time loss of 1.41 h. The decrease in time taken compared to scenario 1 may be because despite being slowed down by a certain percentage at each cell, the vessel is still moving at an average speed higher than that it did in scenario 1. This may allow it to clear the route slightly faster. Scenario 4 took slightly longer by 0.29 h as compared to scenario 3. This is only a 2.9% increase in time taken despite manoeuvring away from BN8 cells. The time loss is not substantial and routing this may be unnecessary. However, this is due to the cells being only a small area. Should the weather conditions of BN8 appear in multiple cells in a large area, a detour may be the preferred method to save on fuel costs as well.

5.2 Fuel Costs

Listed in Table 8 below is the consumption of fuel for each scenario, for each type of fuel.

Table 8. Fuel consumption (MT).

Fuel	Scenario 1	Scenario 2	Scenario 3	Scenario 4
MDO Engine				
MDO	42.35	55.74	72.05	74.15
Dual-fuel 95-5 Ammonia-MDO Engine				
MDO	2.12	2.79	3.60	3.71
Ammonia	92.43	121.64	157.24	161.83

Table 9. Cost of fuel.

Fuel Type	Cost (USD/t)
MDO	579.5
Brown ammonia [6]	230
Blue ammonia [6]	350
Green ammonia [6]	670

Table 10. Cost of each voyage (USD) for each scenario.

	Scenario 1	Scenario 2	Scenario 3	Scenario 4
MDO Engine				
MDO	24,543.06	32,299.46	41,751.03	42,970.35
Dual-fuel 95–5 Ammonia-MDO Engine				
MDO and Brown ammonia	22,486.21	29,592.57	38,252.05	39,369.18
MDO and Blue ammonia	33,577.89	44,189.58	57,120.48	58,788.65
MDO and Green ammonia	63,155.70	83,114.94	107,436.30	110,573.92

As there are different methods of procuring ammonia, the cost for each method is different. Table 9 takes into consideration the cost for each type of ammonia, brown, blue, and green.

Table 10 shows that for each of the scenarios, there is an upward trend for each combination of fuel types. This is expected as the scenarios are simulated with increased resistances as they go along, increasing time taken and fuel consumption.

MDO and brown ammonia have the cheapest cost among the dual-fuel engine variations. Despite using only 5% MDO in the green engine, the cost is only 8.38% cheaper than that of using an MDO-only engine. The large consumption of ammonia required for the engine, makes the cost much higher than one would expect from a green fuel. The expenses only increase as blue ammonia is 36.81% more expensive and green ammonia is 157.33% more expensive as compared to using MDO only. This may be seen as an obstacle in the shipping industry for businesses to use green fuel, as it increases their operational costs substantially.

However, MDO and brown ammonia is the clear winner in terms of cost savings. The drawback is in the method of obtaining the ammonia. As it is classified as brown ammonia, the process is synonymous with being less environmentally friendly. The carbon emissions due to the carbon-intensive nature of the coal-based production process and very high and unfavourable.

5.3 Carbon Emissions

The results of the carbon emissions for the two different types of engines are shown in Table 11:

Table 11. Results of carbon emissions.

Scenario	Power demand, $W_{ME,Ship}$ (kWh)	Operating Speed (knots)	Emissions (MT)					
			CO_2	SO_x	NO_x	CH_4	CO	N_2O
MDO Engine								
1	25586.08	18.80	135.78	0.06	3.68	0.0026	0.0113	0.0077
2	29511.05	18.80	178.69	0.08	4.85	0.0034	0.0148	0.0101
3	43832.18	21.45	230.98	0.10	6.27	0.0044	0.0192	0.0131
4	43832.18	21.45	237.73	0.10	6.45	0.0045	0.0197	0.0134
Dual-fuel 95–5 Ammonia-MDO Engine								
1	25586.08	18.80	6.79	0.0029	7.22	-	-	0.4989
2	29511.05	18.80	8.93	0.0038	9.50	-	-	0.6566
3	43832.18	21.45	11.55	0.0049	12.27	-	-	0.8487
4	43832.18	21.45	11.89	0.0051	12.63	-	-	0.8735

Scenario 1 saw no added wind or wave resistance and serves as the control for this study. Operating at 18.8 knots, which is the optimal speed for a 5000 TEU container vessel, the control completed its journey in 10 h, the average time taken to complete the voyage. The correction factor for wind and waves was set at 1 for this condition to observe the base emission levels across the board.

Scenario 2 had an increase in power demand as the correction factor for wind and waves is set at 0.876, according to [1], to introduce the weather resistance that the ship will face. Ship operating speed remains the same as Scenario 1 at 18.8 knots. The vessel experienced an increase of 14.1% in the time taken to reach its destination, from 10 h to 11.41 h. This led to an increase of 31.6% for all emissions.

Scenario 3 increases its speed by 14.1% to 21.45 knots to compensate for the time loss from the previous condition. An increase in operating speed means that there is an increase in power demand, hence an increase in fuel consumption. Ultimately this leads to an increase in emissions. Compared to Scenario 1, there is an increase of 70.11% for all emissions. The time taken for this scenario is 9.93 h.

Scenario 4 is similar to Scenario 3, but the ship completely avoids cells that experience BN8 or higher. This simulation is run to observe the difference in taking a longer route around certain areas that have a considerable involuntary ship speed loss. When ships pass through BN8, there is a speed loss of 32.39%. Avoiding this area to take a longer route may be favourable if the neighbouring areas are of much calmer sea states. For the case of Scenario 4, the neighbouring cells were BN 6 and 7, still applying a speed loss of 4.99% and 13.6% respectively. Overall, the ship experiences increased travel time, fuel consumption and carbon emissions. The voyage was 10.22 h, 2.9% slower than Scenario 3. Fuel consumption is 75.08% higher than that of Scenario 1.

Several observations can be made with the findings. CO_2 and SO_x emissions for the dual-fuel ammonia engine are significantly lower as expected since there is no carbon and sulphur are produced from the combustion of ammonia. The difference is 5% that of the MDO engine. NO_x emissions are 96.19% higher as compared to the MDO engine as there is naturally more nitrogen content in ammonia. N_2O emissions are 6679% higher than the dual-fuel engine as compared to the MDO engine. As there has not been substantial research done on the methane emission factor for ammonia engines, methane emissions for the ammonia component of the dual-fuel type engine are omitted entirely.

6 Conclusions

This study has observed the carbon emissions, fuel consumption and costs of 5000 TEU container vessels travelling through the roaring forties, while experiencing windspeeds of up to Beaufort Number 8, at 34–40 knots. Values are heavily referenced from the Fourth IMO GHG 2020 Report for the average values of the engine rating, engine output, vessel particulars and characteristics, of a container vessel. The carbon emissions are calculated and compared in a comprehensive view, with various scenarios of increasing resistance. With each differing scenario, windspeeds are considered, engine output is increased for speed loss compensation due to weather effects, and the A* algorithm is utilized to optimize better routes to avoid specific areas of high wind speeds. The outcomes and differences are measured and compared as seen in the following:

- The time taken to complete the journey when alternative routes are considered to avoid high-speed winds, was negligible. The areas that are avoided are too small for the data sample collected. This is due to the limitations of collecting weather data accurately and far into the future. This study is limited to weather data for 2 weeks ahead of time only. Capturing data on wind gusts may not be readily available. However, future work may simulate larger storms with areas of higher BN and wind speeds.
- This study is limited to 2 types of engines, MDO only and a 95-5 ammonia-MDO engine. Emissions that are accounted for are as accurate as some factors such as methane emission factors for ammonia, which is not researched heavily, to provide more accurate emission calculations.
- Further study can compare the differences between various compositions of ammonia and MDO engines to eventually find the perfect ratio for such green engines. As it stands, the 95-5 ratio of ammonia and MDO is far too expensive to be used willingly by ship owners and businesses.
- Carbon emissions are significantly reduced with the use of this green engine simply due to the fact of lack of carbon and sulphur content. However, nitrogen-based emissions are significantly increased due to the natural nitrogen content of ammonia. This fact is up for consideration to ship users. Despite this, overall emission levels are drastically lower, making ammonia-based engines highly environmentally friendly in the grand scheme of carbon emissions.

References

1. "Fourth Greenhouse Gas Study 2020," International Maritime Organization. https://www.imo.org/en/ourwork/Environment/Pages/Fourth-IMO-Greenhouse-Gas-Study-2020.aspx. Accessed 13 July 2023
2. "Greenhouse gases," Transport & Environment. https://www.transportenvironment.org/challenges/ships/greenhouse-gases/. Accessed 13 July 2023
3. Zhang, Y., Wen, Y., Tu, H.: A method for ship route planning fusing the ant colony algorithm and the A* search algorithm. IEEE Access **11**, 15109–15118 (2023). https://doi.org/10.1109/access.2023.3243810
4. Ari, I., Aksakalli, V., Aydogˇdu, V., Kum, S.: Optimal ship navigation with safety distance and realistic turn constraints. Eur. J. Oper. Res. **229**(3), 707–717 (2013). https://doi.org/10.1016/j.ejor.2013.03.022
5. Kwon, Y.J.: Speed loss due to added resistance in wind and waves. Naval Arch. 14–16, March 2008
6. Zincir, B.: Environmental and economic evaluation of ammonia as a fuel for short-sea shipping: a case study. Int. J. Hydrogen Energy **47**(41), 18148–18168 (2022). https://doi.org/10.1016/j.ijhydene.2022.03.281
7. Bomelburg, H.J.: Use of ammonia in energy-related applications. Plant/Oper. Progr. **1**(3), 175–180 (1982). https://doi.org/10.1002/prsb.720010311
8. Chai, W.S., Bao, Y., Jin, P., Tang, G., Zhou, L.: A review on ammonia, ammonia-hydrogen and ammonia-methane fuels. Renew. Sustain. Energy Rev. **147**, 111254 (2021). https://doi.org/10.1016/j.rser.2021.111254
9. Wang, F., et al.: Current status and challenges of the ammonia escape inhibition technologies in ammonia-based CO2 Capture Process. Appl. Energy **230**, 734–749 (2018). https://doi.org/10.1016/j.apenergy.2018.08.116

A Model Validation Method Based on Convolutional Neural Network

Ke Fang(✉) and Ju Huo

Control and Simulation Center, Harbin Institute of Technology, Harbin, China
fangke@hit.edu.cn

Abstract. Conventional model validation methods analyze outputs similarity between simulation and real world with same inputs. However, it is hard to guarantee the condition in practice. In order to solve the problem, a method based on convolutional neural network (CNN) is proposed, including data preprocessing, activation function, loss function, and optimization algorithm. Meanwhile, a CNN is established for model validation training and test. Finally, a case study of model validation is presented. The result shows that, the method can obtain 98.5% validation accuracy under the condition of same inputs, and can discriminate credibility levels with different inputs as well.

Keywords: Model Validation · Convolutional Neural Network · TIC · GRA · Different Inputs

1 Introduction

Model validation is the technique to obtain the simulation credibility, which is critical to the application [1]. The most objective way is comparing the outputs similarity between simulation and real world (or reference) [2]. There are many validation methods based on this point, such as hypothesis test, TIC (Theil Inequality Coefficient) [3], GRA (Grey Relational Analysis) [4], spectral estimation [5] etc. These methods require same inputs between simulation and real world.

However, some inputs to the real world are not possible to manipulate, such as wind direction, atmosphere humidity etc. If these factors are sensitive to the outputs, conventional validation methods may result in wrong credibility conclusions, usually as negative ones. This problem is born with similarity analysis techniques and cannot be solved in the range of goodness-of-fit indicators.

Although outputs vary when inputs are different, the model does not change. There should be some inner way to reveal the similarity between simulation and real world. It requires analysis to model behavior pattern but not output quantity comparison. Convolutional neural network (CNN) [6] provides the ability of deep machine learning and category matching. If we can use experts knowledge or "same input" validation results as priori experience to perform training, it is possible to develop a CNN to evaluate the model credibility level regardless of the input condition.

© The Author(s), under exclusive license to Springer Nature Singapore Pte Ltd. 2024
F. Hassan et al. (Eds.): AsiaSim 2023, CCIS 1911, pp. 194–203, 2024.
https://doi.org/10.1007/978-981-99-7240-1_15

2 Model Validation

The purpose of model validation is to prove the model behaves acceptably the same as the real world. One practical way is to use proportional measurements to compare the outputs between the model and real world. See the formula below:

$$C = 1 - \frac{||O_R - O_S||}{||O_R||}|_{I_R, I_S} \tag{1}$$

Where C is the model credibility, O_R, I_R, O_S, I_S are the real world and simulation outputs and inputs. $||$ is the error norm, which varies from similarity analysis techniques. Conventional model validation methods require $I_R = I_S$.

Another way to obtain model credibility is to use enumerated measurements to fit the simulation outputs into most matched credibility level which is priorly arranged. See the formula below:

$$\begin{cases} L = f(O_R, O_S) = [l_1, l_2, ..., l_n] \\ P = [p_1, p_2, ..., p_n] \to L \\ C = \max\limits_{i=1,2,...,n} l_i \to p_i \end{cases} \tag{2}$$

Where L is the matching vector of P, which is the enumerated measurements of credibility level. L and P are one-to-one mapping. f is the matching function which can derive the vector from real world output O_R and simulation output O_S. C is the model credibility, which is granted as the enumerated level that the maximum matching element is corresponding to.

When Formula (1) is used, it requires the simulation inputs and real world inputs remain the same. While Formula (2) does not require the same inputs, instead of a pattern matching mechanism to fit the credibility into a most suitable level.

3 CNN Method for Model Validation

3.1 The CNN Solution

CNN is a deep neural network which uses specific tagged sample sets for training and application. It can extract the feature relationship between input data, so as to categorize the object. This mechanism allows to discriminate simulation and real world outputs, with no need of same inputs. Figure 1 shows the CNN solution for validation.

The procedure of the model validation method based on CNN is:

(1) Determine the possible credibility levels of the current model validation according to prior experience.
(2) Gather the model output data of each credibility level. For "credible" level, use real world outputs.
(3) Perform pre-processing to the data to meet the CNN training requirement.
(4) Use data of each level to train CNN to produce tagged credibility levels training features.
(5) Run the model and gather output data.

(6) Perform pre-processing to the data to meet CNN recognition.
(7) Calculate the result vector which contains possibility of each credibility level. If there is possibility bigger than the acceptability threshold (e.g. $\mu_t = 0.75$), accept the credibility level which the biggest possibility in the vector maps with as the model credibility. Otherwise, the model is not credible.

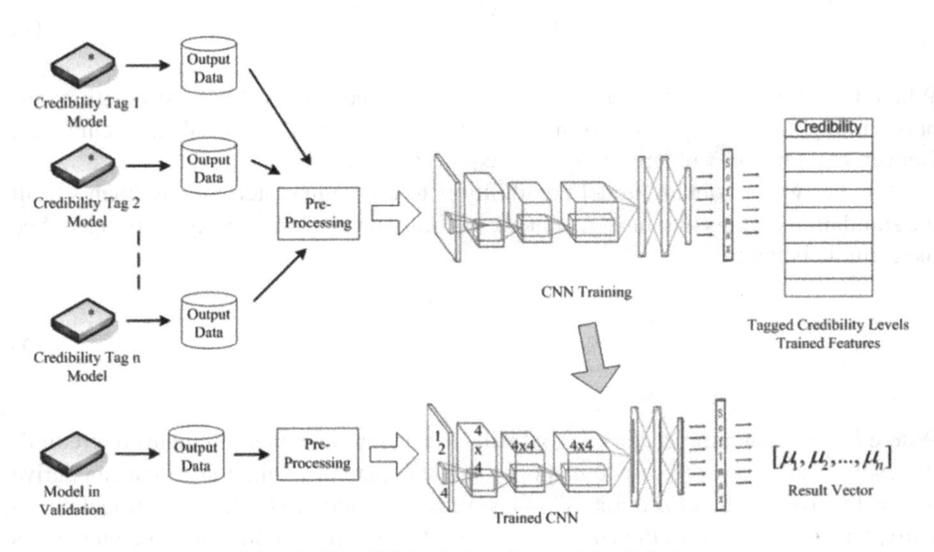

Fig. 1. The CNN solution for model validation

3.2 Data Pre-processing

The main work of data pre-processing is to rebuild the dimension of output data. Usually the model outputs are time series, as Formula (3) shows:

$$X = [x_1, x_2, \ldots, x_q]^T = \begin{bmatrix} x_1(t_1)\ x_1(t_2) \ldots x_1(t_p) \\ x_2(t_1)\ x_2(t_2) \ldots x_2(t_p) \\ \ldots \quad \ldots \quad \ldots \ldots \\ x_q(t_1)\ x_q(t_2) \ldots x_q(t_p) \end{bmatrix} \tag{3}$$

$$0 \le t_1 < t_2 < \ldots < t_p$$

Where X is the set of output time series, q is the dimension of the specific model output (e.g. $q = 3$ for spatial coordinates), $t_1, t_2 \ldots t_p$ is the sampling time stamp. So the dimension of output array is $p \times q$. In order to build the CNN, reshape the input layer to $n \times n \times q$ dimension, where $(n-1) \times (n-1) < p \le n \times n$.

In order to adapt different output dimensions, perform normalization to the reshaped data by Formula (4), which can prevent gradient explosion.

$$x^*(t) = \frac{x(t) - \min\limits_{i=1,2,\ldots,n} x(t_i)}{\max\limits_{i=1,2,\ldots,n} x(t_i) - \min\limits_{i=1,2,\ldots,n} x(t_i)} \tag{4}$$

3.3 Activation Function

Use ReLU function as the activation function for model validation's CNN. Compared with other activation functions (e.g. Sigmoid, Tanh etc.), it avoids the gradient saturation and achieves fast convergence speed.

$$f(x) = \max(0, x) \tag{5}$$

3.4 Loss Function

Use cross entropy function [7] as the loss function for model validation's CNN. It uses the category possibilities from Softmax [8] to calculate the loss of categories.

$$J = -\frac{1}{N}\sum_{k=1}^{N}\sum_{i=1}^{k} y_i \cdot \log(p_i) \tag{6}$$

Where i is the category (credibility level), y_i is the actual tag number of category i, p_i is the matching possibility to category i which is calculated by Softmax, k is the category number, N is the sample number. Formula (7) shows the Softmax function, where z_i is the output of fully connection lay CNN.

$$p_i = \frac{e^{z_i}}{\sum_{j=1}^{k} e^{z_j}} \tag{7}$$

3.5 Optimization Algorithm

Use Adam [9] algorithm as the optimization function for model validation's CNN. It uses first-order and second-order moment estimation of the gradient to dynamically adjust the learning rate for each parameter, as Formula (8) shows.

$$\begin{cases} \theta_t = \theta_{t-1} - \frac{\hat{m}_t}{\sqrt{\hat{n}_t + \varepsilon}} \cdot \eta \\ \hat{m}_t = \frac{m_t}{1-\mu^t}, m_t = \mu \cdot m_{t-1} + (1-\mu) \cdot g_t \\ \hat{n}_t = \frac{n_t}{1-\nu^t}, n_t = \nu \cdot n_{t-1} + (1-\nu) \cdot g_t^2 \end{cases} \tag{8}$$

where η is the learning rate, usually set as 0.001; m_t and n_t are the first and second order moment estimation of the gradient g_t; μ, ν, ε are the hyper parameters which are usually set as 0.9, 0.999 and 1×10^{-6} respectively; t is the iteration time.

4 Case Study

4.1 The Model and Credibility Tag

Use a dual-defense fighter plane's model to verify the method. When the fighter plane detects hostile missile attack, it launches interceptor missiles at the same time of dodging. Use interceptor 1's 2-dimension coordinates (x, y) as the output to evaluate the model credibility. Figure 2 shows the coordinate curve of interceptor missile 1 for tagged simulation model 1/2/3 and the reference.

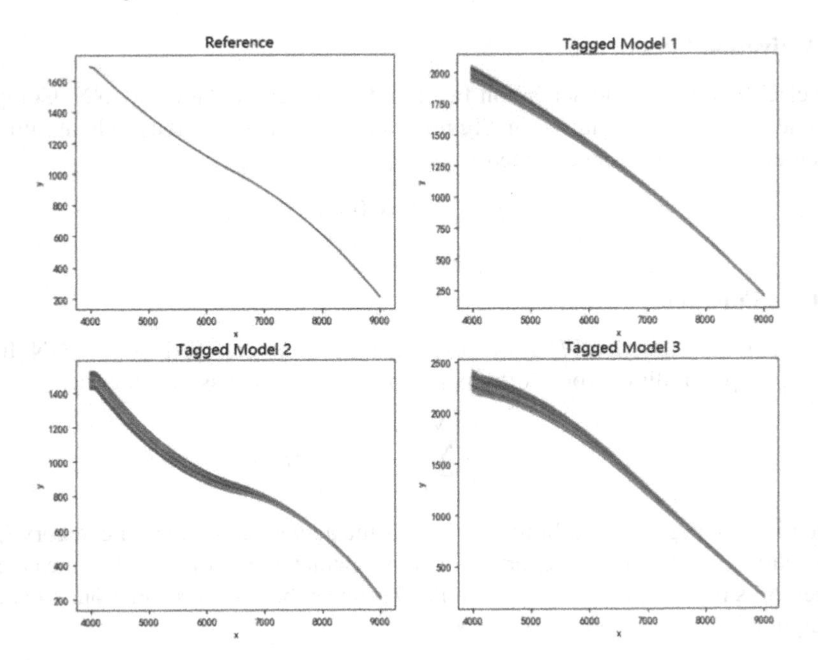

Fig. 2. Output curve of tagged simulation models and the reference

Use TIC [3] and GRA [4] to calculate the similarity between simulation outputs and the reference, and use Formula (9) to transform the results into simulation credibility. Table 1 shows the credibility result.

$$C = \begin{cases} \frac{1-C_t}{\rho_t}(\rho_t - \rho) + C_t, & \rho \in [0, \rho_t] \\ \frac{C_t(1-\rho)}{1-\rho_t}, & \rho \in [\rho_t, 1] \end{cases} \quad C = \begin{cases} \frac{1-C_t}{1-\gamma_t}(\gamma - \gamma_t) + C_t, & \gamma \in [\gamma_t, 1] \\ \frac{C_t\gamma}{\gamma_t}, & \gamma \in [0, \gamma_t] \end{cases} \tag{9}$$

Where C_t is the acceptability threshold of credibility and set as 0.75; ρ_t is the acceptability threshold of the Theil inequality coefficient and set as 0.3; γ_t is the acceptability threshold of the GRA degree and set as 0.8; ρ is the Theil inequality coefficient; γ is the GRA degree; and C is the transformed credibility.

Table 1. Credibility results by TIC & GRA.

Model	TIC ρ	TIC C	GRA γ	GRA C
1	0.113	0.906	0.741	0.695
2	0.069	0.942	0.829	0.786
3	0.184	0.847	0.653	0.612

The result shows that Model 2 has the best credibility, and Model 3 has the worst. So tag the Model 1 as "Good", Model 2 as "Excellent", and Model 3 as "Medium".

4.2 CNN Establishment and Training

Build a CNN with 7-layer structure, which contains 3 convolutional layers, 3 pooling layers, and 1 fully connection layer. The CNN structure is shown in Table 2.

Table 2. The CNN structure for the dual-defense model validation.

No	Layer	Input size	Core size	Core no	Step	Output size
0	Input	$1224 \times 1 \times 2$				$35 \times 35 \times 2$
1	Convolution	$35 \times 35 \times 2$	4×4	16	1	$32 \times 32 \times 16$
2	Pooling	$32 \times 32 \times 16$	2×2	16	2	$16 \times 16 \times 16$
3	Convolution	$16 \times 16 \times 16$	3×3	32	1	$14 \times 14 \times 32$
4	Pooling	$14 \times 14 \times 32$	2×2	32	2	$7 \times 7 \times 32$
5	Convolution	$7 \times 7 \times 32$	4×4	64	1	$4 \times 4 \times 64$
6	Pooling	$4 \times 4 \times 64$	2×2	64	2	$2 \times 2 \times 64$
7	Fully Con	$2 \times 2 \times 64$				3

Group the simulation output data with the proportion of 5:4:1 as the training set, testing set and validation set. Use cross entropy loss function and Adams optimization function, and set iteration time as 50 to train the CNN. Figure 3 shows the loss and accuracy curve of the training.

In Fig. 3, the loss curve descends quickly and progressively stabilized near 0.01. And the accuracy curve ascends quickly and stops fluctuation near 99%. Thus it is clear that the CNN can categorize the 3 tagged models efficiently.

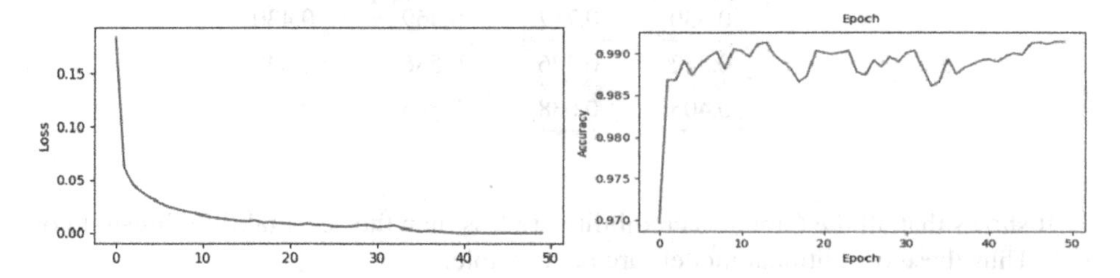

Fig. 3. The loss and accuracy curve of the CNN training.

Use the reference data as the CNN's input and verify the network. The result is [0.0988, 0.8379, 0.0632], which shows that the reference data belongs to "Excellent" category. This matches the real world behavior.

4.3 Same Input Model Validation

Use each model's last 20 series of data and additional 6 series of data in Fig. 4 as the validation data under same input condition. There are 66 series of data in total.

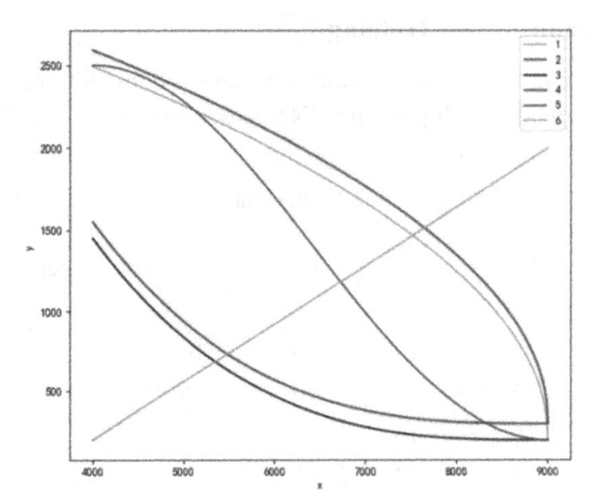

Fig. 4. Output curve of the additional output data.

Use TIC and GRA to calculate the similarity, and use Formula (9) to transform the results into simulation credibility. Table 3 shows the credibility result.

Table 3. Credibility results of additional output data.

Simulation no	TIC ρ	TIC C	GRA γ	GRA C
1	0.335	0.712	0.387	0.363
2	0.358	0.688	0.441	0.413
3	0.342	0.705	0.445	0.418
4	0.330	0.717	0.459	0.430
5	0.322	0.726	0.586	0.549
6	0.405	0.638	0.585	0.548

It shows that all the 6 models credibility are less than the acceptability threshold of 0.75. Thus these 6 additional models are not credible.

Use CNN to evaluate the model credibility by 66 series of data. Put the additional 6 models output in the end. Table 4 shows the result.

According to the results, there is only 1 set of data (Set 65)'s categorized credibility not corresponding to the expected result generated by TIC & GRA. The accuracy of the CNN model validation is 98.5%.

Table 4. CNN results under same input condition.

No	Good	Excellent	Medium	Expected	Actual
1	0.96	0.01	0.03	1	1
…	…	…	…	…	…
41	0.11	0.02	0.87	3	3
…	…	…	…	…	…
60	0.12	0.02	0.86	3	3
61	0.28	0.01	0.71	0	0
62	0.18	0.18	0.64	0	0
63	0.24	0.73	0.03	0	0
64	0.23	0.74	0.03	0	0
65	0.02	0.05	0.93	3	0
66	0.25	0.32	0.43	0	0

4.4 Different Input Model Validation

Adjust the interceptor missile 1's initial velocity in the model, and run the simulation again to generate outputs under different input condition. Table 5 shows the coordinate (x, y) of interceptor missile 1.

Table 5. The model output data for validation under different input condition.

1	2	…	1223	1224
(9000.3, 200.5)	(8994.6, 322.9)	…	(4012.4, 2310.6)	(4012.4, 2310.6)

Use TIC and GRA to calculate the similarity between simulation outputs in Table 5 and the reference, and use Formula (9) to transform the results into simulation credibility. Table 6 shows the credibility result.

Table 6. Credibility results of different inputs.

Simulation no	TIC ρ	TIC C	GRA γ	GRA C
1	0.321	0.728	0.370	0.347

It shows the credibility is less than the acceptability threshold of 0.75. Thus this simulation under different input condition is not credible.

Use the data in Table 7 as the input of the CNN for model validation. Table 7 shows the categorized credibility result.

Table 7. CNN result under different input.

No	Good	Excellent	Medium	Category
1	0.16	0.01	0.83	3

The CNN result shows that the simulation's credibility is "Medium". Since the model did not change, only the input was adjusted in the valid range, the credibility should remain the same as the same input condition. Thus CNN is able to evaluate the model credibility under different input condition, which cannot be done by conventional similarity analysis methods (such as TIC & GRA).

5 Conclusion

The most convincing way to evaluate a model's credibility is to analyze the outputs similarity between the simulation and real world. However, the outputs vary from a system's inputs, which makes model validation difficult under different inputs condition. Since it is not easy to guarantee same inputs between simulation and real world, conventional model validation methods are not as efficient as they look like.

By building a CNN for a target model according to its output data condition, we can perform model validation no matter the inputs are the same or not. Case study proves that, the method can obtain 98.5% validation accuracy when inputs are the same, and can also discriminate credibility levels when inputs are different. The CNN based model validation method may change the embarrassing status of the same input requirements when performing model validation, which is actually not practical most of time.

Meanwhile, CNN method only brings a categorized credibility level but not a quantitative result in [0,1], which reduces the discrimination resolution when comparing multiple simulation implementations to the same object. Possible solutions may use the possibilities of each category of model credibility level, and transform them into a quantitative credibility result.

References

1. Sargent, R.G.: Verification and validation of simulation models. J. Simul. **71**(1), 12–24 (2013)
2. Goldsman, D., Tokol, G.: Output analysis procedures for computer simulations. In: 2000 Winter Simulation Conference Proceedings, Orlando, FL, USA, pp. 39–45 (2000)
3. Khan, N.A., Sulaiman, M., et al.: Numerical analysis of electrohydrodynamic flow in a circular cylindrical conduit by using neuro evolutionary technique. Energies **14**(22), 1–19 (2021)
4. Ning, X., Wu, Y., Yu, T., et al.: Research on comprehensive validation of simulation models based on improved grey relational analysis. Acta Armamentarii **37**(2), 338–347 (2016)
5. Jwo, D.J., Chang, W.Y., Wu, I.H.: Windowing techniques, the Welch method for improvement of power spectrum estimation. Comput. Mater. Continua **6**, 3983–4003 (2021)
6. Kamikawa, S., Sato, T., Terashita, T., et al.: Open data validation of a classification method of eye movement by a convolutional neural network. In: International Forum on Medical Imaging in Asia (2021)

7. Aljohani, N.R., Fayoumi, A., Hassan, S.U.: A novel focal-loss and class-weight-aware convolutional neural network for the classification of in-text citations. J. Inf. Sci. **49**(1), 79–92 (2023)
8. Gao, F., Li, B., Chen, L., et al.: A Softmax classifier for high-precision classification of ultrasonic similar signals. Ultrasonics **112**(1), 106344–106352 (2021)
9. Salem, H.H., Kabeel, A.E., El-Said, E., et al.: Predictive modelling for solar power-driven hybrid desalination system using artificial neural network regression with Adam optimization. Desalination **522**, 115411/1–115411/16 (2022)

A Multiview Approach to Tracking People in Crowded Scenes Using Fusion Feature Correlation

Kai Chen(✉), Yujie Huang, and Ziyuan Wang

Nanjing University of Aeronautics and Astronautics, Nanjing 210016, China
chen_kai@nuaa.edu.cn

Abstract. Most of the current tracking methods for multi-target pedestrian tracking are unable to solve the problem where the tracking targets are blocked and reappears after disappearing in the camera perspectives, which brings great challenges to its practical application. To tackle this problem in dense crowds, we propose a multi-target pedestrian tracking method based on fusion feature correlation under multi-vision: Updating the pedestrian feature pool based on GMM to reduce the feature pollution; Then dynamically calculating the similarity threshold of target features based on K-means algorithm; Use the idea of voting to match pedestrian features and determine the addition and reappearance of pedestrians. The results on open dataset Shelf show that our method improve the accuracy and success rate of tracking under the condition of occlusion and reappearance after disappearance.

Keywords: Multi-vision · GMM · Dynamic threshold · Fusion feature · Reappearance detection

1 Introduction

In recent years, with the continuous expansion of urban population density and the increasing contact between people, the demand for social security management and urban information development has derived the demand for multi-target monitoring and trajectory tracking such as pedestrians. Therefore, it is of great practical significance to use computer vision for multi-target pedestrian tracking task.

In the case of multi-view vision, the multi-object tracking task not only requires to be carried out in the time stream, but also needs to be completed between each view at the same time. Due to the differences in the relative position and background in different images, especially in complex environments such as overlapping, occlusion and background distortion [1, 2], the difficulty of multi-target tracking increased, and the corresponding tracking algorithms also put forward higher requirements. At present, there are lacking related researches on multi-target tracking task. Although the existing algorithms have made different contributions in tracking efficiency and success rate, they still cannot solve the problems when tracking targets are occluded, or targets disappear from the camera views and then reappear [3–5]. In most cases, there are problems

© The Author(s), under exclusive license to Springer Nature Singapore Pte Ltd. 2024
F. Hassan et al. (Eds.): AsiaSim 2023, CCIS 1911, pp. 204–217, 2024.
https://doi.org/10.1007/978-981-99-7240-1_16

such as tracking interruption, tracking drift, and tracking error when occluding [6–10]. Besides, the problems of error-tracking, miss-tracking, and non-correspondence caused by disappearance and reappearance are more difficult to solve [11–13].

Based on this, a multi-target pedestrian tracking method based on fusion feature correlation under multi-vision are designed to solve the problems of long-term occlusion and reappearance of pedestrians.

2 Related Works

In the field of multi-object tracking, many excellent algorithms have been produced, such as DeepSORT[9], ByteTrack[14], MOTDT[1], etc. These algorithms have made excellent innovations in the field of multi-object tracking from the aspects of computational efficiency, tracking accuracy, and solving occlusion problems [15–19].

DeepSORT [9] is an improved algorithm based on SORT target tracking. It extends the core tracking framework in SORT, using Kalman filter and Hungarian algorithm to predict the current position and perform data association, and assigns target IDs to each object. Introducing deep learning models, and a simple CNN network is used to extract and save the appearance features of the detected target after each tracking. The combination of deep neural network and Kalman filter makes the tracking model higher accuracy and stronger robustness, but the Kalman filter cannot accurately predict the state of the targets which are occluded for a long time.

ByteTrack [14] is a method based on the Tracking-by-Detection paradigm. At present, most multi-object tracking methods obtain IDs only by associating detection boxes whose similarity is higher than the threshold, which brings a large number of missed detection and fragmented trajectories due to occlusion. BYTE, a new data association model proposed by [14], not only associates high confidence boxes, but also performs calculations with secondary correlation and background filtering between low confidence boxes and unmatched tracking objects, and then uses the similarity of the trajectories to recover the true target. This method has excellent tracking effect on the low confidence detection box caused by partial occlusion, but it puts forward higher requirements for the detector, and is hard to deal with the correlation matching problem when the tracking objects are completely occluded under monocular vision.

MOTDT [9] combines detection and tracking results on the basis of real-time performance, and selects the best target candidate based on deep neural network, so as to solve the problem of unreliable detection in online tracking. A hierarchical data association strategy is proposed to hierarchically associate target IDs with different candidate boxes using different features to avoid other unwanted objects and background contamination. MOTDT [9] designs a trajectory segment confidence calculation module to evaluate the accuracy of the Kalman filter using time series information. Combined with the hierarchical correlation method, the MOTDT model can run at real-time speed, but since the Kalman filter is not suitable for long-term tracking, this model has great instability for the situation when the targets disappear and then reappear.

The actual experiments show that most of the multi-target tracking algorithms have serious problems such as error-tracking, miss-tracking, target ID hopping [20–22] in the case of complete occlusions, reappearance after long time disappearance and so on. To tackle this, we conduct innovative research on fusion feature extraction and updating, calculation of feature correlation, pedestrian addition and reappearance judgment.

3 Proposed Method

The overall implementation process of the tracking framework is shown in Fig. 1. Firstly, the YOLOv5 [23] algorithm is used to detect pedestrians in multi-view images under the video stream to obtain all the detection boxes. Secondly, the target features are fused based on the correlation of the partition features, and all the pedestrian features to be associated in each frame are extracted. Then, the voting idea is used to calculate the feature similarity matrix, determined the dynamic threshold based on K-means [24], and matched all the target features. Next, combine the pedestrian addition and reappearance judgment to complete the multi-target pedestrian association. All the associated target features are stored in the feature pool, which is updated based on the GMM model. The most representative features of each pedestrian ID are selected as the template for the next feature matching under multi-vision, while the repeated and simple features will be eliminated.

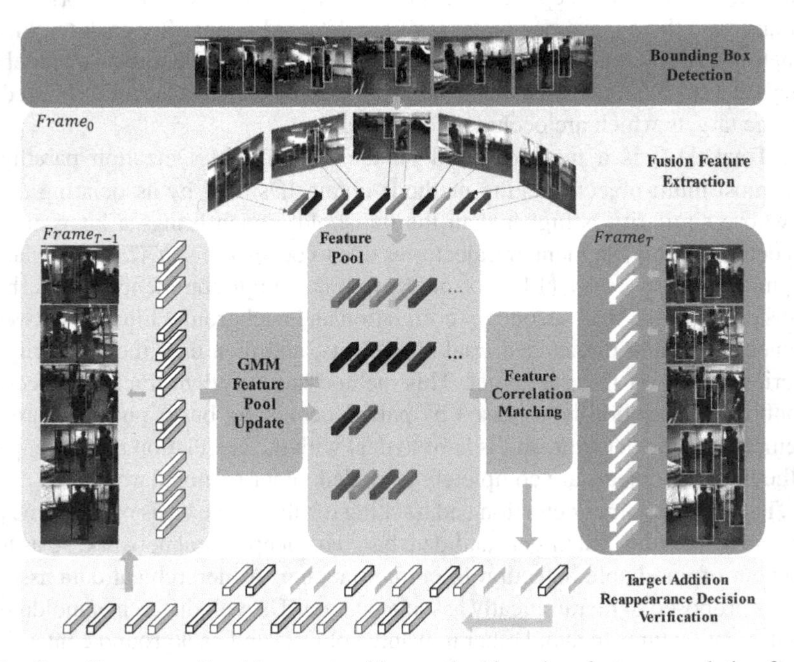

Fig. 1. Overall process of multi-target tracking method based on feature correlation fusion.

3.1 Fusion Feature Extraction and Update

In this section, starting from the feature correlation, feature extraction is carried out based on the partition feature correlation of the target images. Besides, we propose a GMM-based strategy to update the feature templates of all objects respectively.

Fusion Feature Extraction Method. The existing matching methods using pedestrian local features only consider the rough geometric correspondence of people and many parts are isolated from each other, which disperses the feature similarity calculation between different people with similar attributes. The proposed feature extraction method based on partition feature correlation is a feature network that considers the relationship between individual body parts and the rest parts, which provides a more accurate feature representation even in the case of the lack of body parts in the pedestrian images.

Firstly, the target pedestrian image corresponding to the detection box is extracted, and a feature map of size $H \times W \times C$ is extracted from the image based on the ResNet-50 backbone network. Divided it evenly into six horizontal partitions, as shown in Fig. 2. The global max pooling is performed on each partition feature to obtain the partition-level features of size $1 \times 1 \times C$. We propose the partition contrast correlation module (PCC) and the global contrast pooling module (GCP). In PCC, a partition-level feature is extracted from the horizontal partitions to perform different target association using body parts. It is fed into the convolutional network as the main feature, taking partition 1 as an example, the remaining partitions are averaged pooled to aggregate the information of other body parts. After convolution, the main feature $\overline{p}_i$ and aggregate feature $\overline{r}_i$ are integrated to obtain the correlation feature f_1. Similarly, we can obtain $f_2 \sim f_6$ for other partitions respectively. In GCP, the maximum pooling information is continuously added into the average pooling process based on the convolution operation, and the most discriminative part features in the region are aggregated in each horizontal partition to obtain f_0.

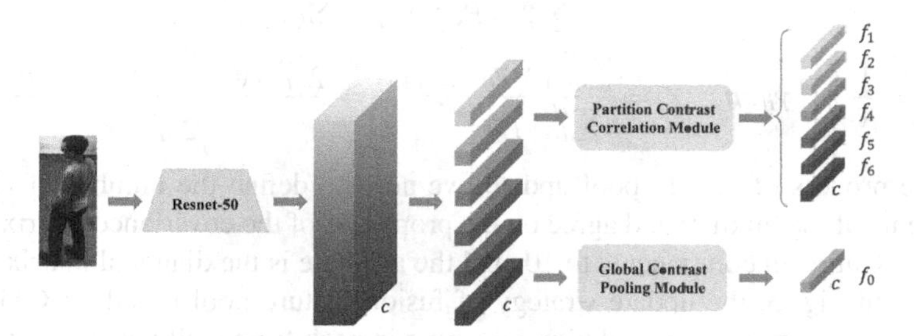

Fig. 2. Fusion feature extraction based on feature correlation.

In this method, f_i can describe the relationship between the information of each specific part and all other parts to adapt to complex occlusion environments, while f_0 can remove the background clutter and effectively reduce the feature pollution. By combining the two, we extract all the pedestrian features to be associated.

Update Strategy of Tracking Feature Pool. In order to ensure the accuracy and success rate of pedestrian tracking relation in multi-view, we propose a feature pool updating strategy based on GMM. In each update, the features under different views and different time points are collected as a mixture of Gaussian components, and each component can express the pedestrian characteristics from different aspects, so that the feature template of each object ID in the feature pool can maintain its generalization ability.

Gaussian mixture model (GMM) is an extension of ordinary Gaussian model. It uses the combination of Gaussian components to describe the data distribution. Computationally GMM is based on the derivation of the EM algorithm, which is mainly divided into expectation (*step-E*) and objective maximization (*step-M*).

GMM is used to calculate the mean and covariance matrix of each Gaussian. Firstly, set the parameters: $X = (x_1, x_1, \ldots, x_N), Z = (z_1, z_2, \ldots, z_N), \theta = (p, \mu, \Sigma)$. Each Gaussian distribution and its mean and covariance matrix can be initialized as $P = (p_1, p_1, \ldots, p_K), \mu = (\mu_1, \mu_2, \ldots, \mu_K), \Sigma = (\Sigma_1, \Sigma_2, \ldots, \Sigma_K)$. θ denotes the cluster center of each class, which contains the probability distribution of the latent variable Z, $\theta^{(i)}$ is the estimated value of the parameter after the ith iteration. N represents the number of samples, and K represents the number of Gaussian distributions. The parameters are substituted into the EM algorithm to obtain the Q-function.

$$Q\left(\theta, \theta^{(i)}\right) = \sum_{i=1}^{N} \sum_{k=1}^{K} (\log P_k + \log \phi(x_i|\mu_k, \Sigma_k)) P(z_i = C_k|x_i, \theta^{(i)}) \quad (1)$$

Thus, the parameters at the next time are solved according to the Q-function. In formula (1), $P(z_i = C_k|x_i, \theta^{(i)})$ is the posterior probability of the hidden latent under the current parameters, and all its internal parameters have been determined, so it is represented by the new variableγ_{ij}. Formula (2) is substituted into the *step-M* to solve the parameters in the next time. Each Gauss and its mean and covariance matrix are obtained, as formula (3).

$$\gamma_{ij} = \frac{P_j \phi(x_i|\mu_j, \Sigma_j)}{\sum_{k=1}^{K} P_k \phi(x_i|\mu_k, \Sigma_k)} \quad (2)$$

$$P_j^{(i+1)} = \frac{1}{N} \sum_{i=1}^{N} \gamma_{ij}, \mu_j^{(i+1)} = \frac{\sum_{i=1}^{N} x_i \gamma_{ij}}{\sum_{i=1}^{N} \gamma_{ij}}, \Sigma_j^{(i+1)} = \frac{\sum_{i=1}^{N} \left(x_i - \mu_j\right)^T (x_i - \mu_j) \gamma_{ij}}{\sum_{i=1}^{N} \gamma_{ij}} \quad (3)$$

In the process of feature pool update, we need to define the number of Gaussian components of the mixture and agree on the properties of the covariance matrix. Set the number of Gaussian components to 10, and the attribute is the diagonal matrix "Diag". As shown in Fig. 3, the update strategy of fusion feature pool based on GMM is as follows: after the multi-target relation matching in each frame, all new related features of each target ID are added to the feature pool. Set the storage limit of a single target in the feature pool to 50.

When the number of features stored by this target ID in the feature pool exceeds the storage limit, the earliest added features are eliminated before adding new features, so that the number of features is maintained at the storage limit. When there is no new associated feature in the frame, the existing features are maintained to complete the feature flow of the feature pool. Finally, when there are more than 15 features of the target ID in the feature pool, GMM is used to select the 10 most representative features among all the features of the target ID as feature template *Query* for association matching in the next frame. When the number of features is less than 15, the existing features are directly used as *Query*.

The feature pool update strategy based on GMM dynamically updates the target ID features stored in the feature pool with the time flow, which ensures low state space

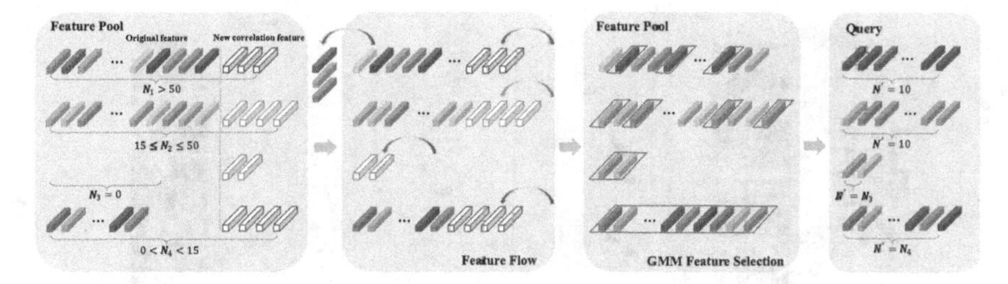

Fig. 3. Feature pool updating process based on GMM.

utilization and improves the computational efficiency of feature matching. Meanwhile, the *Query* always maintains high correlation with the environment and state of the target in the current frame. Besides, for the related target ID, the stored features in the feature pool will not change when the target completely disappears or is completely occluded. Therefore, when the target appears again, it can be re-tracked and associated with the previous ID. The difficult problems such as ID hopping and tracking drift after disappearance and reappearance are effectively avoided [25, 26].

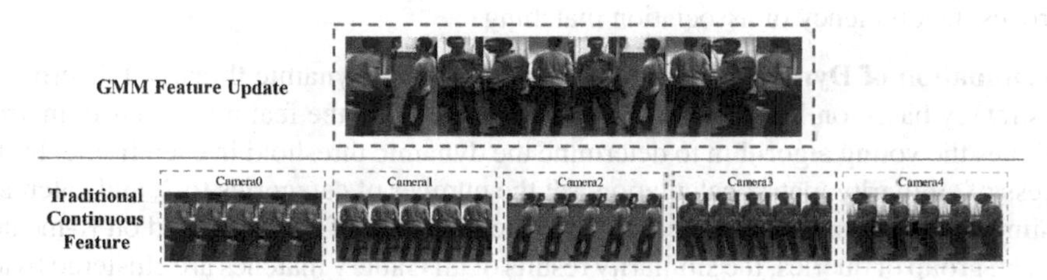

Fig. 4. Using GMM to update feature pool Query and traditional continuous feature maps.

3.2 Calculation of Feature Correlation

After dynamically updating the feature pool, the feature template *Query* selected from the feature pool needs to be matched with the pedestrian detection boxes to be associated in each frame.

Calculation of Similarity Matrix. When obtaining each pedestrian feature to be associated in all views of the current frame, all the features to be related need to be matched by using the feature template selected from the feature pool [27]. As shown in Fig. 4, the template features selected by the feature pool are used as query items, and the 10 features selected for one target ID are taken as a group. All the features to be associated in the current frame form the matching item, Gallery. Through the pairwise distance calculation between the *Query* and the *Gallery*, the similarity matrix of all galleries under each set of queries can be obtained [28, 29]. The smaller the result of each *Query* and *Gallery* is, the more similar the two features are.

The similarity matrix is calculated and judged based on voting. If there is a big difference of the similarity score between a matching item in the *Gallery* and that of

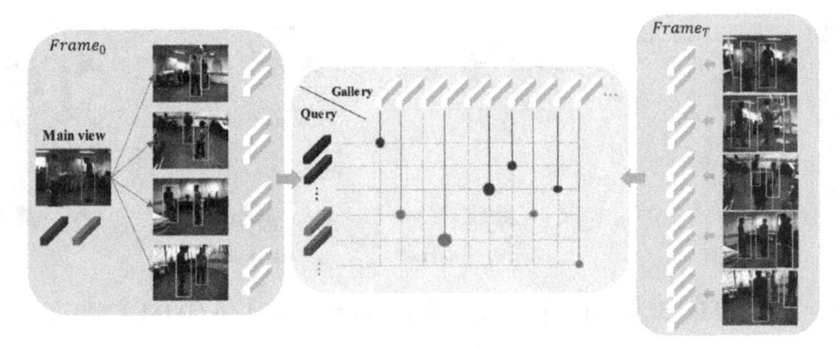

Fig. 5. Similarity matrix calculation of feature correlation.

a group of *Query*, a voting is carried out, and the mean value of the class with more votes will be used as the similarity calculation result. If all the results in a *Query* group have little difference, the average of all the calculated results will be directly taken. This method effectively eliminates the error association results caused by a small amount of feature errors, making the tracking process very good robustness. At the same time, the similarity matrix calculation takes features instead of images as input, effectively improves the efficiency of association matching.

Determination of Dynamic Threshold. We propose a dynamic threshold determination strategy based on K-means, using K-means to cluster the feature similarity matrix, combines the voting algorithm to determine the dynamic threshold in each frame. In the process of similarity matrix calculation, set the number of categories to 2, and calculate the similarity results of each matching item in each group of queries based on K-means. For each group of queries, the similarity results of all Gallery matches are clustered twice based on K-means, and all the matches contained in the category with smaller similarity results (similar features) are found. At this time, the dynamic threshold is determined by combining the clustering results of the similarity matrix, as formula (4) shows.

$$Dy\_threshold = \min\left(avg_{ij}\right) + \left(\max(avg_{ij}) - \min(avg_{ij})\right) \times threshold \tag{4}$$

Where, avg_{ij} is the similarity matrix calculation result of the jth *Gallery* matching item in the ith group of *Query*. *threshold* is the initial fixed threshold.

When the similarity calculation result of matching items *Gallery* included in the secondary clustering is less than the dynamic threshold, the target is considered to be successfully associated with the target ID corresponding to the *Query* group.

Pedestrian Feature Matching and Tracking. When the initial frame feature pool is empty, the association matching needs to be performed separately in each view. In the initial frame, select the view with the largest number of detection boxes as the main view. All features in the main view are used as *Query_0*, and all features in the other views are used as *Gallery_0* to calculate the feature-related similarity matrix. Associate the target features in the other views with the main view to achieve pedestrian tracking between different views in the initial frame, as shown in Fig. 5. Then, add all the associated target features to the feature pool respectively, and start the multi-target tracking process.

To solve the problem caused by the defects of K-means itself, we use IoU and center offset rate to detect the mutation of the pedestrian detection boxes. For the related detection boxes, we calculate the two indicators with the detection boxes of the same person in the same view of previous frame, as shown in formula (5).

$$IoU = \frac{S_{t-1} \cap S_t}{S_{t-1} \cup S_t}, centre\_x = \frac{|x_t - x_{t-1}|}{x_{t-1}} \quad (5)$$

Where $S_{t-1}, S_t, x_{t-1}, x_t$ represent the area of the detection box of the same person in the same view and the X-coordinate of the center points in the previous frame and the current frame respectively. When both IoU and center offset rate are less than the set threshold, the association of target features under multi-vision is considered successfully. In addition, after the object feature association according to the similarity matrix, if the object ID appears in the previous frame in a view but does not appear in the current same view, the miss tracking detection will be triggered. Next, calculate the IoU and center offset rate of the target ID detection box in previous frame and all the unassociated detection boxes in the current frame. If the mutation detection composed of two indicators can be satisfied at the same time, the detection box can be considered to be successfully associated with the target ID.

3.3 Target Addition and Reappearance Determination

In this process, we propose a new addition and reappearance determination method based on feature correlation. If two or more unrelated features still appear in the current frame in more than 60% views, new pedestrian determination will be triggered. As shown in Fig. 6, taking an unassociated feature in each of the five perspectives as an example, all the unassociated features are simultaneously used as *Query_new* and *Gallery_new* to calculate the similarity matrix.

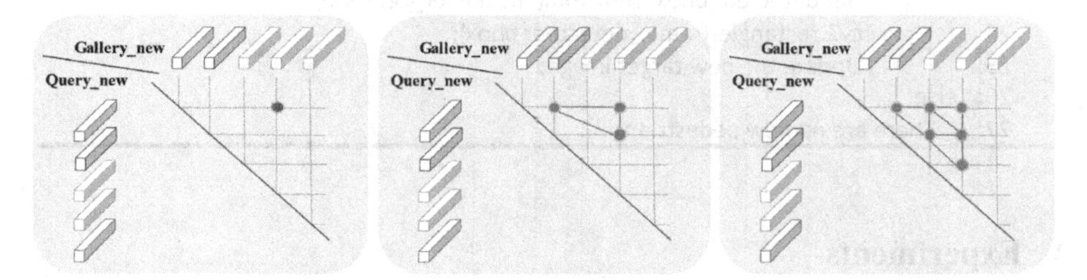

Fig. 6. Pedestrian addition decision diagram based on feature correlation.

For the obtained square matrix, the upper triangular matrix is selected for the target feature correlation. Combined with the dynamic threshold determination method, when at least two features exist in two views and the similarity matrix calculation results are lower than the threshold, the new pedestrians are determined. The specific determination process is shown in Algorithm 1. After determining a new target, the corresponding information will be added into the *Query*, and the new target will be tracked in the next target feature association.

Combined with the feature pool update strategy, if the associated target ID completely disappears in all perspectives, the feature template in its feature pool will still be retained. Therefore, when the target reappears in any perspective, its feature will still be associated with the previous ID.

Algorithm 1

```
1:  INPUT associated pedestrian subscripts in all views;
2:  INPUT all pedestrian Bboxes detected by YOLOv5;
3:    Find unassociated pedestrian subscripts in all views;
4:    Obtain views with unassociated pedestrian as Views;
5:    Extract features of all unassociated pedestrian;
6:    Get frame and bboxes of all unassociated pedestrian;
7:    if len(bbox es) == 0:
8:      return;
9:  if len(Views) > 60% of all views:
10:   Place features of all unassociated pedestrian both into Query_new and Gallery_new;
11:   for query in Query_new:
12:     for gallery in Gallery_new:
13:       Calculate the similarity of two features;
14:   Place all similarity into an array degree to form a similarity matrix;
15:   Obtain the upper triangular matrix of the similarity matrix;
16:   Set num_component = 2;
17:   Obtain the classification results of pedestrian features with K-means as km;
18:   Calculate the Dynamic threshold based on the upper triangular matrix;
19:   Obtain len(min(km.cluster_centers));
20:   if float(min(km.cluster_centers)) < Dynamic threshold
21:     if len(min(km.cluster_centers)) >= 2:
22:       if len(min(km.cluster_centers)) == len(Views):
23:         Get detected_bbox from min(km.cluster_centers);
24:         cv2.rectangle(frame, detected_bbox);
25:         Update the new target image;
26: else:
27:   There are no new pedestrians .
```

4 Experiments

The evaluation indexes used include: center location error (CLE), tracking accuracy, tracking success rate, mismatched number etc. We adopt the success rate evaluation method proposed by [30]. The overlap ratio (IoU) between the tracking prediction frame and the real target frame is calculated to determine the degree of overlap between the two frames, and then the success rate of tracking is determined. Mismatched means under a certain threshold, when the target ID of the same target in the image is matched, the tracking prediction frame does not meet the threshold, and when the tracking prediction frame meets the threshold, the target ID does not match.

We compare the proposed tracking framework with the excellent tracking algorithms mentioned in Sect. 2. To fully analyze the algorithm performance, in Fig. 7(a), the offset distance accuracy of our framework is analyzed with other two tracking algorithms. Here, only CLE less than 50 is evaluated. In this case, the curve represents the proportion of the number of detected targets offset distances less than different offset thresholds to the total number of frames.

In Fig. 7(b), the tracking success rate under different algorithms is compared, as shown in formula (6). As can be seen from Fig. 8(a) and (b), the tracking framework proposed in this paper not only has high tracking accuracy, but also better tracking target success rate than other tracking algorithms when there are complex situations such as target addition, long-term occlusion, and reappearance under multi-vision.

$$S = \frac{R_i \cap R_0}{R_i \cup R_0} \tag{6}$$

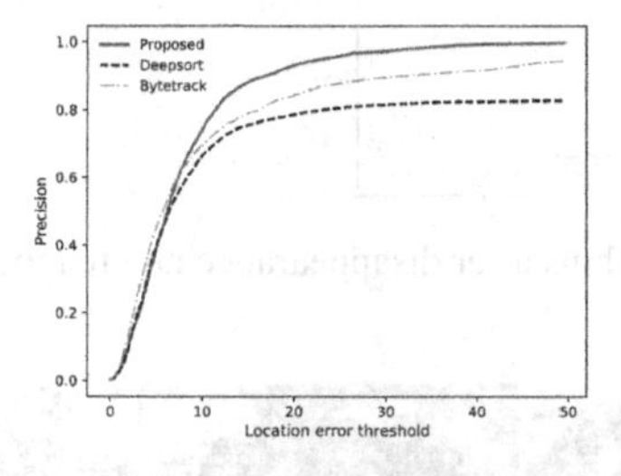

(a) Tracking accuracy comparison

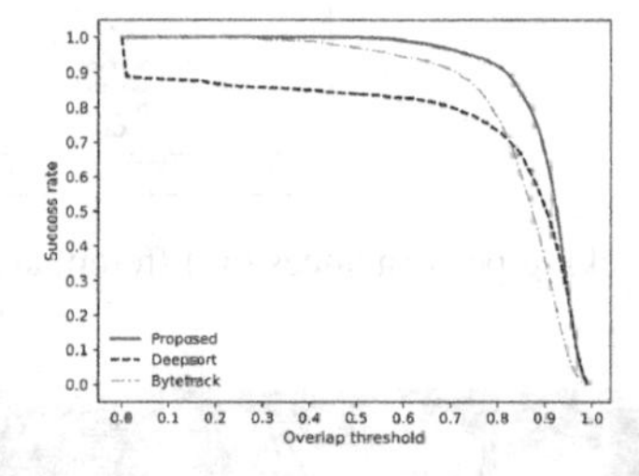

(b) tracking success rate Comparison

Fig. 7. Comparison of multi-view tracking results under different algorithms.

Figure 8 compares the error-tracking of different tracking algorithms with video streams under different intersection ratio thresholds. The curve represents the sum of the number of errors of each algorithm to a certain frame over time under multi-vision. It can be seen that the tracking framework proposed can show better tracking stability and control error performance than other algorithms under different thresholds, especially in the control of the number of wrong tracks. In contrast, other algorithms have persistent ID mismatch in the later stage due to reasons such as reappearance and different views being occluded to different degrees.

Figure 9 shows the tracking results of some keyframes. Each line is a different view of the image in a frame. The second row is the scene where pedestrian No. 2 completely disappears from all perspectives and reappears, where this algorithm can track the same target and predict the same ID as the real value, while other algorithms will have ID matching errors and other situations. In general, the algorithm can track and correlate all targets effectively and predict the correct ID under multi-vision.

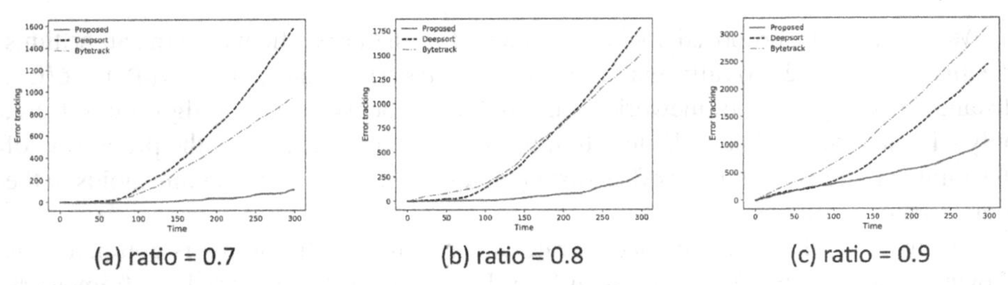

Fig. 8. Mismatches numbers occurring over time by different algorithms under different IoU thresholds.

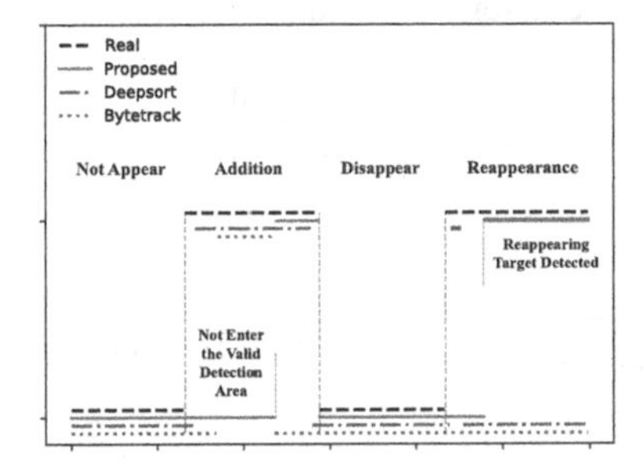

Fig. 9. Tracking performances of different algorithms after disappearance and reappearance

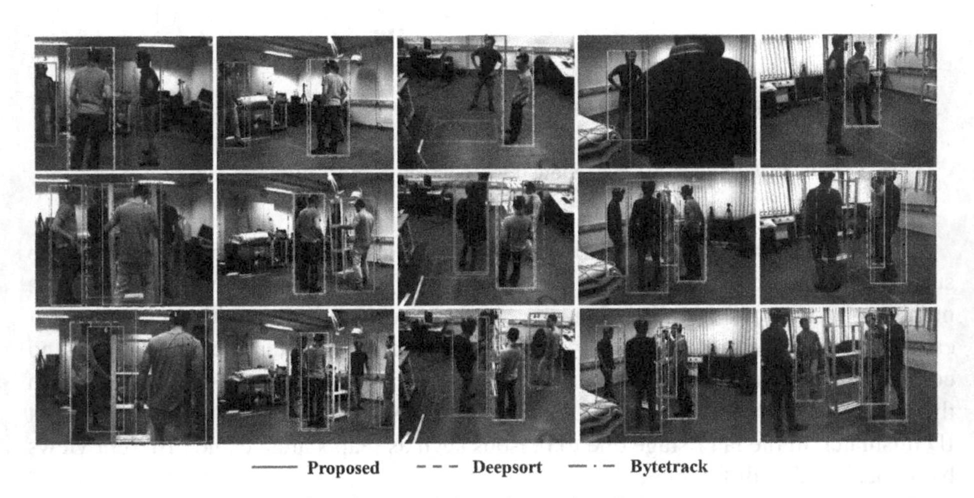

Fig. 10. Illustration of some key frames.

Figure 10 shows the tracking results of some keyframes. Each line is a different view of the image in a frame. The first line is the scene when pedestrian No. 2 is added. In general, all algorithms in this frame achieve good tracking effects. The second row is the scene where pedestrian No. 2 completely disappears from all perspectives and then reappears, where our algorithm can track the same target and predict the right ID, while other algorithms have ID matching errors and other situations. The third line is the scene where everyone is standing in the effective tracking area under multiple perspectives. It can be found that other algorithms will cause ID matching errors, tracking drift and other situations when occlusion and reappearance occur. In contrast, our algorithm can effectively correlate all targets and predict the correct ID under multi-vision.

5 Conclusion

To tackle the problem of multi-target pedestrian tracking under multi-vision in a dense crowd, we propose a multi-target pedestrian tracking method, which is based on the fusion feature correlation to track multi-target pedestrians. Use the voting algorithm to conduct correlation matching of pedestrian features in multi-perspective, reducing the problems such as association errors and tracking errors caused by problems such as occlusion. The dynamic threshold calculation based on K-means ensures the universality of the tracking scene and improves the accuracy and stability of multi-pedestrian tracking. The method ensures the accuracy and success rate of tracking and identification of newly added pedestrians, and solves the problems of target ID hopping and tracking drift after pedestrians' disappearing and reappearing. After feature matching and verification, a feature pool update strategy based on GMM is proposed. The feature pool is dynamically updated every frame to improve the representativeness of the template features in the feature pool in the current environment and effectively reduce the feature pollution problem. Our goal is to find more optimized and efficient multi-target tracking and modeling methods under multi-vision.

References

1. Chen, L., Ai, H., Zhuang, Z., Shang, C.: Real-time multiple people tracking with deeply learned candidate selection and person re-identification. In: 2018 IEEE International Conference on Multimedia and Expo (ICME), pp. 1–6. IEEE (2018)
2. Yao, R., Shi, Q., Shen, C., Zhang, Y., Van Den Hengel, A.: Part-based visual tracking with online latent structural learning. In: PROCEEDINGS of the IEEE Conference on Computer Vision and Pattern Recognition, pp. 2363–2370. IEEE (2013)
3. Khurana, T., Dave, A., Ramanan, D.: Detecting invisible people. In: Proceedings of the IEEE/CVF International Conference on Computer Vision, pp. 3174–3184 (2021)
4. Li, W., Xiong, Y., Yang, S., Xu, M., Wang, Y., Xia, W.: Semi-TCL: semi-supervised track contrastive representation learning. arXiv preprint arXiv: 2107.02396 (2021)
5. Dai, P., Weng, R., Choi, W., Zhang, C., He, Z., Ding, W.: Learning a proposal classifier for multiple object tracking. In: Proceedings of the IEEE/CVF Conference on Computer Vision and Pattern Recognition, pp. 2443–2452 (2021)
6. Zhuang, B., Lu, H., Xiao, Z., Wang, D.: Visual tracking via discriminative sparse similarity map. IEEE Trans. Image Process. **23**(4), 1872–1881 (2014)
7. Comaniciu, D., Ramesh, V., Meer, P.: Real-time tracking of non-rigid objects using mean shift. In: Proceedings IEEE Conference on Computer Vision and Pattern Recognition, vol. 2, pp. 142–149. IEEE (2000)
8. Berclaz, J., Fleuret, F., Turetken, E., Fua, P.: Multiple object tracking using k-shortest paths optimization. IEEE Trans. Patt. Anal. Mach. Intel. **33**(9), 1806–1819 (2011)

9. Wojke, N., Bewley, A., Paulus, D.: Simple online and realtime tracking with a deep association metric. In: 2017 IEEE International Conference on Image Processing (ICIP), pp. 3645–3649. IEEE (2017)
10. Lucas, B.D., Kanade, T.: An iterative image registration technique with an application to stereo vision. In: 7th International Joint Conference on Artificial Intelligence, vol. 2, pp. 674–679 (1981)
11. Brasó, G., Leal-Taixé, L.: Learning a neural solver for multiple object tracking. In: Proceedings of the IEEE/CVF Conference on Computer Vision and Pattern Recognition, pp. 6247–6257 (2020)
12. Lu, Y., Wu, T., Chun Zhu, S.: Online object tracking, learning and parsing with and-or graphs. In: Proceedings of the IEEE Conference on Computer Vision and Pattern Recognition, pp. 3462–3469 (2014)
13. He, K., Zhang, X., Ren, S., Sun, J.: Spatial pyramid pooling in deep convolutional networks for visual recognition. IEEE Trans. Patt. Anal. Mach. Intell. **37**(9), 1904–1916 (2015)
14. Zhang, Y., et al.: ByteTrack: multi-object tracking by associating every detection box. In: Avidan, S., Brostow, G., Cissé, M., Farinella, G.M., Hassner, T. (eds) Computer Vision – ECCV 2022. ECCV 2022. Lecture Notes in Computer Science, vol. 13682. Springer, Cham (2022). https://doi.org/10.1007/978-3-031-20047-2_1
15. Belagiannis, V., Amin, S., Andriluka, M., Schiele, B., Navab, N.: 3D pictorial structures for multiple human pose estimation. In: Proceedings of the IEEE Conference on Computer Vision and Pattern Recognition, pp. 1669–1676. IEEE (2014)
16. Khan, S.M., Shah, M.: A multiview approach to tracking people in crowded scenes using a planar homography constraint. In: Leonardis, A., Bischof, H., Pinz, A. (eds) Computer Vision – ECCV 2006. ECCV 2006. Lecture Notes in Computer Science, vol 3954. Springer, Heidelberg (2006). https://doi.org/10.1007/11744085_11
17. Luiten, J., et al.: Hota: a higher order metric for evaluating multi-object tracking. Int. J. Comput. Vision **129**(548–578), 17 (2021)
18. Chu, P., Ling, H.: Famnet: joint learning of feature, affinity and multi-dimensional assignment for online multiple object tracking. In: Proceedings of the IEEE/CVF International Conference on Computer Vision, pp. 6172–6181 (2019)
19. Meinhardt, T., Kirillov, A., Leal-Taixe, L., Feichtenhofer, C.: Trackformer: Multi-object tracking with transformers. In: Proceedings of the IEEE/CVF Conference on Computer Vision and Pattern Recognition, pp. 8844–8854 (2022)
20. Liu, Z., et al.: Swin transformer: hierarchical vision transformer using shifted windows. In: Proceedings of the IEEE/CVF International Conference on Computer Vision, pp. 10012–10022 (2021)
21. Hager, G.D., Belhumeur, P.N.: Efficient region tracking with parametric models of geometry and illumination. IEEE Trans. Patt. Anal. Mach. Intel. **20**(10), 1025–1039 (1998)
22. Hornakova, A., Henschel, R., Rosenhahn, B., Swoboda, P.: Lifted disjoint paths with application in multiple object tracking. In: International Conference on Machine Learning, pp. 4364–4375. PMLR (2020)
23. Zhu, X., Lyu, S., Wang, X., Zhao, Q.: TPH-YOLOv5: improved YOLOv5 based on transformer prediction head for object detection on drone-captured scenarios. In: Proceedings of the IEEE/CVF International Conference on Computer Vision, pp. 2778–2788 (2021)
24. Hartigan, J.A., Wong, M.A.: Algorithm AS 136: A k-means clustering algorithm. J. Royal Stat. Soc. **28**(1), 100–108 (1979)
25. Chen, K., Song, X., Zhai, X., Zhang, B., Hou, B., Wang, Y.: An integrated deep learning framework for occluded pedestrian tracking. IEEE Access **7**(26060–26072), 13 (2019)
26. Stadler, D., Beyerer, J.: Improving multiple pedestrian tracking by track management and occlusion handling. In: Proceedings of the IEEE/CVF Conference on Computer Vision and Pattern Recognition, pp. 10958–10967 (2021)

27. Sikdar, A., Chatterjee, D., Bhowmik, A., Chowdhury, A.S.: Open-set metric learning for person re-identification in the wild. In: 2020 IEEE International Conference on Image Processing (ICIP), pp. 2356–2360. IEEE (2020)
28. Ren, S., He, K., Girshick, R., Sun, J.: Faster R-CNN: towards real-time object detection with region proposal networks. Adv. Neural Inf. Process. Syst. **28** (2015)
29. Wang, Z., Zheng, L., Liu, Y., Li, Y., Wang, S.: Towards real-time multi-object tracking. In: Vedaldi, A., Bischof, H., Brox, T., Frahm, JM. (eds.) Computer Vision – ECCV 2020. ECCV 2020. Lecture Notes in Computer Science(), vol. 12356. Springer, Cham (2020). https://doi.org/10.1007/978-3-030-58621-8_7
30. Yi, W., Lim, J., Yang, M.: Online Object Tracking: A Benchmark Supplemental Material. IEEE (2013)

Prototype Learning Based Realistic 3D Terrain Generation from User Semantics

Yan Gao[1], Jimeng Li[2], Jianzhong Xu[3], Xiao Song[4], and Hongyan Quan[1](✉)

[1] School of Computer Science and Technology, East China Normal University, Shanghai 200062, China
hyquan@cs.ecnu.edu.cn

[2] School of Software Engineering, East China Normal University, Shanghai 200062, China

[3] Training and Simulation Center, Army Infantry Academy, Nanchang 330103, Jiangxi, China

[4] School of Cyber Science and Technology, Beihang University, Beijing 100191, China

Abstract. Customizing 3D terrain based on user semantics plays an important role in military simulation, but it is difficult to realize realistic results because of the limited ability of some simple Convolutional neural network (CNN) models. In order to meet the personalized needs of users, this article proposes a prototype learning based terrain generation network (ProTG Net). Concretely, it extracts terrain semantics based prototype features from a small number of terrain surface samples, and then transfers the pre-learned features to user customization. Specifically, a prototype learning based framework is designed, including a terrain texture generation module (TGM), prototype feature generation module (PGM), and multiple prototype features matching module (FMM). TGM is designed as the CGAN based Pix2pix (Pixel to Pixel) structure, which can generate realistic terrain textures based on user semantics, providing a reliable terrain texture data source for prototype learning. Based on the semantic terrain texture generated by TGM, multi-features are extracted in PGM including the adaptive super-pixel guided features and the terrain spatial feature. In addition, multiple feature matching strategy is proposed for achieving the better matching between prototype matching and user semantic features. Taking a public dataset of real terrain as an example, it was verified that the prototype based method can generate realistic 3D terrain and achieve user customization to obtain realistic results.

Keywords: Terrain · Prototype learning · Feature · Superpixel

1 Introduction

Realistic 3D terrain generation is a classic hotspot with a long history in computer graphics. Its purpose is to model vivid 3D terrain according to simulated geometric coordinates or sampled real terrain data [1]. The realistic 3D terrain synthesis task has always been a hot topic in the academic and military simulation fields.

In past researches, many terrain generation methods have emerged, with the three main categories being fractal based terrain generation methods [2], height field based terrain generation [3], and texture based terrain generation [4]. The Height field based

© The Author(s), under exclusive license to Springer Nature Singapore Pte Ltd. 2024
F. Hassan et al. (Eds.): AsiaSim 2023, CCIS 1911, pp. 218–229, 2024.
https://doi.org/10.1007/978-981-99-7240-1_17

Terrain Generation (HTG) is one of the mainstream strategies in terrain simulation. Especially as a classic terrain generation method, the elevation map strategy has the characteristics of small storage capacity and easy conversion into mathematical models [5], making it the preferred choice for most researchers in 3D terrain research. In the research of HTG, there have been process based terrain generation (PTG), simulation based terrain generation algorithm (STG), and sample based terrain generation method (SBTG). In PTG based methods, the geometric height of terrain maps requires an iterative process to synthesize terrain, and fractal set theory is its principle. In early research, Degeorgio V and Mandelbrot et al. first proposed the use of fractal set theory for terrain generation tasks [6]. Then, using terrain height for modeling, an interpolation method was proposed to insert a new pixel point between adjacent two points [7]. In addition, PTG based on wavelet transform [8] and PTG based on Berlin noise [9] were also proposed. In STG based methods, the goal is to simulate the impact of physical erosion on terrain, where grid setting methods are used to add control conditions at each vertex to simulate the physical erosion process [10, 11]. Real time STG based methods are also proposed [12]. In the SBTG based method, it generates a new terrain map by selecting duplicate terrain textures. In its early research, texture sample regions were selected by calculating the distance between adjacent points [13], and then terrain texture styles were selected as global features. On this basis, local terrain textures were generated into realistic terrain maps [14]. Overall, there exists the high complexity and computation in terrestrial generation, besides, it is hard to achieve user customized semantics.

Recently, with significant improvements in the performance of CNN, some HTG methods based on CNN have emerged [15–17]. In existing deep learning based prototypes, it obtains feature measurement results by calculating the similarity between the query features and support prototype features, which has been widely applied in many research fields, such as segmentation [18] and object tracking [19].

Although deep neural networks have made significant progress in terrain modeling, there are still typical problems that require a large amount of dense annotations in deep model training. Recently, the rapid development of deep learning analysis has led to the emergence of class specific prototype representations [20–22], which fully utilize knowledge from support set knowledge and provide better generalization ability in alignment regularization between support and queries.

At the IEEE Conference on Computer Vision and Pattern Recognition, superpixel guided clustering (SGC) and guided prototype allocation (GPA) were provided in adaptive prototype learning [21]. SGC extracts more representative prototypes by aggregating similar feature vectors, while GPA can guide the acquisition of more accurate matching prototypes. Inspirited by the previous works, prototype learning principles are applied to the study of terrain generation. Unlike [21] in prototype learning, which only calculates the previous adaptive superpixel guided features, in our method, in addition to the superpixel guided features, the terrain spatial features are also considered in this work.

Our method utilizes a small number of terrain texture maps to construct a sample dataset of terrain textures, and then uses prototype learning to extract terrain features from it. The extracted texture is further matched with the user terrain semantics input to obtain customization. Therefore, we extract the prototype features of the terrain and then query the similarity between the query features and existing terrain prototype support features to obtain the generated results. In summary, the contribution of our method can be described as:

(1) A prototype learning based 3D terrain generation framework is proposed. It transfers the problem of 3D terrain modeling to feature matching in Metric space based on prototype learning. Multiple features were extracted in prototype learning, including adaptive superpixel guided features and terrain spatial features, and multiple features matching for better performance is achieved in terrain generation prototype learning.
(2) A terrain texture generation strategy based on wavelet transform is proposed. Borrowing the pix2pix framework to synthesize terrain textures to extract precise terrain features for further prototype learning.
(3) A terrain generation framework based on user semantics is studied, which takes user customization as the query condition and can meet the needs of user customization in generating terrain.

2 Prototype Learning Based 3D Terrain Generation

The goal of our method is to study a terrain Generative model based on prototype learning. It inputs user-defined semantics and outputs the generated terrain elevation map. Our prototype based learning terrain generation network (ProTG Net) is based on a terrain texture generation strategy, as shown in Fig. 1. On the basis of terrain texture generation, prototype features with finer details can be extracted. Then prototype feature matching can be performed in metric space. In this section, we will first introduce our framework, and then provide a detailed introduction to our strategy, including the design methods of terrain texture generation module (TGM), prototype feature generation module (PGM), and multiple prototype features matching module (FMM).

2.1 The Framework of ProTG-Net

As shown in Fig. 1, our overall architecture of ProTG-Net is based on a prototype learning framework, which includes terrain texture generation, prototype type feature generation, and multiple prototype features matching. In terrain texture generation, we design a TGM module. In prototype feature generation, we provide a prototype generation module denoted as PGM. In prototype feature matching, we designed multiple prototype features matching module denoted as FMM, as shown in Fig. 1.

In TGM, it inputs user semantic and outputs the generated customization of terrain texture map. It is designed as a GAN (Generative Adversarial Network) structure. TGM includes DWT (Discrete Wavelet Transform) generators and patch based discriminators. A generator based on DWT is designed to obtain more detailed terrain information for precise detail. TGM is encouraged by Pix2pix [23], different from the state of art, we design the DWT generator as dual encoding. The first encoding process uses a regular encoding and decoding structure, while the second encoding process is based on DWT theory. Both encoding and decoding structures use 4-layer down-sampling encoding and 4-layer up-sampling decoding calculations.

In PGM, prototype features are extracted in this module. In this work, two kinds features are considered, adaptive superpixel guided feature and the terrain spatial feature,

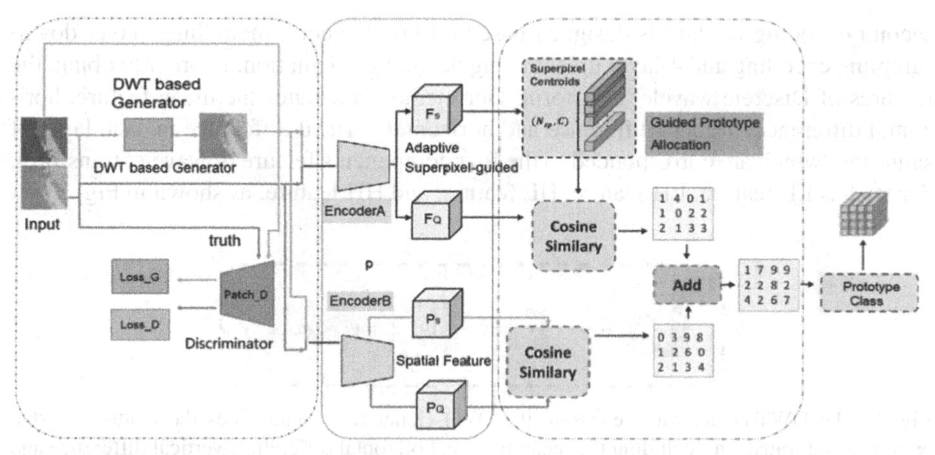

Fig. 1. The framework of ProTG-Net. Our prototype based terrain generation network (ProTG Net) is based on a terrain texture generation strategy. On the basis of terrain feature generation, prototype features are extracted, and then prototype features are matched in metric space. It includes a terrain texture generation module (TGM), a prototype feature generation module (PGM), and multiple prototype features matching module (FMM).

as shown in Fig. 1. In the previous work, adaptive super-pixel guided clustering and guided prototype allocation are introduced by the state of art [21] for multiple prototype extraction and allocation. Due to the fact that the adaptive superpixel guided clustering method is a parameter free and training free method, we apply its ideas to our work. In addition, unlike [21], this work also considers the spatial characteristics of terrain. *Encoder A* is designed for encoding the adaptive superpixel guided feature and *Encoder B* is designed for encoding the terrain spatial feature, respectively.

FMM is designed to select matching prototypes to provide more accurate guidance. Based on the two features extracted from the previous statement, in the prototype matching stage, as in the previous work [21], we considered superpixel guided feature matching in prototype learning. In addition, this work also considered terrain spatial feature matching. On this basis, as shown in Fig. 1, we combine the two path matching result to obtain more stable results.

2.2 The Texture Generation Module TGM

In TGM, we use a small number of terrain texture maps to construct a terrain dataset with a few terrain samples. Then, in prototype learning, we extract prototypes from the constructed dataset and further match them with the input semantic map. Therefore, our method utilizes the rich features of terrain texture maps to extend the corresponding features of terrain segmentation maps to generate realistic 3D terrain.

TGM is designed using the framework Pix2pix [23], where Unet [24] is used as a generator. Unlike existing technologies, we have designed the DWT generator as dual encoding [25]. The first encoding process is a regular encoding and decoding structure [23]. Like the first encoding method, in order to more accurately synthesize terrain, the second encoding method is designed based on DWT theory, including 4-layer down-sampling encoding and 4-layer up-sampling decoding calculation. In order to obtain the features of Discrete wavelet transform, Our method calculates the mean feature, horizontal difference, vertical difference and horizontal-vertical difference in each layer of sampling, which are corresponded to the four components feature of wavelet transform, denoted as LL feature, LH feature, HL feature, and HH feature, as shown in Fig. 2.

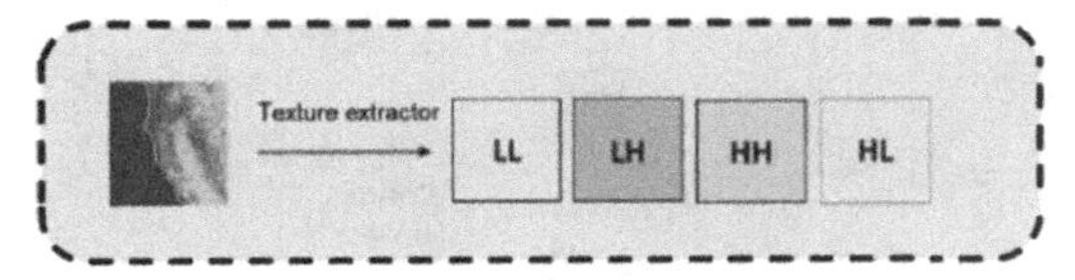

Fig. 2. The DWT-Generator. we design the DWT-Generator to calculates the features of discrete wavelet transform including the mean feature, horizontal difference, vertical difference and horizontal-vertical difference, which are corresponded to the four components features denoted as LL feature, LH feature, HL feature, and HH feature, as shown in this Figure.

In order to obtain more accurate terrain features, our method calculates 4-layer wavelet features and decodes the four component features from the first level of down-sampling to generate texture results.

Inspired by the Pix2pix network [23], we designed a patch based discriminator denoted *patch_ D* (as shown in Fig. 1) used to generate as realistic a terrain texture map as possible. The input of the discriminator is an image pair of terrain semantic map and terrain texture map from the generator, and the discriminator network outputs matrix N with elements sized $n \times n$, each element in the matrix represents a patch, which corresponds to the receptive field of the image pair. The patch based discriminator is designed with 3 convolution operations, with batch normalization processing performed after the first convolution, and only batch normalization and activation processing performed after the second convolution calculation.

2.3 The Prototype Features Extraction and Comparison

In PGM design, inspired by existing technology [21], an adaptive superpixel guided feature is provided in prototype learning. Inspired by existing technology, the proposed method extracts adaptive superpixel guided features. In order to enhance the details of terrain features, we consider using sample terrain textures to extract terrain spatial features.

In the extraction of adaptive superpixel guided features, just like the work [21], our proposed strategy aims to adaptively changing the number of prototypes and their spatial range based on perception content, and enables the prototypes to have content adaptation and spatial perception capabilities. We divide the supporting features into several representative regions based on the similarity for adaptively selecting more important prototypes from the more similar features in the query stage.

We define superpixels as a set of pixels with similar features, and dynamic update the superpixel centroids of representative prototypes by aggregating similar feature vectors.

In this study, as in previous work [21], the distance function D associated with the coordinates of each pixel and the supporting feature map is defined as

$$D = \sqrt{d_p^2 + kd_s^2} \tag{1}$$

where d_p denoted the pixel feature distance and d_s denotes the spatial feature distance, k denotes the weight of spatial feature. In the calculation of pixel feature distance and spatial feature distance, color features and terrain pixel coordinates are considered. On the basis, the correlation mapping Q between each pixel and all superpixels is calculated as

$$Q = e^{-D} \tag{2}$$

where D denotes the distance between every pixel and the superpixels calculated from Formula 1 Then, we treat the correlation mapping Q as the weight of masking query features and update the superpixel centroid at each iteration step.

Different from the previous work [21] that only calculates the previous adaptive super-pixel guided features in prototype learning, in our method, in addition to the adaptive superpixel guided features, this work also considers terrain spatial features. As shown in Fig. 1, the generated terrain texture from TGM is encoded into 4-layers texture features using the convolutional units that is composed of the resNet [26] block, batch normalization, and activation process, and then, the terrain texture support feature P_S is obtained. We take the similar process to obtain the terrain query feature P_Q calculated from the user input semantics data.

In order to obtain a more precise query result in prototype feature matching, we provide a multi-feature matching strategy in prototype learning coupling with the extracted adaptive superpixel guided features and terrain spatial features. In the adaptive superpixel guided features, inspired by existing techniques [21], our method adopts an adaptive prototype allocation strategy. It first calculates the cosine similarity between each prototype and query feature element in the feature space, and the similarity result is denoted as M_A. Similarly, in the terrain spatial feature space, our method calculates the cosine similarity between each prototype and the query feature elements in the spatial feature space, and then the similarity result is denoted as M_B, and further to combine these two similarity matrix to obtain the enhanced prototype query result, see Fig. 1.

2.4 Network Loss Function

We define two kinds of loss, the generating loss L_G and discriminative loss L_D. L_G is defined as:

$$L_G = W_a L_{G\_DT} + W_b L_P + W_c L_{G\_L} \tag{3}$$

where L_{G_DT} and L_{G_DW} denote the discriminative losses. L_{G_DT} is calculated by the terrain semantics map and output of G_DT, L_P is the similarity L1 loss calculated by the similarity between M_A and M_B. $W_c L_{D\_L}$ is the L1 loss calculated by the DEM grayscale image and the ground true DEM image. W_a, W_b and W_c are three constants taking 1, 10 and 100 respectively, to weight the influence of each Loss function on the total loss.

In the prototype network, the specific expression L_P is as follows:

$$L_P = \sum |C_1 - C_2| \tag{4}$$

In this equation, the smaller the result of L_P, the more similar C_1 and C_2 are, indicating that the prototype feature matches the query better.

L_{G_DT} is defined as:

$$L_{G\_DT} = \frac{1}{Z}\sum(-\log(D(Y) + e)) \tag{5}$$

where Z denotes the size of the patch in Generator, Y denotes the DEM grayscale image generated by the generator network, and e is a constant close to 0,which is used to prevent the gradient from disappearing.

L_{G_L} is defined as:

$$L_{G\_L} = \frac{1}{M}|Y - Y_T| \tag{6}$$

where M denotes the number of the number of elements in the output terrain map Y, and Y_T denotes the ground true DEM grayscale map corresponding to the input semantics.

In the discriminant network, the optimization objective function is as follows:

$$L_D = \frac{1}{Z}[-\log(D(Y_T) + e) + \log(1 - D(Y) + e)] \tag{7}$$

The first item is the Loss function of the result of discrimination against the real label, and the second item is the prediction of D discrimination to the output of the generator.

3 Implementation Results and Analysis

3.1 Datasets and the Evaluation Method

Dataset. We evaluated the effectiveness our TSTG-Nets model on the public benchmark dataset and provided by a Kaggle Competitions [27]. The Kaggle dataset, it composes of 5000 groups, and each group consists of terrain texture map and its corresponding semantics. The resolution of each image in the group is both 512 × 512, randomly cropped from a map drawn from global terrain. Each terrain segmentation map is created with random parameters, and the initial obtained segmentation map is filtered by median and abrupt pixels are smoothed to eliminate noise. We take the 5000 groups images and divide them into training, validation, and testing sets according to the ratio of 06:06:0.1 to do the experiments.

Evaluation Method. We use SSIM, PSNR and FID as evaluation indicators. In our experiments, we use some evaluation metrics to evaluate the performance of the model, where SSIM (Structural Similarity) serves as a standard for representing the similarity between two images. The SSIM is calculated as:

$$SSIM = \frac{(2\mu_x\mu_y + c_1)(2\sigma_{xy} + c_2)}{\left(\mu_x^2 + \mu_y^2 + c_1\right)\left(\sigma_x^2 + \sigma_y^2 + c_2\right)} \quad (8)$$

where x is the result graph obtained by the network model, and y is the corresponding label in the dataset. μx is the mean of x, μ_{xy} is the covariance of x and y, μ_{xy} is the variance of x, and c1 and c2 are two constants, similar to about y symbols.

Besides, PSNR (Peak Signal to Noise Ratio) is used to measure the quality of an image, such as clarity, which is based on the MSE (Mean Square Error), which can be calculated as:

$$MSE = \frac{1}{mn}\sum_{i=0}^{m-1}\sum_{j=0}^{n-1}\left[X(i,j) - Y(i,j)^2\right] \quad (9)$$

where X stands for result graph obtained by the network model, Y for the corresponding label in the dataset, m for the height of the image, and n for the width of the image.

Based on MSE, PSNR can be defined as:

$$PSNR = 10 \times \log_{10} \frac{L^2}{MSE} \quad (10)$$

where L is the maximum intensity of the image, here we take 255 for it.

The FID indicator is similar to SSIM in that it measures the vector distance between two images. Unlike SSIM, a smaller FID value indicates a higher quality of generated images. The formula is:

$$FID = \left\|\mu_r - \mu_g\right\|^2 + Tr\left(\sum r + \sum g - 2\left(\sum r \sum g\right)^{\frac{1}{2}}\right) \quad (11)$$

where μ_r denotes the mean vector of image set generated by *r*, and μ_g denotes the mean vector of image set generated by *g*.

3.2 Network Hyperparameters

This study carry on the experiment on the computer with operating system Windows 10, and with a memory of 16GB, a CPU of In-telRCoreTM i5–8400 3.80GHz, and a graphics card of NVIDIA GeForce GTX 1070 Ti 8GB. The deep learning framework used is Python 1.1.0 and Python 3.7, and the experimental running environment for this chapter is Jupyter Notebook. The network hyper-parameter of the experiment is set as the learning rate 2e-4, the optimizer the Adam optimizer, and the batch size 4. For each dataset, the overall training round is 200 epochs.

3.3 Experimental Results and Analysis

We compare the experimental results with the existing related methods on the terrain dataset used in this article. The comparison methods are Pix2pix [23], Oasis [28] and SPADE-GAN [29]. In order to adopt the principle of fairness, the discriminator of the network is changed to PatchGan, and the Learning rate of 2e-4 and the epoch of 200, the and batch size are selected as 4.

Figure 3 shows the visualization results of generating 3D terrain in the experiment. In Fig. 3. In each row, the 3D terrain results are samples from the same sample user semantics showed in the first image in each row, and the other images are comparative experimental results from different network structures. Each row produces different samples one by one. From the results of these 3D terrain generation, it can be seen that our prototype learning method proposed in this article has the most ideal experimental results.

Table 1 shows the comparative results of the same terrain dataset on the Pix2pix [23], SPADE GAN [28], Oasis [29], and the methods used in this paper.

It can be seen that compared with the pix2pix network using Unet as the generator, the method in this paper leads by 0.835 and 22.861 on SSIM and PSNR, respectively, and is 0.107 and 3.61% higher than the pix2pix network; Compared with the Oasis and SPADE in GAN methods, the SSIM index is 0.099 and 0.023% points higher, and the PSNR index is 2.686 and 1.399% points higher, respectively. In terms of FID indicators, the method in this study achieved a minimum value of 20.915.

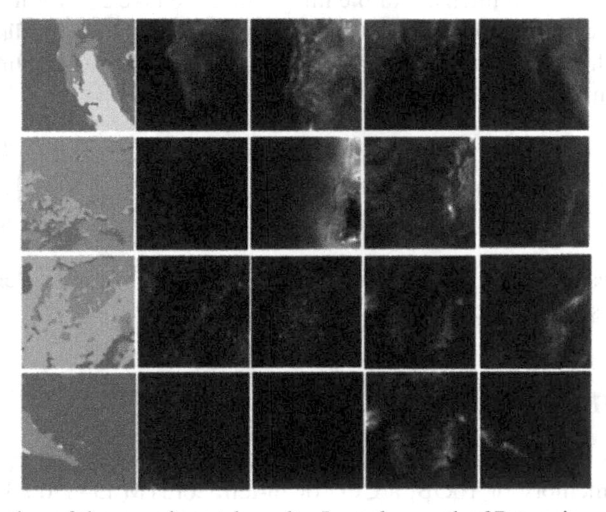

Fig. 3. Visualization of the experimental results. In each row, the 3D terrain results are samples from the same sample user semantics showed in the first image in each row, and the other images are comparative experimental results from different network structures. Each row produces different samples one by one.

In addition, we conducted ablation experiments, take the pix2pix network as the Backbone network, record the prototype generation module as PGM, and record the

Table 1. Comparison results (%)

Methods	SSIM	PSNR	FID
Pix2pix	0.728	19.251	186.925
Oasis	0.736	20.165	30.813
SPADE-GAN	0.812	21.462	96.215
ours	0.835	22.861	20.915

prototype matching model as PAM. Table 2 shows the comparative results of the ablation experiments.

From the table, it can be seen that the network obtained by combining backbone with prototype generation module is 0.053% points higher on SSIM than backbone, 0.98% points higher on PSNR, and the FID index drops to 96.154; The network obtained by combining backbone with prototype matching module is 0.069% points higher on SSIM than backbone, 1.502% points higher on PSNR, and the FID index is reduced to 67.146; The method in this chapter combines PGM and PAM to obtain a network that leads by 0.835 and 22.861 compared to backbone, and the FID index drops to 20.915. Therefore, it can be concluded that PGM and PAM have certain effectiveness.

Table 2. Ablation experiment(%)

Methods	SSIM	PSNR	FID
Backbone	0.728	19.251	186.925
Backbone + PGM	0.781	20.231	96.154
Backbone + PAM	0.797	20.753	67.146
Ours(Backbone + PGM + PAM)	0.835	22.861	20.915

4 Summary

Customizing 3D terrain based on user semantics plays an important role in military simulation, this paper introduces the methods and shortcomings of existing terrain generation techniques, and provides a detailed introduction to the prototype learning algorithm. In order to meet the personalized needs of users, this article proposes a prototype learning based terrain generation network. It extracts terrain semantics based prototype features from a small number of terrain surface samples, and then transfers the pre-learned features to user customization. Specifically, a prototype learning based framework is designed, including a terrain texture generation module (TGM), prototype feature generation module (PGM), and multiple prototype features matching module

(FMM). An effective loss function is designed based on the similarity matching algorithm of prototype learning, which greatly improves the image quality of DEM gray-scale image.

References

1. Trapp, M., Döllner, J.: Interactive close-up rendering for Detail+Overview visualization of 3D digital terrain models. In: Banissi, E., Ursyn, A. (eds.) IV, Part I, 275–280 (2019)
2. Jain, A., Sharma, A., Rajan, K.S.: Adaptive & multi-resolution procedural infinite terrain generation with diffusion models and Perlin noise. In: Biswas, S., Raman, S., Roy-Chowdhury, A.K. (eds.) ICVGIP 2022, pp. 55:1–55:9. ACM (2022)
3. Scott, J.: Realism in Data-Based Terrain Synthesis. Ph.D. thesis, Victoria University of Wellington, New Zealand (2020)
4. Hojo, A., Takagi, K., Avtar, R., Tadono, T., Nakamura, F.: Synthesis of l-band SAR and forest heights derived from tandem-x DEM and 3 digital terrain models for biomass mapping. Remote. Sens. **12**(3), 349 (2020)
5. Liu, L., Xiao, S., Zhou, Z.: Aircraft engine remaining useful life estimation via a double attention-based data-driven architecture. Reliab. Eng. Syst. Saf. **221**(5), 108330 (2022)
6. Degiorgio, V., Mandelbrot, B.B.: The Fractal Geometry of Nature (1984)
7. Fournier, A., Fussell, D.S., Carpenter, L.C.: Computer rendering of stochastic models. Commun. ACM **25**(6), 371–384 (1982)
8. Bruun, B.T., Nilsen, S.: Multiscale representation of terrain models using average interpolating wavelets. In: Bjørke, J.T., Tveite, H. (eds.) ScanGIS 2001, pp. 33–44 (2001)
9. Perlin, K.: An image synthesizer. In: Cole, P., Heilman, R., Barsky, B.A. (eds.) SIGGRAPH 1985, pp. 287–296. ACM (1985)
10. Wang, C., et al.: Model, design, and testing of an electret-based portable transmitter for low-frequency applications. IEEE Trans. Antennas Propag. **2021**(69), 5305–5314 (2021)
11. Roudier, P., Peroche, B., Perrin, M.: Landscapes synthesis achieved through erosion and deposition process simulation. Comput. Graph. Forum **12**(3), 375–383 (1993)
12. Olsen, J.: Realtime procedural terrain generation-realtime synthesis of eroded fractal terrain for use in computer games. Combination perturbation (2004)
13. Dachsbacher, C.: Interactive terrain rendering: towards realism with procedural models and graphics hardware. Ph.D. thesis, University of Erlangen-Nuremberg, Germany (2006)
14. Brosz, J., Samavati, F.F., Sousa, M.C.: Terrain synthesis by-example. In: Braz, J., Jorge, J.A., Dias, M.S., Marcos, A. (eds.) GRAPP 2006, pp. 122–133. INSTICC (2006)
15. Lim, F.Y., Tan, Y.W., Bhojan, A.: Visually improved erosion algorithm for the procedural generation of tile-based terrain. CoRR abs/2210.14496 (2022)
16. Valencia-Rosado, L.O., Guzman-Zavaleta, Z.J., Starostenko, O.: Generation of synthetic elevation models and realistic surface images of river deltas and coastal terrains using cGANs. IEEE Access **9**, 2975–2985 (2021)
17. Emmendorfer, L.R., Emmendorfer, I.B., de Almeida, L.P.M., Leal Alves, D.C., Neto, J.A.: A self-interpolation method for digital terrain model generation. In: Gervasi, O. (ed.) ICCSA 2021. LNCS, vol. 12949, pp. 352–363. Springer, Cham (2021). https://doi.org/10.1007/978-3-030-86653-2_26
18. Shan, D., Zhang, Y., Liu, X., Liu, S., Coleman, S.A., Kerr, D.: MMPL-Net: multi-modal prototype learning for one-shot RGB-D segmentation. Neural Comput. Appl. **35**(14), 10297–10310 (2023)
19. Wang, Q., Zhang, W., Yang, W., Xu, C., Cui, Z.: Prototype-guided instance matching for multiple pedestrian tracking. Neurocomputing **538**, 126207 (2023)

20. Wang, K., Liew, J.H., Zou, Y., Zhou, D., Feng, J.: PANet: few-shot image semantic segmentation with prototype alignment. In: 2019 IEEE/CVF, ICCV 2019, pp. 9196–9205. IEEE (2019)
21. Li, G., Jampani, V., Sevilla-Lara, L., Sun, D., Kim, J., Kim, J.: Adaptive prototype learning and allocation for few-shot segmentation. In: CVPR 2021, pp. 8334–8343. Computer Vision Foundation/IEEE (2021)
22. Li, J., Zhou, P., Xiong, C., Hoi, S.C.H.: Prototypical contrastive learning of unsupervised representations. In: ICLR 2021, OpenReview.net (2021)
23. Isola, P., Zhu, J., Zhou, T., Efros, A.A.: Image-to-image translation with conditional adversarial networks. In: CVPR 2017, pp. 5967–5976. IEEE Computer Society (2017)
24. Ronneberger, O., Fischer, P., Brox, T.: U-net: Convolutional networks for biomedical image segmentation. In: Navab, N., Hornegger, J., Wells, W.M., Frangi, A.F. (eds.) MICCAI 2015. LNCS, vol. 9351, pp. 234–241. Springer, Cham (2015). https://doi.org/10.1007/978-3-319-24574-4_28
25. Feng, X., Zhang, W., Su, X., Xu, Z.: Optical remote sensing image denoising and super-resolution reconstructing using optimized generative network in wavelet transform domain. Remote. Sens. **13**(9), 1858 (2021)
26. He, K., Zhang, X., Ren, S., Sun, J.: Deep residual learning for image recognition. In: 2016 IEEE Conference on Computer Vision and Pattern Recognition (CVPR 2016), pp. 770–778. IEEE Computer Society (2016)
27. https://www.kaggle.com/datasets/tpapp157/earth-terrain-height-and-segmentation-map-images
28. Edgar S., Vadim S., Dan Z., Juergen G., Bernt S., Anna K.: You only need adversarial supervision for semantic image synthesis. In: International Conference on Learning Representations (ICLR) (2021)
29. Park, T., Liu, M.Y., Wang, T.C., Zhu, J.Y.: GauGAN: semantic image synthesis with spatially adaptive normalization. In: ACM SIGGRAPH 2019 Real-Time Live!. ACM (2019)

Research on Matrix Factorization Recommendation Algorithm Based on Local Differential Privacy

Yong Li, Xiao Song(✉), Ruilin Zeng, and Songsong Liu

Beihang University, Beijing 100083, China
songxiao@buaa.edu.cn

Abstract. Mobile Edge Computing (MEC) has gained significant attention in enhancing the efficiency of Recommendation systems. However, the trustworthiness of servers poses a challenge as they can potentially compromise user privacy. To address this issue, we propose a framework for matrix factorization-based recommendation using Local Differential Privacy (LDP). Initially, user data is perturbed using Piecewise Mechanism (a kind of LDP algorithm) and published to an edge server. The edge server performs basic computations on the perturbed data, while the cloud server employs matrix factorization to compute latent factors for users and items, which are then sent back to the edge server. Finally, the edge server computes similarity values and generates personalized recommendations for users. Through extensive simulations, our algorithm ensures recommendation accuracy while preserving user privacy. By comparing with the generalized differential privacy mechanism, the Piecewise Mechanism used in this paper has a better recommendation effect, thereby demonstrating its practical utility.

Keywords: Recommendation Algorithm · Matrix Factorization · LDP

1 Introduction

Recommendation System has become an integral part of our daily lives. Traditional cloud-based Recommendation systems face challenges such as network latency, which hinders users from receiving timely recommendations. With the advancements in wireless communication, Mobile Edge Computing (MEC) has emerged as a new computing paradigm that leverages edge servers to deploy cloud resources, enabling fast computation and storage capabilities. Therefore, reshaping the traditional cloud-centric recommendation framework into a cloud-edge collaborative mode, utilizing edge servers for data collection and partial computation tasks, can effectively reduce the impact of network latency on user experience [1].

One of the core methods in Recommendation systems is matrix factorization (MF) [2], where the recommendation server collects user rating data and uses it to predict items of interest to users. However, rating data often contains sensitive information about users, such as health conditions, political views, or sexual orientation, raising

© The Author(s), under exclusive license to Springer Nature Singapore Pte Ltd. 2024
F. Hassan et al. (Eds.): AsiaSim 2023, CCIS 1911, pp. 230–241, 2024.
https://doi.org/10.1007/978-981-99-7240-1_18

increasing concerns about privacy in personalized recommendations [3]. Traditional privacy protection methods often introduce data distortion, thereby compromising the quality of recommendations. Thus, striking a balance between privacy protection and recommendation accuracy becomes a crucial challenge.

Differential Privacy (DP) [4], an information-theoretic framework for privacy preservation, has gained significant traction in the realm of recommendation systems [5–7]. By injecting controlled levels of noise into the dataset, DP serves as a mechanism to rigorously quantify and limit the adversaries' ability to discern sensitive individual information. However, traditional differential privacy mechanisms assume that the data collection server is trusted, which is not easily achievable in practice. Considering the presence of untrusted servers, users need to protect their privacy themselves. Local Differential Privacy (LDP) [8–13], as an extension of differential privacy, allows each user to independently perturb their own data, while the server aggregates the perturbed data to extract valuable information. LDP provides stronger privacy protection compared to traditional differential privacy but introduces higher levels of noise. Thus, maintaining recommendation accuracy while applying LDP for user privacy protection becomes a technical challenge.

In this paper, we propose a user-based matrix factorization (MF) recommendation framework that leverages LDP techniques for privacy protection. In this framework, each user's original rating data is perturbed by the user themselves and sent to the edge server. The edge server preprocesses the perturbed data and sends it to the cloud computing center. The cloud computing center employs matrix factorization to compute the latent factors of users and items, ultimately calculating item similarity and returning the results to users for predicting ratings on unrated items. The proposed data perturbation method not only provides privacy protection for the original rating values but also reduces the noise introduced by perturbation. We have conducted a series of simulation experiments to demonstrate the performance of the proposed method. The main contributions of this paper can be summarized as follows:

- We design a CF recommendation framework based on Local Differential Privacy, which intelligently utilizes the cloud-edge collaborative framework to enhance recommendation efficiency while protecting user privacy.
- We use Piecewise Mechanism (PM) to perturb the data to protect the user's rating values without significantly degrading the quality of the recommendations and ensuring user privacy protection.
- We evaluate the performance of the proposed method through simulations using real-world data. The results show that the method provides robust privacy protection while preserving recommendation quality.

The rest of the paper is organized as follows. Section 2 presents the background knowledge of LDP and Matrix factorization. Section 3 introduces the proposed recommendation framework. Simulation results are provided in Sect. 4. Finally, conclusions are drawn in Sect. 5.

2 Preliminaries

2.1 Local Differential Privacy

Unlike centralized differential privacy, local differential privacy does not rely on trusted servers but instead performs differential privacy processing on the user side and aggregates the processed data on the server side. The definition of local differential privacy is as follows:

Definition 1. (ε-local differential privacy). Assuming the existence of a random algorithm f, with the domain and range being D and R, respectively, if there are n users and n records, where each user corresponds to a record, and if the algorithm f produces similar output results t^* ($t^* \in R$) for any two records t and $t^{'}$ ($t, t^{'} \in D$), satisfying the following inequality:

$$Pr\left[f(t) = t^*\right] \leq e^{\varepsilon} Pr\left[f\left(t^{'}\right) = t^*\right] \tag{1}$$

then f satisfies ε-local differential privacy. Local differential privacy provides stronger privacy protection for data, preventing data leakage in malicious third-party entities. In the definition of differential privacy, ε is referred to as the privacy budget, which measures the level of privacy preservation. A smaller ε implies less distortion in the data, lower privacy loss, and lower security, while a larger ε results in greater data distortion, higher privacy loss, and higher security.

This paper utilizes the following properties of differential privacy:

Lemma 1. (post-processing). Given random algorithms F_1 and F_2, assuming F_1 satisfies ε -differential privacy, if we define algorithm F' as $F' = F_1(F_2(x))$, then F' also satisfies ε -differential privacy.

The compositionality of differential privacy can be divided into sequential compositionality and parallel compositionality, ensuring that multiple differential privacy models can be used to address the same problem.

Lemma 2. (serial combination). Assuming there are n mutually independent random algorithms $F_i(i = 1, 2, \ldots, n)$, applied to dataset I individually, satisfying ε_i-differential privacy, then applying F_i simultaneously to dataset I provides $\sum_{i=1}^{n} \varepsilon_i$-differential privacy.

2.2 Matrix Factorization Algorithm

Local differential privacy mechanism is used to perturb the dataset in order to protect user privacy. In recommendation systems, matrix factorization algorithms are commonly employed. Let's assume the user-item rating matrix R is of size $m \times n$, where R represents the rank of the matrix. The task of the recommendation server is to predict the rating values for items that haven't been rated by users. However, due to the sparsity of rating data, the number of ratings k is much smaller than $m \times n$, especially in big data scenarios where both n and m are large. Therefore, recommenders can infer unknown rating information of users based on known rating information.

Matrix factorization techniques decompose the user-item rating matrix into two low-rank matrices, namely the user factor matrix P and the item factor matrix Q, for recommendation purposes. The advantage of matrix factorization lies in its ability to capture the underlying relationships between users and items, enabling effective recommendation even in the presence of sparse data. The relationship between users and items can be modeled through the inner product in the latent factor space:

$$R \approx PQ^T \tag{2}$$

Here, P is an $m \times d$ matrix (m denotes the number of users, d denotes the number of factors), and Q is an $n \times d$ matrix (n denotes the number of items). Each row of P corresponds to a user's factor vector p_u, and each row of Q corresponds to an item's factor vector q_i The goal of matrix factorization is to compute the factor vectors p_u for each user and q_i for each item, such that the estimated interest level $\hat{r}_{ui}$ of user u for item i, defined as $\hat{r}_{ui} = p_u {q_i}^T$, closely approximates the true rating r_{ui}.

When dealing with incomplete matrices, overfitting is prone to occur. Therefore, it is recommended to model only the existing rating entries and employ regularization to avoid overfitting issues. The objective function for optimization incorporates a regularization coefficient λ as defined in the model shown in Eq. (3).

$$(P, Q) = \min_{P,Q} \sum_{R} [(r_{u,i} - p_u q_i^{\mathrm{T}})^2 + \lambda(\|q_i\|^2 + \|p_u\|^2)] \tag{3}$$

where λ is the regularization coefficient, k represents the existing rating entries in the training set. The optimization method used in this case is stochastic gradient descent (SGD).

$$p_u^t = p_u^{t-1} - \gamma_t\{\nabla_{p_u} + 2\lambda_u p_u^{t-1}\} \tag{4}$$

$$q_i^t = q_i^{t-1} - \gamma_t(\nabla_{q_i} + 2\lambda_q q_i^{t-1}\} \tag{5}$$

Here, p_u^t and q_i^t represent the values of p_u and q_i at the t-th iteration, and γ_t is a constant representing the learning rate at the t-th iteration. ∇_{p_u} and ∇_{q_i} are the gradients of p_u and q_i.

3 Local Differential Privacy Based Recommendation Algorithm

3.1 Overall Framework

This study focuses on the cloud-edge collaborative recommendation scenario and aims to investigate the impact of the local differential privacy mechanism on the privacy protection process and its influence on the recommendation effectiveness of matrix factorization collaborative filtering algorithms. Traditional recommendation algorithms typically process large volumes of data on trusted cloud servers to achieve high accuracy.

However, centralized differential privacy processing relies on trusted servers to prevent potential privacy breaches by malicious servers. To address this challenge, this paper employs the local differential privacy mechanism, which perturbs user data locally and aggregates the processed rating data at the edge server. The cloud server then utilizes this aggregated data, which has undergone privacy-preserving transformations, to complete the recommendation tasks. The research framework comprises three key components: the client, the edge server, and the cloud server. The client stores the original user data and has the capability to apply local differential privacy perturbations. The edge server performs basic processing on the dataset, while the cloud server analyzes the uploaded data and returns the latent factors of users and items through matrix factorization. The overall framework is illustrated in Fig. 1.

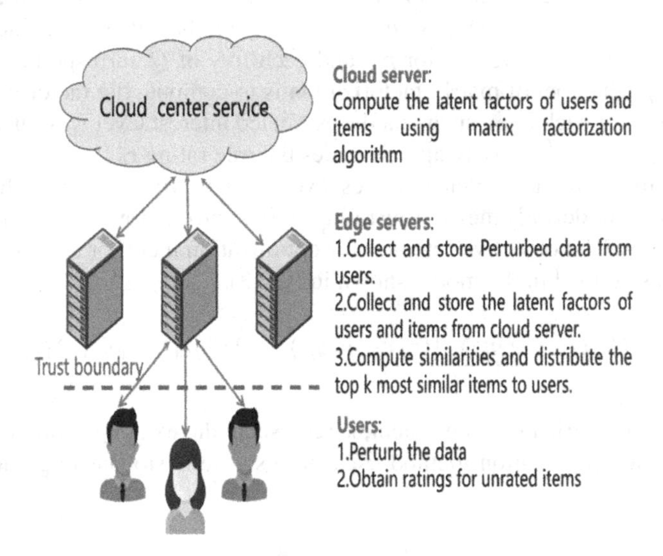

Fig. 1. Local Differential Privacy Recommendation Algorithm Framework.

3.2 Local Differential Privacy-Based Recommendation

The differential privacy matrix factorization collaborative filtering algorithm consists of three phases.

Phases 1: The first phase involves perturbing the dataset using the local differential privacy mechanism and transmitting it to the edge computing center.

Phases 2: In the second phase, the edge computing center performs preprocessing on the perturbed dataset and transmits the processed dataset to the cloud computing center. The cloud computing center conducts matrix factorization on the processed dataset, yielding the latent factors of users and items, which are then sent back to the edge server.

Phases 3: The third phase entails similarity calculations on the edge server to determine the similarity between users and items, thereby generating predicted scores for user-item pairs.

A. Piecewise Mechanism. In the initial phase, the data is subjected to encoding and perturbation using a local differential privacy mechanism [14] to safeguard the privacy of the data. The perturbed data is subsequently analyzed on the server side. For this research, we utilize a Piecewise Mechanism for perturbing the original ratings. The Piecewise Mechanism is advantageous over the Duchi Mechanism [11, 14] and the Harmony Mechanism [14] as it is capable of handling multidimensional data while preserving the continuity of the resulting output data.

The Piecewise Mechanism [14] perturbs the original input values t_i, belonging to the input domain $[-1, 1]$, through differential privacy, resulting in a perturbed value t_i^* within the specified output range $[-C, C]$.

In Piecewise Mechanism, the probability density function (*PDF*) of the perturbed value follows a piecewise constant function, as defined in Eq. (6).

$$pdf\left(t_i^* = x|t_i\right) = \begin{cases} p, & if\ x \in [l(t_i), r(t_i)] \\ \frac{p}{exp(\varepsilon)}, & if\ x \epsilon [-C, l(t_i)] \cup [r(t_i), C] \end{cases} \tag{6}$$

The output values are bounded within the range $[-C, C]$. The parameters in the function are determined by Eqs. (6) and (7).

$$C = \frac{exp\left(\frac{\varepsilon}{2}\right) + 1}{exp\left(\frac{\varepsilon}{2}\right) - 1}, \quad p = \frac{exp(\varepsilon) - exp\left(\frac{\varepsilon}{2}\right)}{2exp\left(\frac{\varepsilon}{2}\right) + 2} \tag{7}$$

$$l(t_i) = \frac{C+1}{2} \cdot t_i - \frac{C-1}{2}, \quad r(t_i) = l(t_i) + C - 1 \tag{8}$$

This algorithm introduces perturbation to the original data, effectively returning a value from the interval $[l(t_i), r(t_i)]$ with probability q, and returning a value from the interval $[-C, l(t_i)] \cup [r(t_i), C]$ with probability *1-q*. The value of q is determined by Eq. (8).

$$q = \frac{e^{\varepsilon/2}}{e^{\varepsilon/2} + 1} \tag{9}$$

In the Piecewise Mechanism, when the algorithm's input domain is $[-1, 1]$ and the rating data domain is $[a, b]$, it is usually necessary to normalize the data that exceeds this input domain before applying differential privacy. Similarly, after processing the data, it is necessary to convert the output data back to the original value domain. Let's assume the perturbed value is denoted as r°, and the returned rating value is denoted as r. The relationship between them satisfies Eq. (10):

$$r = \frac{(r^{\circ}/z) \times (b - a) + a + b}{2} \tag{10}$$

In Eq. (10), the value of z is taken as C in the Piecewise Mechanism. Here, a and b represent the left and right boundaries of the perturbed value's data domain, respectively. This transformation allows the perturbed value to be returned as a rating value. The pseudocode for the operation of the Piecewise Mechanism is provided in Algorithm 1.

Algorithm 1: Piecewise Mechanism

Input: t: Original data, $t \in [-1,1]$

Output: t': Perturbed data, $t' \in [-C, C]$

1: Randomly select a random number x from the interval $[0,1]$

if $x < \frac{e^{\varepsilon/2}}{e^{\varepsilon/2}+1}$ **then**

Choose a value for t' from any point within the interval $[l(t), r(t)]$

else

Choose a value for t' from any point within the interval $[-C, l(t)] \cup [r(t), C]$.

return t'

B. Matrix Factorization. Upon acquiring the perturbed rating matrix, the subsequent phase entails the processing of perturbed ratings via matrix factorization. Utilizing a matrix factorization algorithm on the perturbed rating matrix, the extraction of user latent factors P' and item latent factors Q' is achieved. This computational process is executed on the cloud server.

According to Eq. 2 – Eq. 5, we can finally get the user potential factor and item potential factor after perturbation. Each row of P' corresponds to a user's factor vector p'_u, and each row of Q' corresponds to an item's factor vector q'_i. The whole process is implemented by Algorithm 2.

Algorithm 2: Matrix factorization algorithm with perturbations

Input: t': Perturbed data, $t' \in [a, b]$

d–number of factors,

λ–regularization parameter,

Output: Approximate factor matrices $P_{n \times d}$ and $Q_{m \times d}$

$$(P', Q') = \min_{P', Q'} \sum_{R'} [(r'_{u,i} - p'_u q'^{\mathrm{T}}_i)^2 + \lambda(\| q'_i \|^2 + \| p'_u \|^2)]$$

return P' and Q'

Subsequently, the cloud server sends the derived user latent factors and item latent factors to the edge server.

C. Similarity Calculation. The computation in the third stage takes place in the edge server, where the calculation of distance between row vectors is performed to determine the correlation among rows, representing the similarity between users. Common distance metrics utilized include Pearson correlation coefficient, cosine similarity, and Euclidean distance [30]. Through testing, this study selects Pearson correlation coefficient to assess the similarity between users accurately.

The similarity between user p and user q, measured using Pearson correlation coefficient, is calculated as shown in Eq. (10). The formula is adjusted by considering the average ratings of users to enhance the precision of similarity measurements.

$$sim_{u,v} = \frac{\sum_{i\in I}\left(R'_{u,i} - \overline{R'_u}\right)\left(R'_{v,i} - \overline{R'_v}\right)}{\sqrt{\sum_{i\in I}\left(R'_{u,i} - \overline{R'_u}\right)^2}\sqrt{\sum_{i\in I}\left(R'_{u,i} - \overline{R'_u}\right)^2}} \tag{11}$$

In Eq. (11), I represents the set of all movies, $\overline{R'_u}$ and $\overline{R'_v}$ denote the average ratings of users p and user q, respectively, while $R'_{u,i}$and$R'_{v,i}$ represent the ratings of users users p and user q for item i.

Subsequently, based on the calculated user similarity, the set of K most similar users to the target user is determined. Equation (11) is employed to evaluate the preference level of the target user for movies, and the top N items preferred by the target user are recommended in descending order of preference.

4 Simulation and Analysis

4.1 Simulation Experiment

A. Dataset. The dataset utilized in this study is the MovieLens dataset [15], which is widely recognized as a publicly available movie rating dataset. Although movie ratings are generally considered non-sensitive information, it is essential to acknowledge the potential for attackers to extract users' personal data, such as personal preferences, health status, political inclination, etc., through their ratings. For this study, we selected the smaller subset of the MovieLens Latest Datasets, which comprises more than 100,000 ratings provided by over 600 users for a collection of more than 9,000 movies. Users have the option to rate each movie on a scale of 0.5 to 5, with increments of 0.5.

B. Evaluation Metric. The matrix factorization-based recommendation algorithm employs similarity calculations to determine the level of preference for each movie by each target user. Subsequently, the algorithm ranks the movies in descending order of preference and recommends the top K movies to the user. Upon obtaining the recommendation results, it is imperative to evaluate the efficacy of the recommendation algorithm using the following evaluation metrics.

The first evaluation metric is Precision, which quantifies the ratio of correctly recommended movies to the total number of recommended movies. In other words, it measures the accuracy of the matrix factorization recommendation algorithm by assessing the proportion of accurately recommended movies in the overall recommendation set.

$$Recall = \frac{\sum_u |R(u) \cap T(u)|}{\sum_u |T(u)|} \tag{12}$$

The second evaluation metric is Recall, which measures the proportion of correctly predicted movies out of the total movies that the user has rated. It gauges the comprehensiveness of the matrix factorization recommendation algorithm by evaluating the ratio of correctly recommended movies to the total number of movies that the user has rated.

$$Precision = \frac{\sum_u |R(u) \cap T(u)|}{\sum_u |R(u)|} \tag{13}$$

These evaluation metrics serve as means to assess the recommendation effectiveness of the matrix factorization algorithm in a more scholarly manner.

C. Parameter Setup. The implemented recommendation algorithm includes several parameters. These parameters are the train-test split ratio (*train_rate*) for dividing the dataset into training and testing sets, the number of similar users chosen (K) in the recommendation algorithm, the number of top movies recommended to the user (K), and the privacy budget (ε) in the local differential privacy mechanism. Except for experiment three, the privacy budget remains constant throughout the experiments. The specific settings for these parameters are shown in Table 1.

In the experiments, we vary the privacy budget size in the three local differential privacy mechanisms. We set the privacy budget ranging from 0.1 to 2. For each privacy budget value, we repeat the experiment ten times and take the average to obtain the recommendation performance of the matrix factorization algorithm under each privacy budget.

These adjustments enable us to assess the recommendation performance of the matrix factorization algorithm under different privacy budget sizes in a systematic manner.

Table 1. Parameter table

Parameter	Value
train_rate	0.2
K	30
N	5
ε	1.0

4.2 Simulation Results

This study employed a Piecewise Mechanism (PM) to perturb rating data, and the results are presented below. The method used for comparison is the perturbed matrix factorization algorithm, with the generalized random response mechanism as additional comparisons. Figure 2 shows the results of matrix factorization (MF) and its mean value (MF_AVG), while Fig. 3 shows the experimental results under different privacy budgets (comprising methods such as GRR, PM, and MF_AVG). In both sets of experiments, MF_AVG is employed as the reference benchmark method.

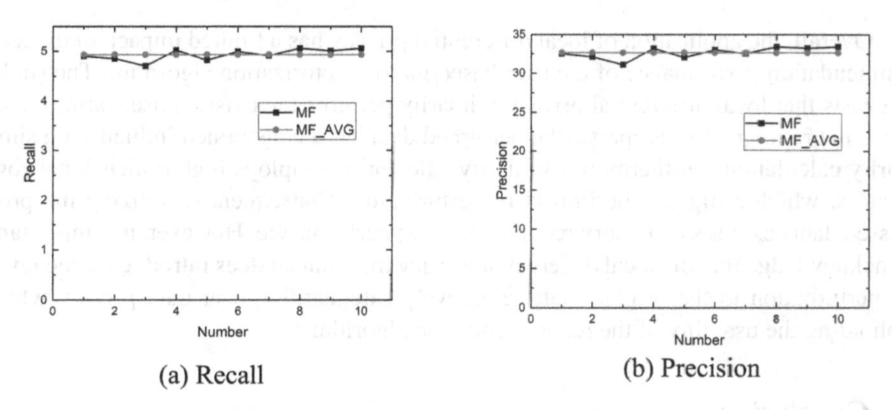

(a) Recall (b) Precision

Fig. 2. The results of matrix factorization and its mean value.

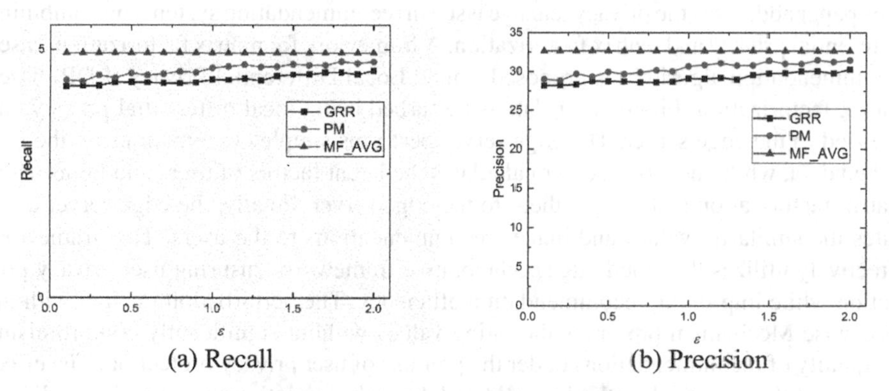

(a) Recall (b) Precision

Fig. 3. The experimental results under different privacy budgets

Based on the empirical observations, it becomes evident that an escalation in the privacy budget parameter (ε) correlates with an augmented probability of user privacy breach. Simultaneously, the amplitude of noise is curtailed, engendering enhancements in the performance of the Recall and Precision metrics. Furthermore, the application of local differential privacy results in a certain degree of degradation in the recommendation performance of the user-based matrix factorization algorithm. However, among the two methods considered, the Piecewise Mechanism exhibits superior performance.

From the analysis conducted, it is suggested that the generalized random response mechanism belongs to the local differential privacy method based on frequency statistics. By randomly perturbing the original data to any candidate value, it is more suitable for count queries. However, in the context of movie recommendation, this mechanism may introduce too many errors. In contrast, the Piecewise Mechanism perturbs the values to any value within the range, resulting in less scattered perturbation values and a smaller impact on similarity calculation (When the privacy budget is set to 2, the accuracy of the PM mechanism decreases by 4%), thus yielding improved recommendation performance.

Overall, the application of local differential privacy has a limited impact on the recommendation performance of the user-based matrix factorization algorithm. The study suggests that local differential privacy primarily perturbs the existing user ratings, and since the rating matrix is sparse, the perturbed data has a constrained influence on similarity calculation. Furthermore, similarity calculation employs high-dimensional row vectors, which mitigates the impact of perturbation. Consequently, utilizing the processed data ensures satisfactory recommendation performance. However, it is important to acknowledge that the local differential privacy mechanism does introduce some level of perturbation to the original data, effectively safeguarding user data privacy while upholding the usability of the recommendation algorithm.

5 Conclusion

This paper addresses the privacy leakage issue in recommendation systems by combining differential privacy and matrix factorization. A framework for matrix factorization-based recommendation algorithm is proposed, called Local Differential Privacy (LDP) based matrix factorization. Firstly, user data is perturbed using local differential privacy and released to the edge server. The edge server performs simple computations on the perturbed data, while the cloud server calculates the latent factors of users and items using matrix factorization and returns them to the edge server. Finally, the edge server computes the similarity values and makes recommendations to the users. This framework effectively utilizes the cloud-edge collaborative framework, ensuring user privacy protection while improving recommendation efficiency. The perturbation method with the Piecewise Mechanism preserves the rating values without significantly compromising the quality of recommendations under the premise of user privacy protection. The effectiveness of the proposed method is validated through simulations using real-world data. The simulation results demonstrate that the method provides strong privacy protection while preserving the quality of recommendations.

Acknowledgements. This work was supported by the National Key Research and Development Program of China (No. 2020YFB1712203).

References

1. Cheng, Y.S., Hsu, P.Y., Liu, Y.C.: Extracting attributes for recommender systems based on MEC theory. In: 2018 3rd International Conference on Computer and Communication Systems (ICCCS), pp. 125–129. Nagoya, Japan (2018)
2. Koren, Y., Bell, R., Volinsky, C.: Matrix factorization techniques for recommender systems. Computer **42**(8), 30–37 (2009)
3. Weinsberg, U., Bhagat, S., Ioannidis, S., Taft, N. BlurMe: inferring and obfuscating user gender based on ratings. In: Conference on Recommender Systems. ACM (2012)
4. Dwork, C., Roth, A.: The algorithmic foundations of differential privacy. Found. Trends Theor. Comput. Sci./ **9**(3–4), 211–407 (2014)
5. Mcsherry, F., Mironov, I.: Differentially private recommender systems: building privacy into the Netflix Prize contenders. In: Knowledge Discovery and Data Mining. ACM (2009)

6. Zhu, X., Sun, Y.: Differential privacy for collaborative filtering recommender algorithm. In: ACM on International Workshop on Security & Privacy Analytics, pp. 9–16 (2016)
7. Chen, Z., Wang, Y., Zhang, S., et al.: Differentially private user-based collaborative filtering recommendation based on k-means clustering. Expert Syst. Appl. **168**, 114366 (2021)
8. Warner, S.L.: Randomized response: a survey technique for eliminating evasive answer bias. J. Amer. Stat. Assoc. **60**, 63–69 (1965)
9. Holohan, N., Leith, D.J., Mason, O., et al.: Optimal differentially private mechanisms for randomised response. IEEE Trans. Inf. Forensics Secur. **12**, 2726–2735 (2017)
10. Duchi, J.C., Jordan, M.I., et al.: Local privacy and statistical minimax rates. In: 2013 IEEE 54th Annual Symposium on Foundations of Computer Science, pp. 429–438. IEEE (2013)
11. Cormode, G., Jha, S., et al. Privacy at scale: local differential privacy in practice. In: Proceedings of the 2018 International Conference on Management of Data, pp. 1655–1658 (2018)
12. Erlingsson, L., Pihur, V., Korolova, A.: RAPPOR: Randomized aggregatable privacy preserving ordinal response. In: Proceedings 23rd ACM SIGSAC Conference on Computer and Communication Security, pp. 1054–1067 (2014)
13. Fanti, G., Pihur, V., Erlingsson, L.: Building a RAPPOR with the unknown: privacy-preserving learning of associations and data dictionaries. Proc. Priv. Enhan. Technol. **3**, 1–21 (2016)
14. Wang, T., Blocki, J., Jha, S. K.: Locally differentially private protocols for frequency estimation. In: USENIX Security Symposium (2017)
15. Harper, F.M., Konstan, J.A.: The MovieLens datasets: history and context. ACM Trans. Interact. Intell. Syst. **5**, 1–19 (2015)

Enhancing Face Recognition Accuracy through Integration of YOLO v8 and Deep Learning: A Custom Recognition Model Approach

Mahmoud Jameel Atta Daasan[1] and Mohamad Hafis Izran Bin Ishak[2](✉)

[1] Universiti Teknologi Malaysia, Skudai Johor, Malaysia
[2] Control and Mechatronics Department, Universiti Teknologi Malaysia, Johor Bahru, Malaysia
hafis@fke.utm.my

Abstract. Student attendance and being present during lectures plays a very important role in the value of the lesson and understanding of each student. There are a number of methods to record the attendance of the students such as signature on paper, QR code, RFID method and finally model based system. The most used method is signature of students even though it is mostly used it suffers from unethical behavior by some students and students forgetting to register their attendance. This research will develop a system to improve attendance registration accuracy by integrating it with custom recognition model. The class will be installed with a device that consists of a microcontroller and camera. The system will take an image of the class and send it to the database to be analyzed and then after recognizing the faces each face will be registered followed by the date and time it has been detected. All data will be stored in a database. The database can be accessed by the lecturers at the end of the class to review the attendance of students without the need to waste time on manually registering it.

Keywords: YOLO v8 · CNN · Deep learning

1 Introduction

The correlation between higher education attendance and academic performance has been carefully investigated for decades. Concurrently, there has been discussion on whether secondary institutions should have mandatory attendance policies. Most of the studies discovered a favorable connection between attendance and academic achievement [1]. For having poor attendance records, certain higher education institutions have penalized students. To discourage poor attendance, most institutions haven't established an automated method for taking attendance; instead, teachers must manually enter each student's attendance information into the system, which may be time-consuming and laborious when there are a lot of students [2]. The most faced problem regarding attendance system is faced due to time spend on taking attendance by passing the list of students or scanning the QR code which could waste precious time that could be spend on lecturers. The current system that has been implemented could be affected by technical

© The Author(s), under exclusive license to Springer Nature Singapore Pte Ltd. 2024
F. Hassan et al. (Eds.): AsiaSim 2023, CCIS 1911, pp. 242–253, 2024.
https://doi.org/10.1007/978-981-99-7240-1_19

issues, maintenance, or non-accurate system recording. Hence a systematic attendance system is required to help save time for lecturers to do their daily job by providing an automated attendance system that records students attendance automatically. Biometric attendance is a common and contemporary method of simplifying attendance. It is possible to attach the biometric device to the database and automatically update attendance. The distinctness of thumb prints allows for the solution of the proxy issue. The students need to wait in line for biometric attendance means that this solution still loses time. This problem can be resolved in class by passing the biometric device around, but it can be upsetting. But none of the approaches to taking: There is still a challenge with attendance: there is no way to guarantee that pupils remain seated for the entire lesson. A student has the option of entering the classroom soon before attendance or leaving it right after it [3]. Face recognition technique that will be described in Sect. 3 will solve all the issues that are being faced. Students are not required to record their attendance manually. Since the camera takes record of attendance, the final attendance will be given after the students have been recorded throughout the session, which can be verified by the lecturer after the class and course that is given.

2 Literature Review

2.1 Convolution Neural Network

The CNN architecture, which automatically creates its own feature extractors from a big data set, can classify and distinguish varied object patterns straight from images without the need for pre-processing. In this research, a framework is established to extract and construct structured low-level properties of an object via CNN architecture. A CNN for structured low-level feature extraction uses additional processing, such as convolutional operation, local sampling, further convolutional operation, and subsampling, as shown in Fig. 1 [4].

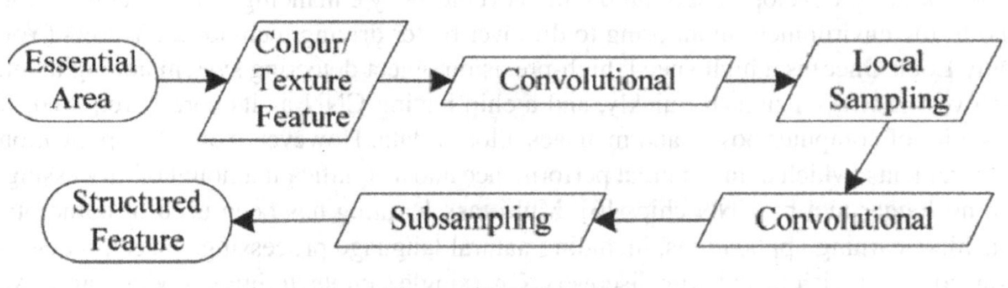

Fig. 1. CNN Structure

A simple convolution neural network model structure diagram shown in Fig. 2. Two convolution layers (C1, C2) and two sub-sampling layers (S1, S2) alternately make up a basic convolution neural network model. First, three trained filters (referred to as convolution kernels) and addable bias vectors convolution the original input image. The C1 layer produces three feature maps, and for each feature map Three new feature maps are produced in the S1 layer by applying a nonlinear activation function after the localized regions are averaged and weighted. The three trained filters of the C2 layer are then convoluted with these feature maps, and three feature maps are output through the S2 layer. The S2 layer's final output is vectorized before being used as an input for the standard neural network's training [5].

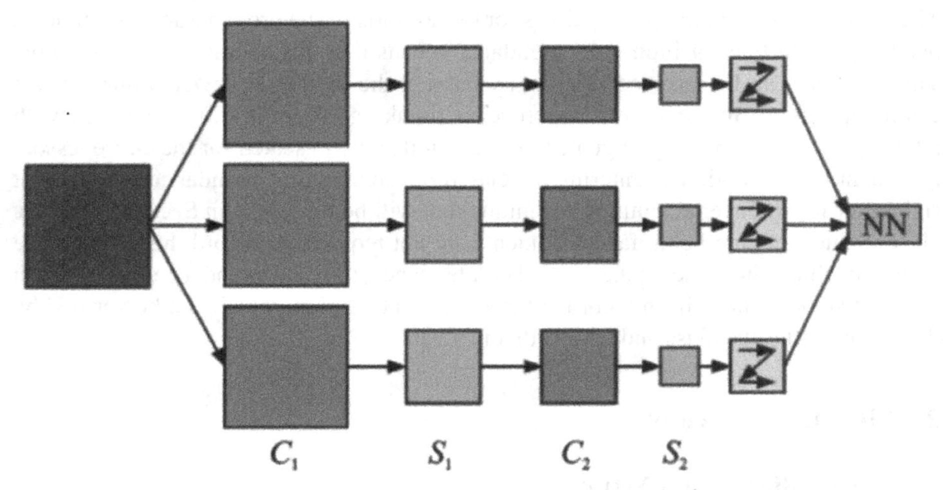

Fig. 2. Simplified CNN Structure [5].

2.2 Deep Learning

These days, object detection systems are used more and more in AI training. Object detection employs a few different strategies. For instance, it can be applied to factories that label their products to help operators classify components or devices. Additionally, it advances the development of autonomous vehicles by enhancing object identification and traffic environment monitoring to discover better driving techniques. Yolov4 (You Only Look Once) is a high-speed, high-precision object detecting system among them. Yolov4 must be calculated quickly, and a chip having CNN as its core is required. It has a lot of computer power and manages a lot of data. However, Yolov4's application requirements, which demand great performance and a significant amount of processing, are no longer met by CNN chips [6]. Multi-task learning has been used in numerous machine learning applications, including natural language processing, speech recognition, computer vision, and drug discovery. On a single sample, training models can make numerous predictions in areas like semantic segmentation and image classification [7].

2.3 Existing Model

YOLO V7. One of the most reliable object identification models currently available, YOLOv7, achieves cutting-edge performance in trade-offs between accuracy and speed. In the range of 5 to 160 frames per second, YOLOv7 outperforms all YOLO versions past and present in both speed and accuracy [8]. However, YOLOv7 is excessively memory heavy since, like past iterations of YOLO, it is based on deep neural networks and has millions of parameters. It is difficult to deploy them on devices with memory and resource limitations because of this [8].

3 Proposed Approach

3.1 Python

Currently, Python is the most popular programming language. Python is popular among programmers because it is simple but versatile. Python, despite being straightforward, can handle challenging tasks. Up till now, Python has been used for the back end. For image identification, Python is still a largely dependable programming language. The Python library provides good support for image recognition. One of the strongest and most useful Python modules is Scimitar Machine Learning, which is recognized for its ability to handle facial recognition and motion detection.

3.2 OpenCV

The most well-liked and probably simplest method for recognizing faces in Python is to use the OpenCV package. Python is currently a part of OpenCV, which was first developed in C/C + +. Bindings. It uses machine learning techniques to identify faces in photos. Faces are incredibly complex, made up of innumerable little patterns and traits that must fit.

3.3 Roboflow

No matter their level of expertise or experience, developers can create their own computer vision apps using Roboflow. It gives all the resources you require to develop a solid computer vision model and implement it in real-world settings.

3.4 Dataset

The incredible collection of images used to make the model was painstakingly selected and put together [8]. The dataset, which consists of a varied collection of photographs, covers a wide range of topics and situations. It has been meticulously gathered to guarantee its representativeness and applicability to the particular task at hand. The vast amount of high-quality photographs in the dataset, each meticulously labelled and annotated to aid the training process, is what makes it special. The model is well-positioned to achieve remarkable performance and generalization abilities with such a large dataset. By making the dataset more easily accessible through Roboflow as shown in Fig. 3, researchers and developers may easily tap into its power and produce cutting-edge AI applications. This dataset serves as evidence of the commitment and work made to create a solid and trustworthy model with the potential to have a big influence in a number of fields.

Fig. 3. Dataset in Roboflow

3.5 YOLO V8

Application areas for the newest and most advanced YOLO model, YOLOv8, include object detection, image categorization, and instance segmentation. YOLOv8 was created by Ultralytics, who also created the famous YOLOv5 model that shaped the sector. YOLOv8 features a few architectural upgrades and improvements over YOLOv5 (Fig. 4).

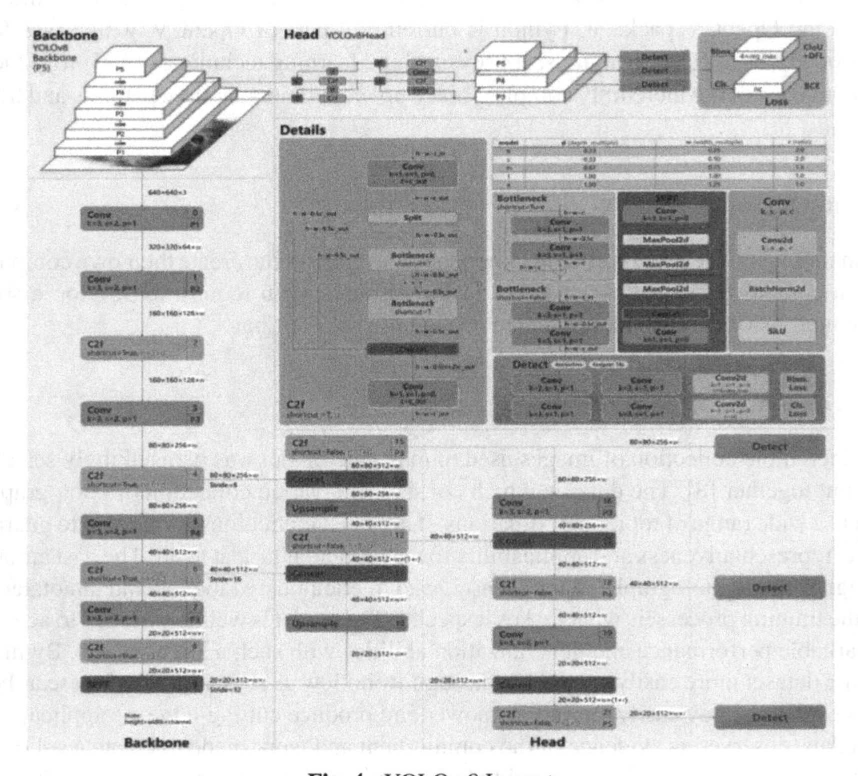

Fig. 4. YOLO v8 Layout

3.6 The Layout of YOLOv8

New Convolutions in YOLOv8. According to the inaugural post from Ultralytics, the YOLOv8 architecture has undergone several modifications and new convolutions:

1. The introduction of C2f, which replaced C3, caused alterations to the system's core. The stem's initial 6x6 convolution was changed to a 3×3 convolution. While just the output from the final Bottleneck was used in C3, C2f combines the outputs from the Bottleneck (which is made up of two 3×3 convs with residual connections).
2. Two convolutions were eliminated (YOLOv5 configuration #10 and #14).
3. The Bottleneck in YOLOv8 is identical to that in YOLOv5, with the exception that the kernel size of the first convolution was increased from 1×1 to 3×3. This modification denotes a move in favor of the ResNet block identified in 2015.

3.7 Reasons to Use YOLO v8

1. YOLOv8 has a good accuracy rate, according the COCO and Roboflow 100 tests.
2. YOLOv8 comes with a tonne of developer-friendly features, such as a clear CLI and a thoughtfully created Python package.
3. Because there is a significant community surrounding YOLO and a developing community surrounding the YOLOv8 model, there are many people in computer vision circles who could be able to assist you when you need advise.

YOLOv8 obtains great accuracy on COCO. For instance, the medium YOLOv8m model yields a 50.2% mAP when assessed on COCO. When assessed against Roboflow 100, a dataset that precisely evaluates model performance on various task-specific areas, YOLOv8 scored significantly higher than YOLOv5. More information about this is provided in the paper's performance study.

YOLOv8's developer-friendly features are also crucial. In contrast to previous models where tasks are dispersed across several executable Python files, YOLOv8 provides a CLI that makes training a model easy. A Python package that offers a more streamlined development experience than earlier versions is also available.

4 Proposed Methodology

4.1 Capturing Images

Using a laptop, a USB camera, or a camera mounted on the ceiling, pictures will be taken of all of the students throughout the first stage of my research. After that, the picture will be stored in the database.

4.2 Face Detection

A multitude of computer applications employ the face detection approach to locate people in digital images. Face detection recognizes many faces in an image and the essential facial characteristics that identify them, such as emotional state, or it uses the expressions on the faces to identify age, gender, and emotions. Face detection is usually the first step in many face-related technologies, such as face identification or verification. In Fig. 5, a workflow diagram is displayed.

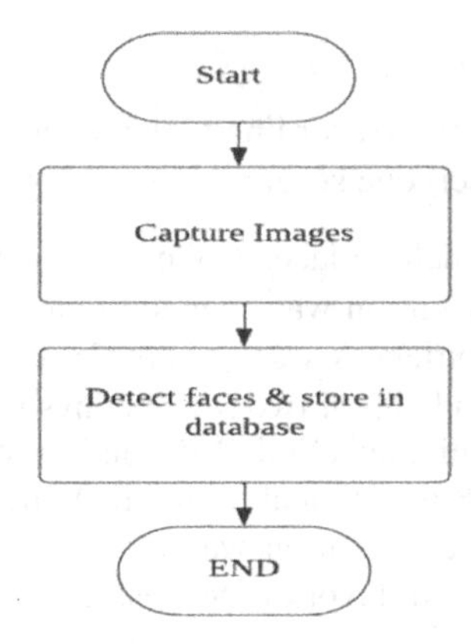

Fig. 5. Workflow Flowchart

4.3 Face Recognition

After getting the images from the database all images will run through the YOLO v8 model to recognize the faces in the images, after that the recognized faces will be stored in the attendance/record table with date and time of detection as shown in Fig. 6.

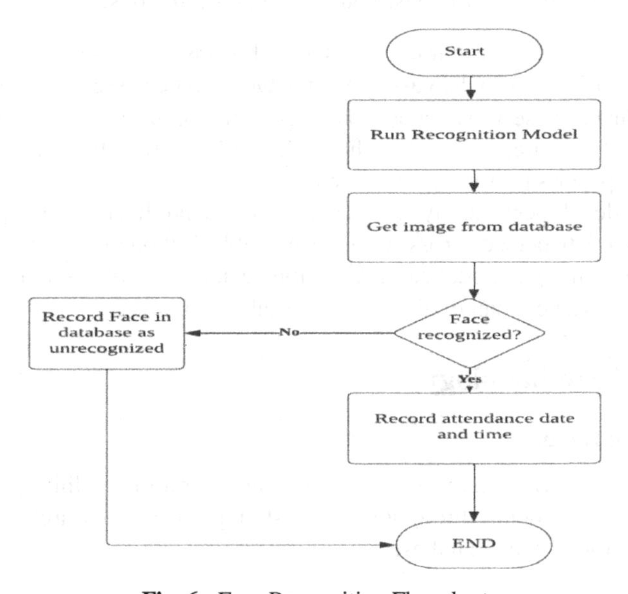

Fig. 6. Face Recognition Flowchart

4.4 Labelling Images

The first step in creating the face recognition model is to collect data for each face and upload it into Roboflow where the images could be annotated as shown in Fig. 7 and then stored into a database.

Finally, the labelled faces will be trained using YOLO V8 on google collab.

Fig. 7. Image Annotation

4.5 Training Phase

As shown in Fig. 8 the training has been set up to run for 100 epochs after uploading the API key for the database from Roboflow as shown in Fig. 7. After the training is done the confusion matrix will be shown with some test images to check the validation of the model (Fig. 9).

```
!mkdir {HOME}/datasets
%cd {HOME}/datasets

!pip install roboflow

from roboflow import Roboflow
rf = Roboflow(api_key="CatVCcTtIIEI5MnAMP1S")
project = rf.workspace("utm-vamfr").project("attendance-tuqb1")
dataset = project.version(5).download("yolov8")
```

Fig. 8. Dataset API

```
box_loss   cls_loss   dfl_loss  Instances       Size
   1.148     0.7756      1.178          6        800: 100% 40/40 [00:35<00:00,  1.13it/s]
  Images  Instances      Box(P          R       mAP50  mAP50-95): 100% 3/3 [00:01<00:00,  1.59it/s]
      91        105      0.828      0.722      0.856      0.574
      91         17      0.176      0.882      0.861      0.568
      91          2          1          0      0.398      0.293
      91          4          1      0.538      0.945      0.577
      91         17      0.882      0.882      0.901      0.589
      91         19      0.902      0.842      0.935      0.602
      91         22      0.894      0.909      0.958      0.646
      91         24      0.939          1      0.993      0.739
```

Fig. 9. Training Dataset

4.6 Model Graph Result

As shown in Fig. 10, the confusion matrix of the model is shown with 7 trained classes as well as the background was included, it is shown that most classes predictions are at 1.0 which is 100% with a small confusion between 2 classes which is needed to be improved.

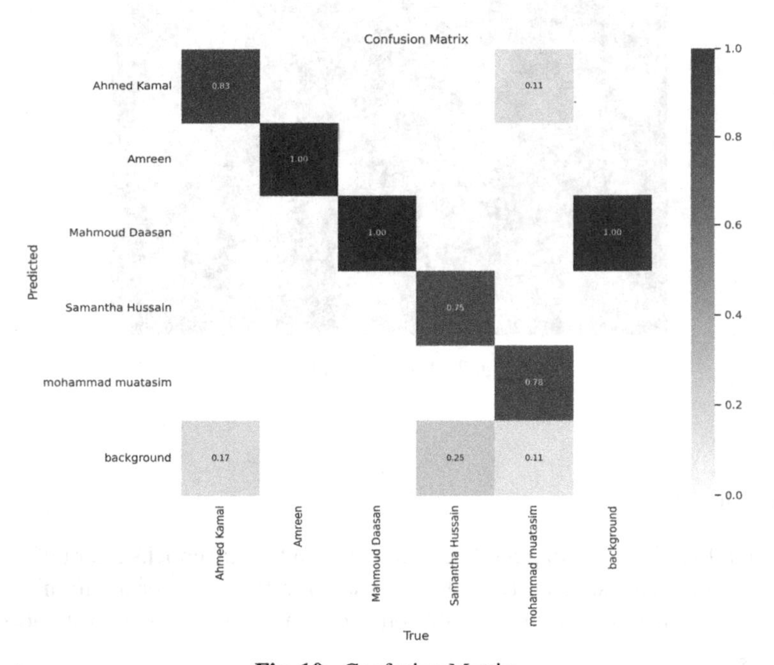

Fig. 10. Confusion Matrix

In a conventional YOLO (You Only Look Once) training graph, the loss value would be represented by the vertical axis. Loss is a gauge of the model's effectiveness during training. Better model performance is shown by lower loss values, and worse performance is indicated by larger values. To reduce the loss value during training is the objective.

"box_loss", "cls_loss", and "dfl_loss" in the context of YOLOv8 as shown in Fig. 11. The following items make up the total loss value:

"box_loss": This is the bounding box regression loss, which calculates the discrepancy between the predicted bounding box dimensions and coordinates and the actual values. The projected bounding boxes are more accurate when the box_loss is lower.

The term "cls_loss" stands for classification loss and refers to the discrepancy between the expected class probabilities and the actual class probabilities for each object in the image. The model is better at classifying the items when the cls_loss is lower.

dfl_loss": New to the YOLOv8 YOLO design is the deformable convolution layer loss. This loss is used to calculate the error in the deformable convolution layers, which are designed to improve the model's capacity to detect objects with various scales and aspect ratios. The model is more capable of handling object deformations and variations in appearance when the dfl_loss is lower.

Typically, the weighted aggregate of these individual losses represents the total loss amount. The precise units of the vertical axis would vary on how it was implemented, but in general, they stand for the size of the mistake or the discrepancy between the values that were predicted and the actual values.

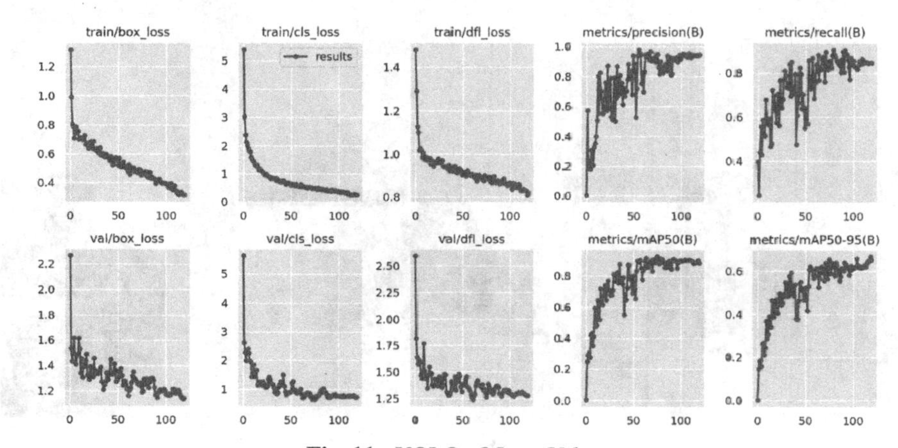

Fig. 11. YOLO v8 Loss Values

5 Results

5.1 Implementation of Face Recognition

The following images illustrates the face recognition using the YOLO v8 model and has been tested into multiple images as shown in Fig. 12a and 12b.

a.

b.

Fig. 12. Face recognition Result 1

6 Conclusion

In this report, python, Ultralytics and YOLO v8 had been used to recognize faces of 7 different people by combining YOLO and facial recognition, we have discovered that the YOLO library is growing more popular and performing better when it comes to finding and recognizing faces.

One of the industries that makes the most use of facial recognition technology is security. Facial recognition technology is being used by software companies to make it simpler for customers to utilize their products. A particularly effective method that could help law enforcement locate criminals is facial recognition. This technology may be enhanced to be used in a variety of settings, including ATMs, confidential information access, or handling other delicate materials. As a result, other security measures like passwords and keys risk becoming obsolete. Instead of forcing you to visit a kiosk to pay for a ticket, facial recognition would scan your face, run it through a system, and charge the account that you've previously registered.

Acknowledgment. The authors would like to express their sincere gratitude to Mohammad Muatasim Siddig Abdelghani, Ahmed Kamal Eldin Abdalla Abdelrahman and Samantha Hussain for their invaluable support and insightful discussions during this research. Their contributions have greatly enriched the quality of this work.

References

1. Chang, L., Cutumisu, M.: Online engagement and performance on formative assessments mediate the relationship between attendance and course performance: Revista de Universidad y Sociedad del Conocimiento. Int. J. Educ. Technol. Higher Educ. **19**(1) (2022). https://vpn.utm.my/scholarly-journals/online-engagement-performance-on-formative/docview/2619964337/se-2. https://doi.org/10.1186/s41239-021-00307-5
2. Liew, K.J., Tan, T.H.: QR code-based student attendance system. In: 2021 2nd Asia Conference on Computers and Communications (ACCC), pp. 10–14 (2021). https://doi.org/10.1109/ACCC54619.2021.00009
3. Rao, A.: AttenFace: a real time attendance system using face recognition. ArXiv.Org (2022). https://vpn.utm.my/working-papers/attenface-real-time-attendance-system-using-face/docview/2736486810/se-2
4. Dong, L., Izquierdo, E.: A knowledge structuring technique for image classification. In: 2007 IEEE International Conference on Image Processing, San Antonio, TX, USA, pp. VI - 377 – VI – 380 (2007). https://doi.org/10.1109/ICIP.2007.4379600
5. Al-Saffar, A.A.M., Tao, H., Talab, M.A.: Review of deep convolution neural network in image classification. In: 2017 International Conference on Radar, Antenna, Microwave, Electronics, and Telecommunications (ICRAMET), Jakarta, Indonesia, pp. 26–31 (2017). https://doi.org/10.1109/ICRAMET.2017.8253139
6. Liu, F.-Y., Liao, C.-L., Chou, P.-W., Fan, Y. -C.: Objects detection deep learning system based on 2-D winograd convolutional neural network. In: 2021 IEEE 10th Global Conference on Consumer Electronics (GCCE), Kyoto, Japan, pp. 454–455 (2021). https://doi.org/10.1109/GCCE53005.2021.9750962
7. Lou, Y., Fu, G., Jiang, Z., Men A., Zhou, Y.: PT-NET: improve object and face detection via a pre-trained CNN model. In: 2017 IEEE Global Conference on Signal and Information Processing (GlobalSIP), Montreal, QC, Canada, pp. 1280–1284 (2017). https://doi.org/10.1109/GlobalSIP.2017.8309167
8. Wang, C.-Y., Bochkovskiy, A., Liao, H.-Y.M.: YOLOv7: trainable bag-of-freebies sets new state-of-the-art for real-time object detectors, arXiv [cs.CV] (2022)
9. Attendance object detection dataset and pre-trained model by mahmouddaasan (no date) Roboflow. https://universe.roboflow.com/mahmouddaasan/attendance-tuqb1. Accessed 10 Aug 2023 1
10. Author, F.: Contribution title. In: 9th International Proceedings on Proceedings, pp. 1–2. Publisher, Location (2010)

NARXNN Modeling of Ultrafiltration Process for Drinking Water Treatment

Mashitah Che Razali[1,2], Norhaliza Abdul Wahab[1](✉), Noorhazirah Sunar[1], Nur Hazahsha Shamsudin[2], Muhammad Sani Gaya[3], and Azavitra Zainal[1,4]

[1] Faculty of Electrical Engineering, Universiti Teknologi Malaysia, Johor Bahru, Johor, Malaysia
norhaliza@utm.my

[2] Faculty of Electrical Engineering, Universiti Teknikal Malaysia Melaka, Hang Tuah Jaya, Durian Tunggal, Melaka, Malaysia

[3] Department of Electrical Engineering, Kano University of Science and Technology, Wudil, Nigeria

[4] Instrumentation and Control Engineering Section, Malaysian Institute of Industrial Technology, Universiti Kuala Lumpur, Johor Bahru, Johor, Malaysia

Abstract. Ultrafiltration (UF) process has gained attention over times, particularly in treating drinking water treatment. A major challenge in achieving high quality of drinking water in UF process is membrane fouling. Membrane fouling has great effects on the performance of filtration process. To overcome fouling accurately, a prediction model is necessary. With prediction model, membrane fouling can be handled in a right way, so that, efficiency of filtration process can be maximize. This paper presents a study on modeling based on non-linear autoregressive with exogenous input neural network (NARXNN) of UF pilot plant specifically from treating drinking surface water. The NARXNN was used to model the permeate flux and transmembrane pressure (TMP). In this work, LM training algorithm was employed. The performance of the model was measured based on mean square error (MSE), root mean square error (RMSE) and correlation of coefficient (R). The simulation results demonstrate that proposed NARXNN modeling able to give high prediction rate with R value of 0.91743. It shows, the prediction values agree well with the actual values. With this model, membrane fouling can be successfully simulated and monitor accordingly.

Keywords: Drinking Water Treatment · NARXNN · Permeate Flux · Transmembrane Pressure · Ultrafiltration

1 Introduction

Drinking water which also recognized as a potable water is a water that is safe and sound to be utilized to endure livelihood. The water must adhere to drinking water quality standard that has been set by each country. It is essential to acquire high quality of drinking water where it can cut down the tendency of many problems especially related to the health issues which eventually give impact to the country development. With reliable

© The Author(s), under exclusive license to Springer Nature Singapore Pte Ltd. 2024
F. Hassan et al. (Eds.): AsiaSim 2023, CCIS 1911, pp. 254–264, 2024.
https://doi.org/10.1007/978-981-99-7240-1_20

treatment, water bone diseases like hepatitis A, cholera and dysentery that conclusively cause death can be overcome. Apprehending the crucial of these issues, numbers of researchers has been studied regarding to the treatment, modeling, monitoring, and prevention strategies for drinking water treatment.

There are variety types of treatment process have been proposed and developed by numerous researchers available to treat the drinking water. The selection of treatment is usually depending on the type and quality of the influent water itself. Others factor like size, materials, and process operating condition as well present significant consequence to the quality of the treatment. Membrane filtration specifically ultrafiltration (UF) treatment has been broadly used in treating drinking water due to the ability to offer high effluent characteristic such as by providing reliable solid-liquid separation, low energy consumption, and requirement of small foot print [1, 2]. Nevertheless, the existence of membrane fouling which arise from the filtration process cause decrement of effluent water and pollutants removal rate [3]. Indirectly influence the efficiency of operation and maintenance process [4]. It is significant to possess efficient approach to lighten membrane fouling.

Over time, modeling of filtration treatment has become very useful for water quality monitoring. This ensures the water quality are easily excess and evaluate by the authorities. The developed model is valuable for water quality assessment and development of control strategies for supervision of water supply. Many studies have been done to predict the filtration properties of drinking water treatment like transmembrane pressure (TMP), permeate flux and membrane permeability by using different modeling methods such as artificial neural network (ANN), fuzzy based method, and genetic algorithm (GA) [5–8].

Study by Matheri et al. [9] proposed ANN in order to discover the relationship between chemical oxygen demand (COD) and trace metal in wastewater treatment plants (WWTP). The proposed method able to predict both parameters with robust compared than conventional modeling with coefficient determination of 0.98 and 0.99 respectively. Study by Naghibi et al. [10] compared multi-layer perceptron adaptive neural network (MLP-ANN) and adaptive neural fuzzy inference system (ANFIS) in prediction of membrane adsorption of WWTP. Regression results shown that ANFIS able to give high prediction rate compared with MLP-ANN. Study by Abdel Daiem et al. [11] proposed non-linear autoregressive with exogenous input neural network (NARXNN) and Seagull optimization algorithm (SOA) in modeling of activated sludge system. Simulation results shown that NARXNN able to predict the digested sample characteristic and biogas production with correlation coefficient close to 1 for each training, validation, and testing data. While study by Mihaly et al. [12] developed NARXNN modeling and optimization based on GA. The study successfully predicts the water line effluents and performance indicators. Study by Yang et al. [13] also found out that modeling based on NARXNN able to predict the effluent quality effectively which circuitously preceding to enhancement in effluent quality.

Recognizing the successful rate of NARXNN in modeling of non-linear system, this study was aims to develop models for drinking water treatment based on NARXNN modeling with the purpose to predict permeate flux and TMP of UF process. The pilot plant of UF process was developed accordingly where it was designed specifically for influent water from river.

2 Ultrafiltration Drinking Water Treatment Pilot Plant

The ultrafiltration (UF) pilot plant for the treatment of drinking water was developed. The plant employed hollow fiber membrane module made from modified polyethersulfone (PES) material with surface area of 0.2 m^2 that act as a filtering mechanism. Figure 1 illustrated the schematic diagram of the designed UF pilot plant. The designed plant has been located at the Process Control Lab, Faculty of Engineering, Universiti Teknologi Malaysia, Johor, Malaysia. The plant was equipped with two tanks, which are for influent and effluent water correspondingly.

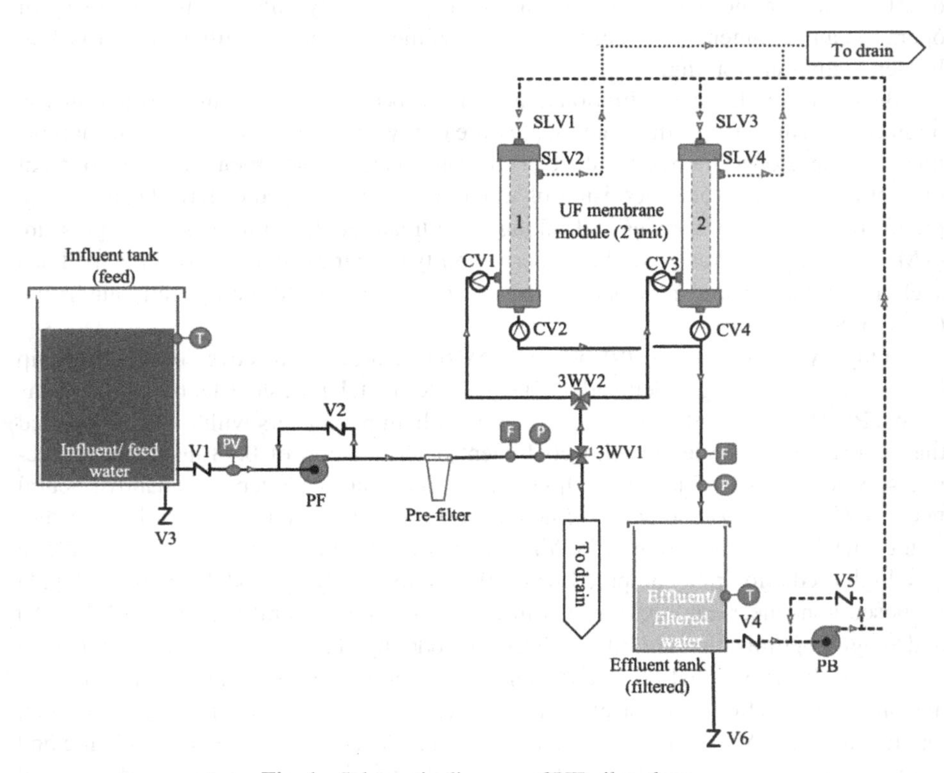

Fig. 1. Schematic diagram of UF pilot plant.

The plant can be assigned into four main parts, which are influent, filtration, backwash, and effluent stream. Each of these parts have their own important role. At the inlet/influent stream, influent is fed by pump to the membrane module where the velocity of the influent is controlled by using proportional valve. Before entering to the membrane module, the influent water is pre-filter by using fabric filter to remove the contaminants. The existing of pre-filter able to reduce the accumulation of particles in the UF plant during the filtration process and increase the long life of the membrane. At the filtration stream, the filtration process occurred where the suspended solid is separated from the fluid and produce the effluent water. At the backwash stream, the membrane will be cleaned up regularly to reduce the fouling propensity. At the outlet/effluent stream, it consists of the effluent water that produced from the filtration process which is filtered water. Here, the flow rate of the effluent water which known as effluent permeate flux is measured by using electronic flow sensor. Table 1 provide the list of instruments that involve in the development of UF pilot plant.

Table 1. Instrument in UF pilot plant.

Tag Number	Description
V1–V6	Ball valve
3WV1–3WV2	Three-way valve
SLV1–SLV4	Solenoid valve
CV1–CV4	Check valve
Pre-filter	Fabric pre-filter
PF	Pump filtration
PB	Pump backwash
PV	Proportional valve
F	Flow sensor
P	Pressure sensor

To collect the desired data for the modeling purposes, the plant necessity be operated correctly with the right selection of process mode. The designed plant was prepared for cross-flow operating mode. The study was utilized river water at the Faculty of Engineering, University Teknologi Malaysia as the influent water. To control and monitor the UF pilot plant, supervisory control, and data acquisition (SCADA) system is developed. The system is mainly used to collect and process the real-time data on the filtration process. These data are used for the modelling purpose based on NARXNN modeling. The SCADA system was utilized LabVIEW software to provide graphical programming approach which visualize the UF pilot plant and measurement data. Figure 2 shows the dataset obtain from the UF plant. This dataset was used both for training and testing of NARXNN modeling.

3 NARXNN Modeling

Non-linear autoregressive with exogenous input neural network (NARXNN) is effective in prediction of non-linear system. It required data from the experimental system. By that, high accuracy in mapping between input and output of non-linear system can be attain. The gather data as shown in Fig. 2 was used to train the NARXNN network.

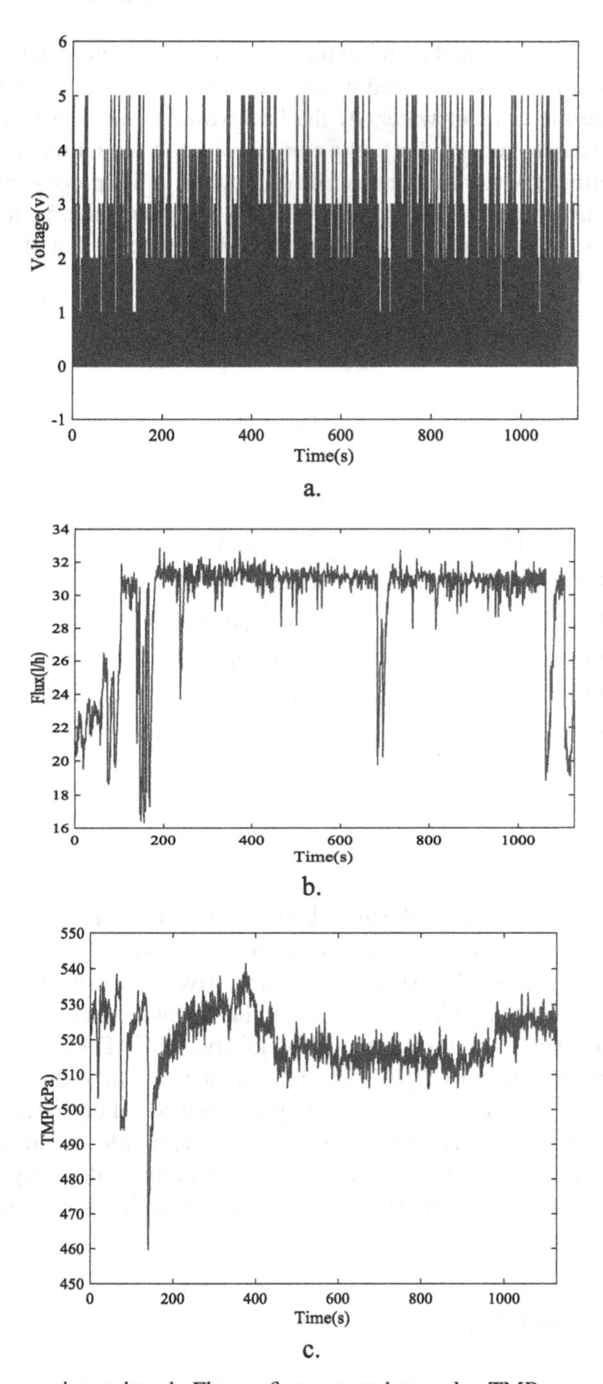

Fig. 2. a. Voltage as input data, b. Flux as first output data and c. TMP as second output data.

The modelling process was executed by using MATLAB Simulink software version R2020b. As the first step, the input and output data with corresponding unit are loaded to the workspace area. Total 3,378 data were loaded for NARXNN models trained. To guarantee stable convergence of the model, the data was being normalize. The resulting normalize data was divided into 2 parts based on cross-validation partition, which are 60% for training and 40% for testing purpose. After data division process and before creating neural network topology, the data are assigned as the input output data in matrix form and that matrix form was undergo transpose process. Once it completed, neural network topology was constructed. Figure 3 shows the architecture of NARXNN in MATLAB that was used for the prediction of flux and TMP.

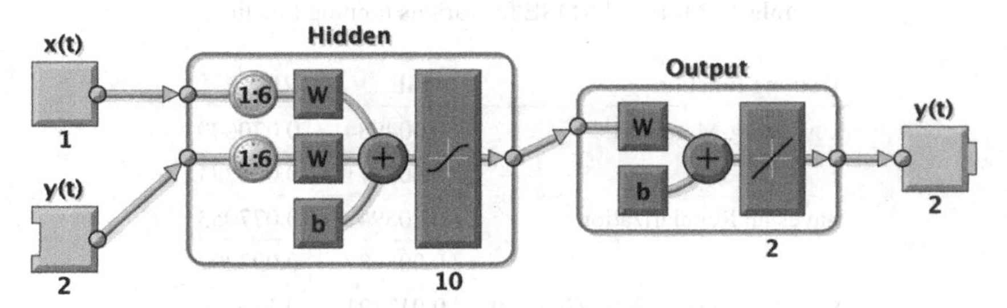

Fig. 3. NARXNN architecture in MATLAB.

Each network comprises of three layers of connected artificial neurons, which are input, hidden, and output layer. Each connection between artificial neuron will transmits a signal from one to another. The number of neurons in the hidden layer were determined using the trial-and-error method until it provides best network architecture. Number of neurons in the hidden layer were finalize based on the least value of root mean square error. In this case, 10 number of hidden neurons was chosen. The input signal which consists of pass output signal with time delay is multiplied by a random number known as connection weights (w) and a constant value known as bias (b). In this work, a Lavenberg Marquardt (LM) was utilized to train the network.

Training of the NARXNNs is an important issue for the accuracy of the develop model. In this study, the performance and accuracy of the flux and TMP prediction was measured based on three criteria which are the correlation of coefficient (R), mean square error (MSE), and root mean square error (RMSE). Equations (1)–(3), shows the formula to calculate R, MSE and RMSE respectively [14].

$$R = \frac{\sum_{i=1}^{n}(x_i - \overline{x})(y_i - \overline{y})}{\sqrt{\sum_{i=1}^{n}(x_i - \overline{x})^2 \sum_{i=1}^{n}(y_i - \overline{y})^2}} \tag{1}$$

$$MSE = \frac{1}{n}\sum (y_i - \overline{y})^2 \tag{2}$$

$$RMSE = \sqrt{\frac{1}{n}\sum (y_i - \overline{y})^2} \tag{3}$$

where n is the number of data, x is the first measured variable, y is the second measured variable, x_i/y_i is the actual value and $\overline{x_i}/\overline{y_i}$ is the predicted value.

4 Result and Discussion

In this study, the MATLAB R2020b is used to create the proposed NARXNN. The NARX prediction model is trained and tested by using actual data from designed UF pilot plant. Many training functions are implemented on the proposed NARXNN Some of them are tabulated as illustrated in Table 2. It was found that, Levenberg-Marquardt (LM) training function provide the best MSE and RMSE with value of 0.004984, 0.009271, 0.070600 and 0.096284 respectively. The proposed NARXNN with LM training function attained lowest value of errors for both training and testing data.

Table 2. MSE and RMSE for various training function.

Training function	MSE	RMSE
Levenberg-Marquardt	0.004984	0.070600
	0.009271	0.096284
Bayesian Regularization	0.005942	0.077083
	0.008687	0.093202
Variable Learning Rate Gradient Descent	0.017121	0.130846
	0.022677	0.150589
Gradient Descent with Momentum	0.124284	0.352540
	0.129276	0.359550

The regression relationship between actual and predicted values is presented as in Fig. 4. Vertical axis of Fig. 4 represents the equation of regression. Based on the obtain correlation of coefficient, R which is 0.91743 that is near to 1, it indicates the high precision and efficiency of proposed NARXNN.

Figure 5 shows the performance of NARXNN in modeling UF drinking water treatment. Attained graph shows minimal errors with a good tracking between predicted and actual data.

In this study, NARXNN is used for prediction of permeate flux and TMP of filtration process. Comparison among actual and predicted values of permeate flux and TMP are illustrated in Fig. 6, 7, 8 and 9 correspondingly. From the figures, it clearly shows that the predicted values of NARXNN is closer to actual values for both permeate flux and TMP. The model of NARXNN able to establish good prediction for both training and testing data. The obtained NARXNN can be used to represent the real physical model of UF pilot plant. Then, it can be used as a prevention, prediction, optimization, and as well as a control tool. For instance, NARXNN can be applied in optimizing filtration variables. At this point, the variables will be optimize accordingly based on the desired output. In UF drinking water treatment, major factor that affect the performance of the process is membrane fouling. Membrane fouling can cause degradation of entire filtration process. It is important to optimize variables that contribute to the development of membrane fouling such as pressure and hydrodynamic conditions. Optimization process will give

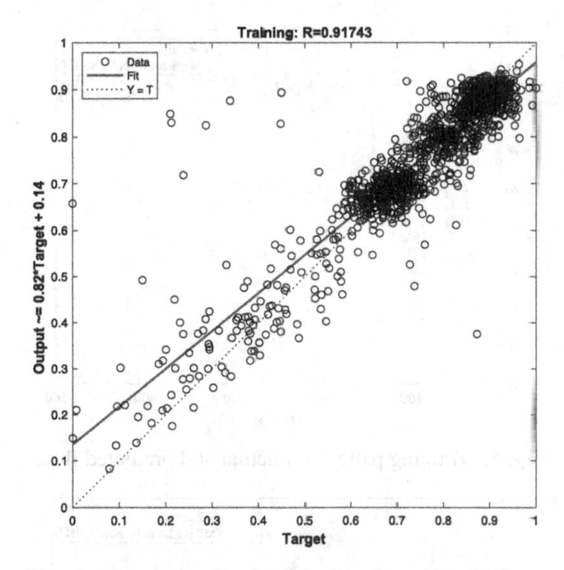

Fig. 4. Regression line of actual and predicted values.

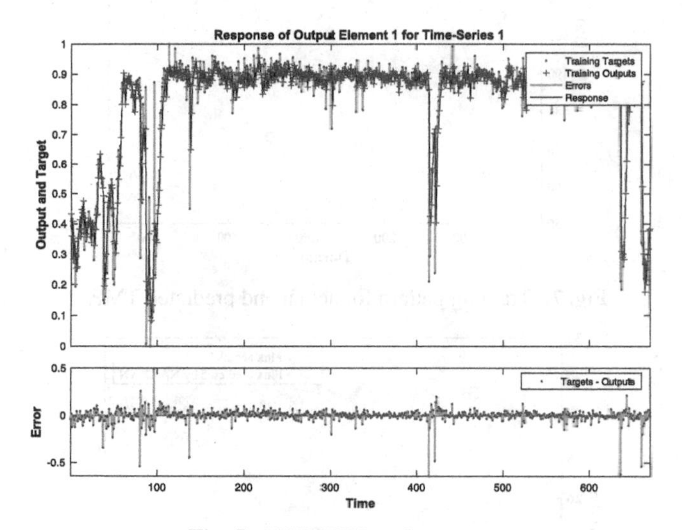

Fig. 5. NARXNN performance.

significant result in removal rate of membrane fouling and indirectly improve the quality of permeate flux.

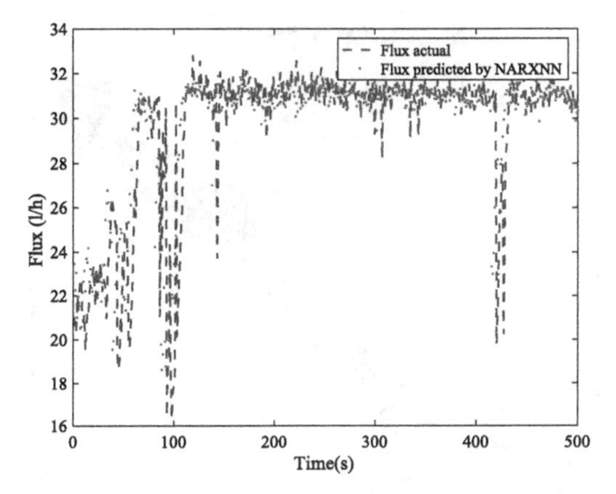

Fig. 6. Training pattern for actual and predicted flux.

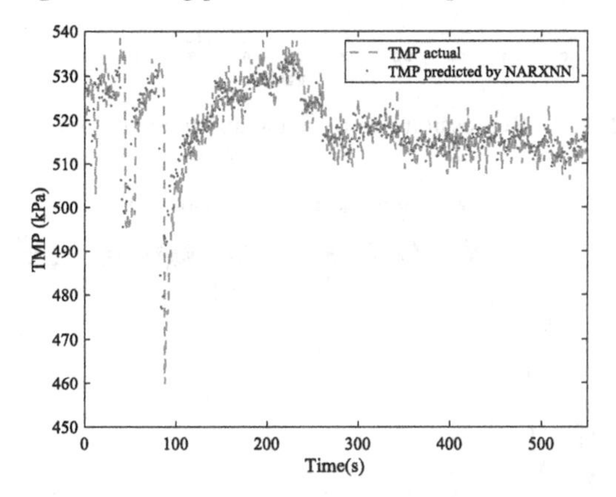

Fig. 7. Training pattern for actual and predicted TMP.

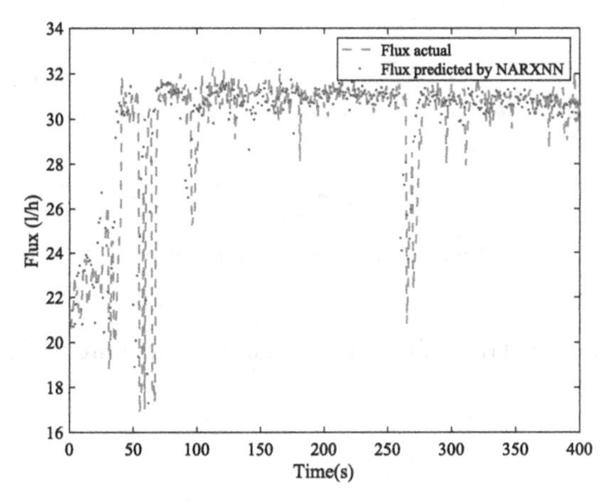

Fig. 8. Testing pattern for actual and predicted flux.

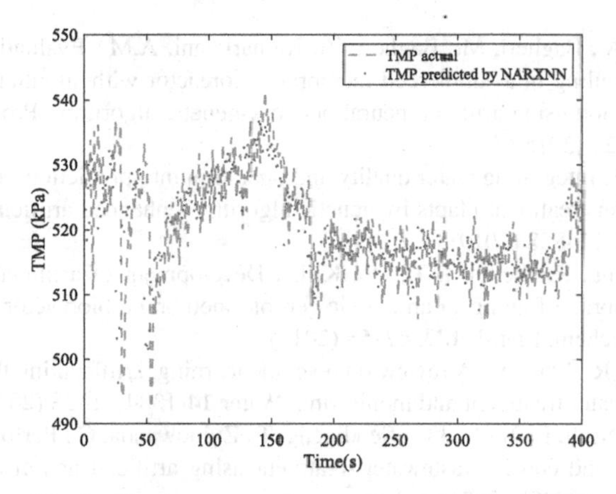

Fig. 9. Testing pattern for actual and predicted TMP.

5 Conclusion

UF process is a treatment that has been widely used in drinking water treatment. A major challenge of this treatment is membrane fouling. To control and manage the occurrences of membrane fouling, model that represent the real system is required. This study presented modeling based on NARXNN for drinking water treatment. The filtration process was done using hollow fiber membrane with river water as influent. In this work, the data was collected from the design UF pilot plant. Collected data was used for training and testing of NARXX. Modeling results shows good prediction of training and testing data for both permeate flux and TMP. The results of this study benefit the safety of the drinking water supply and are useful for the monitoring and control purposes.

Acknowledgments. The authors are grateful to the Universiti Teknologi Malaysia UTM High Impact Research (UTMHR) & No Vot. Q.J130000.2451.08G74 and the Ministry of Higher Education (MOHE), for their partial financial support through their research funds. The first author wants to thank the Universiti Teknikal Malaysia Melaka (UTeM) and the Ministry of Higher Education (MOHE) for the 'Skim Latihan Akademik Bumiputera' (SLAB) scholarship.

References

1. Liu, W., Yang, K., Qu, F., Liu, B.: A moderate activated sulfite pre-oxidation on ultrafiltration treatment of algae-laden water: fouling mitigation, organic rejection, cell integrity and cake layer property. Sep. Purif. Technol. **282**, 120102 (2022)
2. Wang, X., Ma, B., Bai, Y., Lan, H., Liu, H., Qu, J.: The effects of hydrogen peroxide pre-oxidation on ultrafiltration membrane biofouling alleviation in drinking water treatment. J. Environ. Sci. **73**, 117–126 (2018)
3. Chang, H., et al.: Long-term operation of ultrafiltration membrane in full-scale drinking water treatment plants in China: characteristics of membrane performance. Desalination **543**, 116122 (2022)
4. Zhang, L., et al.: The performance of electrode ultrafiltration membrane bioreactor in treating cosmetics wastewater and its anti-fouling properties. Environ. Res. **206**, 112629 (2022)

5. Mirbagheri, S.A., Bagheri, M., Bagheri, Z., Kamarkhani, A.M.: Evaluation and prediction of membrane fouling in a submerged membrane bioreactor with simultaneous upward and downward aeration using artificial neural network-genetic algorithm. Process Saf. Environ. Prot. **96**, 111–124 (2015)
6. Zhang, Y., et al.: Integrating water quality and operation into prediction of water production in drinking water treatment plants by genetic algorithm enhanced artificial neural network. Water Res. **164**, 114888 (2019)
7. Schmitt, F., Banu, R., Yeom, I.-T., Do, K.-U.: Development of artificial neural networks to predict membrane fouling in an anoxic-aerobic membrane bioreactor treating domestic wastewater. Biochem. Eng. J. **133**, 47–58 (2018)
8. Lowe, M., Qin, R., Mao, X.: A review on machine learning, artificial intelligence, and smart technology in water treatment and monitoring. Water **14**(1384), 1–28 (2022)
9. Matheri, A.N., Ntuli, F., Ngila, J.C., Seodigeng, T., Zvinowanda, C.: Performance prediction of trace metals and cod in wastewater treatment using artificial neural network. Comput. Chem. Eng. **149**, 107308 (2021)
10. Naghibi, S.A., Salehi, E., Khajavian, M., Vatanpour, V., Sillanpää, M.: Multivariate data-based optimization of membrane adsorption process for wastewater treatment: multi-layer perceptron adaptive neural network versus adaptive neural fuzzy inference system. Chemosphere **267**, 129268 (2021)
11. Abdel daiem, M.M., Hatata, A., Said, N.: Modeling and optimization of semi-continuous anaerobic co-digestion of activated sludge and wheat straw using nonlinear autoregressive exogenous neural network and seagull algorithm. Energy **241**, 122939 (2022)
12. Mihály, N.-B., Luca, A.-V., Simon-Várhelyi, M., Cristea, V.M.: Improvement of air flowrate distribution in the nitrification reactor of the waste water treatment plant by effluent quality, energy and greenhouse gas emissions optimization via artificial neural networks models. J. Water Process Eng. **54**, 103935 (2023)
13. Yang, Y., et al.: Prediction of effluent quality in a wastewater treatment plant by dynamic neural network modeling. Process Saf. Environ. Prot. **158**, 515–524 (2022)
14. Meng, Y., Yun, S., Zhao, Z., Guo, J., Li, X., Ye, D.: Short-term electricity load forecasting based on a novel data preprocessing system and data reconstruction strategy. J. Build. Eng. **77**, 107432 (2023)

Improvement of Vision-Based Hand Gesture Recognition System with Distance Range

Muhammad Eirfan Mukhtar[1], Noorhazirah Sunar[1](✉), Nur Haliza Abd Wahab[2], Nor Aishah Muhammad[1], and Mohd Fua'ad Rahmat[1]

[1] Faculty of Electrical Engineering, Universiti Teknologi Malaysia, Johor Bahru, Johor, Malaysia
noorhazirah@utm.my

[2] Faculty of Computing, Universiti Teknologi Malaysia, Johor Bahru, Johor, Malaysia

Abstract. Hand gestures are widely explored for the human action interface, reducing the complexity of interaction between humans and computers. Hand gesture-based interaction for computer applications enables touchless operation without the need for other devices such as a keyboard and mouse. It also guarantees sterility and safer interaction for multiple users. The issue of HGR has a low range of detection. This study aims to improve the distance range for HGR from the user to the computer camera using the zoom algorithm that is processed by OpenCV and the Mediapipe library. To verify the proposed system, a Graphical User Interface (GUI) for four applications is developed in HGR mode. The GUI includes the development of cursor mouse event commands and a virtual keyboard display. The proposed method was then tested for some time, and the result indicates that the maximum range of detection increased with the accuracy of the HGR. The proposed method meets the expected outcome and can be used for applications such as slide presentations and watching video applications.

Keywords: Human Computer Interaction · Hand Gesture Recognition · Graphical User Interface · OpenCV · Mediapipe

1 Introduction

Hand gesture recognition (HGR) is an emerging technology that allows computers to interpret and respond to the gestures made by a human hand. With the advancement of technology in computer vision and artificial intelligence, the application of HGR has revolutionized the ability to control and interact with electronic devices remotely. It can be achieved through image processing techniques and machine learning algorithms that are trained to recognize specific hand movements or positions. HGR as a remote device application has been widely used in various applications such as Virtual Reality (VR) [1, 2] and Augmented Reality (AR) [3], smart home automation [4], health and rehabilitation [5], display and presentation [6], and gaming and entertainment [7]. There are several motivations that lie within the usage of current remotes that users use nowadays.

© The Author(s), under exclusive license to Springer Nature Singapore Pte Ltd. 2024
F. Hassan et al. (Eds.): AsiaSim 2023, CCIS 1911, pp. 265–275, 2024.
https://doi.org/10.1007/978-981-99-7240-1_21

Firstly, most users nowadays demand more advanced ways to interact with computers [8]. Secondly, the usage of wired remotes is limited due to cable length limitations, such as keyboards or mouse. It can also improve accessibility for the elderly, serve as assistive technology, and create more ease and independence in the digital environment [9]. It also provides safety and accessibility because it allows touchless interaction, promoting safety and hygiene [10]. It helps reduce the spread of germs and the risk of contamination.

Several methods can be used to detect the hand with the webcam. Firstly, a contour-based method [11] by converting an RGB image frame to grayscale image. Then, contours of colored objects can be found, and after that, find the centroid of the contour; 2) skin color-based methods, where these methods use the color information of skin to segment the hand region from the background; and 3) hand landmark methods. The contour methods are relatively simple and easy to implement, but they are sensitive to variations in lighting and skin color [12]. Meanwhile, the hand landmark method for hand gesture recognition is a technique that involves detecting and tracking the landmarks or key points on a person's hand to recognize and interpret hand gestures [13]. This method relies on computer vision algorithms and machine learning techniques to analyze the hand's shape, pose, and movement.

Several researchers have explored the HGR tracking to improve range of hands and sensors. In [14], the researchers use a depth-based approach to improve the accuracy. The proposed method uses chamber distance to measure the shape of similarity objects with the database. The chamber distances are defined using distance transforms which need advanced depth sensors. Meanwhile, in [15], added two different classifiers based on multi-class MVMs and Random Forest in leap motion and depth sensor approach. Another method, using a HGR using a hand landmark method utilizing the Mediapipe library. This method is a less complex implementation compared to other methods. However, Most of the HGR systems that have been built have a low range of detection [16]. Although Mediapipe can still track the hand in this range, the accuracy of certain mouse events, such as left clicks, decreased rapidly. It is expected that this range's limitations can be improved using magnifying techniques.

2 Methodology

The system has three main parts for system architecture, which consist of input, processing, and output as shown in Fig. 1. The inputs are hand images or frames taken by the webcam. Next, it will be processed using PyCharm as a compiler and supported by OpenCV, Mediapipe and Python. Finally, the mouse and keyboard action were produced depending on the algorithm built. For the virtual keyboard, the Python function was used to run "On Screen Keyboard" and users can use virtual mouse to input the keyboard. For the GUI, python library tkinter, was utilized.

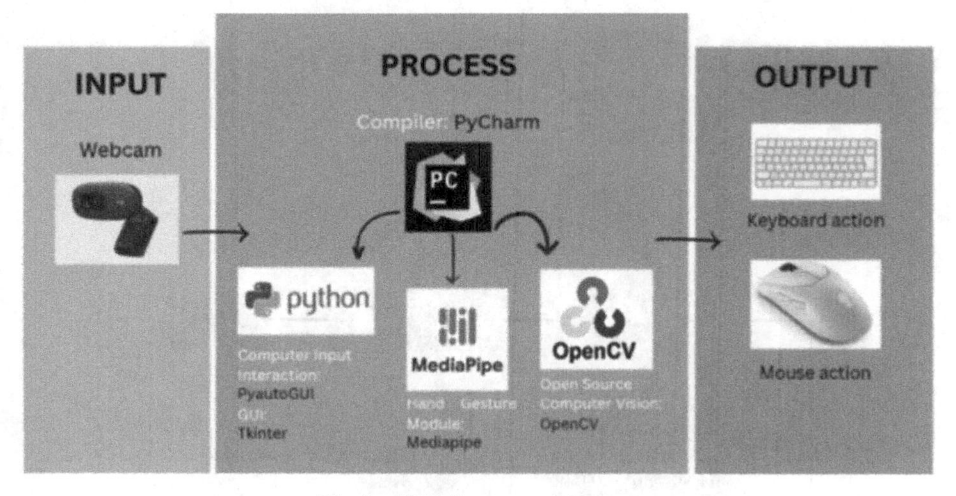

Fig. 1. HGR System architecture

2.1 Hand Tracking Module

Find Hand: This method is used to find the presence of hand in front of the webcam. It first converts the image format from BGR to RGB so that Mediapipe library can process it. If there is the result of hand, then it will draw the hand landmark and boundary box around the hand.

1. Find Hand: This method is used to find the presence of hand in front of the webcam. It first converts the image format from BGR to RGB so that Mediapipe library can process it. If there is the result of hand, then it will draw the hand landmark and boundary box around the hand.
2. Find Position: This method is used to know the coordinates of selected hand landmarks. The coordinates are only in 2D dimension since it's reading it from the frames or images. The coordinates of the hand can be manipulated depending on the application, such as finding distance.
3. Finger Up: This method is used to get information about whether the finger is in an up or down position. It is useful for giving commands to the mouse.
4. Find Distance: This method is used to calculate the distance between two selected hand landmarks in pixels. The distance calculation was done by taking two landmarks' coordinates.

2.2 Virtual Mouse Algorithm

The command for the mouse algorithm was built by capturing hand images from webcam and detecting the hand palm as depicted in Fig. 2. After that, the cursor mouse command was executed based on the landmark position of the hand. Image flip was implemented to flip the image 180 degrees clockwise to get the mirror image from the screen by using OpenCV.

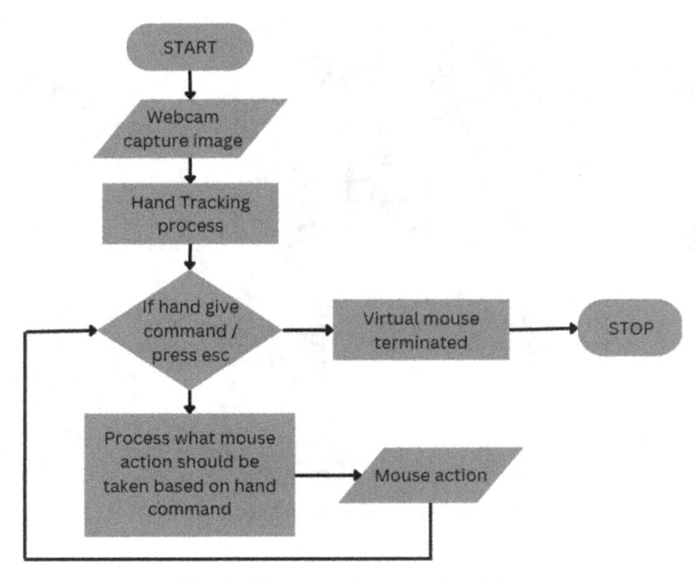

Fig. 2. The virtual mouse algorithm

2.3 Virtual Keyboard Algorithms

Figure 3 summarises the flow of the virtual keyboard algorithm. The virtual keyboard was implemented by directly using On Screen Keyboard provided by Window Operating System. The keyboard automatically runs thanks to the Python function "subprocess. Open". The user does not need to click on the desired input keyboard for writing, instead, they just need to hover over it for 1–2 s.

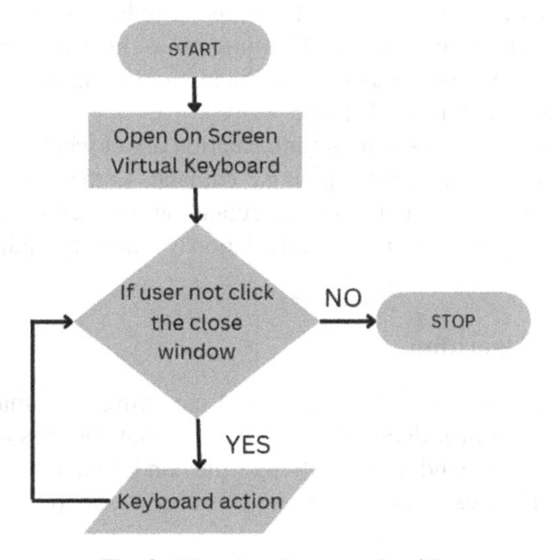

Fig. 3. The virtual mouse algorithm

2.4 GUI Algorithm

From the user's perspective, the process starts with the user selecting the desired application that they want to use in HGR mode. Based on the application chosen, the hand tracking process will display where users can use the mouse and keyboard virtually. The flow of GUI algorithms can be described as in Fig. 4.

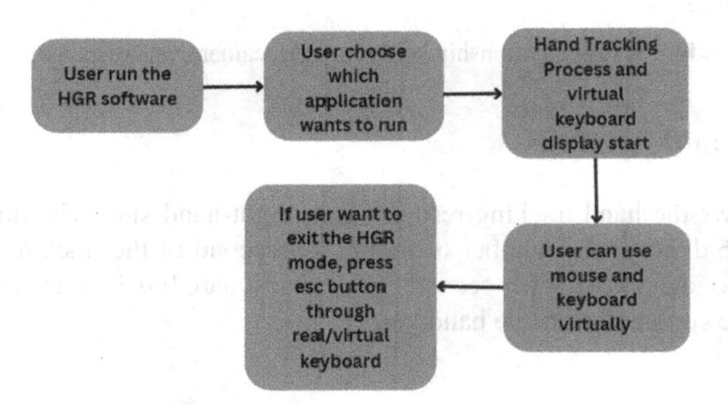

Fig. 4. The GUI Algorithm

2.5 Range Limitation

Magnifying techniques or adding a function zoom to the recognition of the HGR process was proposed to improve HGR recognition at a wide range. By zooming into the region of interest, some parameters, such as the distance between two landmarks, can be manipulated. But before zooming, the distance at which the zooming process starts needs to be known. However, OpenCV and Mediapipe do not support detecting the range between hand and webcam since it is supported up to 2D mode. Thus, the calculation to find the distance between hand and webcam was implemented in the zooming process. The distance can be obtained by calculating the focal length using Eq. (1).

$$f = (w * d)/W \tag{1}$$

where f is the focal length, w is the width in pixels, d is the distance in unit cm, and W is the width in unit cm. Figure 5 labeled the parameters that were used to calculate the focal length. Initially, the value distance d was set to 15 cm. Meanwhile, the width, w can be found using the distance of two landmarks, which are landmark (0) and landmark (9). In the hand tracking module, the algorithm to find distance is added. The width between those two landmarks was measured and is about 10 cm. After that, the formula for focal length was implemented inside Python code with OpenCV support. Then the hand needs to be placed exactly 15 cm from the webcam when the program runs. The focal length was printed on the GUI. After getting focal length, the distance can be calculated by manipulating the same formula. The zooming process was set to start at a distance more than 80 cm. The zoom was set to magnify two times only.

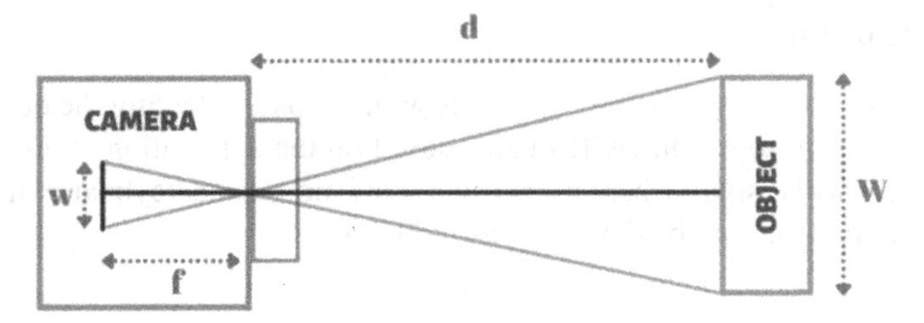

Fig. 5. The relationship between hand, camera, and distance

3 Results and Analysis

Figure 6 shows the hand tracking results on the right-hand side. The number on the top left Fig. 6 displays the number of frames per second of the tracking process that corresponds to images taken per second. The green square box indicated the bounding box where the system detects the hand.

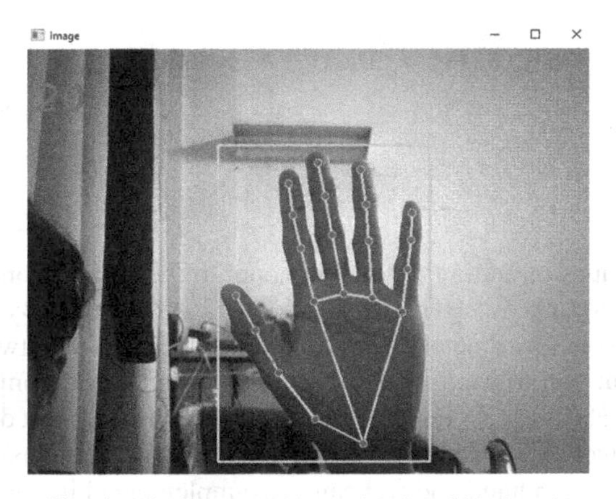

Fig. 6. The hand tracking using hand landmark.

For the system's accuracy, the HGR was tested at 1.5 m and 2.5 m. Using the mouse commands as shown in the command list in Fig. 7, The HGR was tested and repeated 50 times. The result of the HGR accuracy is tabulated in Table 1. From the table, the HGR is highly accurate to detect HGR at 1.5 m from user to the camera. The result from Table 1 also shows that at higher ranges, accuracy was still maintained at a high level. However, the Left click command displayed the highest drop in accuracy among the mouse commands. The left click command was triggered when two hand landmarks from the tip of index and middle finger were very close to each other. Note that the distance between them is in pixel units. As the distance user from webcam is increase, the distance in pixel between index and middle finger is decreased that affects the HGR accuracy.

The left click can be manipulated by adding zoom algorithm in the process of HGR. The left click will trigger when index and middle fingertip distances are below 15 cm.

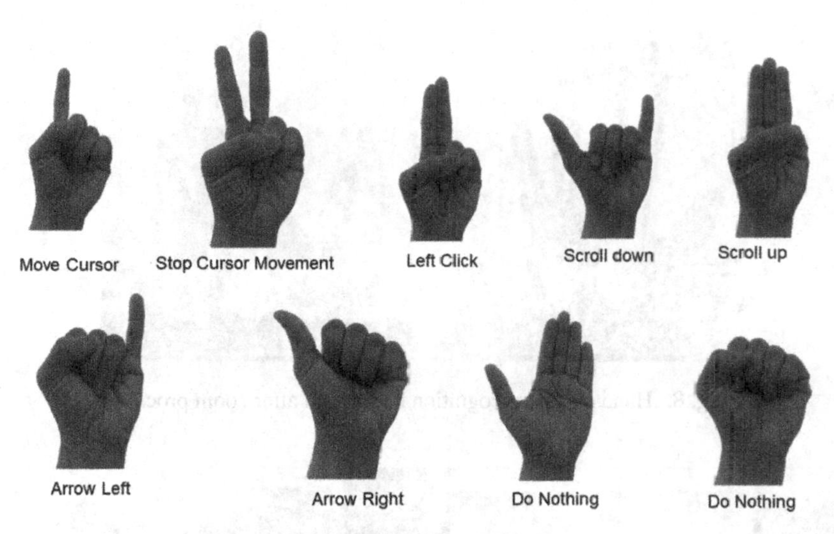

Fig. 7. Command List

Table 1. Mouse command accuracy repeated 50 times from user to camera distance 1.5 m and 2.5 m

Mouse Command	Accuracy percentage at 1.5 m (%)	Accuracy percentage at 2.5 m (%)
Move Cursor	100	
Stop Cursor Movement	100	
Left Click	100	
Scroll down	100	
Scroll up	100	
Arrow Left	100	
Arrow Right	100	
Do Nothing	100	

So, in zoom mode, the value can be changed to below 7 so that it will not trigger itself when the hand does not give the left click command. The zooming algorithm can be stacked further if desired. The result of the zoomed image is shown in Fig. 8. The effect of zoomed image algorithm at distance 2.5 m can be seen in Fig. 9. The left image is the HGR without zoomed image algorithm and the right image is after zoomed image algorithm. From the image, after zoomed image algorithm, the hand landmark is more cleared compared to the without zoomed image algorithm.

The GUI was developed to integrate the HGR control as a function of mouse and keyboard. Figure 10 shows the GUI developed for display and presentation applications. GUI was developed using the Tkinter library which is supported by Python. The GUI was developed to display 4 options of applications namely Microsoft PowerPoint, Canva,

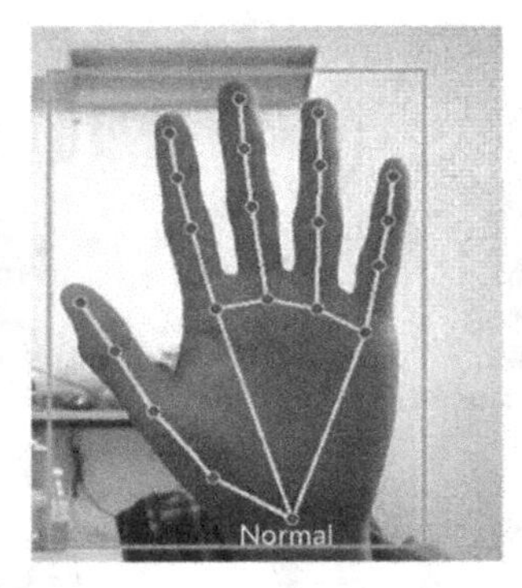

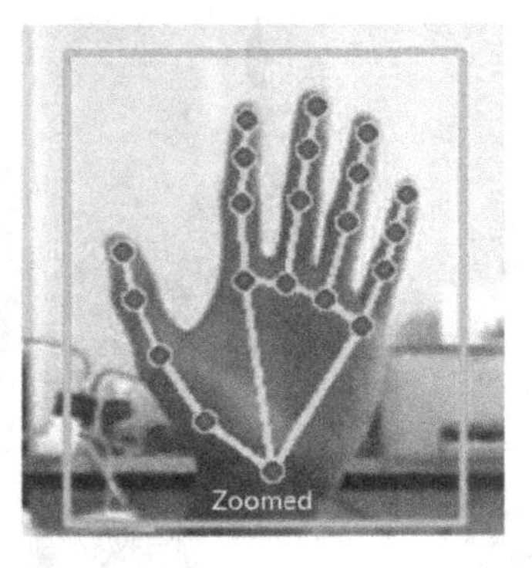

Fig. 8. Hand gesture recognition before and after zoom process

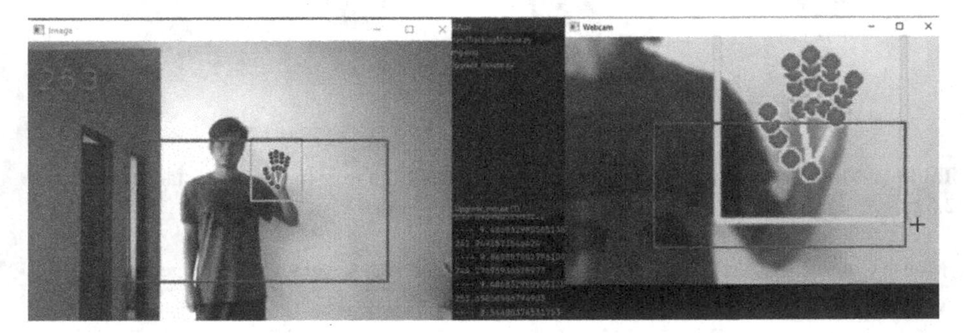

Fig. 9. Hand gesture recognition at 2.5 m without zoom algorithm and with zoom algorithm.

YouTube and Netflix. The GUI also provided a user guide for the first-time user that is unfamiliar with the GUI. The GUI focuses on several usages like presentation and watching video that can control using HGR.

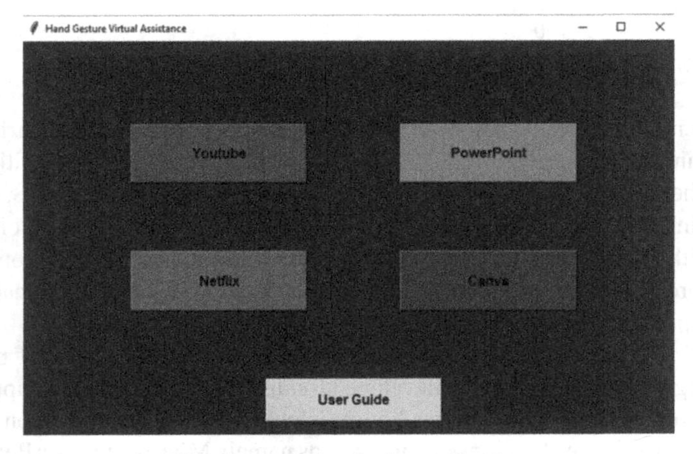

Fig. 10. GUI for display application

Figure 11 shows the GUI display when user choose PowerPoint application. The GUI will display virtual keyboard and webcam images for HGR control. From this point, user can hover the hand as a mouse control and choose the slide to open and make the slide show for the power point. In Fig. 12, display the slide show in power point application. Using the mouse command, the user can move to the next slide or to the previous slide. Meanwhile, Fig. 13 shows when user choose other applications namely You Tube, Netflix and Canva. These applications are web browsers and user can use developed mouse command to use the applications.

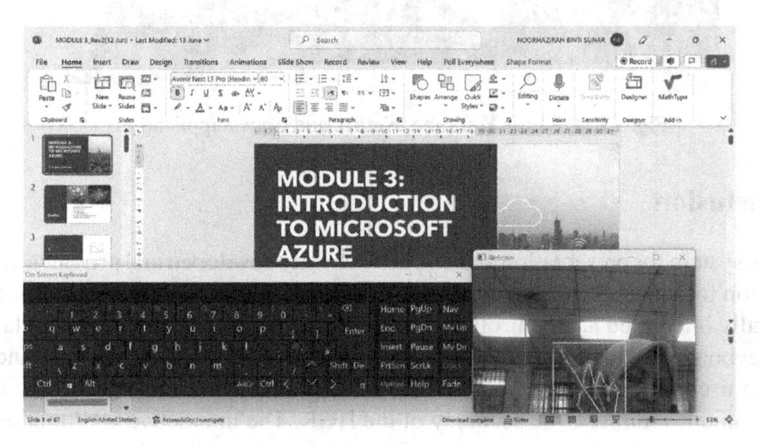

Fig. 11. GUI for display Power Point application

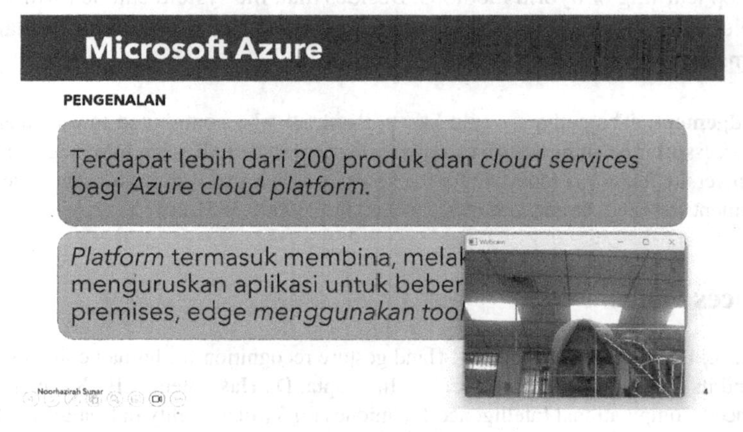

Fig. 12. Display slide show in Power Point Application

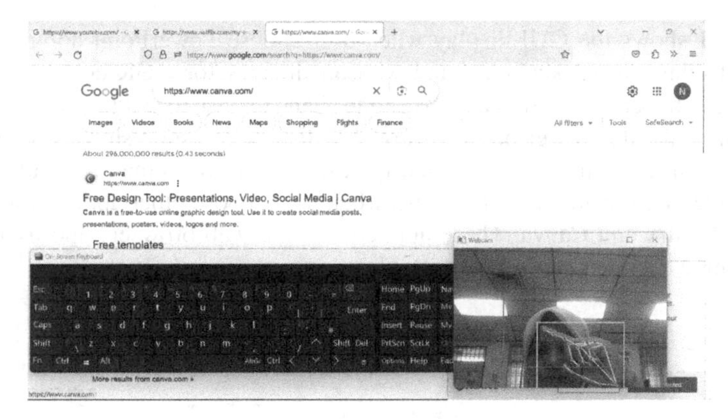

Fig. 13. GUI for YouTube, Netflix and Canva applications

4 Conclusion

The HGR system for presentation and display was fully developed and provides a promising solution for better electronic remote control. The action of mouse and keyboard was successfully integrated and controlled by hand gestures where the system displayed the virtual keyboard and mouse that can be controlled by HGR. The range of distance from camera to user was improved by employing a zoom technique. Results show that the added techniques improve the accuracy of the HGR. The interactive GUI that gives the user application options and guides them has also been established. It is suggested that the system's robustness can be improved by incorporating more advanced techniques such as deep learning or hybrid methods. Besides that, the system can be further applied to other electronic devices such as television and other electronic home appliances for better human machine interaction.

Acknowledgement. The authors would like to thank all who contributed toward making this research successful. The authors wish to express their gratitude to Research Management Center (RMC), Universiti Teknologi Malaysia for the financial support and advice for this project (UTM Encouragement Research Grant with reference no: Q.J130000.3851.20J11.

References

1. Gupta, S., Bagga, S., Sharma, D.K.: Hand gesture recognition for human computer interaction and its applications in virtual reality. In: Gupta, D., Hassanien, A.E., Khanna, A. (eds.) Advanced Computational Intelligence Techniques for Virtual Reality in Healthcare. SCI, vol. 875, pp. 85–105. Springer, Cham (2020). https://doi.org/10.1007/978-3-030-35252-3_5
2. Kang, T., Chae, M., Seo, E., Kim, M., Kim, J.: DeepHandsVR: hand interface using deep learning in immersive virtual reality. Electronics **9**, 1863 (2020). https://doi.org/10.3390/electronics9111863
3. Murhij, Y., Serebrenny, V.: Hand gestures recognition model for augmented reality robotic applications. In: Ronzhin, A., Shishlakov, V. (eds.) Proceedings of 15th International Conference on Electromechanics and Robotics "Zavalishin's Readings." SIST, vol. 187, pp. 187–196. Springer, Singapore (2021). https://doi.org/10.1007/978-981-15-5580-0_15

4. Hakim, N.L., Shih, T.K., Kasthuri Arachchi, S.P., Aditya, W., Chen, Y.-C., Lin, C.-Y.: Dynamic hand gesture recognition using 3DCNN and LSTM with FSM context-aware model. Sensors **19**, 5429 (2019). https://doi.org/10.3390/s19245429
5. Mahmoud, N.M., Fouad, H., Soliman, A.M.: Smart healthcare solutions using the internet of medical things for hand gesture recognition system. Complex Intell. Syst. **7**, 1253–1264 (2021). https://doi.org/10.1007/s40747-020-00194-9
6. Huang, Y., Yang, J.: A multi-scale descriptor for real time RGB-D hand gesture recognition. Pattern Recognit. Lett. **144**, 97–104 (2021). https://doi.org/10.1016/j.patrec.2020.11.011
7. Parekh, P., Patel, S., Patel, N., Shah, M.: Systematic review and meta-analysis of augmented reality in medicine, retail, and games. Vis. Comput. Ind. Biomed. Art. **3**, 21 (2020). https://doi.org/10.1186/s42492-020-00057-7
8. Sharmila, A.: Hybrid control approaches for hands-free high level human–computer interface-a review. J. Med. Eng. Technol. **45**, 6–13 (2021). https://doi.org/10.1080/03091902.2020.1838642
9. Muneeb, M., Rustam, H., Jalal, A.: Automate appliances via gestures recognition for elderly living assistance. In: 2023 4th International Conference on Advancements in Computational Sciences (ICACS), pp. 1–6 (2023). https://doi.org/10.1109/ICACS55311.2023.10089778
10. Khaleghi, L., Artan, U., Etemad, A., Marshall, J.A.: Touchless control of heavy equipment using low-cost hand gesture recognition. IEEE Internet Things Mag. **5**, 54–57 (2022). https://doi.org/10.1109/IOTM.002.2200022
11. Pinto, R.F., Borges, C.D.B., Almeida, A.M.A., Paula, I.C.: Static hand gesture recognition based on convolutional neural networks. J. Electr. Comput. Eng. **2019**, 4167890 (2019). https://doi.org/10.1155/2019/4167890
12. Ansari, M.A., Singh, D.K.: An approach for human machine interaction using dynamic hand gesture recognition. In: 2019 IEEE Conference on Information and Communication Technology, pp. 1–6 (2019). https://doi.org/10.1109/CICT48419.2019.9066173
13. Peral, M., Sanfeliu, A., Garrell, A.: Efficient hand gesture recognition for human-robot interaction. IEEE Robot. Autom. Lett. **7**, 10272–10279 (2022). https://doi.org/10.1109/LRA.2022.3193251
14. Liu, X., Fujimura, K.: Hand gesture recognition using depth data. In: 2004 Sixth IEEE International Conference on Automatic Face and Gesture Recognition, 2004, Proceedings, Seoul, Korea (South), pp. 529–534 (2004). https://doi.org/10.1109/AFGR.2004.1301587
15. Marin, G., Dominio, F., Zanuttigh, P.: Hand gesture recognition with jointly calibrated leap motion and depth sensor. Multimedia Tools Appl. **75**, 14991–15015 (2016)
16. Zhou, L., Du, C., Sun, Z., Lam, T.L., Xu, Y.: Long-range hand gesture recognition via attention-based SSD network. In: 2021 IEEE International Conference on Robotics and Automation (ICRA), pp. 1832–1838 (2021). https://doi.org/10.1109/ICRA48506.2021.9561189

Synthetic Data Generation for Fresh Fruit Bunch Ripeness Classification

Jin Yu Goh[1], Yusri Md Yunos[1], and Mohamed Sultan Mohamed Ali[1,2](✉)

[1] Faculty of Electrical Engineering, Universiti Teknologi Malaysia, 81310 UTM Johor Bahru, Johor, Malaysia
sultan_ali@fke.utm.my

[2] Department of Electrical Engineering, College of Engineering, Qatar University, Doha, Qatar

Abstract. In the field of fresh fruit bunch ripeness classification, the availability of annotated real-world datasets is often limited, posing challenges for developing accurate and robust classification models. To address this limitation, synthetic data generation techniques have emerged as a promising solution, offering the potential to augment dataset sizes and improve model performance. This study investigates the application of synthetic data for enhancing FFB ripeness classification. By leveraging simulation-based approaches, a diverse set of synthetic FFB images was generated, replicating variations in lighting conditions, object appearances, and environmental factors. The synthetic dataset was carefully designed to match the quantity of real data, ensuring a balanced comparison. Through extensive experiments, the performance of classification models trained on synthetic data was evaluated and compared with models trained solely on real data. The results demonstrated the efficacy of synthetic data in improving the accuracy and generalisation capability of the models. Furthermore, performance evaluation metrics, including mean average precision (mAP), were employed to assess the models' performance across different ripeness levels. The findings highlight the potential of synthetic data for fresh fruit bunch ripeness classification and emphasise the importance of leveraging simulation-based techniques for generating high-quality synthetic datasets. This research contributes to the advancement of classification models in scenarios with limited real-world data availability, paving the way for improved accuracy and reliability in FFB ripeness classification.

Keywords: Synthetic data · Fresh fruit bunch · Ripeness classification · Simulation

1 Introduction

As the field of artificial intelligence continues to advance, the application of deep learning techniques for solving complex problems has gained substantial momentum. One such problem that holds significant importance in the agricultural sector is the ripeness classification of oil palm fresh fruit bunches (FFBs). Accurate and timely identification of FFB ripeness is crucial for optimising harvesting operations, improving yield quality, and enhancing overall productivity. Traditionally, the ripeness classification of FFBs

© The Author(s), under exclusive license to Springer Nature Singapore Pte Ltd. 2024
F. Hassan et al. (Eds.): AsiaSim 2023, CCIS 1911, pp. 276–288, 2024.
https://doi.org/10.1007/978-981-99-7240-1_22

has relied on manual inspection by human experts, which is time-consuming, subjective, and prone to inconsistencies [1]. However, with the advent of computer vision and deep learning, there is great potential to revolutionise this process. By leveraging the power of artificial neural networks, the capability to automatically analyse images of FFBs can be harnessed and accurately classify their ripeness levels. Real-time object detection models have predominantly relied on network architecture optimisations, anchor-based predictions, and efficient feature extraction. In addition, these features exhibit invariance to rotation, scale, and partial occlusions, which lack robustness against changes in illumination or partial deformations of objects. Furthermore, the effectiveness of the model is limited to textured objects that possess distinctive interest points and local features [2]. Nevertheless, the performance of convolutional neural networks (CNNs) heavily hinges upon the availability of an extensive and diverse training dataset, which is often scarce for more intricate computer vision tasks like object classification due to the numerous variables involved. This scarcity results in a laborious and time-consuming process of manually preparing annotated datasets [3]. In addition, recent advancements in computer-generated imagery have facilitated the automation of data generation, enabling the creation of fully annotated synthetic datasets. Gao et al. demonstrated the realistic simulation of the pelvic X-ray dataset [4] from human models combined with domain generalisation or adaptation techniques. By leveraging synthetic data for training, the models achieved a comparable performance to those trained on a meticulously matched actual data training set. Notably, this approach provided the added advantage of circumventing the ethical and practical complexities associated with collecting data directly from live human subjects. In addition, Becktor et al. proposed a pipeline for generating simulated synthetic data that matches the target domain for maritime object detection [5] that the real-life maritime data collection is harsh and dangerous. The trained CNNs are shown to outperform conventional methods of training the models with actual images and self-annotation datasets in multiple field examples which require high fidelity and large datasets [6–11]. However, within the domain of FFBs, the challenge lies in the large quantity and availability of high-fidelity datasets that represent real-world data captured by genuine vision sensors in complex environments. This scarcity of reliable datasets hampers the development of effective models for processing FFBs data and extracting valuable insights.

In this paper, an approach for FFB ripeness classification datasets that integrates simulation-based domain randomisation and synthetic data generation was conducted. The generated synthetic data is capable of providing the CNNs model training with the complexity required for dynamic outdoor environment on-tree FFBs characteristics. Through domain randomisation techniques, variations in lighting conditions, backgrounds, textures, and object placement characteristics were introduced to ensure the models can effectively adapt to different environmental factors, especially in dynamic outdoor environments. The core concept of the methodology is to leverage simulation environments that mimic real-world scenarios and generate synthetic data that closely resembles the actual FFBs encountered in real-world practice. The working principle of the simulation approach, the details of the work procedures for conducting the simulation, and the training result on CNN models with discussion will be discussed in the following sections.

2 Working Principle

In the context of FFBs ripeness detection, the conventional approach to data collection and annotation involves the labour-intensive process of physically capturing real-world images or videos of FFBs and manually annotating them with object labels or bounding boxes, as shown in the left column of Fig. 1. This time-consuming method requires meticulous setup, image capture, and individual annotation by human annotators who carefully examine and label each FFB based on their understanding of ripeness. However, synthetic data generation in the simulation provides a more efficient and automated solution for generating training data in this domain, as shown in the right column of Fig. 1. By utilising 3D model assets and adjusting randomisation parameters, synthetic data generation in simulation allows for the creation of diverse FFBs datasets. These datasets include precise annotations and undergo automated validation within the simulated environment, eliminating the need for manual annotation. This approach significantly reduces the time and effort required for conventional manual dataset preparation, making it particularly advantageous when large amounts of diverse data with annotated ground truth are needed for training ripeness detection models. The synthetic data generated closely mimics the characteristics of real FFBs, enabling the creation of training datasets of various sizes, at any given time, and in any desired location within the simulated environment. This flexibility is invaluable for training CNN models to accurately detect and classify the ripeness of FFBs. Synthetic data generation in simulation thus presents a reliable and efficient approach to producing training material specifically tailored for FFB ripeness detection.

The generation of synthetic data for FFBs ripeness detection relies on two fundamental principles: ground truth generation and domain randomisation. The ground truth generation involves associating the automated annotation process, such as bounding boxes or semantic labels. Within the simulation environment, the randomisation of domain parameters is crucial to emulate the complexity of the dynamic outdoor environment found in palm oil estates. Both working principles will be discussed in the following subsections. In this study, Unity®, an open-source game engine, was employed along with a package called Unity Perception to generate synthetic data. Unity Perception offers valuable tools for producing synthetic data with precise annotations and a framework for designing bespoke randomisation techniques for synthetic data. Despite the package providing a few randomisation tools, additional scripts will incorporate the remaining randomisation functionalities. Additionally, the experimental designs used to validate the output images and train a deep-learning model with the generated data will be discussed. The simulation work procedures are categorised into two main parts that contributed to the synthetic data generation of FFBs: 3D modelling that involved the creation of realistic and detailed 3D models of FFBs followed by domain randomisation with simulation setup that introduced the variability into the simulated environment to create diverse and realistic scenarios.

2.1 3D Modelling

The methodology for 3D modelling of a FFB involved the utilisation of Autodesk 3ds Max® software to create two distinct types of 3D models representing different levels of

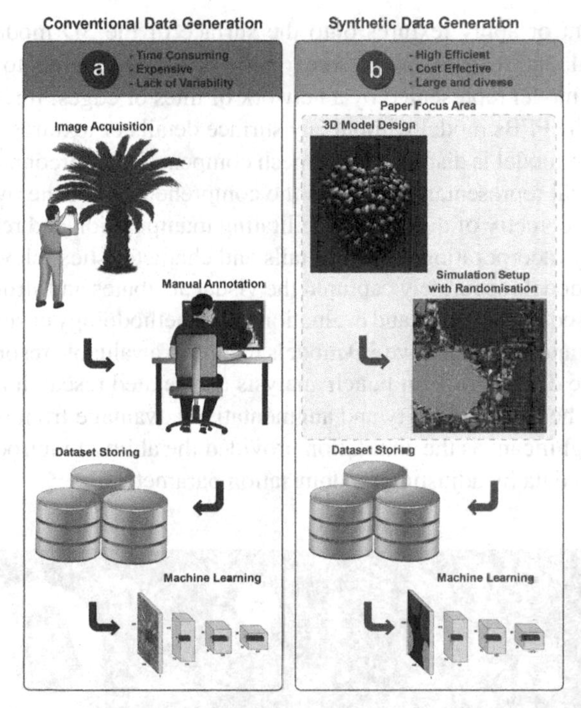

Fig. 1. The overall concept of FFB synthetic data generation and machine learning usage with a) Conventional data generation approach b) Synthetic data generation approach.

ripeness: ripe and unripe. The models were meticulously designed to closely resemble actual FFBs, complete with accurate representations of fruitlets, spikelets, and stalks, as shown in Fig. 2(a). The design of each component within the FFB model was carefully executed, ensuring the inclusion of features that are pertinent for dataset usage. The FFBs 3D model, consisting of 206,404 polygons and 108,447 vertices, exhibits a high level of detail and resolution. Polygons are the building blocks of 3D models, representing the surfaces and contours of the objects. Vertices, on the other hand, are the points where the edges of polygons meet, as shown in Fig. 2(b). Hence, the significant number of polygons indicates that the model employs numerous surfaces to accurately represent the intricate contours and features of the FFBs. This results in smooth curves, intricate fruitlets, spikelets, and stalk, and a faithful representation of the overall shape and form of the FFB. Moreover, the abundance of vertices contributes to the precise definition of the FFB's geometry, ensuring the accurate positioning of edges and corners. The higher vertex count allows for finer details and enhances the fidelity of the model's surface. In 3D modelling, UV mapping and wireframe are fundamental concepts and techniques used to create realistic textures and define the structure of a 3D object. UV mapping is the process of projecting a 2D texture onto a 3D model. It involves creating a flattened 2D representation of the FFBs surface, known as a UV map, which serves as a template for applying textures. The UV map comprises coordinates (U and V) that correspond to the vertices of the 3D model. Texture artists utilise these UV coordinates to accurately paint or apply textures onto the surface of the 3D model, imparting it with colour, detail, and realism. Wireframe, on the other hand, refers to the underlying structure of a 3D model represented by a network of lines or edges. It reveals the shape and topology of the FFBs model without any surface details or textures. In a wireframe representation, the model is displayed as a mesh composed of interconnected edges and vertices. This visual representation enables the comprehension of the overall structure, proportions, and geometry of the model, facilitating manipulation and refinement of the object's shape. By incorporating precise details and characteristics relevant to the FFB, the resulting 3D models accurately captured the visual attributes and structural elements necessary for subsequent analysis and evaluation. This methodology ensured the creation of highly realistic and representative 3D models that form a valuable resource for various applications in the domain of fruit bunch analysis and related research fields. By using 3D model FFBs, the data variability and augmentation advantage from actual photos of FFBs are more significant as the simulation provided the ability to introduce variability into the generated data by adjusting randomisation parameters.

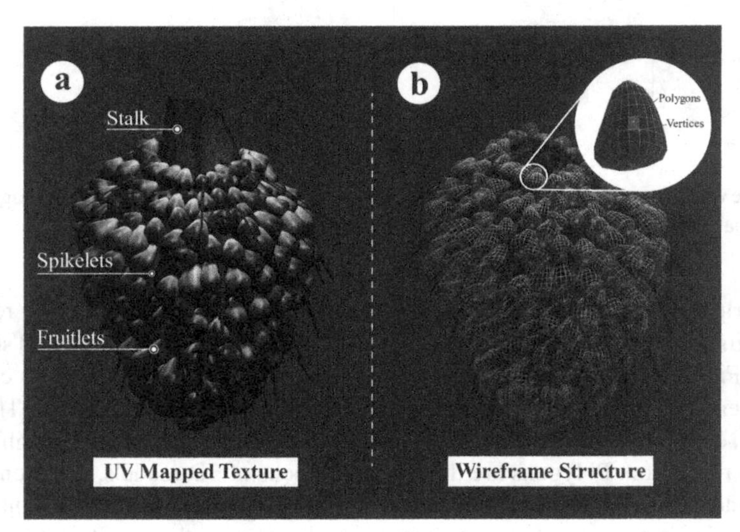

Fig. 2. The 3D Model of FFB created in Autodesk 3ds Max® software. a) The 3D Model with UV Mapped texture that represented the colour features. b) The wireframe of the 3D model that represented the structure geometry of the model.

2.2 Simulation Setup and Domain Randomisation

Rather than aiming for a replication of the reality palm oil estate environment, models can be developed to exhibit robustness in the face of environmental variability. To address the challenges posed by domain mismatch between simulation and the real world, the effective technique for generating synthetic training data versatility is domain randomisation, which introduces synthetic or enhanced data with random variations in environmental properties. A randomisation framework can be employed to introduce variation into synthetic environments, resulting in diverse datasets. This framework involves a control entity, such as a scenario, that coordinates the randomisations in the scene. By defining predetermined orders and schedules for randomisations, the creation of intricate and deterministic variations throughout a simulation will be conducted. By adopting these approaches and utilising suitable frameworks, researchers can develop robust models that exhibit improved performance and generalisation capabilities in the face of environmental variability and domain mismatch. By applying domain randomisation, a range of factors, including lighting conditions, the placement and orientation of the FFB models as foreground objects, and the positioning of background objects featuring distinct fruit images, are subject to randomisation, as shown in Fig. 3. The camera is positioned to capture images at a resolution of 640×640 pixels, enabling the acquisition of high-quality frames for domain randomization within the simulation. The camera setup incorporates a specific field of view of 27° illustrated in Fig. 3, carefully chosen to capture an optimal view of the simulated environment and ensure the desired level of coverage and detail in the captured frames. This dynamic and versatile approach enables the generation of a multitude of diverse scenarios and environments during each simulation run. By introducing variations in these critical aspects, the models are exposed to a comprehensive range of conditions, equipping them with the necessary flexibility and resilience to handle the complexities of real-world scenarios effectively. The integration of domain randomisation ensures that the simulated environment accurately represents the inherent variability present in palm oil estates and fosters the development of robust and adaptable models capable of addressing the challenges posed by real-world environments. To attain a randomised environment utilising the 3D models, a series of randomisers were employed, each responsible for a distinct randomisation task.

In essence, the subsequent elements were subjected to randomisation:

- Foreground FFB object: A randomly chosen subset of the two different ripeness FFB model objects is generated and positioned randomly in front of the camera for each frame, as shown in Fig. 3. The density of these objects is also randomised, resulting in frames where they are placed closer together and more numerous than in others. Moreover, the sizes of these objects are randomised for each frame, and the entire group of objects is given a consistent random rotation. The position of the placement may also exceed the camera frame to be partially seen, as shown in the camera preview in Fig. 3, the purpose is to mimic the actual condition of FFBs on the tree that may be obstructed by the leaves.
- Background objects: To recreate the surrounding environment, a virtual scene is created, incorporating randomised background objects. Different geometric shapes such as circles, rectangles, and triangles are used, and UV mapping techniques are applied to assign different fruit textures to these shapes as shown in Fig. 3. A collection of basic 3D objects is positioned randomly in close proximity to one another, forming an intersecting arrangement akin to a "wall" situated behind the grocery objects. These objects are adorned with diverse textures, randomly selected from a pool of different fruit and vegetable images and applied to them for each frame. Furthermore, the rotations and colour hues of these objects undergo randomisation on a per-frame basis. This approach effectively mirrors the diverse and complex backgrounds found in palm oil estates, enhancing the realism of the simulation.

- Lighting conditions: The lighting conditions in the simulation are based on the average sunlight intensity experienced in the equator region, ranging from 9 klx to 67 klx. Point lights are used to replicate the sunlight source typically found in outdoor palm oil estates shown in Fig. 3 illustrated as the view from behind of the objects, ensuring the simulation reflects real-world environmental conditions accurately. The facing of sunlight was made towards the object to make sure the lighting condition can be reflected on the objects fully and captured by the camera.

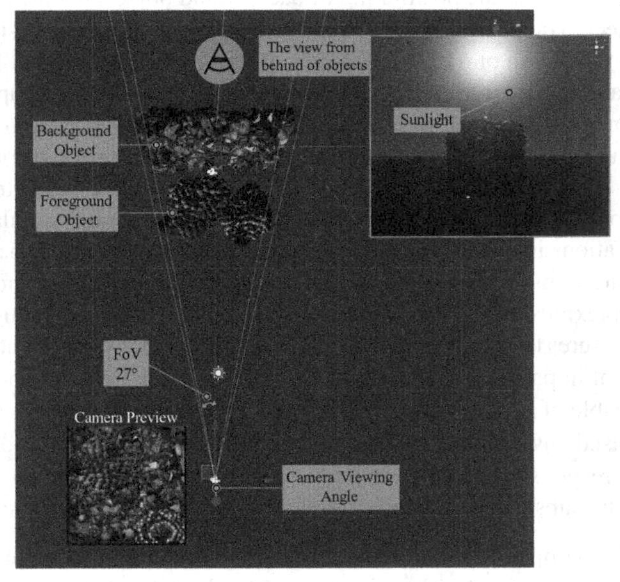

Fig. 3. The overall scene for the simulation, every frame of randomisation was captured by the camera. Foreground objects and background objects generated with different positions and orientations for every frame. The representation of the sunlight in the simulation displayed in the window.

The creation of synthetic data requires automatic annotation labelling of generated images that involves ground truth labelling. Ground truth refers to the accurate and reliable information or data that serves as a reference or benchmark for evaluating the performance of a system or model. In the context of computer vision or machine learning, ground truth data often involves manually annotated or verified data that represents the correct or desired outcomes for a given task. In this case, ground truth data may consist of labels assigned to the ripe or unripe FFBs objects or classes in an image. This ground truth data is used to train and evaluate computer vision models, enabling the learning process and conducting accurate predictions or classifications. In this simulation, both frames and ground truth information will be captured and generated, which the ground truth data will be derived by the labelers. Labelers is a C# customisable component for labels assigned to the 3D assets in the scene. To ensure versatility across different dataset labelling and generation, the labelers are configured with a mapping that translates human-readable semantic labels into numeric canonical class IDs used for training the target model. During simulation, these labelers compute ground truth by analysing the state of the 3D scene and the rendering results, employing a custom rendering pipeline.

2.3 Performance Metrics

In object detection tasks, predictions are made by assigning a bounding box and a class label. The accuracy of these predictions is determined by measuring the overlap between the predicted bounding boxes and the ground truth bounding boxes, which is commonly referred to as intersection over union (IoU). To compute the precision and recall of the object detection model, an IoU threshold is set, establishing a criterion for the IoU value. A significant area under the curve indicates a combination of strong recall and high precision, where high precision corresponds to a low false positive rate, and high recall corresponds to a low false negative rate. The average precision (AP) is determined by calculating the area under the precision-recall curve. In the context of multi-class detection tasks, the most commonly used metric is the mean average precision (mAP) score. AP is derived from precision (p) and recall (r), and mAP is computed by the average is computed of all AP divided by the total number of classes (Q), as shown in the following Eq. (1) and (2):

$$AP = \int_{0}^{1} p(r)dr \tag{1}$$

$$mAP = \frac{1}{Q}\sum_{q=1}^{Q} AP_q \tag{2}$$

with $q = 1 \ldots Q$ and Q is the number of classes. mAP yields a high value close to 1 when the model demonstrates commendable performance in both recall and precision. On the other hand, the minimum value achievable is zero.

3 Result and Discussion

Once the simulation was completed, the synthetic data was generated with the annotated labels in JavaScript Object Notation file format. The dataset was distributed into three categories: Training - 80%, validation - 10% and testing - 10%. The dataset was originally created in Synthetic Optimised Labeled Objects dataset format, which required dataset conversion to You Only Look Once (YOLO) dataset format to be compatible with the training of the CNN model. YOLO v8 real-time object detection model by Ultralytics® was employed with Darknet-53 as the backbone. Several key hyperparameters were selected for the training process, including a learning rate of 0.01, weight decay of 0.0005, Adam as the optimiser, and a batch size of 16 on an NVIDIA® 1660Ti GPU, which is the recommendation value from Ultralytics®. The performance of the manually labelled real dataset and the generated synthetic dataset was assessed in this phase. The CNNs was trained independently using two datasets: (i) the manually labelled dataset with 2D bounding boxes comprising 150 images from the actual palm oil estate, (ii) the simulation-generated synthetic dataset consisting of 30000 images with foreground objects randomly placed on background images. In this experiment, the effectiveness of using synthetic data for the object detection model was analysed, and benchmark comparison was made for dataset quantity optimising and validating with real FFBs datasets.

3.1 Dataset Quantity Optimising

The quantity of data available for training a CNN model plays a pivotal role in determining its performance. This is particularly evident in the context of FFBs ripeness classification, where the limited availability of real data imposes significant constraints on CNN's ability to learn and generalise patterns effectively. By examining the results of an experiment conducted in this regard, it becomes evident that increasing the dataset size yields improved performance, underscoring the importance of large-scale data. The experiment involved exclusively synthetic data, which was generated and utilised in training the YOLOv8 model. To ensure consistency, the validation conditions were maintained with a consistent epoch of 1 for each dataset size. The findings of this experiment were compiled in Table 1 and illustrated in Fig. 4, offering quantitative evidence of the impact of dataset size on the model's performance. The results highlight the necessity of accessing large and diverse datasets to enhance the CNN's ability to discern subtle patterns and make accurate predictions. A larger dataset provides a broader representation of the real-world scenarios and variations that the model may encounter during deployment. This exposure to diverse examples helps the CNN to learn a more comprehensive range of features and patterns, enabling it to make more accurate and robust predictions. Moreover, a larger dataset mitigates the risk of overfitting, which occurs when a model becomes overly specialised to the training data and fails to generalise well to unseen examples. With a limited dataset, the model may memorise specific instances rather than truly understanding the underlying patterns. However, by training on a large dataset, CNN is forced to learn more generalised representations, reducing the likelihood of overfitting and improving its ability to handle new, unseen data, which further justified from this experiment that 30,000 synthetic datasets were deployed in the same training condition, the result showed major improvement in comparison with the previous dataset sizes. By using 30,000 images, the mAP50 reached 0.995, and mAP50–95 obtained 0.814, which is very close to the 1.0 ideal performance. Hence, the use of synthetic data can provide large data with a lesser amount of time to prepare. Generating synthetic data allows researchers to create an almost unlimited number of training samples, surpassing what can be feasibly collected from real-world sources. This large volume of synthetic data enables CNN models to be trained on a diverse range of scenarios, variations, and object configurations that may not be easily accessible or present in real datasets.

3.2 Validation with Real FFBs Dataset

The validation of the synthetic FFBs dataset with the real FFBs dataset is essential to assess the performance and generalisation capability of the CNN model trained on synthetic data. While synthetic data offers advantages in terms of dataset size and controllability, it is crucial to validate the model's performance on real-world data to ensure

Table 1. mAP performance on the validation set for each dataset size.

Size of Dataset	Metrics	
	mAP50	mAP50–95
500	0.092	0.029
1000	0.075	0.035
1500	0.109	0.043
2000	0.292	0.162
2500	0.291	0.155
3000	0.602	0.295
30000	**0.995**	**0.814**

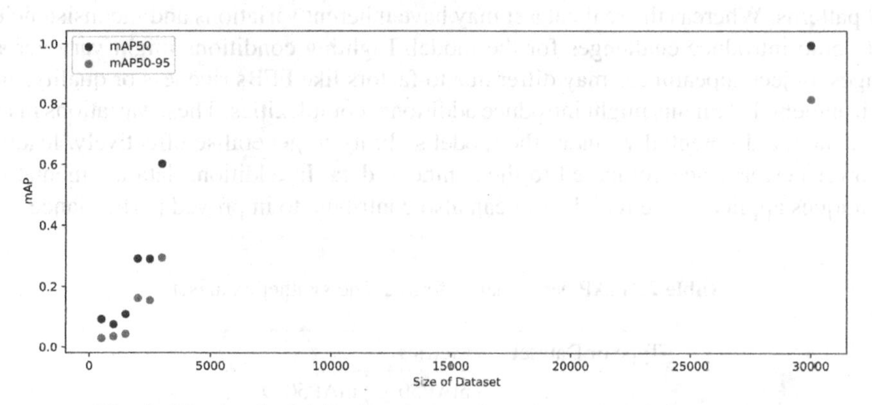

Fig. 4. The graph plotting for mAP performances of each dataset size.

its applicability in practical scenarios. By comparing the model's performance on both synthetic and real FFB datasets, an evaluation of the ability to generalise and accurately classify real FFB instances was carried out. This validation process helps identify any discrepancies or limitations that may arise from the differences between synthetic and real data, such as variations in lighting conditions, object appearances, or environmental factors. The real FBBs dataset consists of 150 actual images of FFBs from primary and secondary data collection with data augmentation performed as shown in Fig. 5. Since the available data is limited, data augmentation is crucial for small real FFBs datasets as it helps to artificially increase the size and diversity of the available data. By applying transformations such as flipping, rotation, cropping, shearing, blurring, Gaussian Noise, cutout and mosaic, data augmentation introduces additional variations and helps prevent overfitting. It allows the model to learn from a larger and more representative set of examples, improving its ability to generalise and make accurate predictions on unseen data. To ensure a balanced comparison, the synthetic data used for training is matched in quantity to the real dataset. The validation is conducted under identical conditions, providing a fair evaluation of the CNN model's performance. By maintaining consistency in dataset size and training conditions, any observed differences in performance between the synthetic and real datasets can be attributed to the inherent characteristics and limitations of each dataset.

The obtained results are tabulated in Table 2, with the synthetic data outperforming the real data by 80.11% which increased from 0.533 to 0.96 for mAP50; whereas the result for mAP50–95 increased from 0.219 to 0.678 by 209.59%. The result indicates a significant performance advantage of the synthetic dataset in this particular experiment. These substantial differences highlight the potential benefits of using synthetic data for training the CNN model. There are several factors contributing to this notable performance gap. Firstly, the synthetic data was carefully generated with specific control over various factors such as lighting conditions, object appearances, and environmental variations. This control allowed for a more consistent and controlled learning experience for the model, enabling it to understand better and capture the essential features and patterns. Whereas the real dataset may have inherent variations and inconsistencies that could introduce challenges for the model. Lighting conditions might vary across images, object appearances may differ due to factors like FFBs ripeness or quality, and environmental elements might introduce additional complexities. These variations in the real data could potentially hinder the model's ability to generalise effectively, leading to lower performance compared to the synthetic data. In addition, data augmentation techniques applied to the real dataset can also contribute to improved performance.

Table 2. mAP performance for real and synthetic dataset.

Type of Dataset	Metrics	
	mAP50	mAP50–95
Real	0.533	0.219
Synthetic	**0.96**	**0.678**

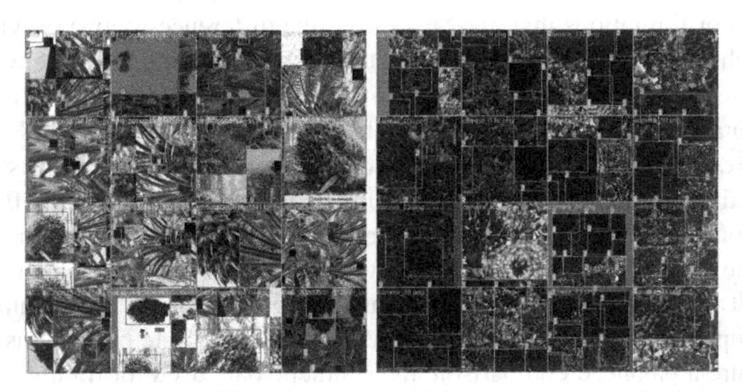

Fig. 5. Training batches for real FFBs dataset on the left and synthetic FFBs dataset on the right.

4 Conclusion and Future Work

In this paper, the significance of synthetic data in addressing the challenges of FFBs ripeness classification is emphasised. By training classification models on this extensive synthetic dataset, substantial improvements in accuracy and generalisation capabilities compared to models trained solely on real data were observed. The first experiment focused on the impact of dataset size, which demonstrated higher mAP values with larger datasets. This highlights the advantages of utilising a larger synthetic dataset, allowing the model to learn from a more comprehensive range of examples and capture a wider variety of ripeness patterns. The controllability and scalability of synthetic data facilitated the generation of a dataset well-suited to the classification task, resulting in superior model performance. The second experiment focused on the comparison between real and synthetic data. By training classification models on synthetic data, significant improvements in accuracy and generalisation capability were observed compared to models trained solely on real data. The synthetic data proved to be a valuable resource for addressing the limitations of limited real-world datasets in FFB ripeness classification, with the synthetic dataset surpassing the real dataset by 80.11% for mAP50 and 209.59% for mAP50–95. Synthetic data can serve as a valuable tool in situations where real data availability is limited, providing a means to generate larger and more diverse datasets for training robust models.

Future research in enhancing synthetic data for FFB ripeness classification involves two key areas. Firstly, validating models in real field conditions with dynamic video to assess their real-world applicability. Secondly, incorporating weather conditions and variations in the simulation process to enhance synthetic data authenticity and improve performance in real-world scenarios.

Acknowledgement. This work was supported by Ministry of Higher Education under Fundamental Research Grant Scheme FRGS/1/2022/TK07/UTM/02/42 and Universiti Teknologi Malaysia under UTM Fundamental Research Grant Q.J130000.3823.22H55.

References

1. Shabdin, M.K., Shariff, A.R.M., Johari, M.N.A., Saat, N.K., Abbas, Z.: A study on the oil palm fresh fruit bunch (FFB) ripeness detection by using Hue, Saturation and Intensity (HSI) approach. IOP Conf. Series Earth Environ. Sci. **37**, 012039 (2016)
2. Bolón-Canedo, V., Sánchez-Maroño, N., Alonso-Betanzos, A.: A review of feature selection methods on synthetic data. Knowl. Inf. Syst. **34**, 483–519 (2013)
3. de Melo, C.M., Torralba, A., Guibas, L., DiCarlo, J., Chellappa, R., Hodgins, J.: Next-generation deep learning based on simulators and synthetic data. Trends Cogn. Sci. **26**, 174–187 (2022)
4. Gao, C., et al.: Synthetic data accelerates the development of generalizable learning-based algorithms for X-ray image analysis. Nat. Mach. Intell. (2023). https://doi.org/10.1038/s42256-023-00629-1
5. Becktor, J., Schöller, F.E.T., Boukas, E., Blanke, M., Nalpantidis, L.: Bolstering maritime object detection with synthetic data. In: IFAC-PapersOnLine. Elsevier B.V., pp. 64–69 (2022)

6. Wu, X., Liang, L., Shi, Y., Fomel, S.: FaultSeg3D: using synthetic data sets to train an end-to-end convolutional neural network for 3D seismic fault segmentation. Geophysics **84**, IM35–IM45 (2019). https://doi.org/10.1190/geo2018-0646.1
7. Manettas, C., Nikolakis, N., Alexopoulos, K.: Synthetic datasets for Deep Learning in computer-vision assisted tasks in manufacturing. Procedia CIRP **103**, 237–242 (2021)
8. Olatunji, J.R., Redding, G.P., Rowe, C.L., East, A.R.: Reconstruction of kiwifruit fruit geometry using a CGAN trained on a synthetic dataset. Comput. Electron. Agric. **177**, 105699 (2020). https://doi.org/10.1016/j.compag.2020.105699
9. Seth P, Bhandari A, Lakara K (2023) Analyzing Effects of Fake Training Data on the Performance of Deep Learning Systems
10. Josip, J., Matthias, K., Christoph, P., Lukas, P., Stefan, W.: Object detection and pose estimation based on convolutional neural networks trained with synthetic data. In: IEEE/RSJ International Conference on Intelligence Robots and Systems (IROS). IEEE (2018)
11. Aayush, P., et al.: Structured domain randomization: bridging the reality gap by context-aware synthetic data. In: International Conference on Robotics and Automation (ICRA) (2019)

Application of Artificial Intelligence for Maternal and Child Disorders in Indonesia: A Review

Diva Kurnianingtyas, Indriati, and Lailil Muflikhah(✉)

Department of Informatics Engineering, Faculty of Computer Science, Universitas Brawijaya, 65145 Malang, Indonesia
{divaku,lailil}@ub.ac.id

Abstract. The development of Artificial Intelligence (AI) technology is used to minimize the risk of maternal disorders during pregnancy. Maternal health needs to be monitored so as not to cause problems during the baby's birth. The purpose of this study is to provide a literature review from 2017 to 2023. The method used is Preferred Reporting Items for Systematic Reviews and Meta-Analyses (PRISMA). This study is a basis for researchers to help solve maternal health problems using AI. This technology is proven to detect and predict problems during pregnancy, thereby reducing maternal and infant mortality and preventing abnormalities in the development process. In addition, AI can help improve medical personnel's performance by minimizing human error. This study also presents trends in AI problems and methods used. However, the rapid development of AI methods has not provided novelty in solving maternal pregnancy issues because researchers rarely implement them on a wider scope of problems. Furthermore, it needs to be developed with other methods, such as simulation.

Keywords: Systematic reviews · Pregnancy disorders · caesarian section · Health status · Preeclampsia

1 Introduction

Pregnancy is a process a maternal goes through to deliver a baby, generally with careful preparation. Maternal health is an essential factor that needs to be considered comprehensively. The unmonitored maternal health history will cause problems at the time of the baby's birth, such as low birth weight (*Berat Badan Lahir Rendah* or *BBLR*), stunted baby growth (*Pertumbuhan Bayi Terhambat* or *PJT*), low immunity, and a high risk of death for the baby, maternal or both [1–3]. The development of Artificial Intelligence (AI) technology is one of the solutions to detect and predict disorders during pregnancy automatically [4].

Based on the previous explanation, AI application has successfully increased efficiency and productivity in helping maternal during pregnancy. The purpose is to present various brief pieces of literature from 2017 to 2023. This study will help medical personnel to monitor and detect early pregnancy disorders in maternal. In addition, it can help researchers to develop AI for health problems to be more optimal.

© The Author(s), under exclusive license to Springer Nature Singapore Pte Ltd. 2024
F. Hassan et al. (Eds.): AsiaSim 2023, CCIS 1911, pp. 289–306, 2024.
https://doi.org/10.1007/978-981-99-7240-1_23

This study has five sections, namely Sect. 2 about the methodology of the literature study, next related to the review results obtained. Afterward, Sect. 4 presents a discussion on the application of AI. Conclusions and potential study follow in the last section.

2 Methodology

The relevant article search strategy in this study used Preferred Reporting Items for Systematic Reviews and Meta-Analyses (PRISMA) flowchart [3] (see Fig. 1). First, determine the queries needed. The query uses two types, namely *"Kecerdasan Buatan untuk Ibu Hamil di Indonesia"* and "Artificial Intelligence for Maternal Health in Indonesia". The query used Bahasa and English because the case study discussed maternal health in Indonesia. Second, the article search was limited to the year of publication from 2017 to 2023. The search was conducted on Google Scholar by identifying titles that match the study. Third, the article obtained were re-filtering based on the abstract. Based on the results of these three stages, 30 articles were obtained, the distribution of which can be seen in Table 1.

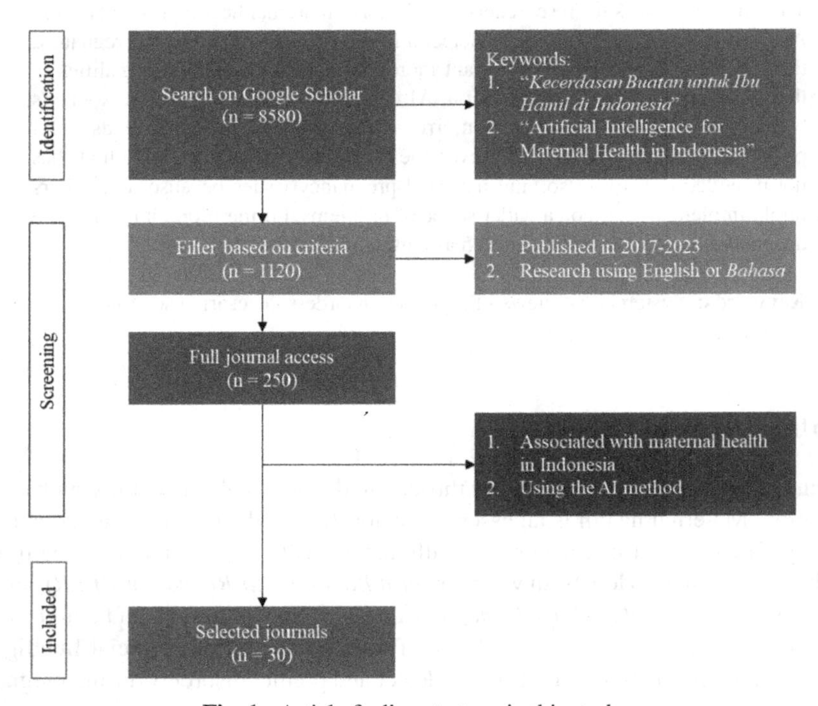

Fig. 1. Article finding strategy in this study

Table 1. Result for query in Google Scholar

Year	AI using	Algorithm	Reference
2017	Detection, Prediction	Artificial Neural Network, Backward Chaining, C45, Certainty Factor Decision Tree, Forward Chaining, Random Forest	[4–7]
2018	Detection	Backward Chaining, Certainty Factor	[8, 9]
2019	Detection, Prediction	Breadth First Search, Forward Chaining, Naïve Bayes, Neural Network	[10–13]
2020	Detection, Prediction	Certainty Factor, Forward Chaining, K-Means, Naïve Bayes, Natural Language Processing, Soft Voting-based Ensemble	[2, 14–18]
2021	Detection, Prediction	Certainty Factor, Decision Tree, Forward Chaining, Linear Discriminant Analysis, Naïve Bayes, Neural Network, Support Vector Machine	[19–25]
2022	Detection	Breadth First Search, Forward Chaining, K-Nearest Neighbour, Naïve Bayes	[26–28]
2023	Detection, Prediction	Forward Chaining, Machine Learning, Deep Learning	[29, 30]

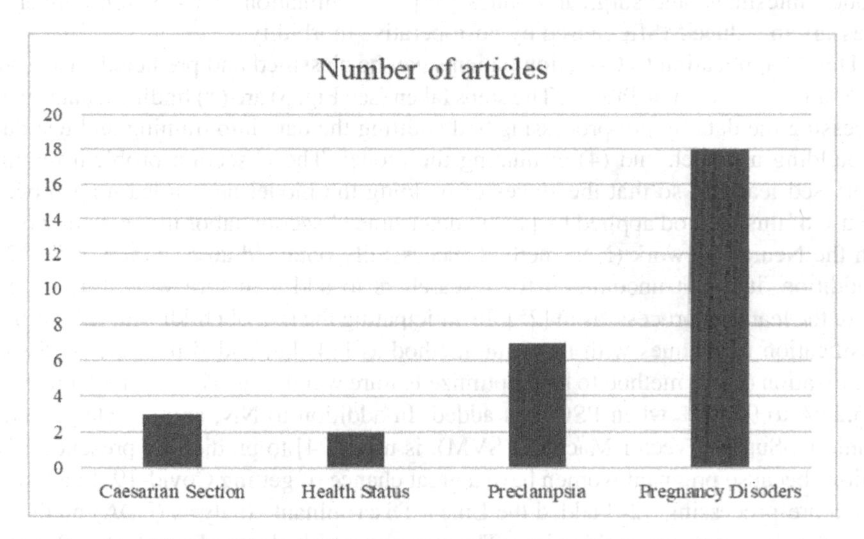

Fig. 2. Number of articles selected by maternal issue type

3 Result

AI has helped find preventive and treatment solutions for maternal health. Based on the findings of the selected articles (see Fig. 2), AI has successfully solved to the problem of Caesarian section, Health status, Preeclampsia, and Pregnancy disorders. The success of AI on each problem will be discussed in the next section.

3.1 Caesarian Section

Caesarian section (C-section) is one way to deliver a fetus to reduce the risk of maternal and infant mortality. This delivery can help achieve Indonesia's Millennium Development Goals (MDGs) targets [31]. Based on data from the Ministry of Health 2022, the maternal mortality rate is around 305/100,000 live births which have not reached the target set in the 2024 National Medium-Term Development Plan (*Rencana Pembangunan Jangka Menengah Nasional* or *RPJMN*) target of around 183/100,000 live births [32]. Therefore, a C-section is one of the efforts that can reduce the Maternal Mortality Rate (MMR or *AKI/Angka Kematian Ibu*).

However, C-section is the last alternative to childbirth as it carries five times more risk than expected delivery. Several factors can increase the risk, including embolism or blockage of blood vessels, endometritis or endometrium inflammation, bleeding during surgery, anesthesia, complication, and recovery of the shape and location of the uterus to be imperfect [33, 34]. One of the impacts is the emergence of infections in the uterus, bladder, intestines, and surgical wounds [35]. Determination of C-section delivery is necessary to reduce MMR caused by postoperative morbidity.

The AI application to C-section patients can be classified and predicted easily with the parameters shown in Table 2. The steps taken (see Fig. 3) are (1) finding a dataset, (2) processing the data by pre-processing and splitting the data into training and test data, (3) building a model, and (4) evaluating the model. The C-section problem requires supervised learning so that the stages of building the model have a learning process. The use of this method applied by [13] to determine C-section labor in pregnant women with the Neural Network (NN) method successfully obtained an accuracy of 74.17%. In addition, it is not uncommon for researchers to add a feature weighting process before the learning process, as in [25]. In anticipating the risk of childbirth, [25] utilized classification techniques with the same method as [14] but added the Particle Swarm Optimization (PSO) method to help optimize feature weighting. The accuracy increased by 6.25% to 93.75% when PSO was added. In addition to NN, another classification technique, Support Vector Machine (SVM), is used [24] to predict the presence of C-sections because pregnant women have a great chance of getting Covid-19. In addition, in data pre-processing, [24] added the Linear Discriminant Analysis (LDA) method to obtain data by the research objectives. The accuracy result obtained was 100%. Based on the three studies, a Confusion Matrix, Receiver Operating Characteristic (ROC) Curve, and Area Under Curve (AUC) are commonly used to measure the performance of models applied to C-section problems.

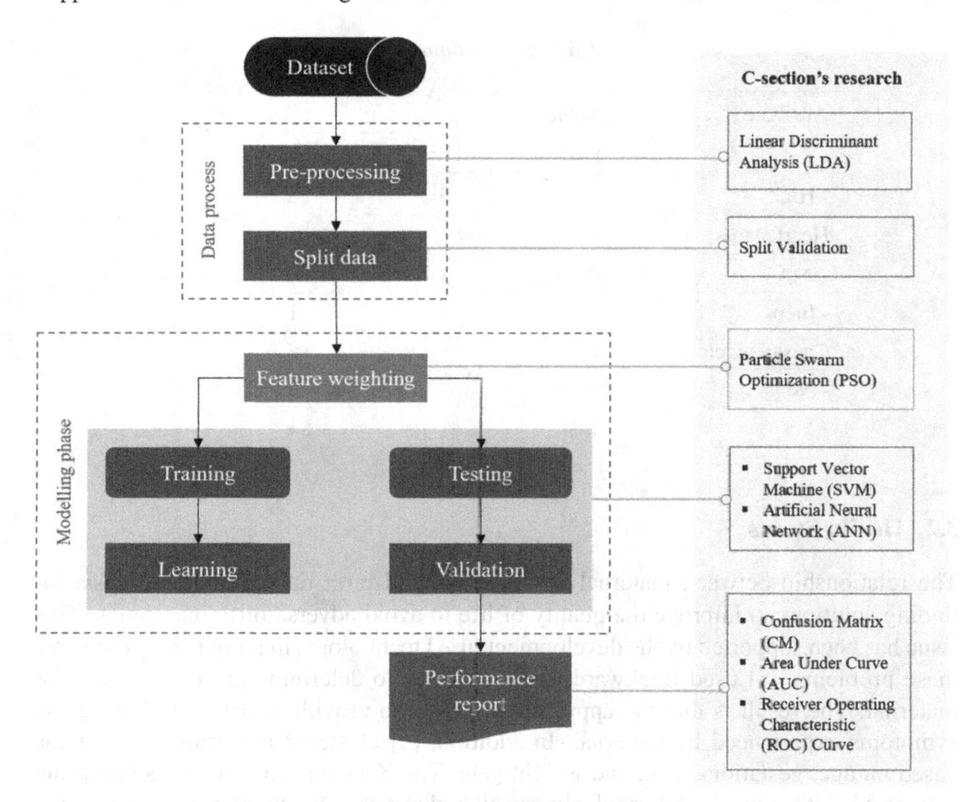

Fig. 3. Research framework of C-section

Table 2. C-section dataset features

Attribute	Value		
	[13]	[24]	[25]
Age (years)	17–41	17–40	17–40
Delivery number	1, 2, 3, 4	1, 2, 3, 4	1, 2, 3, 4
Delivery time			
- Timely	0	1	0
- Premature	1	2	1
- Latecomer	2	3	2
Blood pressure			
- Low	0	1	0
- Normal	1	2	1

(*continued*)

Table 2. (*continued*)

Attribute	Value		
	[13]	[24]	[25]
- High	2	3	2
Heart status			
- Apt	0	1	0
- Inept	1	2	1
Caesarean class			
- No	0		0
- Yes	1		1

3.2 Health Status

The relationship between maternal health status and birth outcomes is the basis for finding solutions to improve the quality of life to avoid adverse birth out-comes. This issue has been supported by the development of AI technology, making it easier to solve these problems. [8] used Backward Chaining (BC) to determine proper nutrition for maternal. The result is that the approach is proven to provide nutrition following the symptoms experienced by maternal. In addition, [2] clustered maternal health status based on age, gestational age, and weight gain. The K-means method helps group the maternal health status into normal, abnormal, and at-risk. The parameters to determine the maternal health status can be used in Table 3. The results are used as evaluation material to improve maternal quality in pregnancy and reduce the risk of disorders in infants.

Table 3. Maternal nutritional status features based on [2, 8]

Attribute	Value
Status	
Age (years)	19–40
Age of pregnancy (months)	2–9
Weight gain (kg)	1–30
Nutrition needs	
Carbohydrate	0, 1
Proteins	0, 1
Vitamin	0, 1

(*continued*)

Table 3. (*continued*)

Attribute	Value
Zinc	0, 1
Folic acid	0, 1
Symptoms	
The body is easily sluggish, tired, and tired quickly	0, 1
Lack of energy	0, 1
Irritable	0, 1
Stressed and always have a headache	0, 1
Sleepiness	0, 1
Abnormal body weight	0, 1
Sleepiness	0, 1
Abnormal body weight	0, 1
Appetite is not good	0, 1
Loose bowel movements and frequent constipation	0, 1
Dry and chapped lips	0, 1
Gums often bleed	0, 1
Dark eye bags	0, 1
Swollen feet	0, 1

3.3 Preeclampsia

Preeclampsia is one of the causes of maternal and fetal mortality and morbidity due to hypertension, proteinuria, and edema in pregnancy. Based on World Health Organization (WHO) predictions, preeclampsia cases have the potential to occur in developing countries seven times higher than in developed countries. Delays in recognizing the emergence of dangerous conditions and taking preventive measures are important factors in increasing maternal and fetal deaths. The government has established an obstetric-related policy in Ministry of Health (*Kementerian Kesehatan* or *Kemenkes*) No. 369/Menkes/SK/III/2007 to support prevention by optimizing health during pregnancy [22, 27]. However, the policy needs to be supported by AI technology to make the prevention process effective.

Several studies have proposed effective and efficient methods for preventing preeclampsia during pregnancy. The parameters to identify preeclampsia in maternal can be used in Table 4. [12] successfully used Long Short-Term Memory (LSTM) to prevent the risk of preeclampsia through the classification and prediction process shown with an accuracy of 90.22%. Afterward, [15] could obtain an accuracy of 96.66% with the Soft Voting-based Ensemble method. [16] using Natural Language Processing (NLP), the Confidence Interval (CI) value can reach 0.88–0.89 in predicting preeclampsia in maternal.

In addition, some researchers have created early detection systems for preeclampsia, such as an early detection system to monitor maternal to get immediate and appropriate treatment for preeclampsia [22]. The use of the Certainty Factor (CF) method in the system has an accuracy of 91.45%. In addition, [27] built a system using the Forward Chaining (FC) method proven to help increase knowledge in minimizing the risk of preeclampsia.

Table 4. Preeclampsia features based on [12, 16, 22]

Attribute	Value
Demographic	
Age (years)	19–50
Role in family [wife, child, primary member, additional member]	0, 1
Work [company-paid labor, government-paid labor, entrepeneur, non-labor]	0, 1
Status	
Case history	
[hereditary preeclampsia, hereditary hypertension (PHT), hereditary pregnancy with hypertension (PPHT), herditary diabetes pregestational, immune disorders, birth control (Keluarga Berencana/KB) acceptor]	0, 1
First pregnancy	0, 1
Pregnancy distance	0, 1
Twin pregnancy	0, 1
Blood pressure	0, 1
Body mass index (BMI)	0, 1
Glucose (mg/dl)	90–110
Proteinuria (mg/dl)	<10
Symptoms	
Severe headache	0, 1
Heartburn	0, 1
Visual disturbances	0, 1
Depression/stress	0, 1
Difficulty breathing	0, 1
Weak baby movements	0, 1

3.4 Pregnancy Disorders

One of the changes that occur in maternal is a decreased immune system. The risk of diseases increases if maternal do not maintain their health. Various kinds of research to reduce the risk with AI both the application of the Naïve Bayes (NB) method [10, 21],

FC [7, 14, 19, 20, 29], Breadth First Search (BFS) [26], and the combination of FC and BFS [11] can help early detection of the risk of disorders during pregnancy. In addition, the application of the Artificial Neural Network (ANN) method [5, 6] is used to predict the risk of pregnancy based on the detection results, and the Genetics Algorithm (GA) method [36] to optimize the method so that the detection and prediction obtained are accurate.

The maternal's lack of information and knowledge regarding the risk of disorders that will occur in the pregnancy process can result in miscarriage. This problem happens because the mother is late in knowing her disease during pregnancy. Table 5 is an example of a symptom matrix in maternal disease. [37] developed a website-based system to prevent the risk of abnormalities during pregnancy in order to reduce the Infant Mortality Rate (IMR or *AKB/Angka Kematian Bayi*). This system applies the FC method, which has been proven to help experts and mothers consult during pregnancy.

Table 5. Symptoms matrix in the disease [14]

Symptoms		Disease								
		D1	D2	D3	D4	D5	D6	D7	D9	D10
S1	Discharge of spots	✓								
S2	Bloody discharge	✓								
S3	Prolonged bloody discharge	✓								
S4	Nausea and excessive vomiting	✓								
S5	Pain in the placenta	✓								
S6	Prolonged dizziness	✓								
S7	Pain during urination	✓								
S8	Vaginal bleeding	✓								
S9	Drastic changes in pregnancy symptoms	✓								
S10	Pain and tenderness in the pelvis and abdomen	✓								

(*continued*)

Table 5. (*continued*)

Symptoms		Disease								
		D1	D2	D3	D4	D5	D6	D7	D9	D10
S11	Dysminorrhea (pain during menstruation)		✓							
S12	Irregular menstruation		✓							
S13	Vaginal discharge and odor		✓							
S14	Abdominal pain		✓							
S15	Pain during intercourse		✓							
S16	Heavy bleeding		✓							
S17	Swelling in the groin			✓						
S18	Lump or swelling in the vagina			✓						
S19	Pain and tenderness in the pelvis			✓						
S20	Intense aches and pains during exertion			✓						
S21	Abdominal cramps before or during menstruation				✓					
S22	Difficulty sleeping				✓					
S23	Back pain				✓					
S24	Dizziness and headache				✓					
S25	Loss of appetite					✓				
S26	Chills					✓				

(*continued*)

Table 5. (*continued*)

Symptoms		Disease								
		D1	D2	D3	D4	D5	D6	D7	D9	D10
S27	Heartburn					✓				
S28	No menstruation for about two months					✓				
S29	Dehydration					✓				
S30	Hypotension or low blood pressure					✓				
S31	Heart palpitations					✓				
S32	Excessive salivation					✓				
S33	Feeling stressed, confused, anxious					✓				
S34	Highly sensitive to smells					✓				
S35	Discharge of fluid						✓			
S36	Pain in the lower abdomen						✓			
S37	Itching in the external genital area						✓			
S38	Vaginal discharge, inflammation, and redness of the external genitalia						✓			
S39	Intense pain						✓			
S40	Water discharge at the surgical site							✓		

(*continued*)

Table 5. (*continued*)

Symptoms		Disease								
		D1	D2	D3	D4	D5	D6	D7	D9	D10
S41	Pain on the left side							✓		
S42	Prolonged back pain							✓		
S43	No period and positive pregnancy test							✓		
S52	Body fatigue								✓	
S53	Fever								✓	
S54	Itching in the abdomen and legs								✓	
S55	Nausea and vomiting								✓	
S56	Sudden abdominal pain								✓	
S57	Frequent feeling of dizziness								✓	
S58	Chest tightness								✓	
S59	Problems urinating and defecating								✓	
S60	Pain in hips, stage, and thighs								✓	
S61	Heavy menstruation									✓
S62	Intense blood discharge									✓
S63	Period pain									✓
S64	Pain when pressing on the pelvis									✓
S65	Pelvic pain after intercourse									✓

(*continued*)

Table 5. (*continued*)

Symptoms		Disease								
		D1	D2	D3	D4	D5	D6	D7	D9	D10
S66	Feeling of pressure in the lower part of the colon									✓
S67	Abdominal fullness and bloating									✓

D1: Abortion; D2: Adenomyosis; D3: Bartholinitis; D4: Menstrual disorders; D5: Hyperemesis gravidarum; D6: Vaginal infection; D7: Ectopic pregnancy; D8: Ovarian cyst; D9: Uterine myoma

4 Discussion

Indonesia is the most populous country in Southeast Asia, with 277.43 million people by 2023. In connection with this, the government seeks to improve the population's quality by monitoring maternal health to give birth to a healthy and quality generation and reduce MMR and IMR [38] because Indonesia is one of the ten countries with the highest postpartum mortality rate globally [39]. Disorders that occur during pregnancy will cause severe impacts if not monitored properly. AI has a function to help early detection and prediction of these problems. This technology is done to help the performance of medical personnel. The increasing number of patient visits to health services will reduce the performance of medical personnel. AI technology can reduce human error when medical personnel experience fatigue due to many patients requesting health services. In addition, AI helps the performance of medical personnel so that work is completed quickly and with good and precise accuracy. The basis of AI is to learn in order to solve problems. In maternal problems, AI performs the detection process and then can make predictions [1–4]. Various AI methods can be used for detection, prediction, or both. Based on Fig. 4 (a), the most commonly used methods for detection and prediction are ANN and Decision Tree (DT).

One branch of technology that AI belongs to is Expert System (ES). This method adopts the working principles of experts to solve problems to find conclusions and decisions based on existing facts. The system is designed to help laypeople solve complex problems that experts can only solve. ES has four important components, including the knowledge base, database, user interface, and inference engine. These four components cannot be separated [10, 17, 19, 26, 27]. The problem of pregnancy disorders is in dire need of ES to improve the accuracy of interpretation, diagnosis, prediction, monitoring and recommendations needed to minimize the risk of disorders during pregnancy. Following Fig. 4 (b), the methods most often used by researchers to develop ES include FC, BC, BFS, and CF.

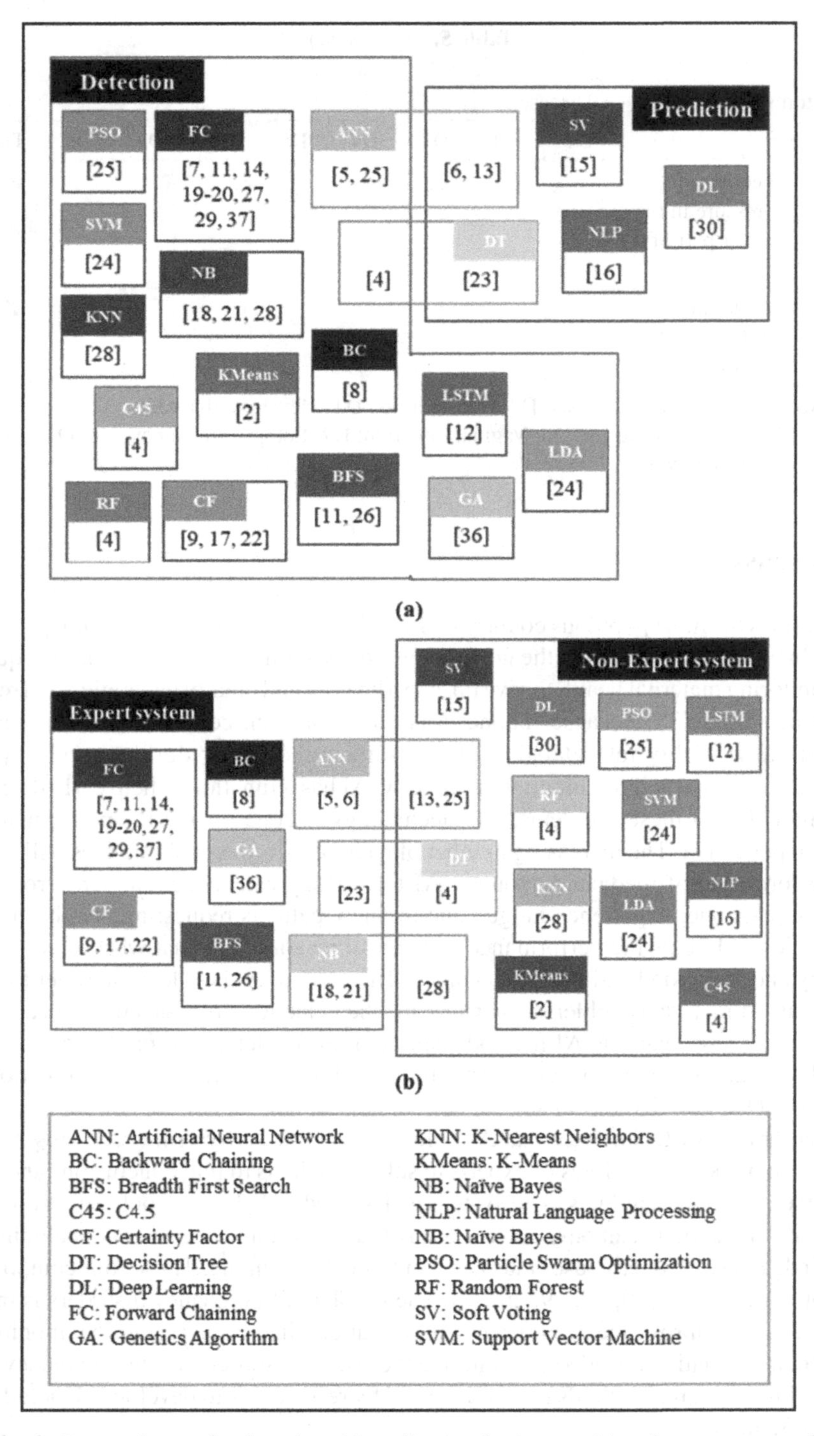

Fig. 4. Research mapping based on AI application for its uses (a) and expert system (b)

The development of AI technology in the health sector felt in Indonesia, especially in diagnosis. However, several challenges must be faced, such as AI technology development. Indonesia is a developing country that generally has relatively low income, so it has yet to be able to design its technology. Next, AI technology based on learning many data increases the risks of data security and privacy violations. In addition, implementation of information and communication technology in Indonesia is uneven. Most people have difficulty accessing technology, especially AI. This constraint causes the use of technology to be suboptimal. Efforts to reduce MMR and IMR cannot be instantaneous. AI technology can help with obstacles, such as the limited number of medical personnel, unreachable health facilities, and the need for maternal knowledge [39–41]. Resolving these challenges requires collaborating with the government, the community, and other interested parties. Addressing this challenge requires collaboration with the government, communities, and other interested parties to find the proper regulations to maximize AI technology in Indonesia.

5 Conclusion

In this study, we discuss AI's application to maternal health problems. Problems are divided into four categories: Caesarian section, Health status, Preeclampsia, and Pregnancy disorders. AI is implemented to minimize the risk of complications during pregnancy. This prevention helps reduce maternal and child mortality and prevents abnormal growth and development of babies. This study helped stakeholders find ways to solve maternal and child health problems using AI. In addition, researchers with similar expertise found research gaps in AI methods and maternal and child health issues.

AI proves its ability to do learning, problem-solving, and pattern recognition. In addition, AI has a branch of knowledge, namely ES. ES is proven to be able to solve complex problems with the help of experts. This system is considered suitable for accelerating the detection and prediction of maternal disorders with the help of a doctor's knowledge. The number of patient visits that continues to increase does not interfere with ES performance. Instead, it speeds up the detection and prediction process. Thus, events in mothers and children during pregnancy can be minimized. Methods of FC, BC, BFS, and CF are often used. The progress of AI methods is relatively rapid, but researchers tend to choose AI methods that are commonly known. These developments can be used as opportunities so that the process of detecting and predicting maternal and child disorders is more accurate by increasing the scope of the problem. In addition, the problems are more than just a matter of detection and prediction. Therefore, this limitation can be developed in future research by combining it with other methods, such as the simulation approach. In addition, because there are still obstacles to implementation in Indonesia, it is necessary to determine priority needs in the health sector. This obstacle can allow researchers to conduct studies using various approaches to assist stakeholders in determining policies to use AI technology optimally.

Acknowledgement. This study was supported by Universitas Brawijaya Indonesia [Number of grant: 611.76/UN10.C200/2023].

References

1. Pratiwi, L., KM, M.: Kesehatan Ibu Hamil. CV Jejak (Jejak Publisher) (2021)
2. Pa, A.L.B.: Penerapan Metode K-Means Clustering dalam Pengelompokkan Status Kesehatan Ibu Hamil. In: Seminar Nasional Informatika (SENATIKA), pp. 759–766 (2022)
3. Widodo, A.W., Kurnianingtyas, D., Mahmudy, W.F.: Optimization of healthcare problem using swarm intelligence: a review. In: 2022 IEEE International Conference on Industrial Engineering and Engineering Management (IEEM), pp. 747–751. IEEE (2022)
4. Mambang, M., Byna, A.: Analisis perbandingan Algoritma C. 45, random forest Dengan Chaid Decision Tree Untuk Klasifikasi Tingkat Kecemasan Ibu Hamil. Semnasteknomedia Online. 5, 1–2 (2017)
5. Anggraeny, F.T., Muttaqin, F., Munir, M.S.: Modeled early detection of pregnancy risk based on Poedji Rochjati score card using relief and neural network. In: Proceedings. pp. 519–525 (2018)
6. Maylawati, D.S., Ramdhani, M.A., Zulfikar, W.B., Taufik, I., Darmalaksana, W.: Expert system for predicting the early pregnancy with disorders using artificial neural network. In: 2017 5th International Conference on Cyber and IT Service Management (CITSM), pp. 1–6. IEEE (2017)
7. Afiana, F.N., Hariawan, A., Setiyadi, H.: Perancangan Metode Forward Chaining Untuk Mendeteksi Dini Gangguan Masa Kehamilan. In: Conference on Information Technology, Information System and Electrical Engineering (CITISEE). pp. 78–82 (2017)
8. Yuvidarmayunata, Y.: Sistem Pakar Berbasis Web Menggunakan Metode Backward Chaining Untuk Menentukan Nutrisi Yang Tepat Bagi Ibu Hamil. INTECOMS J. Inf. Technol. Comput. Sci. **1**, 231–239 (2018)
9. Aji, A.H., Furqon, M.T., Widodo, A.W.: Sistem pakar diagnosa penyakit ibu hamil menggunakan metode Certainty Factor (CF). J. Pengemb. Teknol. Inf. Dan Ilmu Komputer 2, 27–36 (2017)
10. Mustafa, W.F., Kusrini, K.: Sistem Pakar Diagnosa Penyakit Pada Ibu Hamil Menggunakan Teorema Bayes Di Apotek Rumah Sederhana Jayapura. INFOS J. Inf. Syst. J. **1**, 33–39 (2019)
11. Widiastuti, S.H., Imansyah, N.: Implementasi Forward chaining dan breadth first pada sistem pakar diagnosa gangguan kehamilan. JSAI. **2**, 154–158 (2019)
12. Sakinah, N., Tahir, M., Badriyah, T., Syarif, I.: LSTM with adam optimization-powered high accuracy preeclampsia classification. In: 2019 International Electronics Symposium (IES), pp. 314–319. IEEE (2019)
13. Suwarno, P.A.S.: Performance evaluation of artificial neural network classifiers for predicting cesarean sections. Evaluation **59**, 66–67 (2019)
14. Gunawan, A., Defit, S., Sumijan, S.: Sistem Pakar dalam Mengidentifikasi Penyakit Kandungan Menggunakan Metode Forward Chaining Berbasis Android. J. Sistim Inf. dan Teknol. 15–22 (2020)
15. Simbolon, O., Widyawati, M.N., Kurnianingsih, K., Kubota, N., Ng, N.: Predicting the risk of preeclampsia using soft voting-based ensemble and its recommendation. In: 2020 International Symposium on Community-centric Systems (CcS), pp. 1–6. IEEE (2020)
16. Sufriyana, H., Wu, Y.-W., Su, E.C.-Y.: Artificial intelligence-assisted prediction of preeclampsia: Development and external validation of a nationwide health insurance dataset of the BPJS Kesehatan in Indonesia. EBioMedicine **54**, 102710 (2020)
17. Saripurna, D., El, E.: Sistem Pakar Mendiagnosa Penyakit Pre-eklamsia pada Ibu Hamil dengan Menggunakan Metode Certainty Factor pada Rumah Sakit Ibu dan Anak Medan Johor. J. Cyber Tech. 4 (2022)
18. Ginting, M.A.P.F., Azlan, A., Halim, J.: Sistem Pakar Dalam Mediagnosa Penyakit Plasenta Previa Dengan Menggunakan Metode Teorema Bayes. J. Cyber Tech. 3 (2020)

19. Maulana, A.: Sistem Pakar Untuk Mendeteksi Gangguan Kehamilan Menggunakan Metode Forward Chaining. J. Inform. Polinema. **8**, 17–24 (2021)
20. Kuncoro, D.: Metode Forward Chaining Untuk Diagnosa Gangguan Kehamilan
21. Handoko, M.R., Neneng, N.: Sistem Pakar Diagnosa Penyakit Selama Kehamilan Menggunakan Metode Naive Bayes Berbasis Web. J. Teknol. Dan Sist. Inf. **2**, 50–58 (2021)
22. Fitrilina, F., Albbi, M., Agustian, I., Herawati, A., Massardi, N.A.: Sistem Peringatan Awal Resiko Preklamsia pada kehamilan menggunakan metoda Certainty Factor dan Android. J. Nas. Tek. Elektro. 45–54 (2021)
23. Wiyanto, W., Maulida, M.I., Fauziah, S.: Penerapan Sistem Pakar Berbasis Android Dengan Metode Decision Tree Untuk Memprediksi Postpartum Haemorrhage Pada Wanita Hamil. Pelita Teknol. **16**, 29–40 (2021)
24. Abdillah, A.A., Azwardi, A., Permana, S., Susanto, I., Zainuri, F., Arifin, S.: Performance evaluation of linear discriminant analysis and support vector machines to classify cesarean section. Eastern-European J. Enterp. Technol. **5**, 113 (2021)
25. Setia, I.C., Arifin, T.: Penentuan Penanganan Persalinan Caesar dengan Neural Network dan Particle Swarm Optimization. SISTEMASI. **10**, 346–356 (2021)
26. Al Ayubi, F., Indriyanti, A.D.: Perancangan Sistem Pakar untuk Mendiagnosis Kelainan pada Ibu Hamil menggunakan Metode Breadth First Search. J. Emerg. Inf. Syst. Bus. Intell. **3**, 18–26 (2022)
27. Lestari, D., Nawang, E.: Aplikasi Sistem Pakar Deteksi Peringatan Awal Resiko Preeklamsia Pada Kehamilan Menggunakan Metode Forward Chaining: Application Of Expert System Detection Of Early Warning Risk Of Preeclamsia In Pregnancy Using Forward Chaining Method. Media Publ. Penelit. Kebidanan. **5**, 38–41 (2022)
28. Rinanda, P.D., Delvika, B., Nurhidayarnis, S., Abror, N., Hidayat, A.: Perbandingan Klasifikasi Antara Naive Bayes dan K-Nearest Neighbor Terhadap Resiko Diabetes pada Ibu Hamil: comparison of classification between Naive Bayes and k-nearest neighbor on diabetes risk in pregnant women. Malcom. Indones. J. Mach. Learn. Comput. Sci. **2**, 68–75 (2022)
29. Agave, S., Ulum, M.B., Kom, S., Kom, M.: Aplikasi Sistem Pakar Untuk Diagnosa Penyakit Ibu Hamil Menggunakan Metode Forward Chaining Berbasis Website. J. Komputasi. **11**, 1–10 (2023)
30. Aljameel, S.S., et al.: Prediction of preeclampsia using machine learning and deep learning models: a review. Big Data Cogn. Comput. **7**, 32 (2023)
31. Subekti, S.W.: Indikasi persalinan seksio sesarea. J. Biometrika dan Kependud. **7**, 11 (2018)
32. Indonesia, R.: Rencana pembangunan jangka menengah nasional 2020–2024. Peratur. Pres. Republik Indones 303 (2020)
33. Batubara, A.R., Fitriani, F.: Faktor-Faktor Yang Berhubungan Dengan Risiko Kematian Bayi 0–28 Hari Di Kabupaten Bireuen. J. Healthc. Technol. Med. **5**, 308–317 (2019)
34. Dila, W., Nadapda, T.P., Sibero, J.T., Harahap, F.S.D., Marsaulina, I.: Faktor yang Berhubungan dengan Persalinan Sectio Caesarea Periode 1 Januari-Desember 2019 di RSU Bandung Medan. J. Healthc. Technol. Med. **8**, 359–368 (2022)
35. Juliathi, N.L.P., Marhaeni, G.A., Mahayati, N.M.D.: Gambaran Persalinan dengan Sectio Caesarea di Instalasi Gawat Darurat Kebidanan Rumah Sakit Umum Pusat Sanglah Denpasar Tahun 2020. J. Ilm. Kebidanan. (J. Midwifery) **9**, 19–27 (2021)
36. Nafi'iyah, N.: Sistem Penentuan Keluhan Ibu Hamil dengan Algoritma Genetika Algoritma. In: SEMNASKIT 2015 (2018)
37. Manganti, A.: Sistem Pakar Diagnosa Penyebab Keguguran Pada Ibu Hamil Menggunakan Metode Forward Chaining. J. Sist. Inf. dan Sains Teknol. **3**, 491998 (2021)
38. Kemenkes, R.I.: Rencana strategis kementerian kesehatan tahun 2015–2019. Jakarta Kementeri. Kesehat. RI (2015)

39. Damayanti, N.A., Wulandari, R.D., Ridlo, I.A.: Maternal health care utilization behavior, local Wisdom, and associated factors among women in Urban and Rural Areas, Indonesia. Int. J. Womens. Health. 665–677 (2023)
40. Nantabah, Z.K., Effendi, D.E., Agustina, Z.A., Ipa, M., Laksono, A.D., Laksono, A.D.: Hospital accessibility in Indonesia. Medico-Legal Updat. **21**, 125–133 (2021)
41. Nawabi, F., Krebs, F., Lorenz, L., Shukri, A., Alayli, A., Stock, S.: Understanding determinants of pregnant women's knowledge of lifestyle-related risk factors: a aross-sectional study. Int. J. Environ. Res. Public Health **19**, 658 (2022)

Up Sampling Data in Bagging Tree Classification and Regression Decision Tree Method for Dengue Shock Syndrome Detection

Lailil Muflikhah[1(✉)], Agustin Iskandar[1], Novanto Yudistira[1], Bambang Nur Dewanto[2], Isbat Uzzin Nadhori[3], and Lisa Khoirun Nisa[1]

[1] Brawijaya University, Malang, East Jawa, Indonesia
lailil@ub.ac.id
[2] Merdeka University, Malang, East Java, Indonesia
[3] Politeknik Elektronika Negeri Surabaya, Surabaya, Indonesia

Abstract. Dengue virus, DENV, is the cause of dengue fever and carries the risk of developing dengue hemorrhagic fever and dengue shock syndrome (DSS), which can lead to death. The high mortality rate is due to the blockage of blood circulation resulting from delayed patient management, particularly in cases of advanced-stage DSS. Intensive examinations and new treatments are only provided when the patient has been admitted to the hospital. Therefore, we proposed to develop an early detection for DSS risk based on the clinical data using ensemble method through bagging Tree-CART, namely Tree-Bag algorithm. We explored information from patient's clinical data including temperature, vomiting, pain, as well as the levels of erythrocytes, leukocytes, creatine level, hemoglobin, and time series data on the patient's condition. In the preprocessing data stage, we applied oversampling data due to imbalanced class in dataset. The experimental results show that the Tree-Bag achieved high performance with accuracy of 0.91 and the AUC of 0.84. It is dominant when compared to single machine algorithms including: KNN, Naïve Bayes, CART decision tree, and SVM.

Keywords: Dengue Shock Syndrome · Tree-Bag · Ensemble method

1 Introduction

Dengue Fever is an acute febrile illness lasting for 2–7 days with two or more of the following manifestations: headache, abdominal pain, nausea, vomiting, retro-orbital pain, myalgia, arthralgia, skin rash, hepatomegaly, bleeding manifestations, and leukopenia. Dengue Hemorrhagic Fever (DHF) is a case of dengue fever with a tendency for bleeding and manifestations of plasma leakage. Dengue Hemorrhagic Fever (DHF), also known as Dengue Fever with hemorrhagic manifestations, is dengue fever accompanied by liver enlargement and bleeding manifestations. Dengue Hemorrhagic Fever (DHF) is a disease caused by the Dengue Family Flaviviridae virus, with its genus being Flavivirus. The virus has four known serotypes, DEN-1, DEN-2, DEN-3, and DEN-4. Clinically, it has different levels of manifestations depending on the dengue virus serotype [1, 2].

© The Author(s), under exclusive license to Springer Nature Singapore Pte Ltd. 2024
F. Hassan et al. (Eds.): AsiaSim 2023, CCIS 1911, pp. 307–318, 2024.
https://doi.org/10.1007/978-981-99-7240-1_24

In Indonesia, Dengue Hemorrhagic Fever (DHF) remains one of the primary health problems. The increasing mobility and population density contribute to the growing number of cases and the spread of this disease. In 2015, data from the Ministry of Health recorded a total of 126,675 DHF cases in 34 provinces in Indonesia, with 1,229 deaths. These numbers were higher compared to the previous year, where 100,347 DHF cases were reported in 2014, with 907 deaths. The number of outbreaks of DHF also increased from 1,081 cases in 2014 to 8,030 cases in 2015. The reporting of DHF outbreaks expanded from 5 provinces and 21 districts in 2014 to 7 provinces and 69 districts in 2015 [3, 4].

Fig. 1. Development of Dengue Fever Leading to Dengue Shock Syndrome

Dengue Shock Syndrome (DSS) is a severe form of dengue hemorrhagic fever (DHF) characterized by circulatory failure/shock and the progression of the disease is shown in Fig. 1. This condition occurs in individuals with widespread and sudden manifestations of Dengue Hemorrhagic Fever or Dengue Fever (DF), but it is also a clinical concern. Approximately 30–50% of dengue fever patients will experience shock and potentially fatal outcomes, especially if not promptly and appropriately treated. Managing shock in DHF is a critically important issue, as the mortality rate increases if shock is not addressed early and adequately. The fundamental approach to managing DHF shock is volume replacement or intravascular fluid replacement to compensate for the loss of fluid due to capillary wall damage, leading to increased permeability and resulting in plasma leakage. The death may occur when there is severe bleeding, uncontrolled shock, significant pleural effusion and ascites, or seizures. Currently, there is no commercially available vaccine for dengue fever caused by flaviviruses. The primary prevention of dengue fever lies in eliminating or reducing the mosquito vector responsible for transmitting the disease.

Related studies have explored dengue fever detection using supervised learning methods with a single classifier including SVM, Decision Tree, Naïve Bayes, and Back Propagation Neural Network algorithm. The kinds of data set were used including sequence data, spatial data, and clinical data [5–9]. Several studies on dengue using data analysis are conducted using conventional machine learning with single classifier and the performance evaluation are including Logistic regression (59.1%), Linear Regression (17.4%), and General Linear Model with 70% [10]. Also, another research related risk-stratification of dengue on clinical data based using Artificial Neural Network (ANN) achieved an AUROC of 0.82. Specificity of 0.84, and sensitivity of 0.66 [11]. Therefore, this research is proposed an ensemble method by bagging of decision tree with up-sampling data due to imbalanced data distribution in class.

2 Research Method

In general, we proposed Tree-Bag using resampling dataset. First stage, we applied pre-processing data including to handle missing attribute values and normalization. Before applying the machine learning method, we resampled data sets to avoid any bias in defined class. Then, we applied an ensemble method by bagging CART algorithms, and the information detail is shown in Fig. 2.

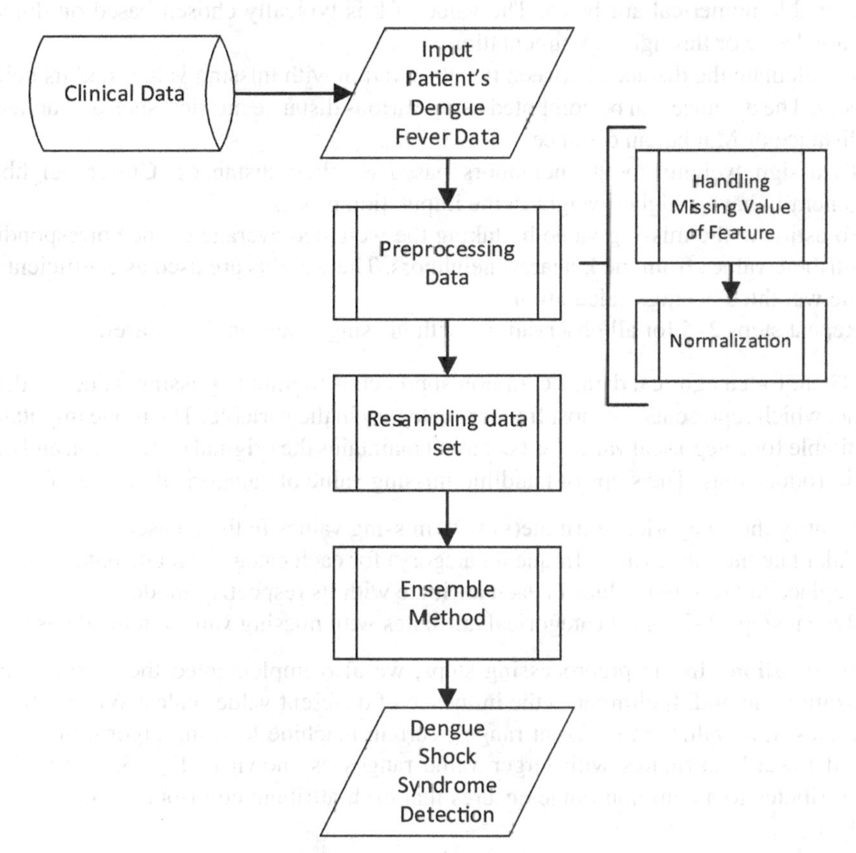

Fig. 2. General Steps of Dengue Shock Syndrome Detection Using Ensemble Method

2.1 Preprocessing Data

Handling Missing Attribute-Value Data. Missing values refer to the absence of data or information for certain variables or observations in a dataset. They can occur due to various reasons, such as data collection errors, data entry issues, or intentional omissions. Handling missing values is an essential step in the data preprocessing stage to ensure accurate and reliable analysis.

In this research, we use two techniques to solve the problem depending on the type of data. In the case of numerical attribute values, one effective method is K-Nearest Neighbors Imputation (KNNI). The KNNI is a data imputation technique that estimates missing values based on the values of its nearest neighbors [12]. The stages of KNN imputation are as follows:

1. To identify the numerical attribute(s) with missing values in the dataset.
2. For each observation with missing values, find its k nearest neighbors based on the available numerical attributes. The value of k is typically chosen based on domain knowledge or through experimentation.
3. To calculate the distance between the observation with missing values and its neighbors. The distance can be computed using various distance metrics, such as Euclidean distance or Manhattan distance.
4. To assign weights to the neighbors based on their distances. Closer neighbors generally have a higher weight in the imputation process.
5. To estimate the missing value by taking the weighted average of the corresponding attribute values from the k nearest neighbors. The weights are used as coefficients in the weighted average calculation.
6. Repeat steps 2–5 for all observations with missing values in the dataset.

Then, for categorical data, a common approach is to impute missing values with the mode, which represents the most frequent category in the variable. The mode imputation is suitable for categorical variables because it maintains the original distribution and does not introduce bias. The steps of handling missing value of categorical data are follows:

1. Identify the categorical attribute(s) with missing values in the dataset.
2. Calculate the mode (most frequent category) for each categorical attribute.
3. Replace the missing values in each attribute with its respective mode.
4. Repeat steps 2–3 for all categorical attributes with missing values in the dataset.

Normalization. In the preprocessing steps, we also implemented the minimax normalization method. It eliminates the influence of different value scales: When different attributes have values in different ranges, certain machine learning algorithms can be biased towards attributes with larger value ranges as shown in Fig. 3. Normalizing the attributes to a common range ensures that each attribute contributes equally to the analysis.

Minimax normalization is a common technique used to transform numerical attribute values into a specific range. It rescales the values of a variable to a fixed range, typically between 0 and 1, to avoid the impact of disparate value ranges on certain machine learning algorithms. Here is a step-by-step description of how minimax normalization works:

1. To identify the numerical attribute(s) that you want to normalize.
2. To determine the minimum (min) and maximum (max) values of the attribute in the dataset. The minimum value is the smallest observed value, while the maximum value is the largest observed value.

3. To apply the following formula to normalize each value in the attribute:

$$normalized\ value = \frac{(original\ value - \text{minimum value})}{(maximum - minimum)} \quad (1)$$

This formula scales the original value based on the range of values in the attribute.

4. Repeat step 3 for all values in the attribute.
5. After normalization, the attribute values will fall within the range of 0 to 1. The minimum value in the attribute will be mapped to 0, the maximum value will be mapped to 1, and all other values will be scaled proportionally between these two endpoints.

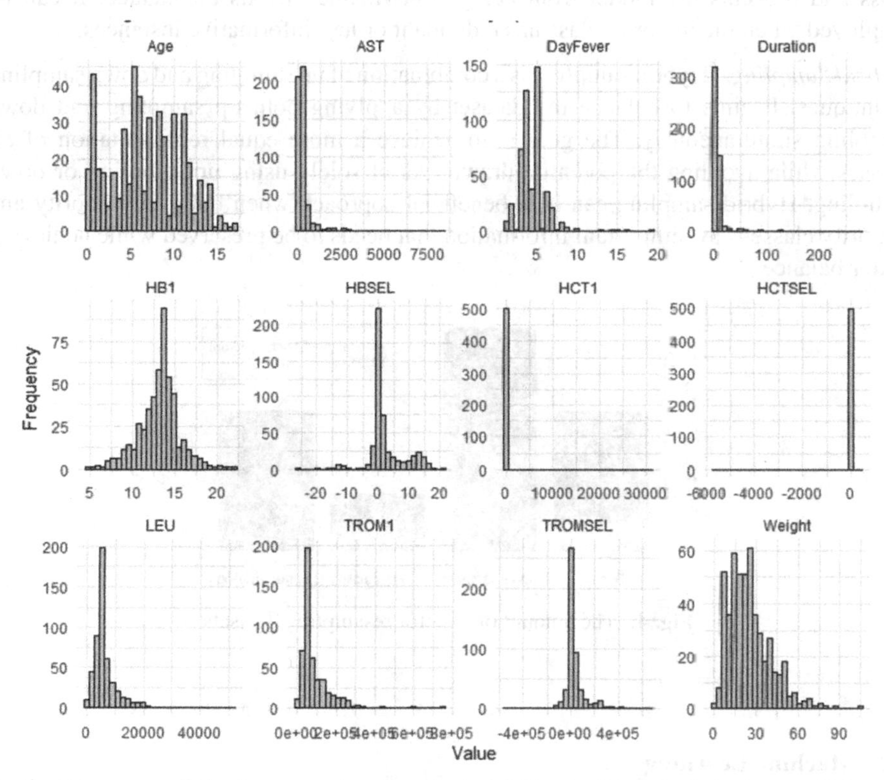

Fig. 3. Histogram of the frequency in numerical variables in Several clinical data of Dengue Shock Syndrome

Resample Data. To improve input data quality, we apply resampling data due to imbalanced class distribution of data sets as shown in Fig. 4. Resampling data is a technique used in data preprocessing to address class imbalance or uneven distribution of data points across different classes or categories. When the number of observations in one class is significantly higher or lower than the number of observations in other classes, class imbalance arises. Resampling methods such as up sampling, down sampling, and hybrid sampling can be used in such instances to create a more balanced dataset.

Up Sampling. Up sampling involves increasing the number of instances in the minority class (the class with fewer observations) to match the majority class. This is typically done by randomly replicating instances from the minority class until a balance is achieved. Up sampling helps to provide more representation to the minority class and prevents the model from being biased towards the majority class. It can be particularly useful when the minority class contains important or critical information.

Down Sampling. Down sampling, on the other hand, involves reducing the number of instances in the majority class (the class with more observations) to match the minority class. It is achieved by randomly removing instances from the majority class until a balance is attained. Down sampling helps in reducing the dominance of the majority class and prevents the model from being overwhelmed by its abundance. It can be employed when the majority class has redundant or less informative instances.

Hybrid Sampling. Hybrid sampling is a combination of up sampling and down sampling techniques. It aims to balance the dataset by applying both up sampling and down sampling simultaneously. The goal is to achieve a more equal representation of all classes while avoiding the potential drawbacks of solely using up sampling or down sampling. Hybrid sampling can be a beneficial approach when both the minority and majority classes have important information that needs to be preserved while achieving better balance.

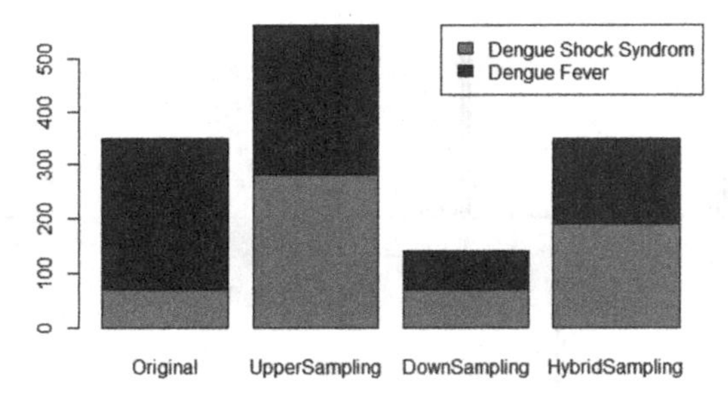

Fig. 4. The amount of data for resampling datasets

2.2 Machine Learning

A method for learning from directed data is machine learning. This suggests that the machine learning algorithm is given a set of instructions on what to look for and how to interpret it. Unsupervised learning is the practice of allowing the machine learning algorithm to operate independently of human supervision. In machine learning, we utilize supervised learning to determine how to map input (X) to output (Y). This is accomplished by employing algorithms to locate a function that can convert a set of input values into a corresponding set of output values. It is necessary for the output variable to be forecastable [13].

Supervised learning is a machine learning technique that learns from labeled training data to assist users in predicting outcomes for unexpected data. In supervised learning, the machine is trained using properly "labelled" data that in fact is tagged with the correct answer or prediction result. Supervised learning can be imagined by learning in the presence of a supervisor or teacher who always directs the correct answer as illustrated in Fig. 5 [5]. In this study, we applied the representative machine learning including: KNN, Naïve Bayes, SVM and CART. We also involved an ensemble method by bagging Tree-CART algorithm.

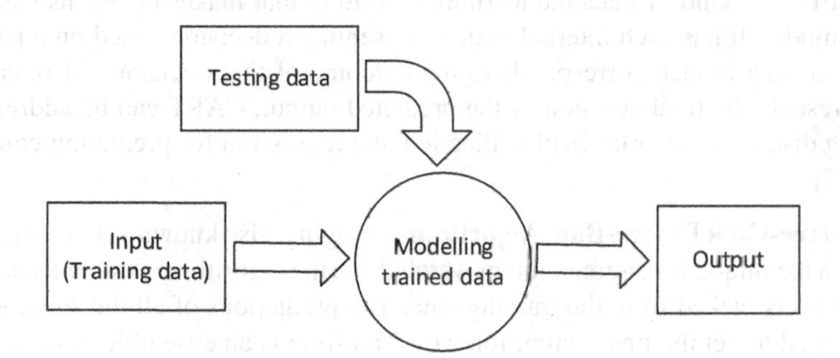

Fig. 5. Illustration of Supervised Learning Method

K-Nearest Neighbor (K-NN). K-Nearest Neighbor or also be called as K-NN classifier is a supervised learning algorithm for classification that uses based on comparison for the characteristics' objects that are divided into their categories to predict the label or category of the new information object. This method is recognized as a method of classification that is considered lazy due to construction for modelling retrieved from training data is not necessary. The majority vote of the training data object class in the k nearest neighbor data to determine the class value of the new data [14].

Naïve Bayes Classifier. Naïve Bayes is a supervised learning method that can classify data. The method is using unconditional and conditional probability to construct the classifier model. The principal concept is the independence of each event (unconditional probability) with very strong (naïve) hypothesis [15]. The model is constructed based on (2).

$$P(C|X) = \frac{P(C)P(X|C)}{P(X)} \tag{2}$$

where X is attributed, and C is class.

The posterior probability value of each class is maximized in the Bayes classifier. It is defined as in the stated formula in (3).

$$H_{MAP} = \text{argmax} P(C|x_1, x_2, \ldots x_n) \text{argmaxPnaïve} \prod_{i=1}^{n} P(x_i|C) \tag{3}$$

Support Vector Machine (SVM). SVM classifier is another supervised learning method for classification [9]. The function first divides the input space into two classes. The goal of this strategy is to find the best hyperplane function for separating different classes. This method's developed function can handle non-linear classification. The kernel trick function is designed to convert a vector space with a suitably high dimension. The SVM classifier uses a variety of kernel functions, including Linear, Polynomial, Sigmoid and the Radial Basis Function (RBF) kernel function as used in this research [16].

CART Decision Tree Algorithm. Classification and Regression Decision Tree (CART) decision tree is a kind of machine learning algorithm that makes decisions based on a tree-like model. It has each internal node representing a decision based on a particular feature and each branch corresponds to the outcome of that decision. Also, each leaf node represents the final decision or the predicted output. CART can be addressed for predicting discrete categories in classification and regression for predicting continuous values [17].

Bagging Tree-CART (Tree-Bag) Algorithm. Bagging, also known as bootstrap aggregation, is a technique for aggregation in which the same algorithm is trained again using various subsets picked from the training data. The predictions of all the submodels are then averaged to get the final output forecast. Tree-Bag is an ensemble of decision tree method by aggregation for output values. Bagging tree CART (Classification and Regression Tree) refers to a specific ensemble learning technique that combines the concepts of bootstrap aggregating (bagging) and decision trees based on the CART algorithm.

In the bagging tree CART ensemble, multiple decision trees are trained using the bagging technique. Each decision tree is built independently on a different bootstrap sample of the training data. The final prediction of the ensemble is obtained by aggregating the predictions of individual trees, either by majority voting (for classification) or averaging (for regression).

3 Result and Discussion

This study utilized 501 clinical data of dengue fever patients, with 401 categorized as fever and 100 patients experiencing shock syndrome. To know the correctness of detection method, then there are various measurements, including accuracy, sensitivity, specificity, and the Area Under the Curve (AUC) as this below Table 1 [18].

Remark:

- True positive (tp): the cases are predicted shock syndrome and, they are shock syndrome.
- True negative (tn): the cases are predicted dengue fever and, they are same condi-tion (no shock)
- False-positive (fp): the cases are predicted shock syndrome, but they are dengue fever.
- False-negative (fn): the cases are predicted dengue fever, but they are shock syndrome.

The proposed method was evaluated using performance evaluation measurements including accuracy, sensitivity, and specificity. Then it was compared to other machine learning algorithms as shown in Table 2, Table 3, Table 4, and Table 5 in various resample

Table 1. The Performances measure metrics

Measure	**Formula**	**Definition**
Accuracy	$\frac{tp+tn}{tp+fn+fp+tn}$	The fraction of successfully identified examples over the total number of occurrences in the dataset
Sensitivity	$\frac{tp}{tp+fn}$	A classifier's ability to recognize the positive label
Specificity	$\frac{tn}{tp+fn}$	A classifier's ability to recognize the negative label
AUC	$\frac{1}{2}\left(\frac{tp}{tp+fn}+\frac{tn}{tp+fn}\right)$	The classifier's capacity to avoid mistaken classification

data. The bagging Tree-CART method is stable in any resample and achieves high performances. In general, this method has superior performance compared to other methods, especially using up sampling of training data sets.

Table 2. Performance Evaluation Without Resample Data Sets

Algorithms	Accuracy	Sensitivity	Specificity	P-Value
Bagging Tree-CART	0.8	0.806	0.5	1
Naïve Bayes	0.78	0.853	0.444	0.878
SVM	0.78	0.953	0.472	0.00183
CART	0.8	0.806	0.5	1
KNN	0.23	0.67	0.187	1

Table 3. Performance Evaluation Using Up Sampling Data Sets

Algorithms	Accuracy	Sensitivity	Specificity	P-Value
Bagging Tree-CART	**0.91**	**0.72**	**0.973**	4.31E−05
Naïve Bayes	0.8	0.6	0.875	0.5595
SVM	0.83	0.552	0.944	0.004095
CART	0.79	0.484	0.928	0.01747
KNN	0.67	0.367	0.961	0.0008747

Furthermore, we also got the AUC measurement is a combination between sensitivity and specificity as shown in Fig. 6. By resample datasets including up sampling, down sampling, and hybrid sampling, it applied to all algorithms. The results showed that Tree-Bag using up sampling achieved the highest performance of AUC. Otherwise, the lowest AUC is prediction using KNN algorithm by hybrid resample (up-down sampling).

Table 4. Performance Evaluation Using Down Sampling Data Sets

Algorithms	Accuracy	Sensitivity	Specificity	P-Value
Bagging Tree-CART	0.8	0.5	0.984	8.50E−05
Naïve Bayes	0.79	0.333	0.804	1
SVM	0.76	0.444	0.938	0.007013
CART	0.79	0.484	0.928	0.0175
KNN	0.72	0.389	0.906	0.057123

Table 5. Performance Evaluation Using Hybrid Up-Down Sampling Data Sets

Algorithms	Accuracy	Sensitivity	Specificity	P-Value
Bagging Tree-CART	0.8	0.806	0.5	1
Naïve Bayes	0.78	0.853	0.444	0.878
SVM	0.78	0.953	0.472	0.00183
CART	0.8	0.806	0.5	1
KNN	0.23	0.67	0.187	1

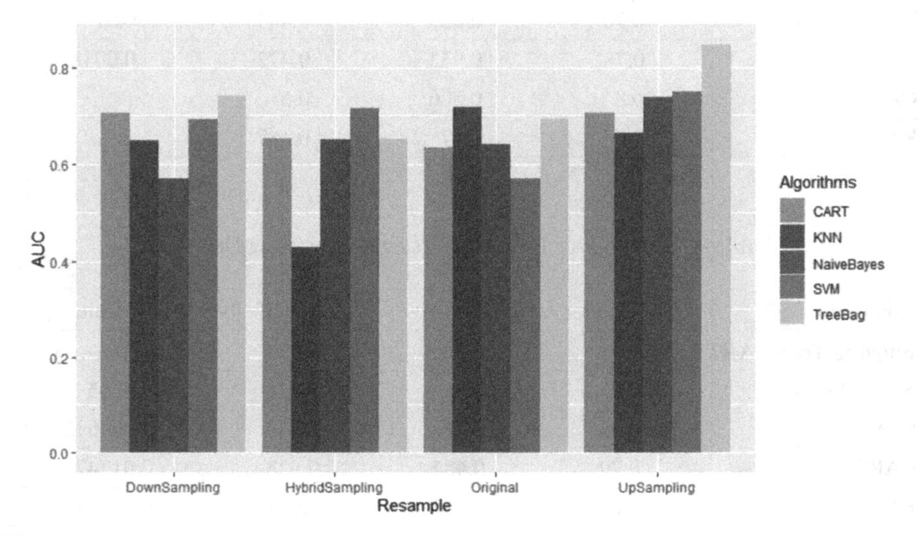

Fig. 6. AUC Comparison for Machine Learning Algorithm in Several kinds of Resample Data

4 Conclusion

Research on detection of Dengue Shock Syndrome based on clinical data was applied using either single learning classifier or ensembled method classifier. In terms of imbalanced class distribution, we applied resample data sets. The experimental result showed that the ensemble method, Tree-Bag by up sampling achieved the highest performance evaluation including accuracy, sensitivity, specificity, and AUC.

Acknowledgement. This research was financially supported by the Indonesian Government through the Ministry of Research and Technology/National Agency for Research and Innovation (RISTEK/DIKTI) on a program of Regular Fundamental Research Grant in 2023 under contract no: 119/E5/PG.020.00.PL/2023 was dated on June 19, 2023.

References

1. Sarma, D., Hossain, S., Mittra, T., Bhuiya, Md.A.M., Saha, I., Chakma, R.: Dengue prediction using machine learning algorithms. In: 2020 IEEE 8th R10 Humanitarian Technology Conference (R10-HTC), pp. 1–6 (2020). https://doi.org/10.1109/R10-HTC49770.2020.9357035
2. Yuan, K., Chen, Y., Zhong, M., Lin, Y., Liu, L.: Risk and predictive factors for severe dengue infection: a systematic review and meta-analysis. PLoS ONE **17**(4), e0267186 (2022)
3. World Health Organisation. Dengue and severe dengue. WHO Fact Sheet [Internet] **117**, 1–4 (2014). Tersedia pada: https://www.who.int/mediacentre/factsheets/fs117/en/index.html
4. WHO | Epidemiology [Internet]. https://www.who.int/denguecontrol/epidemiology/en/ Accessed 16 Sept 2020
5. Marimuthu, T., Balamurugan, V.: A novel bio-computational model for mining the dengue gene sequences. In 2015. https://www.semanticscholar.org/paper/A-NOVEL-BIO-COMPUTATIONAL-MODEL-FOR-MINING-THE-GENE-Marimuthu-Balamurugan/aa6c9fd853eeb246e49355f00050f55e4f501cfc. Accessed 12 Apr 2023
6. Nynalasetti, K.K.R., Varma, G., Rao, M.: Classification rules using decision tree for dengue disease. IJRCCT **1**(3), 340–343 (2014)
7. Husin, N.A., Salim, N., Ahmad, A.R.: Modeling of dengue outbreak prediction in Malaysia: a comparison of neural network and nonlinear regression model. In: 2008 International Symposium on Information Technology, pp. 1–4 (2008)
8. Sudsom, N., Thammapalo, S., Pengsakul, T., Techato, K.: A spatial clustering approach to identify risk areas of dengue infection after insecticide spraying. Jurnal Teknologi. **8**, 78 (2016)
9. Mohd Sharef, N., Husin, N.A., Kasmiran, K.A., Ninggal, M.I.: Temporal trends analysis for dengue outbreak and network threats severity prediction accuracy improvement. J. Digit. Inf. Manag. **17**(3), 122 (2019)
10. Hoyos, W., Aguilar, J., Toro, M.: Dengue models based on machine learning techniques: a systematic literature review. Artif. Intell. Med. **119**, 102157 (2021)
11. Ming, D.K., et al.: Applied machine learning for the risk-stratification and clinical decision support of hospitalised patients with dengue in Vietnam. PLOS Digital Health **1**, e0000005 (2022)
12. Muflikhah, L., Hidayat, N., Hariyanto, D.J.: Prediction of hypertension drug therapy response using K-NN imputation and SVM algorithm (2019). https://doi.org/10.11591/ijeecs.v15.i1.pp460-467
13. Jordan, M.I., Mitchell, T.M.: Machine learning: trends, perspectives, and prospects. Science **349**(6245), 255–260 (2015)
14. Sutton, O.: Introduction to k nearest neighbour classification and condensed nearest neighbour data reduction. University lectures, University of Leicester, vol. 1 (2012)

15. Naive Bayes Classifier - an overview | ScienceDirect Topics. https://www.sciencedirect.com/topics/engineering/naive-bayes-classifier. Accessed 26 Nov 2020
16. Zang, L., Luo, L., Hu, L., Sun, M.: An SVM-based classification model for migration prediction of Beijing. Eng. Lett. vol. **28**(4) (2020)
17. Mienye, I.D., Sun, Y., Wang, Z.: Prediction performance of improved decision tree-based algorithms: a review. Procedia Manuf. **35**, 698–703 (2019). https://doi.org/10.1016/j.promfg.2019.06.011
18. Evaluating a Classification Model. ritchieng.github.io. http://www.ritchieng.com/machine-learning-evaluate-classification-model/. Accessed 01 Nov 2019

Real-Time Crack Classification with Wall-Climbing Robot Using MobileNetV2

Mazleenda Mazni[1,2](✉), Abdul Rashid Husain[1], Mohd Ibrahim Shapiai[3], Izni Syahrizal Ibrahim[4], Riyadh Zulkifli[1], and Devi Willieam Anggara[5]

[1] School of Electrical Engineering, Faculty of Engineering, Universiti Teknologi Malaysia, Skudai, Johor, Malaysia
mazleenda@graduate.utm.my

[2] Faculty of Mechanical Engineering, Universiti Teknologi MARA Cawangan Johor, Kampus Pasir Gudang, Jalan Purnama, Masai, Malaysia

[3] Centre for Artificial Intelligence and Robotics, International Institute of Technology, Universiti Teknologi Malaysia, Kuala Lumpur, Malaysia

[4] Forensic Engineering Centre, Institute for Smart Infrastructure and Innovative Construction, Faculty of Civil Engineering, Universiti Teknologi Malaysia, Johor, Malaysia

[5] School of Computing, Faculty of Engineering, Universiti Teknologi Malaysia, Skudai, Malaysia

Abstract. Detecting cracks on concrete surfaces is a crucial task in civil engineering inspections, but it poses significant challenges due to the small and concealed nature of cracks. Visual detection is particularly difficult on uneven or rough concrete surfaces. To overcome these challenges, our research focuses on developing an automated system that utilizes a wall-climbing robot for crack classification.Our main objective is to introduce a crack classification technique using MobileNetV2, enabling real-time classification without human intervention. The Convolution Neural Network (CNN) model used for crack classification is based on MobileNetV2, which is fine-tuned by adjusting the sensitivity of its hyperparameters. Through extensive experiments, we evaluate the performance of this CNN approach specifically designed for embedded systems. After evaluating our proposed approach of crack-detection on publicly available datasets, we have found that out of all the pre-trained CNN models MobileNetV2 yields the best performance with 99.56% detection accuracy, precision of 99.65%, recall of 99.48%, and F1-Score of 99.56%. However, it is important to note that the training time for this model is relatively high, taking 25,500 s. Future study of the study should focus on optimizing the computation time to improve efficiency.

Keywords: Crack classification · MobileNetV2 · Wall-climbing robot

1 Introduction

The image-based techniques are used to capture the texture and geometry of objects in the scene [1]. In image-based object detection, a typical approach involves training a machine learning or deep learning model on a labeled dataset. The model learns to

© The Author(s), under exclusive license to Springer Nature Singapore Pte Ltd. 2024
F. Hassan et al. (Eds.): AsiaSim 2023, CCIS 1911, pp. 319–328, 2024.
https://doi.org/10.1007/978-981-99-7240-1_25

recognize and differentiate between different objects based on the texture and geometry features extracted from the training images. In real-time object detection, the first step is image acquisition, where convert real-world data captured by cameras into a suitable format that can be efficiently processed by computers or smart devices. This conversion involves transforming the image data into a manageable array of numerical values that can be easily manipulated and analyzed [2].

Examining cracks in extensive concrete structures like tall buildings involves inherent risks when humans attempt access. It also requires substantial investments of time, money, and the installation of extra structures. In addition, maintaining the impartiality of visual inspections in this scenario poses a significant challenge. Substantial research efforts have been dedicated to investigating automated crack detection methods utilizing robotic systems. To overcome this constraint, this research paper focuses on detecting cracks on walls using Convolution Neural Network (CNN) techniques implemented in a wall-climbing robot. Equipped with a powerful embedded platform and a camera, the robot is capable of efficiently identifying cracks on both vertical and horizontal surfaces. By securely attaching itself to vertical structures, the wall-climbing robot demonstrates the ability to classify concrete cracks while effectively navigating indoor environments and bridges.

In recent times, there has been a surge in research efforts aimed at enhancing the accuracy of crack detection through the application using Convolution Neural Network (CNN). Liu et al. [3] introduced a method for concrete crack detection that exhibits remarkable precision even with a smaller training dataset. Their approach, based on the U-Net architecture, proved to be robust, effective, and highly accurate. Zhang et al. [4], drawing inspiration from Full Convolutional Networks (FCN), proposed a full convolutional network utilizing dilated convolutions. This network, composed of encoders and decoders, demonstrated faster convergence and improved generalization on concrete crack test datasets. Zou et al. [5] devised DeepCrack, a crack detection method built upon the SegNet architecture. By intelligently merging convolutional features generated by encoder and decoder networks at the same scale, they successfully infer crack presence. These methods, employing VGG16, ResNet18, and similar networks, were experimented with on GPU-equipped desktop computers, exhibiting impressive accuracy. However, the computational demands of these approaches hinder their application in mobile environments, where real-time processing is crucial.

2 Research Framework

As shown in Fig. 1, the framework for the potential methodology consists primarily of two steps: The initial objective is to build an enhanced crack detecting system using a wall-climbing robot, wireless data transfer, and a monitoring board for the screen. The wall-climbing robot called as has been proposed by Riyadh et al. in [6] which using hybrid method adhesion mechanism as shown in Fig. 2.

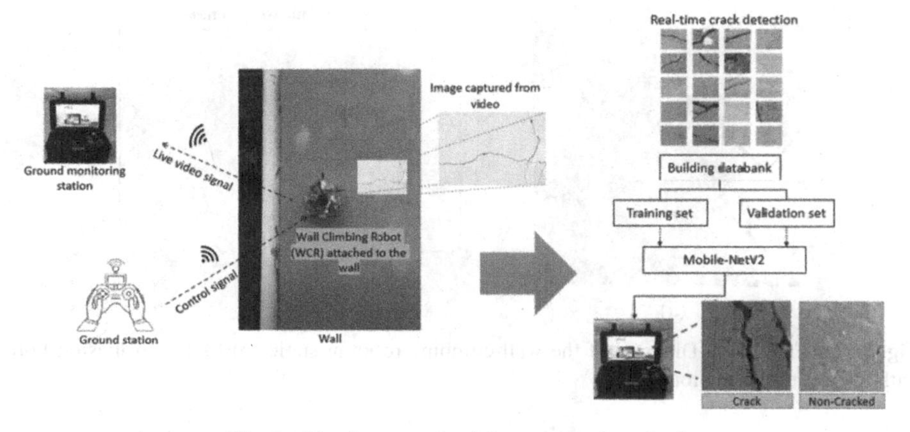

Fig. 1. The framework of the proposed method

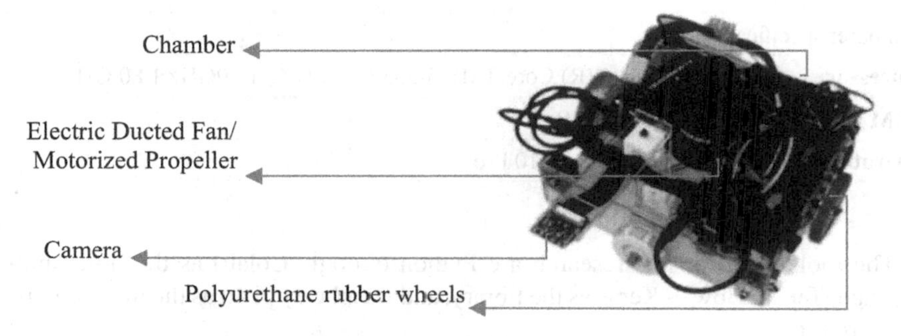

Fig. 2. Wall-climbing robot

2.1 Mechanics of Wall Climbing Robot

As shown in Fig. 3(a), this is the free body diagram of the robot at static to maintain the sticking. From the free body diagram of the robot, Eq. (1) is derived from the summation of force on x and y axis to understand and plot the minimum condition for sticking.

$$F_a \geq mg\left[\frac{sinsin\theta}{\mu_s} - coscos\theta\right], 45 \leq \theta \leq 180 \quad (1)$$

From Eq. (1), F_a is the total adhesion force, m is the total mass of the wall-climbing robot, g is the gravitational acceleration $9.81\frac{m}{s^2}$, θ is the Inclination angle and μ_s is the static friction coefficient between wheels & surface. When the designed robot have $m = 1$ kg and $\mu_s = 0.5$, the plot of the threshold F_a with respect to the θ to maintain sticking is shown in Fig. 3(b). From the plot, it is known that the maximum threshold F_a is at $\theta = 118°$.

2.2 Tools and Hardware

The assessment of classification metrics such as recall, precision, and accuracy, as well as the F1 score, use a personal computer with the specifications detailed in Table 1.

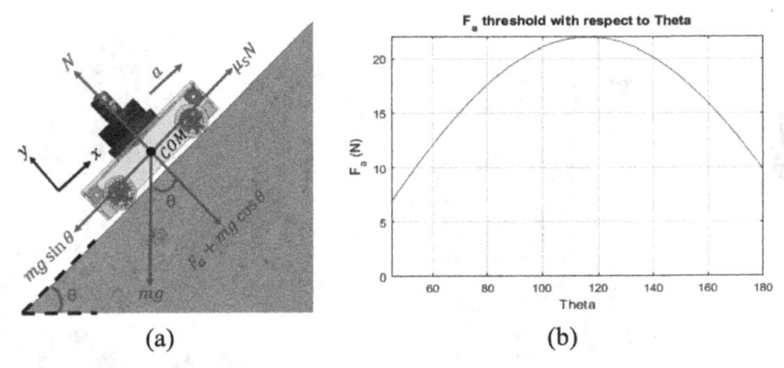

(a) (b)

Fig. 3. (a) Free Body Diagram of the wall-climbing robot at static (b) Plot of Adhesion Force with respect to Inclination Angle

Table 1. Computer specification

Computer specification	
Processor	Intel(R) Core(TM) i5-3337U CPU @ 1.80GHz 1.80 GHz
RAM	4.00 GB
Operating system	Windows 10 Pro

The tools used for this research are Python (Google Colab) as the programming language, TensorFlow + Keras as the library, and Raspberry pi 4b as the microcontroller for the robot.

3 Methods and Definitions

3.1 Dataset Preparation

The databases include images of various concrete surfaces, both with and without cracks and the data is split into two categories: negative (without crack) and positive (with crack). This dataset was obtained from the Mendeley Data - Crack Detection website, which was provided as a helpful resource for crack detection research by Çağlar Fırat Özgenel et al. in [7]. Using the strategy suggested by Zhang et al. in [8], 458 high-resolution photos (4032 × 3024 pixels) were to develop a dataset that takes into account the high variance in illumination conditions and surface finish. The dataset does not make use of any data augmentation techniques like random rotation, flipping, or tilting in any way.

3.2 Data Processing

The images are resized to 224 × 224 pixels for uniformity. The dataset is split into input and testing sets, which are used for learning and evaluation. The input set has two parts: training and validation. The training set is used for learning, while the validation set adjusts the hyperparameters. The input set has more images than the testing set.

However, the specific ratio for dataset splitting is not universally defined and may vary depending on the particular scenario [9]. Based on dataset, 11,429 files belonging to 2 classes (crack or non-cracked). Out of the total files, 9,144 files are being used for training the model. Training data is used to teach the model to recognize patterns and make predictions. The remaining 2,285 files are being used for validation. Validation data is used to assess the model's performance during training and tune its hyperparameters.

3.3 CNN Classifier Model Configuration

In this research, we utilized MobileNetV2 as a CNN classifier. The MobileNetV2 architecture [10], an impactful CNN model is utilised for feature extraction and classification purposes. MobileNetV2 makes use of inverted residuals and linear bottlenecks, which allows it to strike a good balance between the accuracy and computing efficiency of its results. The model has demonstrated successful performance in various image classification tasks, including object recognition and fine-grained classification.

MobileNetV2 uses depthwise separable convolutions to make a small and fast neural network. This method splits the filters into depth and spatial parts and applies one filter to each input channel. Then, it uses 1×1 convolutions to mix the results. This way, it extracts features efficiently with less computation and model size. The networks were implemented using Keras applications, which have ready-made CNN models for training and prediction. Python and TensorFlow were used for programming and backend. The CNN classifier application was built and tested using Google Colab, a platform for collaborative coding and experimentation which also been utilized in various other studies, such as [11, 12], and [13], for collaborative coding and experimentation purposes.

3.4 Analysis of the Sensitivity of Hyper-Parameters

In order to find the top-performing models and train the hyperparameters, a sensitivity analysis is also carried out. The study focuses on several crucial hyperparameters, such as batch size, activation function, optimization function, loss function, and learning rates. In this work, pre-trained ImageNet weights are used to start the training of CNN models. After that, a process of iterative refinement is used to get the hyperparameters to their optimal state. The details of these hyper-parameters are presented in Table 2.

The method of feature extraction in CNNs is nonlinear and uses activation functions where the activation functions introduce nonlinearity to the model. In this study, the Rectified Linear Activation (ReLU) function is employed for the CNN model, as depicted in Eq. (2). ReLU is a linear function; if the input is negative, it generates zero, which essentially deactivates the neuron. This approach offers computational advantages as not all neurons are activated simultaneously. It is crucial to calculate the derivatives of both the actual values and the expected values in order to update the model parameters efficiently. This calculation is performed using the loss function, which measures the discrepancy between the predicted and actual values. The loss function plays a crucial role in updating the model variables during the training process.

$$f(x) = max(0, x) \tag{2}$$

Table 2. Details of hyper-parameters

Parameter Names	Parameter values
Epoch	100
Batch size	32
Learning rate	0.01
Activation function	ReLu
Loss function	Binary Cross-entropy
Optimization function	Adadelta
Callbacks	Model Checkpoint

In this study, the binary cross-entropy (BCE) loss function is chosen for classification purposes. This specific form of cross-entropy is utilized when making a decision between two possibilities, namely crack or non-cracked. It is commonly used in conjunction with the sigmoid activation function to achieve accurate predictions. By incorporating this loss function into the CNN model, it ensures the highest compatibility when applied to new datasets. The mathematical formulation of the binary cross-entropy loss function (LBCE) is presented in Eq. (3), where yj represents the scalar output value, yi denotes the corresponding target value, and n signifies the output size. This equation enables the calculation of the average loss, facilitating effective model training.

$$Loss = -\frac{1}{n}\sum\sum_{i=1}^{n}[yi * log(yj) + (1 - yi) * log(1 - yj)] \tag{3}$$

The goal of optimization in a neural network is to find the best output for a given input by changing the parameters during the training. The loss function measures how well the model predicts the expected results and helps the optimizers to reduce the error. The hyperparameters for CNN classifiers are chosen by analyzing the learning process. A batch size of 32 and 50 epochs are used in this study. A model checkpoint callback saves the best model if the validation loss improves. This helps prevent overfitting and allows us to select the best model based on validation loss. When training the MobileNetV2 model, the Adadelta optimizer is employed, and a custom learning rate is calculated based on best practices from past research on CNNs optimization.

4 Results and Discussion

The results of crack classification were evaluated by using a confusion matrix. A confusion matrix (Fig. 4) is a commonly used tool in classification problems to assess how well a classifier performs. The model's accuracy represents the proportion of images that are correctly classified according to their crack type. On the other hand, the recall and precision scores measure the model's ability to correctly identify concrete cracks among all images classified as containing cracks and among all images, respectively, regardless of crack presence. By combining precision and recall, the F1 score provides a balanced summary of the model's performance. The specific formulas used for evaluation can be found in Eqs. (4) to (7).

$$Accuracy = \frac{TP + TN}{TP + TN + FP + FN} \tag{4}$$

$$Recall = \frac{TP}{TP + FN} \tag{5}$$

$$Precision = \frac{TP}{TP + FP} \tag{6}$$

$$F1_{score} = 2 \times \frac{Precision \times Recall}{Precision + Recall} \tag{7}$$

TP is the number of crack images classified right, TN is the number of non-crack images classified right, FP is the number of non-crack images classified wrong as crack, and FN is the number of crack images classified wrong as non-crack. The confusion matrix shows these outcomes (TP, TN, FP, and FN). This research uses a binary confusion matrix because there are only two classes: "crack" and "non-crack". These metrics measure the accuracy and reliability of the model.

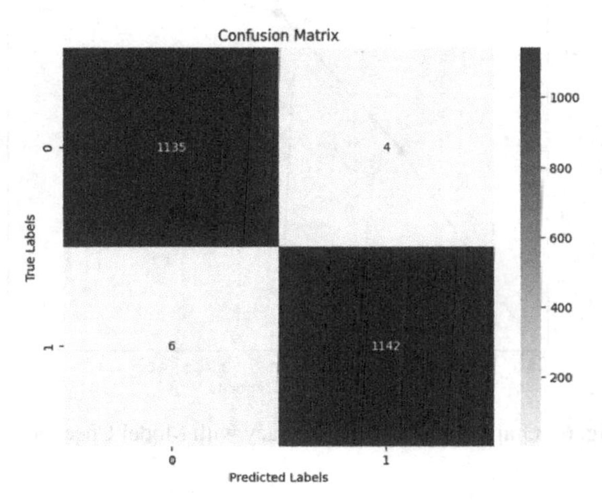

Fig. 4. Confusion matrix

As results, MobileNetV2 attained the performance by adopting Adadelta optimizer with an accuracy of 99.56%, precision of 99.65%, recall of 99.48%, and F1-Score of 99.56%. Next, according to Fig. 5, several sample results of crack identification using the MobileNetV2 classifier are presented. Each result includes the image's identification number, size, predicted class, and the corresponding prediction outcome for the defect. The probability percentages associated with each damage case represent the output generated by the CNN models developed in this study. Based on Fig. 6, the model checkpoint reveals that the overall crack classification accuracy, achieved through parameter tuning, reaches an impressive 99.56%. However, it should be noted that the total computation time to attain this level of accuracy is relatively high, amounting to 25,500 s. Efforts are required to further optimize the computation time and reduce it in future iterations of the study.

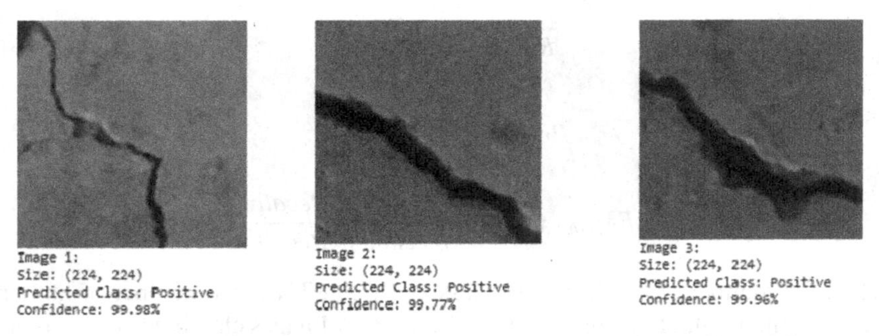

Fig. 5. Sample images for crack prediction using MobileNetV2 model

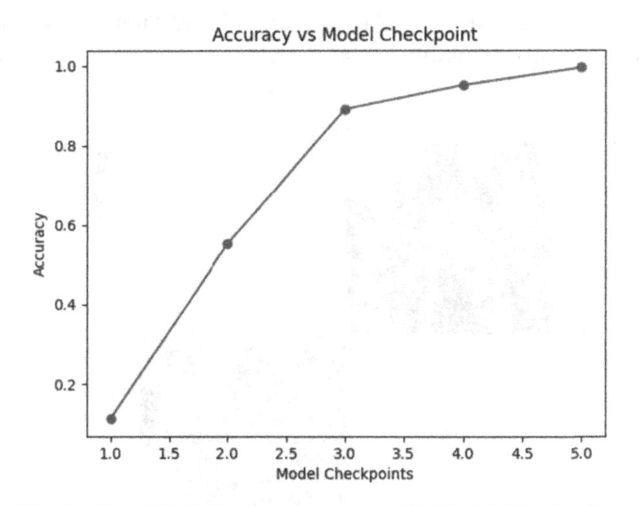

Fig. 6. Graph Performance Accuracy with Model Checkpoints

5 Conclusion

In this research, we utilize a binary classification technique to categorize cracks as either "cracked" or "non-cracked" using a wall-climbing robot. The robot is securely attached to walls and has the capability to detect even micro cracks while traversing both indoor and outdoor surfaces, including vertical and horizontal orientations. We evaluate the performance of a CNN technique that excels in detecting cracks in real-time, even on non-smooth concrete surfaces. The employed CNN model for crack detection is based on MobileNetV2, which is fine-tuned by adjusting its hyperparameters' sensitivity.

The achieved accuracy, after parameter tuning, is an impressive 99.56%. However, it is important to note that the total computation time to achieve this level of accuracy is relatively high, totaling 25,500 s. Future research can improve crack classification by using crack segmentation. This method finds and outlines the crack regions in an image, showing their shape, size, and distribution. This helps to understand crack features and patterns better. Crack segmentation gives a complete view of cracks by capturing their details and locations. This can help to develop better crack monitoring systems that assess the structural health of different assets. Accurate evaluation of crack extent helps to make informed decisions in infrastructure management, such as prioritizing repairs and allocating resources. By using crack segmentation in crack analysis, researchers can identify and visualize cracks precisely. This leads to more effective maintenance strategies and better decision-making processes in infrastructure health monitoring.

Acknowledgement. The authors would like to acknowledge that this research work is funded by Ministry of Higher Education under FRGS with Registration Proposal No: FRGS/1/2020/TK02/UTM/02/4.

References

1. Anggara, D.W., et al.: Grayscale image enhancement for enhancing features detection in marker-less augmented reality technology. J. Theor. Appl. Inf. Technol. **98**(13), 2671–2683 (2020)
2. Anggara, D.W., et al.: Integrated Colormap and ORB detector method for feature extraction approach in augmented reality. Multimed. Tools Appl. **81**(25), 35713–35729 (2022). https://doi.org/10.1007/s11042-022-13548-x
3. Liu, Z., Cao, Y., Wang, Y., Wang, W.: Computer vision-based concrete crack detection using U-net fully convolutional networks. Autom. Constr. **104**, 129–139 (2019). https://doi.org/10.1016/j.autcon.2019.04.005
4. Zhang, J., Lu, C., Wang, J., Wang, L., Yue, X.G.: Concrete cracks detection based on FCN with dilated convolution. Appl. Sci. **9**(13) (2019). https://doi.org/10.3390/app9132686
5. Zou, Q., Zhang, Z., Li, Q., Qi, X., Wang, Q., Wang, S.: DeepCrack: learning hierarchical convolutional features for crack detection. IEEE Trans. Image Process. **28**(3), 1498–1512 (2019). https://doi.org/10.1109/TIP.2018.2878966
6. Zulkifli, R., Husain, A.R., Ibrahim, I.S., Mazni, M., Fauzan, N.H.A.M.: Analysis of the hybrid adhesion mechanism of the wall climbing Robot BT - control. In: Instrumentation and Mechatronics: Theory and Practice, pp. 155–169. Springer, Cham (2022). https://doi.org/10.1007/978-981-19-3923-5_14
7. Özgenel, F., Gönenç Sorguç, A.: Performance comparison of pretrained convolutional neural networks on crack detection in buildings. In: ISARC 2018 - 35th International Symposium Automatic Robotics in Construction International AEC/FM Hackathon Future Building Things, no. Isarc (2018). https://doi.org/10.22260/isarc2018/0094
8. Zhang, L., Yang, F., Daniel Zhang, Y., Zhu, Y.J.: Road crack detection using deep convolutional neural network. In: Proceedings - International Conference on Image Processing ICIP, vol. 2016-Augus, pp. 3708–3712 (2016). https://doi.org/10.1109/ICIP.2016.7533052
9. P. Arafin, A. Issa, and A. H. M. M. Billah, "Performance Comparison of Multiple Convolutional Neural Networks for Concrete Defects Classification," Sensors, vol. 22, no. 22, 2022, doi: https://doi.org/10.3390/s22228714

10. Sandler, M., Howard, A., Zhu, M., Zhmoginov, A., Chen, L.C.: MobileNetV2: inverted residuals and linear bottlenecks. In: Proceedings of the IEEE Computing Social Conference on Computer Vision Pattern Recognition, pp. 4510–4520 (2018). https://doi.org/10.1109/CVPR.2018.00474
11. Falaschetti, L., Beccerica, M., Biagetti, G., Crippa, P., Alessandrini, M., Turchetti, C.: A lightweight CNN-based vision system for concrete crack detection on a low-power embedded microcontroller platform. Procedia Comput. Sci. **207**(Kes), 3948–3956 (2022). https://doi.org/10.1016/j.procs.2022.09.457
12. Li, C., Pan, W., Su, R.K.L., Yuen, P.C.: Multiple structural defect detection for reinforced concrete buildings using YOLOv5s. HKIE Trans. Hong Kong Inst. Eng. **29**(2), 141–150 (2022). https://doi.org/10.33430/V29N2THIE-2021-0033
13. Islam, M.M., Hossain, M.B., Akhtar, M.N., Moni, M.A., Hasan, K.F.: CNN based on transfer learning models using data augmentation and transformation for detection of concrete crack. Algorithms **15**(8) (2022). https://doi.org/10.3390/a15080287

Estimation of Remaining Useful Life for Turbofan Engine Based on Deep Learning Networks

Nurul Hannah Mohd Yusof(✉), Nurul Adilla Mohd Subha, Nurulaqilla Khamis, Noorhazirah Sunar, Anita Ahmad, and Mohamad Amir Shamsudin

Universiti Teknologi Malaysia, Johor Bahru, Johor, Malaysia

nhannah2@live.utm.my

Abstract. Having accurate prediction on the health of machines in manufacturing can lead to a profitable organization if the operations and maintenance decisions are appropriately performed. This hinges on making well-informed operational and maintenance decisions. Incorporating condition monitoring and predictive maintenance strategies can significantly contribute to achieving this goal. By continuously monitoring the real-time condition of machines, organizations can gather valuable data that offers insights into the performance and health of the equipment. However, dealing with a scarce dataset, which is common in real world applications, makes any prognostics on the maintenance system intricate. This is further exacerbated by the unavailability of failure data within the system which makes degradation model is best suited for the said situation. Since there is no extensive study discussing computational time under similar settings of two different networks for the degradation model in estimating RUL, this study investigates a simple Long Short-Term Model (LSTM) method for prognostics, which is compared to a two-dimensional Convolutional Neural Network (CNN) under the same training options. The networks are trained using the popular Commercial Modular Aero-Propulsion System Simulation (C-MAPSS) dataset from the National Aeronautics and Space Administration (NASA). The aim of this study is to estimate the remaining useful life (RUL) of a turbofan engine in the most effective way. With carefully designed and defined network architectures, better performance can be attained, enabling proper foreseen of the RUL of an engine as soon as it is more likely to be close to failure. Based on the comparison, it is noted that the simple LSTM method for RUL prediction outperforms the two-dimensional CNN with better RUL prediction, Root Mean Square Error (RMSE), and computational time. For future improvement, this study can be further explored for a more sophisticated hybrid model that might produce better prediction in various sectors such as manufacturing, automotive, and military applications.

Keywords: Remaining Useful Life · Deep Learning · Prognostics

© The Author(s), under exclusive license to Springer Nature Singapore Pte Ltd. 2024
F. Hassan et al. (Eds.): AsiaSim 2023, CCIS 1911, pp. 329–340, 2024.
https://doi.org/10.1007/978-981-99-7240-1_26

1 Introduction

Maintenance in aerospace, automotive, and manufacturing is very crucial and designing strategies to overcome breakdowns is always intricate and challenging. Traditionally, corrective model in maintenance is scarred due to the fact that the concept of 'run-to-failure' is adapted. This approach, also known as reactive maintenance, that refers to the procedure of addressing and fixing equipment or system faults after they have occurred. It entails taking steps to restore the functionality of the apparatus or system to the state required for regular operation. When performing maintenance tasks like corrective maintenance, unanticipated faults or malfunctions are addressed. When a piece of equipment breaks down, sounds an alarm, or users/operators report a problem, maintenance is carried out as needed. It is relatively unplanned maintenance, commonly associated with unscheduled downtime and increased costs due to emergency repairs. In order to minimize downtime and restore the system or piece of equipment to its intended functioning state, the root cause of the failure must be immediately identified and corrected [1, 2]. Back then, the machines were simpler and specialized expertise was unnecessary. However, this type of maintenance is no longer convenient as the complexity of machines has increased. Major breakdowns can take a significant amount of time to be resolved, leading to temporary halting production. As the industrial revolution progressed, mass production and electrically powered lines took center stage in discussions. Therefore, specific timely planned maintenance started to be adopted, focusing on the practice of carefully attending to every repair based on its necessity and discovery, with the aim of boosting profitability.

Compared to reactive maintenance, preventive or proactive maintenance can be performed in a more strategic way that usually involves performing regular checks, maintenance work, and servicing on equipment or assets to avoid breakdowns, increase reliability, and extend the life time of machine tools as addressed in [3–6]. Essentially, preventive maintenance can be performed based on either periodic cycles or equipment condition. Periodic cycles maintenance can unreasonably costly, which is why device-based maintenance is favored. In device-based maintenance, replacement of parts and alteration of machines only occur when deviations from its normal operation start to appear, making it more efficient and cost effective. The primary goal of preventive maintenance is to identify possible problems and rectify them before they result in equipment failure or performance degradation. However, the drawback of preventive maintenance is that it requires more resources and is less adaptable to changing conditions. Hence, to overcome these issues, predictive maintenance (PdM) can be adopted.

1.1 Predictive Maintenance

As technology of manufacturing evolving, PdM has emerged as the highest level of maintenance that has been a hot topic discussed among researchers and manufacturers as stated in [7–10]. It is a technique of foreseeing issues before any fault arise, achieved through the analysis of production data trends to prevent failure of assets. It is also referred to as prognostics. To achieve this, integration of big data analytics and artificial intelligence is utilized to generate insights, spotting trends and abnormalities. Training a predictive maintenance algorithm can be categorized into three categories: anomaly detection, fault identification or diagnostics, and remaining useful life (RUL) estimation or prognostics. Eventually, the objective of predictive maintenance algorithms is to transform sensor data into actionable maintenance decisions.

To evaluate proper functioning of machines, anomaly detection can be adopted. This algorithm identifies any occurrences or patterns that deviate from the anticipated behavior of a particular machine. Engineers and data scientists often utilize this approach to identify faults in machinery or defects in manufacturing production lines, as seen in [11–13]. However, even though anomaly detection is useful for pinpointing faults, it may be unable to provide insight about the current and future health state of the machines. For that purpose, condition monitoring could be more suitable.

Condition monitoring particularly addresses on diagnosing or assessing the current health state of machines during operation. It achieves this by capturing and analyzing data from sensors on the equipment, which is then used for further development into PdM. Several case studies on fault diagnostics of rotating machinery have been discovered in [14]. This algorithm enables manufacturers to reduce unexpected downtime by expecting occurrence of abnormalities or faults and swiftly identifying their sources. Furthermore, this algorithm also prevents unnecessary scheduled maintenance costs by performing servicing only when it is necessary. Not only just data collection, but the condition monitoring can also leverage the collected data to determine the maintenance status of equipment. To further forecast the health of machine for upcoming event, prognostics of remaining useful life (RUL) prediction can be implemented, as shown in [15–17].

In prognostic algorithm, through tracking sensor data from machines, the failure event occurrence can be predicted, allowing manufacturers to prevent equipment failure. This capability enables the modification of maintenance schedules based these projections, presenting an alternative to traditional preventative maintenance, which relies on predetermined timescale. Additionally, manufacturers can reduce downtime of equipment by foreseeing the issues early, thereby extending the lifetime of the equipment. By determining the root cause of impending failures, manufacturers can speed up bringing the equipment back online.

For predictive maintenance to be successful, prognostics algorithms are essential. Sensors are used to detect temperature, pressure, voltage, noise, and vibration. To extract characteristics known as condition indicators from this data, several statistical and signal processing approaches are used. Using data clustering and classification or other machine learning approaches, these status indicators can be compared with recognized markers of problematic conditions, allowing for effective monitoring of the equipment's health. To train prognostics algorithms, the condition indicators serve as inputs to RUL prediction models. The type of data supplied can influence the choice of RUL model, which can be similarity, survival, or degradation model. The ultimate outcome is a prognostics algorithm that capable of categorizing, forecasting, and providing a confidence bound on the next failure occurrence.

Based on the available information, there are three ways to estimate RUL: survival model, similarity-based, and degradation model. By having a complete series of datasets, similarity model can be performed as in [18], which is a very rare case since data is often scarce. On the other hand, survival model exemplified in [19] is performed when only failure data is available, without complete histories. However, this method has a disadvantage due to its data censoring characteristic, which means a lot of assumption must be made. When no data is available, degradation model can be used, which is advantageous because getting a set of flawless data would be challenging. In a degradation model [20–22], under some circumstances, no failure information from related equipment is accessible but with knowledge of a safety limit, the machine operations should not exceed the threshold, as surpassing it could lead to failure. Prior data from the machine can be utilized to fit a degradation model to the condition indicator that forecasts future changes in the condition indicator. RUL can be calculated using statistical, machine learning, or deep learning methods to estimate the number of cycles before the condition indicator passes the threshold.

1.2 Deep Learning System Model

Convolutional Neural Network (CNN) was an inspiration of a cat visual center and has arisen after multiple effort generations which yields good findings in image recognition and speech analysis. The structure of weight sharing adopted in this method is almost alike biological neural networks with a highly simplified networks and amount of weight. Furthermore, to prevent an intricate data reconstruction and feature extraction, CNN can consider an input of a two dimensional array. Two dimensional CNN makes it possible for the proposed method to be used in the field of industry while requiring no prior knowledge of prognostics or signal processing. Despite of good prognostics on machine health, this type of deep learning model might computationally burden the training due to more layers (5 layers) adopted such in [23]. Apart from that, without dropout technique, the model might face overfitting issue especially when the dataset is small resulting in poor performance of the testing dataset. Hence, to overcome the intricacy of the architecture, this study implementing a simple Long Short-Term Model (LSTM) which is in fact more suitable for sequential data such as in [24, 25]. LSTM is very well-known Recurrent Neural Network (RNN) that is invented to solve the vanishing gradient issue while performing back-propagation in training a model. The gates (input gate, output gate, forget gate) in LSTM has the ability to control the flow of data into and out of the cell with a memory capability which makes it holds the advantage of capturing long-term dependencies between time steps of input. With all these advantage, several metrics must be meticulously considered to measure the performance of these two networks.

Performance Evaluation. The predicted and actual values of RUL need to be compared and analyzed. Appropriate metric is required to evaluate the performance of the trained model.

Computational Time. In deep learning techniques, computational time is considered as the amount of time taken to develop the training progress of the network model. The layers of interconnected neurons may be at variance and usually have need of large data streams and resources to be effectively computed and trained. It is worth noted that computational time may vary depends on the complexity of the network model, size of dataset, capacity of the processor, and optimization techniques employed. Fast computational time is among the utmost important in PdM due to timely decision to be made, minimal downtime, cost effective, safety measures, and reliability of overall system. it also important as an aid to a fast availability of system overall insights. However, PdM selection does not solely depend on computational time, but a trade-off among several factors such as accuracy of the model, availability of dataset, and maintenance requirements.

Root Mean Square Error (RMSE). This RMSE is commonly used as a metric to evaluate machine learning and deep learning tasks. It is an averaged measure of difference between prediction of a model and actual values at hand. The average of the difference will be taken under a square root and the equation will be,

$$RMSE = \sqrt{\frac{1}{N}\sum_{i=1}^{N} d_i^2} \quad (1)$$

The value of different between predicted values and actual values are presented as d and N is the number of prediction values available. This metric delivers a viable measure for evaluating a prediction model in deep learning. RMSE with lower values point toward a better performance of a specific model due to the small differences between predicted and actual values. Both models considered in the next subsection will be evaluated using these two evaluation measures.

In this study, the prognostics of RUL prediction is developed based on a degradation model with a predefined threshold due to the type of dataset available. This paper examines the effects of having deep learning models, specifically using LSTM and two dimensional CNN on RUL predictions in term of computational time and RMSE. A summary of PdM and deep learning networks research is provided in Sect. 1. The developments of these trained networks to generate RUL estimations are shown in Sect. 2. Section 3 consists of the findings within the scope of this research. Finally, some discussion on future directions are narrated in Sect. 4.

2 Methodology

In this study, the procedures of obtaining RUL estimations were set up (Fig. 1) using a 64-bit operating system with Intel(R) Core (TM) i5-9400F CPU @ 2.90 GHz processor and 8 Gb RAM. The Windows edition was Windows 10 Pro version 21H2. There are several components need to be available beforehand:

- MATLAB R2023a
- Deep Learning Toolbox
- Turbofan Engine Degradation Simulation data set (C-MAPSS) [26]

Dataset. This study considered on a prior benchmark investigation available on degradation model of NASA's turbofan engine in [26]. The dataset is developed by NASA by that derived a model-based simulation program named Commercial Modular Aero-Propulsion System Simulation (C-MAPSS) dataset which includes four division of sub-datasets retrieved from 21 sensors. All subsets of data are partitioned into training and testing data. The training data consists of sensors' measurements on run-to-failure mode of several aero-engines that includes multiple operating and fault conditions. To verify

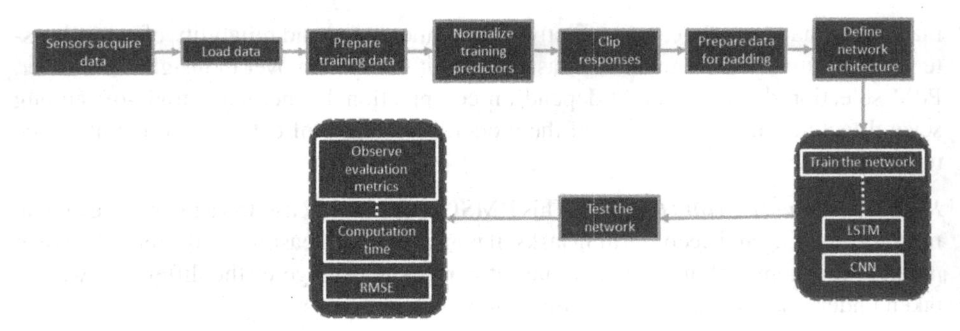

Fig. 1. The flowchart for training CNN and LSTM networks

the testing data, actual values of RUL are provided. In this study, the first data subset named FD001 is specifically considered due to simplicity of the dataset. There are 100 engine units available for training and testing, one operating condition and one fault mode with 17,731 of training samples and 100 testing samples. In this study, four units (unit 67, 69, 92, and 96) out of 100 engines are randomly picked for thorough observations on its RUL estimation.

Networks Architecture. The data to be trained are processed and sorted in sequence format. Furthermore, the data are divided evenly in mini-batches.

CNN. Two dimensional CNN is adopted, where the number of selected features are presented in the first dimension and the length of time sequence is obtainable in the second dimension. The convolutional layers are bundled along with batch normalization layer, activation layer and are stacked together for feature extraction. Finally, the output can be obtained by utilizing the fully connected layers and regression layers in order to achieve the final value of RUL.

LSTM. For a simple LSTM network, the data is trained under a fully connected layer of size 50. The sequence input layer is set equally as the sequential number of features.

To make both networks equally specified, similar training options are adopted using the Deep Learning Toolbox. These networks are defined with a fully connected layer, 200 number of hidden units, and 0.5 dropout probability. The responses are clipped with a predefined threshold 150, to pick up the patterns of sequence data as soon as the engines are likely to failing. Both networks are trained by using the 'adam' solver for 60 epochs. These epochs had mini-batches of size 20 and the learning rate is specified to 0.01. The gradient threshold is set to one in order to avoid the gradient exploding effect. The training data are sorted according to sequence length by setting the 'Shuffle' mode to 'never'. The training progress is turned on to monitor the computational time. After the training is over, then, the performance of the RUL estimations will be analyzed and evaluated in results section.

3 Results

In this section, the comparison between CNN and LSTM networks' performances are presented by addressing the computational time, test engine observation, and RMSE. The performance is monitored under similar condition of training options. The observations of test engines are conducted for engine number 67, 69, 92, and 96.

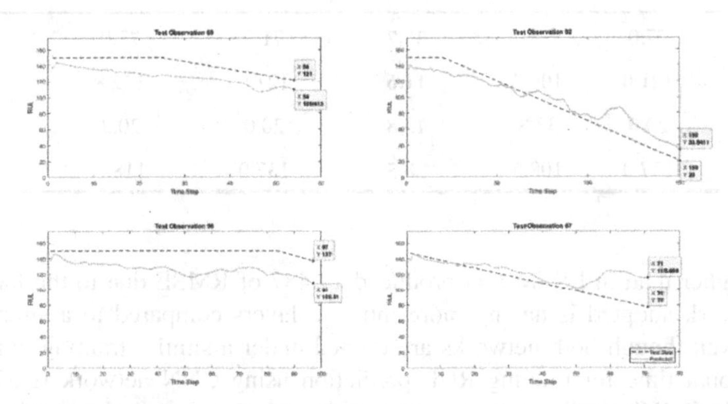

Fig. 2. CNN test observation for different engine number (Color figure online)

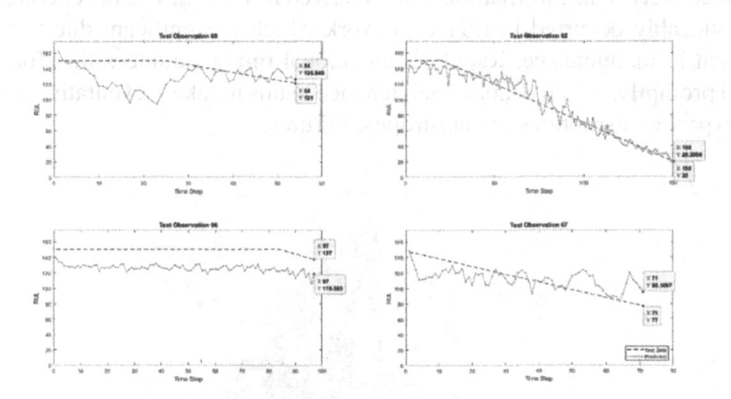

Fig. 3. LSTM test observation for different engine number (Color figure online)

By referring to Fig. 2 and Fig. 3, the predicted RULs are presented for the dataset FD001 by utilizing the CNN and LSTM, respectively. It can be observed that all different four testing engine units produced different RULs but evolving around the predefined threshold. Noticeable difference can be seen between the predicted RULs and the real RUL values indicate by the red solid line and dashed blue line, respectively. The difference is considered as error and it can be seen that the error in RUL prediction of CNN is slightly higher than the one that is LSTM-trained and the numerical values are summarized in Table 1. To verify this, the distributions of the error can be quantified by calculating the RMSE.

Figure 4 and Fig. 5 described the distribution of errors that occurred in the RUL prediction of CNN and LSTM networks. The value of RMSE in CNN is 19.8332 which

Table 1. RUL of test and predicted data for CNN and LSTM networks

Test Observation	CNN			LSTM		
	Test Data	Predicted Data	Error	Test Data	Predicted Data	Error
67	77.0	115.7	**38.7**	71.0	77.0	**6.0**
69	121.0	109.4	**11.6**	121.0	125.8	**4.8**
92	20.0	33.8	**13.8**	20.0	20.2	**0.2**
96	137.0	108.5	**28.5**	137.0	118.6	**18.4**

slightly higher than in LSTM that produced 19.487 of RMSE due to the fact that the CNN network adopted is having more intricate layers compared to a simple LSTM network even though both networks are trained under a similar training options. The computational time for training RUL prediction using CNN network is 37 s, while LSTM-trained RUL prediction took only 9 s to be completed as can be seen in Fig. 6 and Fig. 7, respectively. This information is summarized in Table 2. The faster computational time is noticeably occurred in LSTM network which is significant due to fast-paced environment in maintenance. Rapid computational time guarantees that forecasts can be created promptly, which enables maintenance teams to take preventative measures to avoid unexpected disruptions or catastrophic failures.

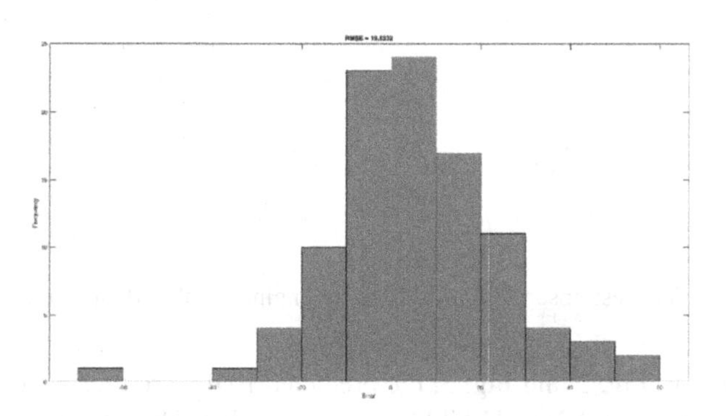

Fig. 4. CNN RMSE

Overall, it can be seen that the RUL prediction using a simple LSTM has outperformed the one using CNN in term of RMSE and computational time as the complexity in a simple LSTM is reduced with a single LSTM layer compared to the two dimensional CNN with five layers as in [23]. The estimations of RUL in prognostic is very crucial to determine the time of engine units most likely to failure or damage because an accurate RUL estimations can aid in maintenance activities to be performed on precise schedule which will allow the maintenance team to conduct precaution steps exactly before any

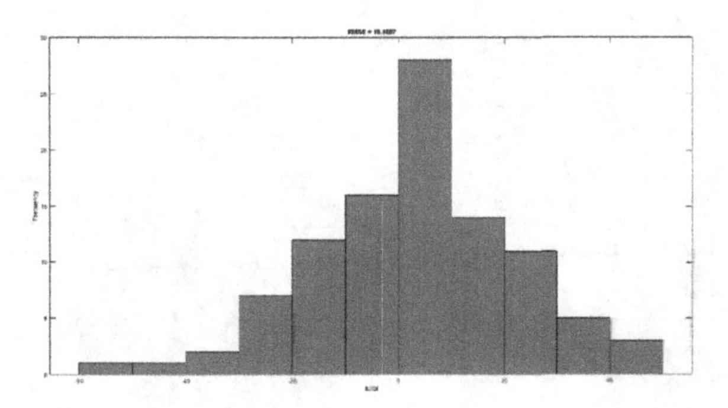

Fig. 5. LSTM RMSE

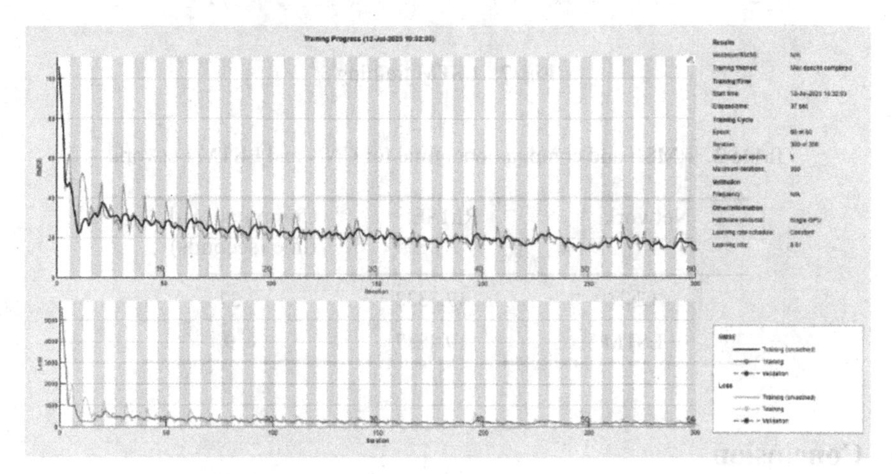

Fig. 6. CNN training

damage occur and prevent unnecessary costly disruptions. Additionally, the selection of a predictive maintenance approach should take into account more than just computational time, taking into account evaluations like model accuracy, interpretability, data accessibility, and the particular needs of the maintenance application. Good evaluation on the status of engines can save a fortune for an organization indirectly, it can enhance the reliability and safety of overall operation, avoid unexpected downtime, and reduce maintenance costs.

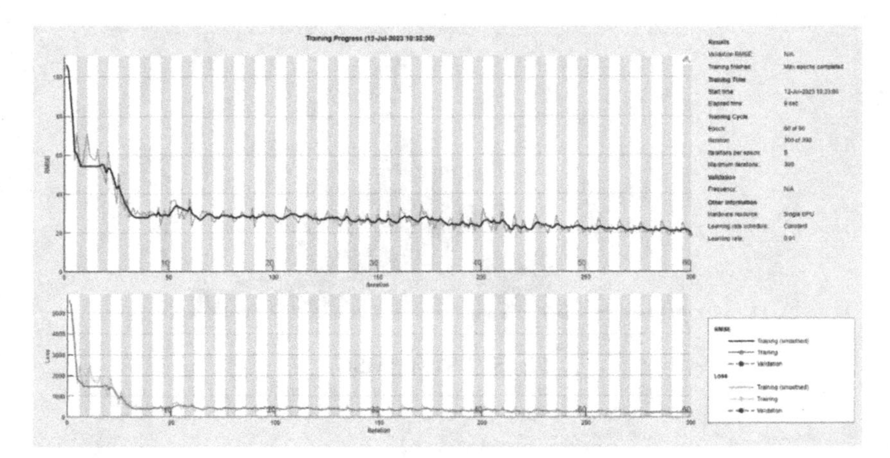

Fig. 7. LSTM training

Table 2. RMSE and computational time for CNN and LSTM networks

Network	RMSE	Computational time (seconds)
CNN	19.8332	**37**
LSTM	19.4887	**9**

4 Conclusion

In this study, a simple LSTM method for prognostics is compared to two dimensional CNN under the same training options. The training for the networks are conducted using the popular C-MAPSS dataset from NASA. The aim of this study is to estimate the RUL of a turbofan engine with the most effective way. With the networks architecture that is carefully designed, a better performance can be attained. The RUL of an engine can be properly foreseen as soon as the engine is more likely close to any failure. Based on the comparison, it is noted that a simple LSTM method for RUL prediction has outperformed the two dimensional CNN with better RUL prediction, RMSE, and computational time. For future improvement, this study can be further explored for a more sophisticated hybrid model that might produce a better prediction in various sectors such as in manufacturing, automotive, and military applications.

Acknowledgement. This work was supported by the Ministry of Higher Education (MOHE), Malaysia under Fundamental Research Grant Scheme (FRGS) (FRGS/1/2020/TK0/UTM/02/36).

References

1. Gan, S., Song, Z., Zhang, L.: A maintenance strategy based on system reliability considering imperfect corrective maintenance and shocks. Comput. Ind. Eng. **164**, 107886 (2022)
2. Özgür-Ünlüakın, D., Türkali, B., Aksezer, S.Ç.: Cost-effective fault diagnosis of a multi-component dynamic system under corrective maintenance. Appl. Soft Comput. **102**, 107092 (2021)
3. Hagmeyer, S., Mauthe, F., Dutt, M., Zeiler, P.: Preventive to Predictive Maintenance (2021)
4. Abdelhadi, A.: Preventive Maintenance Grouping using Similarity Coefficient Methodology (2010)
5. Udoh, N., Ekpenyong, M.: A knowledge-based framework for cost implication modeling of mechanically repairable systems with imperfect preventive maintenance and replacement schedule. J. Appl. Sci. Eng. **26**, 221–234 (2023)
6. You, M.Y., Meng, G.: A framework of similarity-based residual life prediction approaches using degradation histories with failure, preventive maintenance, and suspension events. IEEE Trans. Reliab. **62**, 127–135 (2013)
7. May, G., et al.: Predictive maintenance platform based on integrated strategies for increased operating life of factories. IFIP Adv. Inf. Commun. Technol. **536**, 279–287 (2018)
8. Cao, Q., et al.: KSPMI: a knowledge-based system for predictive maintenance in industry 4.0. Robot. Comput. Integr. Manuf. **74**, 102281 (2022)
9. Rieger, T.: The Application of Data Analytics Technologies for the Predictive Maintenance of Industrial Facilities in Internet of Things (IoT) Environments
10. Gutschi, C., Furian, N., Suschnigg, J., Neubacher, D., Voessner, S.: Log-based predictive maintenance in discrete parts manufacturing. Conf. Intell. Comput. Manuf. Eng. **79**, 528–533 (2019)
11. Cai, X., Xiao, R., Zeng, Z., Gong, P., Ni, Y.: Itran: a novel transformer-based approach for industrial anomaly detection and localization. Eng. Appl. Artif. Intell. **125**, 106677 (2023)
12. Shen, H., Wei, B., Ma, Y., Gu, X.: Unsupervised industrial image ensemble anomaly detection based on object pseudo-anomaly generation and normal image feature combination enhancement. Comput. Ind. Eng. **182**, 109337 (2023)
13. Pota, M., De Pietro, G., Esposito, M.: Real-time anomaly detection on time series of industrial furnaces: a comparison of autoencoder architectures. Eng. Appl. Artif. Intell. **124**, 106597 (2023)
14. Wang, Y., Zhou, J., Zheng, L., Gogu, C.: An end-to-end fault diagnostics method based on convolutional neural network for rotating machinery with multiple case studies. J. Intell. Manuf. **33**, 809–830 (2020)
15. Li, T., Wang, S., Shi, J., Ma, Z.: An adaptive-order particle filter for remaining useful life prediction of aviation piston pumps. Chin. J. Aeronaut. **31**, 941–948 (2018)
16. Feng, K., Ji, J.C., Ni, Q.: A novel gear fatigue monitoring indicator and its application to remaining useful life prediction for spur gear in intelligent manufacturing systems. Int. J. Fatigue **168**, 107459 (2023)
17. Zhou, J., Qin, Y., Chen, D., Liu, F., Qian, Q.: Remaining useful life prediction of bearings by a new reinforced memory GRU network. Adv. Eng. Informatics **53**, 101682 (2022)
18. Aburakhia, S., Tayeh, T., Myers, R., Shami, A.: Similarity-based predictive maintenance framework for rotating machinery. In: International Conference on Communications, Signal Processing, and their Applications (Institute of Electrical and Electronics Engineers Inc.) (2022)
19. Ruppert, T., Csalodi, R., Abonyi, J.: Estimation of machine setup and changeover times by survival analysis. Comput. Ind. Eng. **153**, 107026 (2021)

20. Kang, J., et al.: Remaining useful life prediction of cylinder liner based on nonlinear degradation model. Eksploat. i Niezawodn. – Maint. Reliab. **24**, 62–69 (2022)
21. Yang, C., et al.: A novel based-performance degradation indicator RUL prediction model and its application in rolling bearing. ISA Trans. **121**, 349–364 (2022)
22. Feng, K., Ni, Q., Zheng, J.: Vibration-based system degradation monitoring under gear wear progression. Coatings **12**, 892 (2022)
23. Li, X., Ding, Q., Sun, J.Q.: Remaining useful life estimation in prognostics using deep convolution neural networks. Reliab. Eng. Syst. Saf. **172**, 1–11 (2018)
24. Li, J., Li, X., He, D.: A directed acyclic graph network combined with CNN and LSTM for remaining useful life prediction. IEEE Access **7**, 75464–75475 (2019)
25. Wang, C., Jiang, W., Yue, Y., Zhang, S.: Research on prediction method of gear pump remaining useful life based on DCAE and Bi-LSTM. Symmetry **14**, 1111 (2022)
26. Saxena, A., Goebel, K., Simon, D., Eklund, N.: Damage propagation modeling for aircraft engine run-to-failure simulation. In: International Conference on Prognostics and Health Management (2008)

Simulation Design of Reinforcement-Based Migration System in Software-Defined Networking Using Q-Learning

Avinash Imanuel Gana Raj and Tan Saw Chin(✉)

Faculty of Computing and Informatics, Multimedia University, Cyberjaya, Malaysia
sctan1@mmu.edu.my

Abstract. An innovative strategy, Software-Defined Networking (SDN), provides improved network programmability, flexibility, and scalability. However, a lot of recent material on the transition from traditional network designs to SDN has recommended providing static heuristic algorithms for determining the node migration sequence in legacy networks. The approach is deemed unfeasible for use in real-world circumstances and has issues managing the changing nature of network traffic. Reinforcement learning (RL) has therefore been proposed for use in SDN domains such flow entry management, controller placement, and routing selection. In the study, we first examine the relevant research on reinforcement learning applications on SDN conducted by previous researchers and identify the difficulties in SDN migration. Finally, using Q-learning in this context, we proposed a reinforcement learning technique to get around these problems and carefully transition a network's legacy nodes to SDN nodes. Thus, we have discussed designs for applying reinforcement learning to a hSDN deployment in a legacy network in this article.

Keywords: Reinforcement-learning · Software-defined networking · Migration

1 Introduction

Machine Learning (ML) is a subfield of artificial intelligence research that trains computers to learn from experience. Numerous complicated issues occurring in traffic engineering and route optimization have been successfully solved using machine learning approaches [1–3]. The area of machine learning known as reinforcement learning can learn from its actions and adapt to its surroundings based on both positive and negative reinforcements [4]. The Q-Learning, SARSA, and Greedy algorithms are just a few of the algorithms available for creating this reinforcement learning agent [5]. Maximizing the cumulative reward in the context at hand is the fundamental objective of the learning process. Numerous aspects of Software-Defined Networking have used reinforcement learning [1–4, 6]. The network architecture known as "Software-Defined Networking" (SDN) removes itself from the control and data planes and unites everything under a single centralised controller. Different applications of reinforcement learning to SDN exist,

© The Author(s), under exclusive license to Springer Nature Singapore Pte Ltd. 2024
F. Hassan et al. (Eds.): AsiaSim 2023, CCIS 1911, pp. 341–351, 2024.
https://doi.org/10.1007/978-981-99-7240-1_27

including resource allocation-focused flow entry management [2], dynamic switch and controller clustering [3], drone management framework [7], etc. Reinforcement learning is used in these SDN applications because it optimizes network performance and increases resource utilization. Users and businesses have reaped significant benefits from the development of SDN technology, including the ability to optimize traffic flow inside an SDN network and assign enough resources depending on network traffic and user requests.

It is therefore highly feasible to implement reinforcement learning in SDN migration to optimize the migration process and assure less disturbance to network services [1]. Since converting the current architecture to SDN has both technical and economical limitations, one option to get around this problem is to swiftly and progressively migrate SDN while only choosing the essential nodes. Hybrid SDN Architecture [5], the name given to this migration, allows traditional and SDN-capable nodes to coexist in the network environment shown in Fig. 1.

The article's overall focus will be a well-researched overview of Reinforcement Learning's use in Software-Defined Networks. The following sections are as follows: The related research on reinforcement learning applications on SDN is examined in Sect. 2; the difficulties encountered in other studies on SDN migration are covered in Sect. 3; and the proposed design of the reinforcement-learning based SDN migration system is presented in Sect. 4.

2 Reinforcement-Learning Related Works in SDN

In order to create an effective rules placement algorithm that dynamically gathers the optimal path from the DRL agent and predicts future traffic demands using the well-known prediction method Long Short-Term Memory (LSTM), the authors in [1] used SDN in conjunction with Deep Reinforcement Learning (DRL) in routing optimisation. The Knowledge-Defined Networking (KDN) method, which is separated into three planes, including the Knowledge plane, Data plane, and Control plane, was used to create the framework. The proactive forwarding module uses a DRL agent to select a way to distribute incoming traffic, while the network measurement module in the control plane is in charge of storing information about the traffic on the network. Together, these two modules help the Knowledge plane estimate network congestion using techniques like LTSM by giving it information. The Deep Reinforcement Learning agent, which assures the optimum routing path, optimises the routing. The DRL agent interacts with the environment through three signals: State, Action, and Reward. The DRL is represented as follows: G(V, E, C), where V, E, and C are, respectively, the vertex, edge, and link capacity sets, and "|V | = N" denotes the number of network nodes. The (NxN) Traffic Matrix, or State, represents the current network load. The "Action" that the agent does is to decide which path to follow for the new incoming flow, and the agent's "Reward", (r), is connected to the QoS parameters, which are primarily: Latency (L), Rate (W), and Packet Loss (P L). The Reward r is determined as follows:

$$r = \alpha \cdot W - \beta \cdot L - \gamma \cdot PL \tag{1}$$

where the variable weights $\alpha, \beta, \gamma \in [0, 1]$ are decided by the routing method. The ideal policy, π, for matching the set of states to the set of actions in order to maximise reward, r, is what the DRL agent seeks to find.

In addition, it is stated in [4] that they have proposed an SDN Flow Entry Management Using Reinforcement Learning and modelled using a Markov Decision Process (MDP), where $S = \{s1, s2, \ldots, si\}$ is the state-space, $A = \{a1, a2, \ldots, aj\}$ is the action-space, Pij defines the transition probability from state i to state j, and R defines the reward associated with various actions $a \in A$. State-space, action-space, and reward function definitions are used to categorise the issue:

- State set: Every potential combination of values for the flow match frequency and flow recentness parameters is represented by the state set in our model.
- Action set: In our method, there are three possible courses of action: (i) taking no action; (ii) raising the value associated with the selected parameter; and (iii) lowering the value associated with the selected parameter.
- Reward Function: The amount of communications required between the controller and switches in order to appropriately transmit the input packets is known as the network setup overhead. Based on the assessed setup overhead and the switch's installed forwarding rules, the reward is calculated.

Researchers concentrated on employing reinforcement learning for route selection in accordance with the work done by [8]. They proposed a network architecture known as SDN-enabled wireless-Power Line Communication (PLC) Power Distribution-Internet of Things (PD-IoT), which combines the benefits of PLC and wireless communications to adapt to the expansion of PD-IoT network scale and the continuously rising number of devices. A SARSA-based Delay-aware Route Selection (SDRS) algorithm is the method put out in this research for SDN-enabled wireless-PLC PD-IoT. On the basis of the network condition data and performance feedback supplied by the SDN controller, the SDRS may learn the ideal route selection technique. To regulate the route that data packets take when being sent, SDRS also creates a Q-table. It is capable of integrating wireless communications and PLC in dynamic network situations. SARSA has the benefit of learning the best route selection approach, which can decrease transmission time and increase dependability. Additionally, it can adjust to the complicated and dynamic communication environment. Numerical studies demonstrate that SDRS performs superbly in terms of transmission latency and reliability when compared to other heuristic algorithms, such as the shortest route selection (SRS) algorithm and random route selection (RRS) method. The list of Reinforcement-Learning-based SDN-related works is shown in Table 1.

We may draw the following conclusion from the facts above: Reinforcement learning has demonstrated a number of benefits, including effective routing rules in dynamically changing networks, a decrease in transmission delays and an increase in dependability, as well as the capacity to adapt to the dynamic and complicated communication environment. It has numerous and common applications in the areas of controller placement, flow management, and routing optimisation, among others.

Table 1. Summary of SDN using Reinforcement-learning

Reference	Application Area	Techniques	Description
(Bouzidi et al., 2019) [1]	Routing	Deep Reinforcement Learning (DRL) using LTSM with an ILP approach	• without needing to know the network architecture and traffic patterns in advance, effective routing strategies in dynamic networks • DRL can create new network states, speed up processing, and reduce storage requirements for Q-tables
(Shi et al., 2022) [8]	Route Selection	SARSA-based Delay-aware Route Selection (SDRS)	• discovers the best route-selection technique that may lower transmission delay and increase dependability • become used to the complicated and dynamic communication environment
(Li et al., 2022) [3]	Controller Placement	multi-agent deep Q-learning networks (MADQN)	• appropriate for use in massive combinatorial optimisation • due to the information gained from feedback, may obtain appropriate results by exploring the solution space and be adaptable to the fast changes in network state in SD-IoV • Decreases delay • Balances load • Promotes dependability

(*continued*)

Table 1. *(continued)*

Reference	Application Area	Techniques	Description
(Mu et al., 2018) [4]	Flow Entry Management	Q-Learning	• When compared to the Multiple Bloom Filters (MBF) technique, emulation results utilising the RL algorithm indicate around a 60% reduction in the long-term control plane overhead and about a 14% improvement in the table-hit ratio • When RL is used, the packet delivery ratio is greatly increased • RL was suggested as a way to increase the effectiveness of network routing • Quality of Service (QoS)-aware Adaptive Routing (QAR) outperforms previous learning systems and offers a higher convergence rate in QoS provisioning thanks to an RL-based methodology

3 Research Analysis and Challenges of SDN Migration

For SDN migration, the author [5] developed a static method based on a cost-location aware heuristic algorithm. The method has performed well in terms of network improvement, but it has shortcomings when it comes to handling the dynamic nature of traffic in a specific network. To accomplish load balancing and reduce the amount of new switch-controller assignments, the authors of another work [9] suggested a paradigm for dynamic switch-controller mapping. Although the authors encountered a similar issue to authors [5] during the testing phase of their model since their suggested solution offers static mapping between switches and controllers, which according to the results cannot handle changes in network traffic.

The authors have used Deep Reinforcement Learning (DRL) in the deployment of SDN switches based on the study by [10]. The choice to employ DRL is a result of the limitations of static heuristic methods, which do not take into account the dynamic nature of increasing network traffic, such as the Traffic Matrix (TM). The authors suggest employing SEED (SDN deployment using DRL), an intelligent algorithm, to create a hybrid SDN with an adaptive SDN deployment method that can better respond to dynamic traffic. However, because it makes use of deep neural networks, more computing power is required, which might restrict the scope and complexity of the issues that can be resolved.

Table 2 shows the summary of limitations found in the techniques for SDN migration. The table consists of three columns, the authors, the techniques and its limitation explanation.

Table 2. Summary of Technique Limitations Used for SDN Migration

Authors	Techniques used in SDN migration	Limitations
Wei et al., 2021 [4]	Algorithm for multi-periods cost-location aware SDN migration	A static heuristic approach cannot adequately address the dynamic nature of a particular network, such as traffic increase or topology change, throughout the migration time
(Csikor et al., 2020) [10]	To encourage SDN migration for smaller businesses, HARMLESS, a Hybrid ARchitecture for Migrating Legacy Ethernet Switches to SDN, was developed	HARMLESS increases network latency, which might have an impact on the performance of a task with a high latency need
(Guo et al., 2022) [10]	SEED (SDN deployment using deep reinforcement learning) uses several Traffic Matrices (TMs) to intelligently learn the SDN deployment method	The need to implement it is complex as it uses deep neural network, and is heavy in term of computational resources as well
(AL-Tam & Correia, 2019) [9]	Networking with several controllers using a fractional switch migration	Both flat and hierarchical multi-controller SDN designs have the potential to result in large traffic fluctuations that the controller may not be able to handle, which can cause failure and even cascade failure

We can see several drawbacks of each technique from the restrictions indicated in the research above, particularly in terms of the static approach in SDN Migration. We can observe how poorly it performs when it comes to dealing with the dynamic nature of network traffic because it can only be used in set circumstances. As for DRL, we can observe that it makes extensive use of computer resources because it uses deep neural networks to train its model. This occurs as a result of the sheer volume of parameters that must be understood.

4 The Proposed Design of Reinforcement-Based SDN Migration

We describe our strategy for software-defined networking migration utilising reinforcement learning (RL) in this section. We begin by outlining the broad foundation. The RL approach used for SDN migration will then be explained. The topology of our network can be depicted by the equation G = (V, E), where V signifies the network's nodes and E defines the links. We and Ce, signifies the connection's weight and capacity, respectively, where e ∈ E. The traffic matrix, abbreviated as TMij, which depicts the traffic demand from node i to node j where i,j ∈ V, represents the network's traffic. The migration of SDN nodes occurs over a period of time of t = 1,.., T steps with a budget of B for each step. Each node that is upgraded to an SDN node costs btn, where t ∈ T and n ∈ V. S = {Stn ∈{0, 1}: t ∈ T, n ∈ V }, represents a set of SDN nodes at step t when Stn = 1, n ∈ V. A group of nodes that are needed to be excluded from migration is represented by xS. In(v) signifies the ingress link established into node v where v ∈ V. Out(n) signifies the egress link established from node n at step t where Stn = 1, n ∈ V, t ∈ T. Furthermore, fe indicates the flow of traffic on link e where e ∈ E and g_e^{ij} shows the flow of traffic for link e from node i to j where i,j ∈ V. x^texte is used to represent the splitting flow on the link e ∈ Out(n) at time step t ∈ T and Stn = 1. Table 3 are the summary of the notations used in SDN migration.

The goal of the SDN migration in hybrid SDN is to minimise the MLU for all time periods.

$$\theta = minimize \Sigma_{t=1}^{T} U_t \tag{2}$$

Furthermore, there are a set of constraints to follow when migrating the node. The objective function is subject to a number of restrictions, which are given below:

$$0 \leq U_t \leq 1, t \in [1, n] \tag{3}$$

- The value of MLU is between 0 and 1

$$\forall t \in T, S_{tn} = 1, n, v \in V$$

$$\sum\nolimits_{e \in In(n)} f_e + TM_{nv} = \sum\nolimits_{e \in Out(n)} x_{e\prime}^{t} \tag{4}$$

- Flow conservation is the concept that the traffic flow leaving a node is equal to the traffic flow inside the node plus the traffic coming from the same node.

Table 3. Notation for SDN Migration

Notation/Variable	Description
V	Set of legacy nodes in a network to be migrated to SDN-enabled node
E	Set of links in a network
T	Set of the time step for SDN nodes migration in a network
S_{tv}	1 when node v at time t is a SDN node, 0 otherwise
xS	Set of nodes to be exclude in SDN migration in a network
b_i	The cost for migrating node i to SDN-enabled node
U_t	Maximum link utilization for time t
TM_{ij}	Traffic demand from node i to j
f_e	Traffic flow on link e
x_e^t	Splitting traffic flow on the link e at time t
g_e^{ij}	Traffic flow on link e from node i to node j
C_e	Capacity of link e.
B	Budget per period to migrate SDN node
$In(v)$	Set of ingress link to node v
$Out(n)$	Set of egress link from node n
W_e	Weight setting for each link e

$$\forall t \in T, e \in E, \sum\nolimits_{i,j \in V} g_e^{ij} \leq U_t C_e \tag{5}$$

- A link's capacity can never be exceeded in terms of total flow.

$$\forall t \in T, n \in V, S_{tn} = 0, n \in xS \tag{6}$$

- Upgrades shouldn't be made to the nodes that won't be migrated

$$\forall t \in T, \sum_{n \in V} S_{tn} b_n \leq B \tag{7}$$

- For each time step there is a budget cap for the expense of upgrading a traditional node to an SDN capable node.

The following shows the environment, states, action and the reward with a more broken down explanation of our proposed framework for the SDN migration problem. We have the environment, which our reinforcement learning agent will be interacting with. The state set would be the calculations related to the present environment such as the traffic demand, link capacity, traffic flow and budget. The action space describes the decision taken to migrate a node or not to migrate it and lastly the reward given to the learning agent based on the action taken.

Environment: An agent continuously interacts with the environment by informing it of the observed state P_t by using Reinforcement Learning. After an action, a_t the environment returns the reward, r_t and the next state, P_{t+1}. Finding the best course of action to maximise the expected cumulative discounted reward, $J = \mathbb{E}[\sum_{t=1}^{\infty} y^t r_t]$ is the aim of RL, where $\gamma \in [0, 1]$ is the discount factor and y^t is the discount factor that is multiplied by the quantity of time steps to reflect how important future rewards will become as the learning agent iteratively explores the environment. The environment in this scenario is the SDN migration network.

State set: Based on the topology G, the state, P_t, contains information about the current network, such as the nodes that needed to be migrated which is denoted as S_{tn}= [0 | 1], where 1 means it's an SDN node while 0 means it's a legacy node. There is also the traffic matrix (TM_{ij}) that determines the demand of a certain link flow. The observation consists of the environment's state S_t from the set of states $P = \{P^1, P^2, \ldots\}$. . The (N x N) Traffic Matrix, which represents the current network load, is the state. A learning agent takes a state $P_t = TM_{ij}$, as an input at time step, t. Furthermore, the current state also consist of the reward r_t, link capacity C_e from node i to node j. The budget B is the overall cost of migration for a given network and the learning agent should always ensure the cost of migration, b_i do not exceed the given budget.

Action: The action, a_t includes the following options: (i) Do not migrate the node, (ii) Migrate the node to SDN node. To migrate a node it follows the conditions from Eqs. (5) and (6), where the node to be migrated is indicated by, Stn =[0 | 1]. The nodes shall only be migrated if the conditions are met from the equations stated above, if they don't, the learning agent ignores the node and checks the following node on whether it fulfils the conditions. Furthermore, the learning agent is responsible to determine the proper splitting of traffic flow, f_e on each link to ensure that the traffic demand (TM) is fulfilled while link capacity never exceeded.

Reward: According to Eq. (8), the reward, r_t is computed. When the environment takes the action, a_t at in accordance with the state P_t, we utilise U_t to represent the MLU of the current network. The formulae $U_{ij} = f_e/C_{ij}$, where $i, j \in V$ are the traffic flow variables and f_e is the traffic flow on link e, yields the MLU. The reward's value informs the agent of how effective the action is. The agent will receive a larger compensation and a lesser MLU Ut if the decision made was the best.

$$r_t = 1/U_t \tag{8}$$

Q-Learning is the policy employed by the reinforcement learning algorithm. The Q-Learning technique was chosen because it updates the Q-values at each time step based on the estimated action-value function and does not require a model for the environment. 2018 (Mu et al.). Based on the biggest expected payoff, the Q-learning algorithm assists in determining the best course of action to take in each circumstance. The SDN migration agent uses this outcome afterwards to decide how to proceed and adjust its behaviour over time. The policy is written as:

$$\mathrm{Q(p, a)} = \mathrm{Q(p, a)} + \alpha(\mathrm{r} + \gamma([\mathrm{Q}(p', a')] - \mathrm{Q(p, a)}) \tag{9}$$

Q(p,a) is defined as the pairing of a state and an action. α represents the learning rate weight, determining how much importance is given to new information versus the existing Q-value. It is a value between 0 and 1, while, p', is the following state after taking an action, a, in the state, p, a' is a potential action in state, p', and $max_{a'}$ Q(p', a') represents the maximum Q-value among all actions a' in the next state p'. The algorithm selects the action that yields the highest Q-value in order to estimate the maximum expected future reward. In SDN migration, the state, Pt, consist the traffic demand (TM_{ij}), link capacity (C_e), budget (B) and the reward (r_t).

Furthermore, the action, a, consist of S_{tn}, the choice between migrating a node to an SDN node when $S_{tn}= 1$ or to not when $S_{tn}= 0$ following the conditions mentioned in the Eqs. (5) and (6) above, and deciding the path to split traffic flow (f_e) to different links, e, in the network so it can ensure the traffic flow fulfills the traffic demand (TM_{ij}). Figure 1 depict the proposed design of reinforcement-based SDN migration using Q-learning.

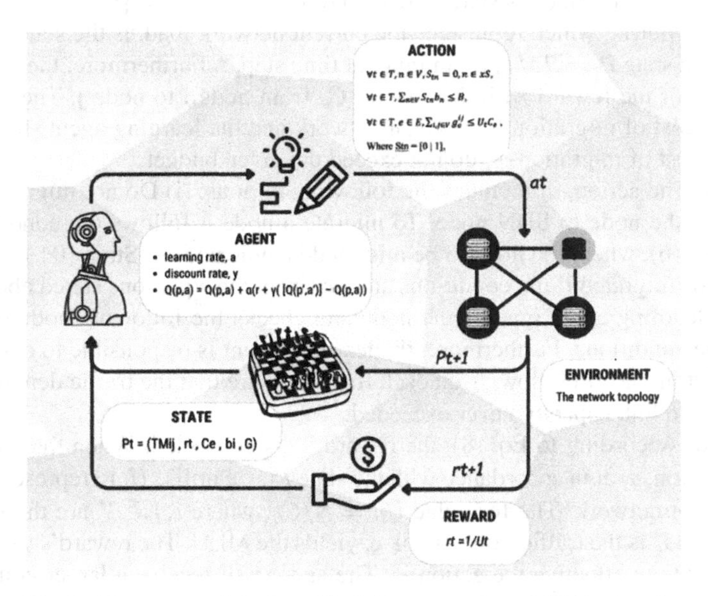

Fig. 1. The proposed design of Reinforcement-based SDN migration

5 Conclusion

Based on the identified research challenges, the prevailing research perspective on SDN migration has primarily concentrated on static methods. These methods often overlook factors like traffic growth and solely rely on historical traffic patterns, failing to capture the dynamic nature of network topology during deployment. As a result, the existing literature on SDN migration, which employs static heuristic algorithms, is impractical for real-world implementation. In response, a novel approach for achieving dynamic SDN migration is proposed. This approach leverages reinforcement learning, specifically Q-Learning, to address the dynamic environment. The method is framed as a Markov Decision Process (MDP) tailored for hSDN. It also enhances performance metrics such as maximum link utilization (MLU), while adhering to various constraints like link capacity and budget. Consequently, the migration process becomes more adept and efficient, empowering the agent to select actions that enhance network performance and tackle SDN migration challenges. The paragraph also outlines forthcoming research directions, which are currently in progress.

References

1. Bouzidi, E.H., Outtagarts, A., Langar, R.: Deep reinforcement learning application for network latency management in software defined networks. In: 2019 IEEE Global Communications Conference (GLOBECOM) (2019). https://doi.org/10.1109/globecom38437.2019.9013221
2. Kim, G., Kim, Y., Lim, H.: Deep reinforcement learning-based routing on software-defined networks. IEEE Access **10**, 18121–18133 (2022). https://doi.org/10.1109/access.2022.3151081
3. Li, B., Deng, X., Chen, X., Deng, Y., Yin, J.: MEC-based dynamic controller placement in SD-IoV: a deep reinforcement learning approach. IEEE Trans. Veh. Technol. **71**(9), 10044–10058 (2022). https://doi.org/10.1109/tvt.2022.3182048
4. Mu, T.-Y., Al-Fuqaha, A., Shuaib, K., Sallabi, F.M., Qadir, J.: SDN flow entry management using reinforcement learning. ACM Trans. Auton. Adapt. Syst. **13**(2), 1–23 (2018). https://doi.org/10.1145/3281032
5. Siew, H.W., Tan, S.C., Lee, C.K.: Hybrid SDN deployment using machine learning. In: Alfred, R., Iida, H., Haviluddin, H., Anthony, P. (eds.) Computational Science and Technology. LNEE, vol. 724, pp. 215–225. Springer, Singapore (2021). https://doi.org/10.1007/978-981-33-4069-5_19
6. Binlun, J.N., Chin, T.S., Kwang, L.C., Yusoff, Z., Kaspin, R.: Challenges and direction of hybrid SDN migration in ISP Networks. 2018 IEEE International Conference on Electronics and Communication Engineering (ICECE) (2018). https://doi.org/10.1109/icecome.2018.8644812
7. Yazdinejad, A., Rabieinejad, E., Dehghantanha, A., Parizi, R.M., Srivastava, G.: A machine learning-based SDN controller framework for drone management. In: 2021 IEEE Globecom Workshops (GC Wkshps) (2021). https://doi.org/10.1109/gcwkshps52748.2021.9682027
8. Shi, Z., Zhu, J., Wei, H.: Sarsa-based delay-aware route selection for SDN-enabled Wireless-PLC power distribution IOT. Alex. Eng. J. **61**(8), 5795–5803 (2022). https://doi.org/10.1016/j.aej.2021.11.029
9. AL-Tam, F., Correia, N.: Fractional switch migration in multi-controller software-defined networking. Comput. Netw. **157**, 1–10 (2019). https://doi.org/10.1016/j.comnet.2019.04.011
10. Guo, Y., Chen, J., Huang, K., Wu, J.: A deep reinforcement learning approach for deploying sdn switches in isp networks from the perspective of traffic engineering. In: 2022 IEEE 23rd International Conference on High Performance Switching and Routing (HPSR), Taicang, Jiangsu, China, 2022, pp. 195–200. https://doi.org/10.1109/HPSR54439.2022.9831203
11. Csikor, L., Szalay, M., Rétvári, G., Pongrácz, G., Pezaros, D.P., Toka, L.: Transition to SDN is HARMLESS: hybrid architecture for migrating legacy ethernet switches to SDN. IEEE/ACM Trans. Network. **28**(1), 275–288 (2020). https://doi.org/10.1109/TNET.2019.2958762

A Preliminary Study on the Possibility of Scene Captioning Model Integration as an Improvement in Assisted Navigation for Visually Impaired Users

Atiqul Islam(✉), Mark Kit Tsun Tee, Bee Theng Lau, and Kazumasa Chong Foh-Zin

Swinburne University of Technology Sarawak Campus, Kuching, Malaysia
aislam@swinburne.edu.my

Abstract. This research introduces a new approach to augment image captioning for visually impaired individuals by integrating depth data with RGB images. An overview of existing assistive tools and technologies indicates their limited adoption due to high costs and various constraints. The proposed model, designed to tackle these challenges, has the potential to enhance scene comprehension and navigation for the visually impaired. The model, which includes stages from data collection to its integration with assistive tools, uses a unique neural network architecture to process both image types, merge their outputs, and generate more detailed and practical descriptions of the environment. However, specific challenges exist, such as securing an appropriate RGB-D image dataset and creating an efficient neural network. Ongoing research efforts are vital to refine the model and evaluate its real-world applicability.

Keywords: Visually Impaired · Captioning · RGB-D

1 Introduction

The global prevalence of visual impairment is alarming, with the Globe Health Organization (WHO) estimating a staggering 285 million people suffering from some form of visual impairment. Of this demographic, 39 million are completely blind. Such individuals grapple daily with an array of challenges, one of the most significant being interior navigation. The ramifications are profound: their autonomy and overall life quality are detrimentally impacted. In fact, falls, a significant consequence of impaired navigation, are alarmingly prevalent among those aged 65 years and above [1]. Statistics calculated by Moreland et al. [1, 2] revealed that among the surveyed respondents who either experience blindness or have difficulty seeing, 42.1% reported having a fall, and an average of 1,500 self-reported falls per 1,000 respondents over the course of a year.

The inability to comprehend visual information in the surroundings, such as signs, landmarks, and impediments, is one of the main challenges experienced by visually impaired people while navigating [2]. This often culminates in disorientation and a severe

© The Author(s), under exclusive license to Springer Nature Singapore Pte Ltd. 2024
F. Hassan et al. (Eds.): AsiaSim 2023, CCIS 1911, pp. 352–361, 2024.
https://doi.org/10.1007/978-981-99-7240-1_28

erosion of confidence in their navigation abilities, particularly in unfamiliar terrains. Ideally, navigation should be an accident-free endeavor, minimizing potential hazards to both the individual and others in the vicinity. Although accidents in real-world contexts can sometimes be inevitable, successful navigation is typically gauged by one's ability to safely and independently reach their intended destination within a reasonable timeframe [3].

Historically, there has been an evolution in the tools and technologies developed to aid those with visual impairments. Initial solutions were simple yet effective, ranging from the traditional white cane, the invaluable assistance of guide dogs, to the aid of fellow humans. However, with technological advancements, there has been a shift towards more sophisticated solutions such as virtual white canes, advanced navigational software, and intelligent robots. Modern iterations often incorporate Bluetooth low-energy beacons and diverse sensors, and cutting-edge research now delves into the possibilities of LIDAR scanning [4].

Despite significant progress and a wide range of technical solutions, navigation assistance devices for the visually impaired are still not extensively utilized, and user approval is poor. Many of these devices are restricted in scope and have limitations [5]. For example, the white cane, which is the most important tool used by individuals with visual impairments, can only detect near-ground obstacles, leaving obstacles below ground level or above knee level undetected [6].

Navigation is a complex task that has multiple precise stages. In order to successfully navigate, users with visual impairment must be aware of their physical location, their relation to the surrounding environment (context) and the route they must follow to navigate to a desired destination [7]. This research will not cover the entire process of navigation, the primary focus will be on defining the relationship between the surrounding environments.

In the realm of technology, image captioning is emerging as a bridge between visual inputs and linguistic outputs. It aims to transform visual data into textual descriptions, with potential applications ranging from augmenting multi-modal search platforms to aiding visually impaired individuals in navigation [8]. Despite the commendable progress in producing succinct descriptions for 2D images, there's a prevalent focus on generating singular sentences. This study, in contrast, seeks to explore if blending depth information with conventional RGB images can refine the captioning task, producing richer descriptions to enhance visually impaired individuals' understanding of their environment.

2 Related Work

Several current navigation aid techniques and technologies have been developed to assist visually impaired people in navigating their surroundings. The aids can be broadly categorized as traditional, such as guide dogs, or human assistance; no-tech solutions like white cane and lastly, software/hardware-based solutions, for example, navigation systems for blinds (Dot Waker, Nearby Explorer, Get There, and Google Maps) [6], portable blind navigation devices such as Canetroller by Microsoft [9] or Smart Cane by The Robust Intelligence Program of the National Science Foundation [10]. The most recent assistive technologies are observed to use visual deep learning techniques and LIDAR scanning [4, 7] for localization.

Traditional supportive technologies for the visually impaired, such as white canes and guide dogs, present certain challenges. The range of detection offered by a white cane is rather limited, reaching most 1.5 m, thus only allowing users to identify imminent obstacles at ground level. Guide dogs, while proficient in assisting navigation, are burdened by the extensive time and significant financial investment required for their training. Moreover, the task of caring for these animals can pose considerable difficulties for those with visual impairments.

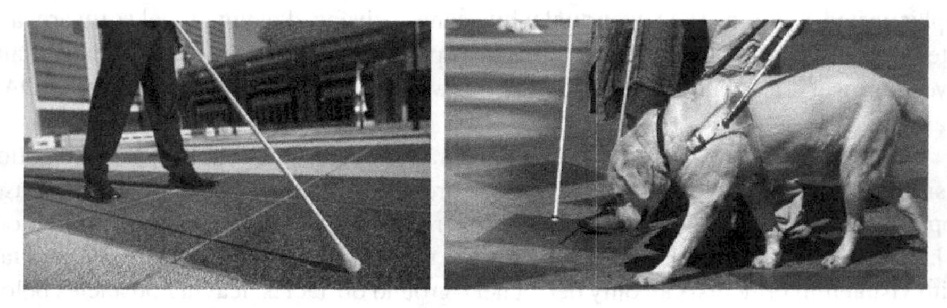

Fig. 1. Example of traditional support for blind people

Amy et al., in their systematic survey [4] examined 35 scholarly articles published between 2016 and February 2021 to categorize and depict the types of navigation aids used by blind, visually impaired, or deafblind individuals in dynamic indoor and outdoor

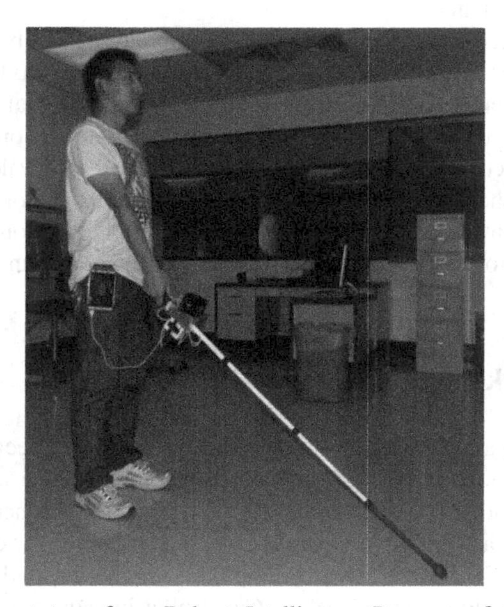

Fig. 2. Smart cane prototype from Robust Intelligence Program of the National Science Foundation

environments. The findings revealed that smart device technology was incorporated in 33 of the 35 studies that fulfilled the selection criteria. Most of these devices made use of low-energy Bluetooth beacons and other sensors, such as haptic sensors. The minimum price for a cane equipped with sensors is $200, while a complete assistive system can cost as much as $6000 [11].

Another comprehensive review [6] by Vahid et al., reported the most commonly used mobile navigation services. Dot Waker, Nearby Explorer, Get There, and Google Maps are the most commonly used navigation systems by visually impaired individuals. All of these are mobile applications addressing the specific needs of sight-free trips. It is supposed that the Talkback screen reader will be used with this application. Among the apps, Google Maps is capable of voice guidance [12].

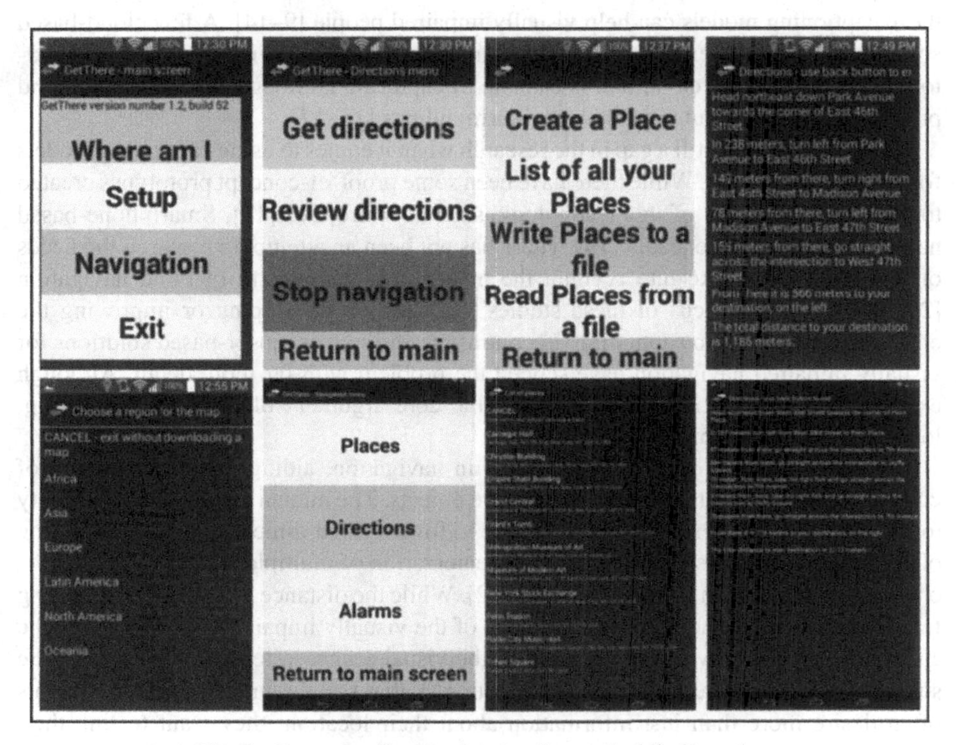

Fig. 3. Example of navigational software for blind people.

In addition to these technologies, there has also been research on navigation prototypes that are AI-based. For example, Zhang et al. conducted a bibliometric narrative review [3] on modern navigation aids for people with visual impairment. Another study by Șipoș et al. proposed a complete, portable, and affordable smart assistant [4] for helping visually impaired people to navigate indoors, and outdoors, and interact with the environment. These AI combined with sensor-based solutions while providing precise location measurements, come at a substantial cost, with prices starting from $300 [4].

Several image and video recognition models have also been developed in recent years, with outputs and capabilities for navigation. For example, a study by Rane et al. [5] proposed an implementation of smart spectacles based on Image Captioning and Optical Character Recognition (OCR) to ease the navigation process. Another study by Ahsan et al. proposed altering AoANet [6], a state-of-the-art image captioning model, to leverage the text detected in the image as an input feature.

Image or video captioning models, in general, have advanced in recent years with the emergence of transformer-based deep learning models. Transformer-based models have proven to be effective in a range of natural language processing tasks, including image and video captioning. By using attention mechanisms to concentrate on the most relevant parts of the input image or video, these models have been able to generate high-quality captions [7, 8]. There has been a multitude of researchers suggesting that these captioning models can help visually impaired people [9–14]. A few cloud-based solutions have also been observed in the literature that combined image-to-text and text-to-speech using AWS cloud that is capable of capturing an image of indoor space and providing speech output explaining the surroundings [15–17].

However, there is still a gap in the research when it comes to using captioning models for assisted navigation. While there have been some proof-of-concept prototypes created to show the use cases of this technology, such as Third-Eye [15], Smartphone-based navigation [17, 18], Concadia [19]. There has not been an attempt to research the needs of blind people to take into account the nuance and complexity of blind navigation [20, 21]. The core focus of these studies has been on developing or improving the core technique of video captioning. In contrast, traditional or sensor-based solutions for visually impaired navigation have scoped the problem with far more detail. Although expensive, these solutions have followed the core argument of navigation: tracking, localization, and planning.

Spatial awareness plays a crucial role in navigation, aiding in the prevention of collisions and the identification of proximate objects. The human visual system's ability to interpret varying perspectives of a scene to form a three-dimensional understanding of the environment has motivated the implementation of multiple cameras to construct or identify the world in three dimensions [22]. While the distance information is missing from the captioning model for the purpose of the visually impaired assistive tools, the captioning model provides the context of the visual scene, in terms of features of the surroundings. It has been reported in the surveys that people with visual impairments often desire more than just information about their location; they want to link their current position to the features of their surrounding environment [15, 23, 24].

Keeping the definition of navigation in mind and the need for contextual and distance information, it can be argued that current approaches built for indoor navigation using captioning systems are not yet ready for navigational use for the visually impaired, because crucial elements for navigation, such as distance of certain hazardous items or optimal safest path are not part of the captioning models. Therefore, iteration is required to incorporate depth and utilization of contextual information into the traditional captioning model for visually impaired assistive tools.

3 Proposed Model

In this study, a novel method is introduced with the objective of amplifying the effectiveness of image captioning to reinforce scene comprehension for visually impaired individuals. This is achieved by integrating depth information with RGB inputs.

The process will commence with the pivotal stage of data collection and preprocessing. An expansive dataset of RGB-D images will be compiled, each of which will be complemented by a corresponding depth map. The subsequent preprocessing will ensure the alignment and normalization of RGB and depth images, laying the foundation for the effective consolidation of these distinct visual data sources.

A transformer-based neural network will form the core of the model architecture, designed to simultaneously process both the RGB and depth images. The transformer architecture has been chosen for this research due to its ability to handle long-range dependencies with ease and its state-of-the-art performance in sequence-to-sequence tasks such as caption generation [25]. The transformer architecture has shown state-of-the-art performance in sequence-to-sequence tasks such as caption generation [26] that is the core component of this research.

The proposed dual-branch architecture will process the RGB image and the depth image in parallel. Once processed independently, the outputs from these branches will be amalgamated through a fusion layer. The fused output will then be funnelled into a caption generation module to produce the descriptive text, encapsulating the visual content of the images.

Post-architecture development, rigorous training of the model will be undertaken using optimal loss function, such as cross-entropy loss. This training will fine-tune the caption generation process, with an aim to optimize the accuracy and relevance of image descriptions.

An evaluation phase will follow to assess the performance of the model on a reserved test set, using established captioning metrics such as BLEU, METEOR, ROUGE, and CIDEr. This evaluation will provide insight into the model's capability to generate precise and comprehensive captions.

Ultimately, the trained and refined model will be integrated with existing assistive tools intended for visually impaired individuals. This fusion is expected to significantly enrich scene comprehension by delivering more comprehensive captions, thereby providing a detailed understanding of the surroundings.

In offering elaborate environmental descriptions, this proposed model will potentially improve the navigation capabilities of visually impaired individuals. It will foster a greater degree of interaction with and understanding of their environment, opening doors for more independent and efficient navigation.

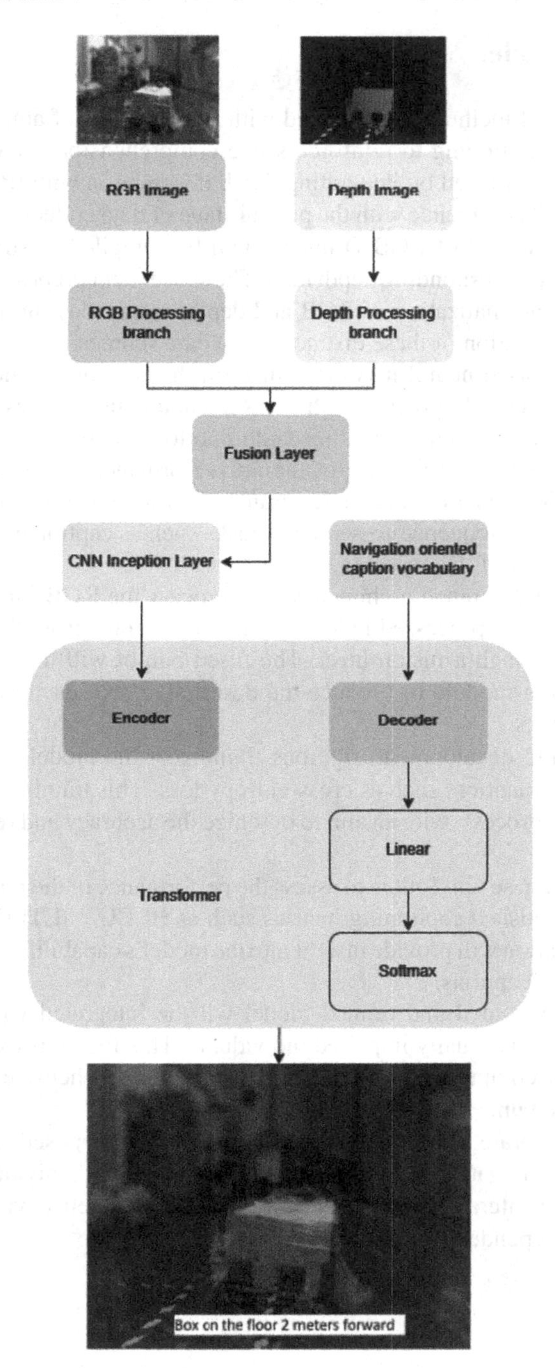

Fig. 4. Proposed Depth-Aware Image Captioning model using Transformer architecture.

4 Discussion

In the literature review section of this study, an examination of existing techniques and tools for aiding navigation in individuals with visual impairments is conducted. These methods span from traditional aids such as guide dogs and white canes to advanced technologies, including AI-infused navigation prototypes and image or video recognition models. Despite substantial strides in developing assistive tools and technologies, their widespread adoption and user acceptance remain deficient. A notable issue is that these advanced solutions often come with a high price tag, making them inaccessible for many. Furthermore, user acceptance of these tools is relatively low due to their limited functional scope and various inherent limitations, such as the inability to describe related contexts or features of any scene.

The model proposed in this study aims to mitigate some of these constraints by fusing depth information with RGB images to enrich image captioning, thus improving scene comprehension for visually impaired individuals. By generating more detailed and informative depictions of surroundings, the model aims to enhance the navigational challenge of individuals with visual impairments.

However, developing and implementing the proposed model is not without its potential hurdles. One such challenge is the collection and preprocessing of an adequate dataset of RGB-D images, a process that requires meticulous alignment and normalization of RGB and depth images. Another obstacle lies in creating an efficient neural network architecture capable of processing both image types, amalgamating their outputs via a fusion layer, and channelling them into a caption generation module. It becomes evident that additional research efforts are required to navigate these obstacles and further refine and optimize the proposed model.

5 Conclusion

In this study, we present a new approach to enhance image captioning for the visually impaired by combining depth data with RGB images. The model proposed here includes various stages such as data collection and preprocessing, the design of the model's architecture, the training phase, performance evaluation, and its fusion with existing assistive tools.

With the provision of more comprehensive and practical depictions of their environment, this method is expected to incrementally improve how visually impaired individuals navigate. Nevertheless, further refinement and optimization of the model are required, and its practical efficiency in real-world scenarios needs to be assessed. The ongoing enhancement of this model could potentially result in a significant breakthrough in assistive technology for the visually impaired.

References

1. Burns, E., Kakara, R.: Morbidity and mortality weekly report deaths from falls among persons aged ≥65 years—United States, 2007–2016 (2007)
2. Moreland, B., Kakara, R., Henry, A.: MMWR, trends in nonfatal falls and fall-related injuries among adults aged ≥65 years—United States, 2012–2018 (2020)
3. Zhang, X., Yao, X., Hui, L., Song, F., Hu, F.: A bibliometric narrative review on modern navigation aids for people with visual impairment. Sustainability **13**, 8795 (2021). https://doi.org/10.3390/SU13168795
4. Şipoş, E., Ciuciu, C., Ivanciu, L.: Sensor-based prototype of a smart assistant for visually impaired people—preliminary results. Sensors **22**, 4271 (2022). https://doi.org/10.3390/S22114271
5. Rane, C., Lashkare, A., Karande, A., Rao, Y.S.: Image captioning based smart navigation system for visually impaired. In: Proceedings - International Conference on Communication, Information and Computing Technology, ICCICT 2021 (2021). https://doi.org/10.1109/ICCICT50803.2021.9510102
6. Ahsan, H., Bhalla, N., Bhatt, D., Shah, K.: Multi-modal image captioning for the visually impaired. In: NAACL-HLT 2021 - 2021 Conference of the North American Chapter of the Association for Computational Linguistics: Human Language Technologies, Proceedings of the Student Research Workshop, pp. 53–60 (2021). https://doi.org/10.18653/v1/2021.naacl-srw.8
7. Cornia, M., Baraldi, L., Cucchiara, R.: Explaining transformer-based image captioning models: an empirical analysis. AI Commun. **35**, 111–129 (2022). https://doi.org/10.3233/AIC-210172
8. Wang, Y., Xu, J., Sun, Y.: End-to-end transformer based model for image captioning. In: Proceedings of the 36th AAAI Conference on Artificial Intelligence, AAAI 2022, vol. 36, pp. 2585–2594 (2022). https://doi.org/10.1609/aaai.v36i3.20160
9. Sharma, H., Agrahari, M., Singh, S.K., Firoj, M., Mishra, R.K.: Image captioning: a comprehensive survey. In: 2020 International Conference on Power Electronics and IoT Applications in Renewable Energy and its Control, PARC 2020, pp. 325–328 (2020). https://doi.org/10.1109/PARC49193.2020.236619
10. Ayoub, S., Gulzar, Y., Reegu, F.A., Turaev, S.: Generating image captions using Bahdanau attention mechanism and transfer learning. Symmetry **14**, 2681 (2022). https://doi.org/10.3390/SYM14122681
11. Ghandi, T., Pourreza, H., Mahyar, H.: Deep learning approaches on image captioning: a review (2022)
12. Jain, V., Al-Turjman, F., Chaudhary, G., Nayar, D., Gupta, V., Kumar, A.: Video captioning: a review of theory, techniques and practices. Multimed. Tools Appl. (2022). https://doi.org/10.1007/s11042-021-11878-w
13. Verma, V., Saritha, S.K., Jain, S.: Automatic image caption generation using ResNet & torch vision, pp. 82–101 (2022). https://doi.org/10.1007/978-3-031-24367-7_7
14. Moctezuma, D., Ramírez-delReal, T., Ruiz, G., González-Chávez, O.: Video captioning: a comparative review of where we are and which could be the route. Comput. Vis. Image Underst. **231**, 103671 (2023)
15. Guravaiah, K., Bhavadeesh, Y.S., Shwejan, P., Vardhan, A.H., Lavanya, S.: Third eye: object recognition and speech generation for visually impaired. Procedia Comput. Sci. **218**, 1144–1155 (2023). https://doi.org/10.1016/j.procs.2023.01.093
16. Ahsan, H., Bhalla, N., Bhatt, D., Shah, K.: Multi-modal image captioning for the visually impaired (2021)

17. Makav, B., Kılıç, V.: Smartphone-based image captioning for visually and hearing impaired (2019)
18. Zaib, S., Khusro, S., Ali, S., Alam, F.: Smartphone based indoor navigation for blind persons using user profile and simplified building information model. In: 1st International Conference on Electrical, Communication and Computer Engineering, ICECCE 2019 (2019). https://doi.org/10.1109/ICECCE47252.2019.8940799
19. Kreiss, E., Fang, F., Goodman, N.D., Potts, C.: Concadia: towards image-based text generation with a purpose (2021)
20. Plikynas, D., Zvironas, A., Gudauskis, M., Budrionis, A., Daniusis, P., Sliesoraityte, I.: Research advances of indoor navigation for blind people: a brief review of technological instrumentation. IEEE Instrum. Meas. Mag. **23**, 22–32 (2020). https://doi.org/10.1109/MIM.2020.9126068
21. Kandalan, R.N., Namuduri, K.: Techniques for constructing indoor navigation systems for the visually impaired: a review. IEEE Trans. Hum. Mach. Syst. **50**, 492–506 (2020). https://doi.org/10.1109/THMS.2020.3016051
22. Jirawimut, R., Prakoonwit, S., Cecelja, F., Balachandran, W.: Visual odometer for pedestrian navigation. IEEE Trans. Instrum. Meas. **52**, 1166–1173 (2003). https://doi.org/10.1109/TIM.2003.815996
23. Fernandes, H., Costa, P., Filipe, V., Paredes, H., Barroso, J.: A review of assistive spatial orientation and navigation technologies for the visually impaired. Univer. Access Inf. Soc. **18**, 155–168 (2019). https://doi.org/10.1007/S10209-017-0570-8/FIGURES/2
24. Nair, V., Olmschenk, G., Seiple, W.H., Zhu, Z.: ASSIST: evaluating the usability and performance of an indoor navigation assistant for blind and visually impaired people. Assist Technol. **34**, 289–299 (2020). https://doi.org/10.1080/10400435.2020.1809553
25. Moro, G., Ragazzi, L., Valgimigli, L., Frisoni, G., Sartori, C., Marfia, G.: Efficient memory-enhanced transformer for long-document summarization in low-resource regimes. Sensors **23**, 3542 (2023). https://doi.org/10.3390/S23073542
26. Rohde, T., Wu, X., Liu, Y.: Hierarchical learning for generation with long source sequences (2021)

Smart Agriculture: Transforming Agriculture with Technology

Pattharaporn Thongnim[1,2](✉), Vasin Yuvanatemiya[1,3](✉), and Phaitoon Srinil[1,2]

[1] Burapha University, Chanthaburi, Thailand
[2] Data Center,Burapha University, Chanthaburi, Thailand
pattharaporn@buu.ac.th
[3] Unmanned Aircraft Systems Training Center,Burapha University, Chanthaburi, Thailand

Abstract. The widespread adoption of digital technologies has brought momentous changes to all economic sectors, and the agriculture sector is no exception. As one of the oldest and most vital professions, agriculture and farming in Chanthaburi, the East of Thailand are also being transformed by the digital revolution. The digital revolution in agriculture, often referred to as "smart farming" or "smart agriculture" involves the integration of various digital tools and technologies into traditional farming practices. These technologies are aimed at enhancing productivity, efficiency, and sustainability in agricultural operations. This study conducts comprehensive various technologies proposed for the agriculture sector. Some of the technologies that could have been included in the study are: Data Center in Chanthaburi (DCC), Internet of Things (IoT), Unmanned Aerial Vehicles (UAVs)) and Smart Agriculture Management System (SAMs). The study also aims to showcase the potential of integrating DCC, and SAMs in agriculture and how these technologies can revolutionize farming practices. The findings could have significant implications for the agricultural industry, encouraging the adoption of these technologies to optimize agricultural processes, increase productivity, and contribute to sustainable farming practices.

Keywords: Smart Agriculture · Internet of Things · Big Data

1 Introduction

According to the Food and Agriculture Organization (FAO), it has been projected that the global population would reach 9.73 billion by the year 2050, with a further growth expected to reach 11.2 billion by the year 2100. Food scarcity and population growth are the greatest obstacles to global sustainable development. Advanced technologies such as artificial intelligence (AI), the Internet of Things (IoT), UAVs and mobile internet can offer practical solutions to the

Supported by Burapha University Chanthaburi Campus.

© The Author(s), under exclusive license to Springer Nature Singapore Pte Ltd. 2024
F. Hassan et al. (Eds.): AsiaSim 2023, CCIS 1911, pp. 362–376, 2024.
https://doi.org/10.1007/978-981-99-7240-1_29

global problems [1]. The potential of smart agriculture lies in its utilization of AI, IoT, UAVs and cyber-physical systems in farm management. This integration of advanced technologies can lead to transformative outcomes in agricultural practices, enabling farmers to optimize their operations for enhanced productivity and sustainability. Moreover, smart agriculture tackles numerous crop production challenges through the monitoring of climate factors, soil characteristics, and soil moisture. IoT technology plays a vital role by seamlessly linking various remote sensors, including robots, ground sensors, and drones [2]. These interconnected devices operate automatically, facilitating efficient and real-time data collection and analysis [3].

Over the past few years, farming has witnessed several technical revolutions, leading to increased industrialization and reliance on technology. With the adoption of intelligent agricultural technologies, farmers now have enhanced control over crop cultivation, resulting in greater predictability and efficiency. The integration of smart farming practices has been driven by the growing consumer demand for farm products, contributing to the global proliferation of these advanced technologies in agriculture [4]. The drive is built upon a diverse range of digital technologies, encompassing Big Data, and digital behaviors like collaboration, mobility, and open innovation [5]. These technological components work together to create a dynamic and innovative approach, fostering the effectiveness in achieving its goals. As a result, the data-driven approach empowers farmers to optimize their agricultural practices, leading to improved overall productivity while ensuring sustainability and environmental stewardship.

Farmers can completely use relevant data sources to extract significant insights, regularities, patterns, and knowledge from accumulated data in order to create qualitative goods, enhance revenues, and make well-informed judgments. Big data is important in smart farming because it allows us to properly exploit this abundance of information. Agricultural practices are currently utilizing comprehensive data analysis tools to intelligently and cost-effectively utilize smart farming data. Therefore, the adaptability of big data in smart farming analytics is demonstrated by a variety of common smart-farming applications [6]. These examples show how big data may help farmers acquire useful insights and improve their farming methods.

In addition, big data encompasses vast amounts of information characterized by its high volume, generated at high velocity, and exist in a variety of formats and types [7]. It is difficult to handle and analyze agricultural data using traditional procedures because the sheer volume of data exceeds the capabilities of standard data processing technologies. Specialized technology and cutting-edge analytical techniques like data mining, machine learning, and artificial intelligence are used to extract valuable insights and value from farming data. With the help of these technologies, businesses and industries can turn raw data into useful knowledge, improve processes, and gain a competitive edge in the current data-driven environment [8]. Despite the extensive number of data mining-related studies published, there is a notable scarcity of literature reviews specifically focusing on smart farming. Hence, it is imperative to address the significance of big data in the field of agriculture.

Thus, the primary focus of this study is on the implementation and utilization of smart agriculture techniques. In current smart farming scenarios, academics are progressively directing their attention towards comprehending and executing agricultural data-oriented methodologies in conjunction with data center in Chanthaburi, Thailand. The provided material possesses significant worth for practitioners seeking to include and use big data within their smart farming solutions with farming application. Futhermore, the primary aim of this study is to address the existing gap and provide a comprehensive examination of modern technology and applications that focus on big data in the field of smart farming such as IoT and drone. The contributions of this study are as follows:

1. Filling the Knowledge Gap: By providing a detailed review of agricultural data techniques in smart farming, the study addresses the existing lack of comprehensive information in this area.
2. Insights for Practitioners: The study equips practitioners with valuable insights and knowledge, enabling them to make informed decisions when implementing big data technologies in their smart farming endeavors.
3. State-of-the-Art Analysis: Through a thorough examination of current practices and applications, the study presents the latest advancements and trends in big-data-focused smart farming data analysis.

Therefore, this study serves as a valuable resource for those interested in the intersection of data and smart farming, offering a comprehensive overview of the techniques and applications that drive innovation and efficiency in modern agriculture. The research questions considered as the main criteria for the selection of the research are as follows: What are the various types of data generated by smart agriculture? What are the favored agricultural data applications in smart agriculture? What are the techniques used for smart agriculture big data analysis? By investigating these research questions, the study aims to provide comprehensive insights into the use of big data in Data Center in Chanthaburi (DCC) and Smart Agriculture Management System (SAMs) in Chanthaburi, Thailand, contributing to a better understanding of how data-driven technologies can transform modern agriculture and improve overall agricultural management.

2 Methodology

In the methodology employed for the agricultural data analysis and smart farming practices, the initial step is to gain a comprehensive understanding of the data that will be utilized in the process. This involves exploring the nature of the data, its sources, and the technologies that can be leveraged for effective data management and analysis.

2.1 Types of Data and Data Resources

The objective of this inquiry is to investigate the different data resources and types of data generated within smart farming practices. The focus is on identifying various categories of data collected, encompassing:

Environmental Data: This category includes information related to climate conditions, soil properties, and water characteristics [9]. Data on temperature, humidity, rainfall, soil nutrients, and water availability are some examples that fall under this category [10]. Environmental data is crucial for making informed decisions about irrigation, fertilization, and pest control.

Sensor Data: Derived from IoT devices, robotics, satellite and drones, sensor data provides real-time information on various aspects of the farm [11]. This data can include measurements of soil moisture, temperature, crop growth, and atmospheric conditions. IoT sensors play a significant role in monitoring and managing the farm environment efficiently.

Agricultural Data: This category involves data concerning both crop and livestock aspects of the farm. It includes information on plant growth, crop health, soil characteristics, as well as data related to animal well-being, weight, activity levels, and feeding patterns [12]. Monitoring agricultural data helps optimize crop management practices, detect early signs of crop diseases, ensure proper nutrient management, and improve overall animal husbandry practices. Integrating and analyzing both crop and livestock data enables farmers to make well-informed decisions, enhance productivity, and achieve sustainable and efficient agricultural practices.

Crop Data: This category involves data specifically related to the cultivation and management of crops on the farm. It includes information on plant growth, crop health, soil characteristics, weather conditions, irrigation schedules, fertilization practices, and pest control measures. Monitoring crop data helps farmers optimize planting schedules, assess soil fertility, detect and address crop diseases and pests, and make data-driven decisions to enhance crop yields and overall agricultural productivity [13]. Analyzing and utilizing crop data play a crucial role in implementing precision agriculture practices, conserving resources, and achieving sustainable crop production.

Other Relevant Information Streams: Smart farming may involve the collection of additional data from diverse sources. This can include data from market trends, supply chain logistics, financial records, and other relevant data streams that impact farm operations and decision-making [14].

Indeed, as the volume of data in agriculture continues to grow, proper categorization becomes essential to manage and make sense of the vast amounts of information generated [15]. Categorization helps organize the data in a structured manner, making it easier to access, analyze, and derive meaningful insights.

2.2 Network Organization

The concept of the network organization pertains to how stakeholders within the agricultural ecosystem interact and collaborate to achieve data process objectives effectively. The behavior of stakeholders within the network organization can significantly impact the success of data center processes and the overall outcomes of data-driven agriculture. Understanding and influencing this behavior are crucial for optimizing data center operations and achieving the objectives of data-driven initiatives in agriculture [16]. In this context, stakeholders can include farmers, researchers, data center operators, technology providers, policymakers, and other relevant entities involved in the collection, management, and analysis of agricultural data. The emergence of Big Data and Smart Farming has led to significant technical changes in the agricultural sector, prompting a need to comprehend the stakeholder network surrounding the farm [17].

The landscape of agriculture is evolving, with stakeholders from diverse backgrounds collaborating to leverage data-driven approaches for enhanced efficiency, sustainability, and productivity in agriculture. Indeed, open data sets in the agricultural domain are typically owned and managed by government institutes responsible for collecting and generating the data. In addition, government organizations and corporations, particularly those in the technology and data sectors, have acknowledged the significant opportunities presented by agricultural data center. Consequently, they are making substantial investments in the exploration, advancement, and application of big data technologies within this field. These organizations possess significant resources and specialized knowledge, which empowers them to create advanced data analytics platforms, Internet of Things (IoT) solutions [18], and cloud-based services specifically designed for the agriculture industry [19].

2.3 Data Center

The surge in demands for data processing, data storage, and digital telecommunications has resulted in a significant expansion of the data center industry [20]. Data centers play a crucial role in modern information technology (IT) infrastructure, serving as specialized facilities designed to house and operate IT equipment used for data processing, storage, and communication networking [21]. As depicted in Fig. 1, the data process commences with identifying the sources from which valuable data is extracted [22]. Subsequently, the data is stored in a suitable data model based on its structure - either structured or unstructured (primary data, secondary data, real time data). The next step involves classifying and filtering the data, depending on the specific type of analysis required [23]. The method of processing is then determined, whether it be data cleaning, data transformation, data integration, or data aggregation. Once the data is classified, appropriate tools are employed for analysis. These tools encompass machine learning (ML), regression, deep learning and other data science techniques. The insights obtained from the analysis are then presented using visualization tools. Lastly, precision agriculture involves the application of various technologies such as crop management software and Irrigation systems.

In addition, the ability of data center networks is determined by the effective communication between devices and the responses from data center networks. Data centers serve as critical industrial infrastructure for dynamic computing and storage needs. The Data Center network is tasked with managing an enormous number of elements within the network. This robust infrastructure enables the storage and processing of large amounts of data in a highly efficient and reliable manner. Furthermore, cloud computing facilitates data accessibility and collaboration. Stakeholders in different locations can access, analyze, and share agricultural data seamlessly through cloud-based platforms [24]. This fosters collaboration between farmers, researchers, and other industry players, leading to innovative solutions and improved agricultural practices.

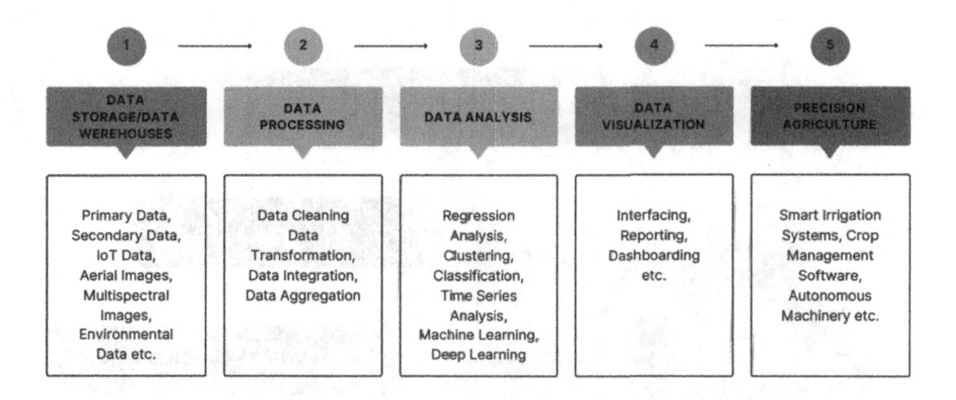

Fig. 1. The framework architecture of data center.

2.4 IoT in Smart Agriculture

In Fig. 2 shows that smart Agricultural Applications concentrate on conducting a thorough examination of current agricultural applications, encompassing a wide range of aspects, including irrigation management, soil quality assessment, weather forecasting, price prediction, and monitoring plant health. For examples, smart irrigation systems are designed to optimize water usage in agriculture by providing the right amount of water to crops at the right time [25]. These systems use soil moisture monitoring devices to measure the moisture content in the soil. Based on the readings, the irrigation system can automatically adjust the irrigation schedule, ensuring that crops receive the appropriate amount of water without overwatering or underwatering [26]. The data collected from smart agriculture applications can be shared with a data center, creating a symbiotic relationship between smart agriculture and data centers. The integration of data centers in smart agriculture allows for centralized storage, management, and analysis of the vast amount of data generated by various IoT devices, sensors, and other smart farming technologies. For fertilization, the data center takes into account factors like soil nutrient levels and crop growth stage. Based on this information, it determines the appropriate amount of fertilizer and other supplements required to support healthy plant growth. The data center then instructs the system to mix and distribute the specified amount of fertilizers to the crops [27].

Therefore, the collaboration between smart agriculture and data centers results in improved agricultural practices, increased productivity, resource efficiency, and enhanced sustainability. By harnessing the power of data analytics and advanced technologies, farmers can make informed decisions to optimize their operations and contribute to the transformation of the agricultural industry.

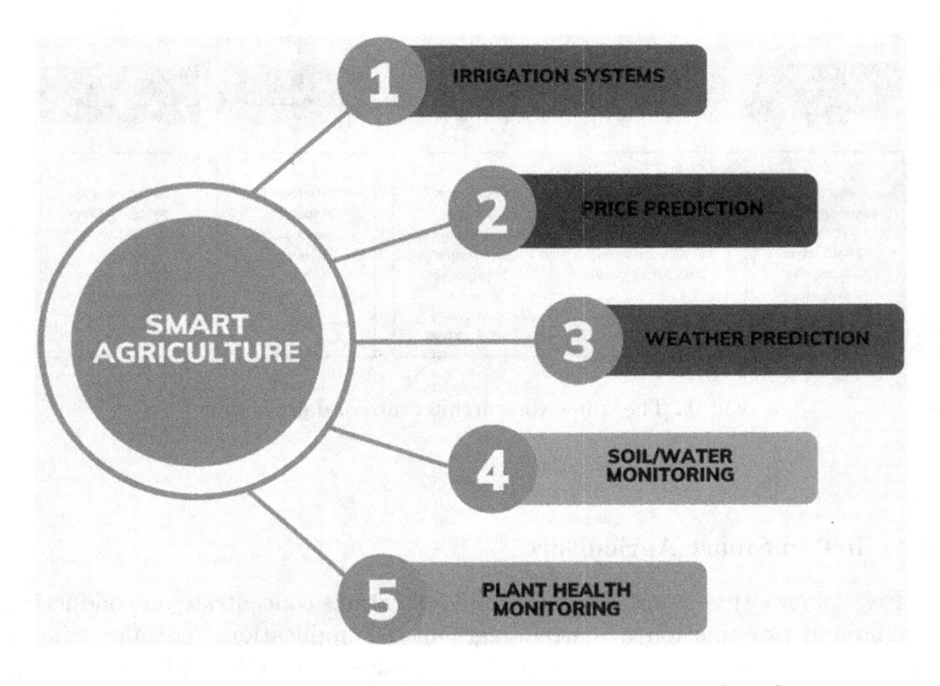

Fig. 2. The implementation of IoT technology in the agricultural sector.

2.5 Machine Learning Models

In the field of smart agriculture, machine learning (ML) algorithms have emerged as powerful tools for analyzing the vast amounts of agricultural data generated today. These techniques have been widely applied to analyze various types of smart farming data, including environmental data, sensor data, and crop data, among others. For example, linear regression is a fundamental ML technique used for predicting numerical values based on input features. In smart agriculture, it can be applied to analyze climate data and predict crop yields, helping farmers make informed decisions about irrigation, fertilization, and pest control [28]. Similarly, time series analysis techniques are used to analyze data that changes over time. In smart agriculture, time series analysis can be employed to study weather patterns, monitor crop growth, and predict market trends. Artificial Neural Networks (ANNs) and their variants have indeed become one of the most widely utilized techniques in the field of smart agriculture. In smart agriculture, ANNs have been applied to various applications, and one of the significant areas of use is yield prediction [29]. Yield prediction models based on ANNs take into account various factors such as climate data, soil characteristics, crop variety, and management practices [30]. They are also used for image recognition in precision agriculture, enabling identification of pests, diseases, and weed species from images captured by drones or cameras. The machine learning algorithm and applications are listed in Table 1.

Table 1. Machine learning models and applications in smart agriculture.

Machine learning Algorithm	Application	Reference
Linear Regression	Minimization of fertilizer and water	[28]
Linear Regression	Prediction of yield	[31]
Gaussian	Detection of leaves	[32]
ANN	Prediction of yield	[29]
ANN	Optimization of water	[30]
Random Forest	Prediction of yield	[33]
Support vector machines	Detection of fruit	[34]

3 Application of Smart Agriculture

Chanthaburi, being one of the agricultural provinces in the east, is an excellent starting point for the smart agriculture, considering its significance in the agricultural sector. By starting with Chanthaburi and gradually expanding to cover other provinces, smart agriculture will play a crucial role in transforming the agricultural landscape in the region. It will empower farmers with data-driven tools and knowledge, fostering a culture of continuous learning and improvement in the agricultural community.

3.1 Data Center in Chanthaburi

Having a dedicated agricultural data center in Chanthaburi, Thailand, in the east of the country, will reflect a commitment to harnessing the power of data and technology to drive agricultural advancements, sustainability, and productivity. It will serve as a vital resource for the entire agricultural community, helping them adapt to the challenges of modern farming and contribute to the growth and development of the region's agricultural sector.

The data center will also foster collaboration and knowledge-sharing among different stakeholders within the agricultural ecosystem. This data center will be a valuable repository of time series data, particularly secondary data, obtained from various government organizations. This data will encompass a wide range of agricultural-related information, including crop yield data, labor statistics, economic indicators, weather data, and more.

As the data center evolves and expands its scope to cover more provinces in the east of Thailand, it can become a regional hub for agricultural information and innovation. Farmers across the region will have access to valuable insights and historical data, aiding them in adapting to the challenges posed by modern farming and making informed choices for their agricultural operation.

After classifying the data, a variety of appropriate tools are employed for analysis, such as data visualization, machine learning, and other data science techniques. The data can be exported as a csv file, enabling users to select and visualize the information. The results and insights derived from this analysis are then presented through visualization tools, as depicted in Fig. 3, facilitating users' comprehension and interpretation of the data effectively.

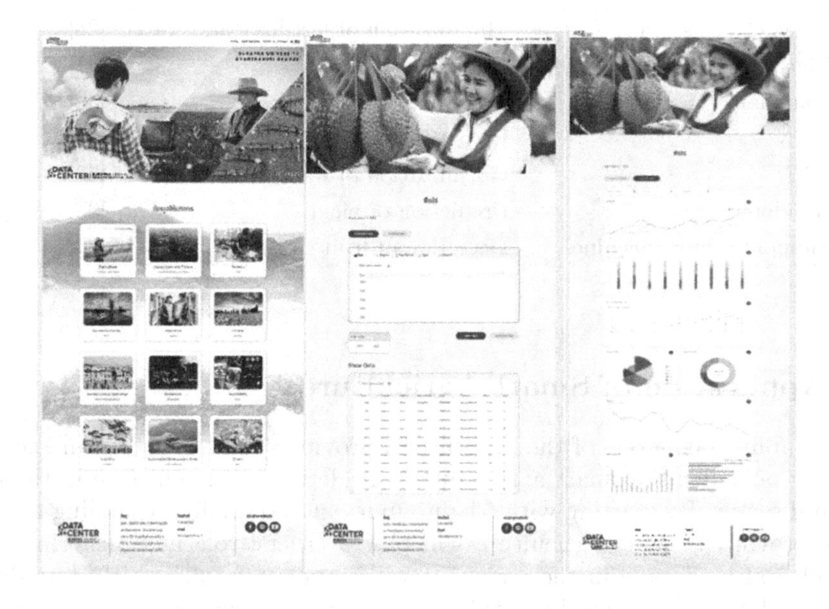

Fig. 3. Functions of data center in Chanthaburi.

For primary data, IoT will continuously capture and store the data in real-time. IoT devices and sensors are equipped to gather data directly from the source, such as agricultural fields, livestock, weather conditions, and other relevant aspects. The data collected by IoT devices is transmitted in real-time to a centralized system or data center, where it is processed, analyzed, and made available for further use. Moreover, the data collected by drones is then stored in the data center for future analysis. As the data center accumulates a significant amount of image data over time, it becomes a valuable repository for assessing crop yield, monitoring changes in vegetation health, and conducting comprehensive analyses to improve agricultural practices. By combining the real-time data collected by IoT devices with image data from drones, the data center can provide a comprehensive view of the farm's performance. This integrated approach to data collection and analysis empowers farmers to make informed decisions, optimize resource allocation, and enhance overall crop yield and farm productivity. Additionally, it facilitates research and innovation in the agriculture sector, leading to the development of more efficient and sustainable farming practices.

3.2 Smart Agriculture Management System (SAMs)

SAMs, which stands for "Smart Agricultural Management System" is a blockchain-based solution designed to enhance traceability, transparency, and security in the agricultural supply chain [35]. By integrating blockchain technology into the data center, SAMs can significantly improve the management and sharing of agricultural data. Figure 4, the context of SAMs application demonstrates that the traceability of durian refers to the capability of tracking the journey of durian fruits from their point of origin (e.g., the farm where they were grown) through various stages in the supply chain until they reach the end consumer or market [36]. Therefore, the traceability of durian through SAMs empowers stakeholders along the supply chain, including farmers, distributors, retailers, and consumers, to have greater confidence in the origin, quality, and

Fig. 4. Traceability of durian of SAMs application.

safety of durian products. This enhanced transparency and trust contribute to a more efficient and sustainable agricultural ecosystem.

SAMs will initially be implemented and tested with durian farms. How SAMs can enhance the traceability of durian? Information about the durian variety, harvest date, farming practices, and quality control measures can be documented on the blockchain. Consumers can access this data by scanning a QR code on the product packaging, gaining insight into the product's authenticity and quality. In addition, each durian's origin, handling, processing, and transportation details are securely recorded on the blockchain. By providing transparent information about the durian's journey from farm to table (Fig. 5), SAMs fosters consumer trust. Consumers can make informed decisions, supporting sustainable and ethical practices in the agriculture sector.

Fig. 5. SAMs application on mobile.

4 Discussion

The research questions formulated in the Introduction and provide answers to each of them based on the discussion presented: What are the various types of data generated by smart agriculture? The data center and SAMs will serve as a central repository for storing agricultural data collected from various organizations and stakeholders in the agricultural sector [37]. Initially, the focus of data storage and management will be on Chanthaburi, and subsequently, it will expand to cover other provinces in the east of Thailand. This approach ensures a systematic and scalable implementation of data storage and analysis capabilities to cater to the agricultural needs of the entire region Data center and SAMs will generate a wide variety of data [38], including environmental data (climate factors, soil characteristics), crop health data (disease, pests, nutrient levels), farm equipment data (operational status, fuel consumption), market and price data, satellite, drone and remote sensing data, energy consumption data, water usage data, farm operations data, and financial data. This diverse range of data enables precision agriculture, efficient resource utilization, and informed decision-making for optimal agricultural practices [39].

What are the favored agricultural data applications in smart agriculture? In smart agriculture, various agricultural data applications are favored for optimizing farm management, increasing productivity, and ensuring sustainability. The integration of a data center and Smart Agricultural Management System (SAMs) further enhances the capabilities of these applications. Some of the favored agricultural data applications in smart agriculture, empowered by the data center and SAMs, include: precision agriculture, crop monitoring, weather and climate prediction, market analysis and price prediction, and sustainable farming practices [40]. It highlights how data center-supported data applications play a pivotal role in promoting sustainable farming practices by enabling farmers to monitor and optimize resource usage. Specifically, IoT-based irrigation systems efficiently manage water usage, leading to reduced water wastage and minimizing environmental impacts [41].

What are the techniques used for smart agriculture big data analysis? Smart agriculture utilizes various techniques for big data analysis to derive meaningful insights and optimize agricultural practices such as machine learning, data visualization and internet of things (IoT) analytics. Together, data center and SAMS create a robust ecosystem that empowers smart agriculture big data analysis. By harnessing the power of advanced technologies, they drive agricultural advancements, optimize resource utilization, and contribute to sustainable and data-driven farming practices [42].

Therefore, the integration of a data center and SAMs application will plays a crucial role in facilitating the successful implementation and operation of this advanced agricultural technology. First of all, the data center acts as a central hub for managing, storing, and processing the vast amount of agricultural data collected. The data center enhances the efficiency, reliability, and effectiveness of the entire agricultural ecosystem. Secondly, the SAMs application complements the data center by providing specialized functionalities and insights tailored to the unique needs of smart agriculture. SAMs serves as an intelligent decision support system that leverages the data center's data resources to offer real-time monitoring, analysis, and recommendations for farm management. Thirdly, SAMs plays a vital role in tracking and ensuring traceability throughout the agricultural supply chain. By integrating data center capabilities with SAMs' tracking functionalities, smart agriculture can achieve enhanced traceability and transparency in various stages of the agricultural process.

5 Conclusion

In conclusion, the agriculture sector's importance in the global economy cannot be overstated, as it drives economic growth and employment opportunities, especially in rural regions. In developing countries like Thailand, agriculture holds a pivotal role in poverty reduction and economic advancement, contributing to the overall prosperity of the nation. The continuous development and enhancement of the agricultural sector in Thailand, fueled by technological advancements and data-driven approaches, further strengthen its potential for sustainable growth and improved livelihoods. Embracing smart agriculture practices, facilitated by the integration of data center and Smart Agricultural Management System (SAMs), empowers farmers with valuable insights, real-time monitoring, and efficient resource management.

The data center acts as a central hub for storing and processing vast amounts of agricultural data, while SAMs offers intelligent decision support and automation, enabling precise and sustainable farming practices. Together, they foster a more productive, resilient, and environmentally conscious agricultural ecosystem, addressing modern challenges and ensuring the well-being of farmers and communities. By making a plan for the future and putting it into action, Chanthaburi's agriculture industry will be able to reach its full potential and become a model of data-driven, sustainable, and resilient agriculture. Integrating new technologies, building more data centers, and using smart farming methods will help create a prosperous and environmentally friendly agricultural setting in the future.

Acknowledgements. We would like to express our heartfelt gratitude to the dedicated and hardworking teams at the Data Center and UAVs (Unmanned Aerial Vehicles) in Chanthaburi, Thailand. Their unwavering support and collaboration have been instrumental in the success of our endeavors in the field of smart agriculture.

References

1. Mohamed, E.S., Belal, A.A., Abd-Elmabod, S.K., El-Shirbeny, M.A., Gad, A., Zahran, M.B.: Smart farming for improving agricultural management The Egyptian. J. Rem. Sens. Space Sci. **24**(3), 971–981 (2021)
2. Adamides, G., et al.: Smart farming techniques for climate change adaptation in Cyprus. Atmosphere **11**(6), 557 (2020)
3. AlMetwally, S.A.H., Hassan, M.K., Mourad, M.H.: Real time internet of things (IoT) based water quality management system. Proc. CIRP **91**, 478–485 (2020)
4. Javaid, M., Haleem, A., Singh, R.P., Suman, R.: Enhancing smart farming through the applications of Agriculture 4.0 technologies. Int. J. Intell. Netw. **3**, 150–164 (2022)
5. De Alwis, S., Hou, Z., Zhang, Y., Na, M.H., Ofoghi, B., Sajjanhar, A.: A survey on smart farming data, applications and techniques. Comput. Ind. **138**, 103624 (2022)
6. Othman, M.M., Ishwarya, K.R., Ganesan, M.: A study on data analysis and electronic application for the growth of smart farming. Alinteri J. Agric. Sci. **36**(1), 209–218 (2021)
7. Wolfert, S., Ge, L., Verdouw, C., Bogaardt, M.J.: Big data in smart farming-a review. Agric. Syst. **153**, 69–80 (2017)
8. De Mauro, A., Greco, M., Grimaldi, M.: A formal definition of Big Data based on its essential features. Libr. Rev. **65**(3), 122–135 (2016)
9. Sun, J., Zhou, Z., Bu, Y., Zhuo, J., Chen, Y., Li, D.: Research and development for potted flowers automated grading system based on internet of things. J. Shenyang Agric. Univ. **44**(5), 687–691 (2013)

10. Li, P., Wang, J.: Research progress of intelligent management for greenhouse environment information. Nongye Jixie Xuebao= Trans. Chin. Soc. Agric. Mach. **45**(4), 236–243 (2014)
11. Faulkner, A., Cebul, K., McHenry, G.: Agriculture gets smart: the rise of data and robotics. Cleantech Agriculture Report (2014)
12. Cole, J.B., Newman, S., Foertter, F., Aguilar, I., Coffey, M.: Breeding and genetics symposium: really big data: processing and analysis of very large data sets. J. Anim. Sci. **90**(3), 723–733 (2012)
13. Bauer, J., Aschenbruck, N.: Design and implementation of an agricultural monitoring system for smart farming. In: 2018 IoT Vertical and Topical Summit on Agriculture-Tuscany (IOT Tuscany), pp. 1–6. IEEE (2018)
14. Miller, H.G., Mork, P.: From data to decisions: a value chain for big data. It Professional **15**(1), 57–59 (2013)
15. Hartmann, P.M., Zaki, M., Feldmann, N., Neely, A.: Capturing value from big data-a taxonomy of data-driven business models used by start-up firms. Int. J. Oper. Prod. Manag. **36**(10), 1382–1406 (2016)
16. Glaroudis, D., Iossifides, A., Chatzimisios, P.: Survey, comparison and research challenges of IoT application protocols for smart farming. Comput. Netw. **168**, 107037 (2020)
17. Carolan, M.: Acting like an algorithm: Digital farming platforms and the trajectories they (need not) lock-in. In: Desa, G., Jia, X. (eds.) Social Innovation and Sustainability Transition, pp. 107–119. Springer Nature Switzerland, Cham (2022). https://doi.org/10.1007/978-3-031-18560-1_8
18. Navarro, E., Costa, N., Pereira, A.: A systematic review of IoT solutions for smart farming. Sensors **20**(15), 4231 (2020)
19. Idoje, G., Dagiuklas, T., Iqbal, M.: Survey for smart farming technologies: challenges and issues. Comput. Electr. Eng. **92**, 107104 (2021)
20. Ren, J., Zhang, D., He, S., Zhang, Y., Li, T.: A survey on end-edge-cloud orchestrated network computing paradigms: transparent computing, mobile edge computing, fog computing, and cloudlet. ACM Comput. Surv. (CSUR) **52**(6), 1–36 (2019)
21. Huang, P., et al.: A review of data centers as prosumers in district energy systems: renewable energy integration and waste heat reuse for district heating. Appl. Energy **258**, 114109 (2020)
22. Santos, M.Y., et al.: A big data analytics architecture for industry 4.0. In: Rocha, Á., Correia, A.M., Adeli, H., Reis, L.P., Costanzo, S. (eds.) WorldCIST 2017. AISC, vol. 570, pp. 175–184. Springer, Cham (2017). https://doi.org/10.1007/978-3-319-56538-5_19
23. Cravero, A., Pardo, S., Sepúlveda, S., Muñoz, L.: Challenges to use machine learning in agricultural big data. Syst. Lit. Rev. Agron. **12**(3), 748 (2022)
24. Bo, Y., Wang, H.: The application of cloud computing and the internet of things in agriculture and forestry. In: 2011 International Joint Conference on Service Sciences, pp. 168–172. IEEE (2011)
25. Alex, N., Sobin, C.C., Ali, J.: A comprehensive study on smart agriculture applications in India. Wirel. Pers. Commun. **129**(4), 2345–2385 (2023)
26. Shock, C.C., Pereira, A.B., Feibert, E.B., Shock, C.A., Akın, A.İ., Ünlenen, L.A.: Field comparison of soil moisture sensing using neutron thermalization, frequency domain, tensiometer, and granular matrix sensor devices: relevance to precision irrigation (2016)

27. Dai, B., Xu, G., Huang, B., Qin, P., Xu, Y.: Enabling network innovation in data center networks with software defined networking: a survey. J. Netw. Comput. Appl. **94**, 33–49 (2017)
28. Muhammed, S.E., Marchant, B.P., Webster, R., Whitmore, A.P., Dailey, G., Milne, A.E.: Assessing sampling designs for determining fertilizer practice from yield data. Comput. Electron. Agric. **135**, 163–174 (2017)
29. Akbar, A., Kuanar, A., Patnaik, J., Mishra, A., Nayak, S.: Application of artificial neural network modeling for optimization and prediction of essential oil yield in turmeric (Curcuma longa L.). Comput. Electron. Agric. **148**, 160–178 (2018)
30. Muangprathub, J., Boonnam, N., Kajornkasirat, S., Lekbangpong, N., Wanichsombat, A., Nillaor, P.: IoT and agriculture data analysis for smart farm. Comput. Electron. Agric. **156**, 467–474 (2019)
31. Lobell, D.B.: The use of satellite data for crop yield gap analysis. Field Crop Res **143**, 56–64 (2013)
32. Motokura, K., Takahashi, M., Ewerton, M., Peters, J.: Plucking motions for tea harvesting robots using probabilistic movement primitives. IEEE Robot. Autom. Lett. **5**(2), 3275–3282 (2020)
33. Bocca, F.F., Rodrigues, L.H.A.: The effect of tuning, feature engineering, and feature selection in data mining applied to rainfed sugarcane yield modelling. Comput. Electron. Agric. **128**, 67–76 (2016)
34. Ramirez-Paredes, J.P., Hernandez-Belmonte, U.H.: Visual quality assessment of malting barley using color, shape and texture descriptors. Comput. Electron. Agric. **168**, 105110 (2020)
35. Zhang, Y., et al.: Blockchain: an emerging novel technology to upgrade the current fresh fruit supply chain. Trends Food Sci. Technol. **124**, 1–12 (2022)
36. Yang, X., Li, M., Yu, H., Wang, M., Xu, D., Sun, C.: A trusted blockchain-based traceability system for fruit and vegetable agricultural products, pp. 36282–36293. IEEE (2021)
37. Coble, K.H., Mishra, A.K., Ferrell, S., Griffin, T.: Big data in agriculture: a challenge for the future. Appl. Econ. Perspect. Policy **40**(1), 79–96 (2018)
38. Weersink, A., Fraser, E., Pannell, D., Duncan, E., Rotz, S.: Opportunities and challenges for big data in agricultural and environmental analysis. Ann. Rev. Resour. Econ. **10**, 19–37 (2018)
39. Weiss, M., Jacob, F., Duveiller, G.: Remote sensing for agricultural applications: a meta-review. Remote Sens. Environ. **236**, 111402 (2020)
40. Wu, B., et al.: Challenges and opportunities in remote sensing-based crop monitoring: a review. Natl. Sci. Rev. **10**(4), p.nwac290 (2023)
41. Martínez, R., Vela, N., El Aatik, A., Murray, E., Roche, P., Navarro, J.M.: On the use of an IoT integrated system for water quality monitoring and management in wastewater treatment plants. Water **12**(4), 1096 (2020)
42. Yousif, J.H., Abdalgader, K.: Experimental and mathematical models for real-time monitoring and auto watering using IoT architecture. Computers **11**(1), 7 (2022)

WebGIS Visualization of Infectious Disease Clustering with a Hybrid Sequential Approach

Elly Warni[1(✉)], Christoforus Yohanes[1], Zahir Zainuddin[1], Tyanita Puti Marindah Wardhani[1], Andi Rusmiati[1], and Muhammad Rizal H[2]

[1] Department of Informatics, Universitas Hasanuddin, Gowa, Indonesia
elly@unhas.ac.id

[2] Department of Informatics, Universitas Teknologi Akba Makassar, Makassar, Indonesia

Abstract. Infectious diseases are diseases that can be transmitted through various media. Based on the results of research that has been conducted at several Health Center in Gowa Regency, it is found that from 2020 to 2022, Health Center of Bontomarannu is one of the Health Center in Gowa Regency that handles many cases of infectious diseases, namely 212 cases of tuberculosis, 224 cases of dengue fever, and 427 cases of typhoid fever. This makes it difficult for the Health Center to classify what factors cause the increase in these diseases every day for one reason, namely the limitations of medical personnel to manually detect the indicators that cause the emergence of these diseases through medical record data. Based on data obtained through questionnaires given to patients of the Bontomarannu Health Center, there are several variables associated with infectious diseases. Therefore, this study aims to visualize the results of clustering infectious disease data (tuberculosis, dengue fever, and typhoid) per village in Gowa Regency in the form of WebGIS. The method used is K-means clustering which will be optimized using the Particle Swarm Optimization (PSO) algorithm to obtain better results. Moreover in the WebGIS visualization section, the frontend will be made using NextJS, the backend using Flask Py-thon, and for DBMS using SQLAlchemy. This WebGIS visualization will display cluster information for each village in Gowa Regency. Through quantitative analysis and clustering, the study aims to visualize the data and identify patterns and trends associated with infectious diseases in Gowa Regency, ultimately aiding in better decision-making and resource allocation for disease prevention and control.

Keywords: *K-means Clustering* · PSO · infectious diseases · WebGIS

1 Introduction

Infectious diseases are diseases that can be transmitted through various media. This type of disease is a major health problem in almost all developing countries because of its relatively high morbidity and mortality rates in a relatively short period of time [1]. Based on the results of research that has been conducted at several Health Center of in the district of Gowa, it was found that from 2020 to 2022, Health Center of Bontonompo II handled around 50 cases of tuberculosis, 20 cases of dengue, and 258 typhoid fever

© The Author(s), under exclusive license to Springer Nature Singapore Pte Ltd. 2024
F. Hassan et al. (Eds.): AsiaSim 2023, CCIS 1911, pp. 377–389, 2024.
https://doi.org/10.1007/978-981-99-7240-1_30

cases. Health Center of Gowa handled 212 cases of tuberculosis, 140 cases of dengue, and 377 cases of typhoid fever, while Health Center of Bontomarannu handled 212 cases of tuberculosis, 224 cases of dengue, and 427 cases of typhoid fever. From these data, it can be seen that the Bontomarannu Health Center is one of the health centers in Gowa Regency that handles many cases of infectious diseases such as tuberculosis, dengue, and typhoid fever. The high number of cases of these diseases makes it difficult for the health center to classify the factors that cause the increase in these diseases every day because one of the reasons is the limitation of medical personnel to manually detect what indicators can be used as a reference through patient information affected by these diseases. However, there is actually potential that can be obtained from data, namely how the data can be processed to support a case resolution [2]. Therefore, a visualization is needed related to the grouping of medical record data of infectious disease patients to minimize the spread of infectious diseases.

K-means is one of the popular clustering methods used to group data into groups based on the closest distance to the cluster center (centroid). However, K-means has a weakness, namely its sensitivity to the initial position or initialization of the centroid [3]. To overcome this problem, K-means optimization can be done using the Particle Swarm Optimization (PSO) algorithm. PSO is able to optimize centroid placement adaptively by searching in the search space based on particle performance [4]. With PSO, centroid initialization will be more efficient and accurate so as to avoid bad local solutions. The use of PSO in K-means helps to improve the accuracy and consistency of data clustering, thus obtaining better results in the analysis of infectious disease data such as tuberculosis, typhoid fever, and dengue hemorrhagic fever at Health Center of Bontomarannu in Gowa Regency.

Geographic Information System (GIS) is a website-based information system that can visualize information to users, namely recording, storing, writing, and analyzing geographic data or mapping of an area [5]. In the case of infectious diseases, data management will be carried out from the list of patients affected by tuberculosis, typhoid fever, and dengue hemorrhagic fever at the Bontomarannu Health Center. This research will optimize K-means clustering using Particle Swarm Optimization (PSO) algorithm. The clustering results will be displayed in the form of a WebGIS visualization that will display cluster information for each village in Gowa Regency.

Therefore, this research will develop research that has been done before by optimizing K-means clustering by optimizing the value of the cluster distance using the Particle Swarm Optimization (PSO) algorithm. The utilization of this clustering can be a solution in helping the analysis of infectious diseases, namely tuberculosis, dengue fever, and typhoid fever in the community based on the research variables. It is hoped that this research can provide information as a basis for taking the necessary actions for treatment, control, and prevention.

2 Related Work

Research related to K-means has been conducted by [6] who proposed the combination of K-means algorithm and genetic algorithm to improve the final result of data clustering with the results showing that the approach successfully reduces the impact of random initialization in K-means. Researcher [7] also tried to combine the K-means algorithm with another algorithm, namely the Particle Swarm Optimization (PSO) algorithm with results showing that the combination helped overcome convergence to a local optimum and improve the accuracy of the clustering results. In addition, researchers [8] also proposed a variation of the K-means algorithm to overcome its weaknesses by introducing a new initialization method called K-means++ that selects the initial cluster centers to reduce the possibility of getting stuck on a bad local solution. From the study, it was found that K-means++ significantly improved the performance of K-means on the dataset. Regarding WebGIS, researchers [9] conducted research related to visualizing the spread of the Covid-19 pandemic using WebGIS with the aim of presenting information related to the spread of the disease visually. The results of this visualization help the government in making decisions in pandemic prevention and also this research is useful for developing effective visualization systems in understanding and overcoming pandemic outbreaks that are very high in transmission.

3 Research Methodology

This research will build a WebGIS-based visualization system using NextJS on the frontend and Flask Python on the backend. This study will visualize the results of data clustering obtained from the results of optimizing the K-means algorithm using the Particle Swarm Optimization (PSO) algorithm. As shown in Fig. 1, the system design in this study from start to finish.

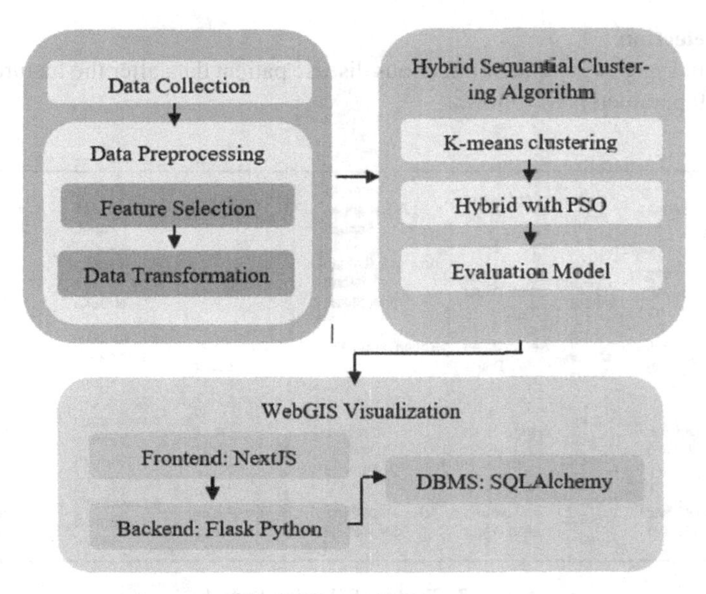

Fig. 1. Research Process

3.1 Data Collection

The first step is to input the dataset into the system. In this research, the user must input data with the Comma Separated Values File (.csv) extension. As shown in Fig. 2, the sample data collection with a total of 21 parameters.

	Umur	Alamat	Nafsu Makan	Kebiasaan Jajan	Kondisi Pembuangan Sampah	Ketersediaan Air Bersih	Sistem Ventilasi Rumah	Riwayat Pemakaian Jamban/Toilet	Diagnosa/Tes Darah
0	60.0	Maccini Baji (Gowa)	Nafsu Makan Normal	Jarang Jajan Diluar	Ada sampah dan jarang dibersihkan	Tersedia air bersih	Ada ventilasi tapi jarang terbuka	Sering Pakai Jamban/Toilet	tbc
1	29.0	Maccini Baji (Gowa)	Nafsu Makan Normal	Sering Jajan Diluar	Bersih tanpa ada sampah	Tersedia air bersih	Ada ventilasi tapi hanya sedikit	Sering Pakai Jamban/Toilet	tbc
2	18.0	Maccini Baji (Gowa)	Nafsu Makan Kurang	Sering Jajan Diluar	Ada lalat/nyamuk di sekitar tumpukan sampah	Tersedia air bersih	Ada ventilasi disemua ruangan	Sering Pakai Jamban/Toilet	dbd
3	9.0	Maccini Baji (Gowa)	Nafsu Makan Kurang	Jarang Jajan Diluar	Ada sampah tapi tidak banyak	Tersedia air bersih	Ada ventilasi tapi hanya sedikit	Jarang pakai jamban/toilet	tifoid

Fig. 2. Data Collection Sample

3.2 Data Preprocessing

Feature Selection

As shown in Fig. 3, a sample of infectious disease patient data after the feature selection stage into 9 parameters as follows.

Umur	Alamat	Nafsu Makan	Kebiasaan Jajan	Kondisi Pembuangan Sampah	Ketersediaan Air Bersih	Sistem Ventilasi Rumah	Riwayat Pemakaian Jamban/Toilet
60.0	Maccini Baji (Gowa)	Nafsu Makan Normal	Jarang Jajan Diluar	Ada sampah dan jarang dibersihkan	Tersedia air bersih	Ada ventilasi tapi jarang terbuka	Sering Pakai Jamban/Toilet
29.0	Maccini Baji (Gowa)	Nafsu Makan Normal	Sering Jajan Diluar	Bersih tanpa ada sampah	Tersedia air bersih	Ada ventilasi tapi hanya sedikit	Sering Pakai Jamban/Toilet
18.0	Maccini Baji (Gowa)	Nafsu Makan Kurang	Sering Jajan Diluar	Ada lalat/nyamuk di sekitar tumpukan sampah	Tersedia air bersih	Ada ventilasi disemua ruangan	Sering Pakai Jamban/Toilet
9.0	Maccini Baji (Gowa)	Nafsu Makan Kurang	Jarang Jajan Diluar	Ada sampah tapi tidak banyak	Tersedia air bersih	Ada ventilasi tapi hanya sedikit	Jarang pakai jamban/toilet

Fig. 3. Feature Selection Sample

Data Transformation

In the data transformation, a Multilabel Binarizer process is performed to convert the features to a form of data that can be accepted by the algorithm used. As shown in Fig. 4, displays sample data before and after Multilabel Binarizer as follows.

Nafsu Makan	Kebiasaan Jajan	Kondisi Pembuangan Sampah	Ketersediaan Air Bersih	Sistem Ventilasi Rumah	Riwayat Pemakaian Jamban/Toilet	Diagnosa/Tes Darah
Nafsu Makan Normal	Jarang Jajan Diluar	Ada sampah dan jarang dibersihkan	Tersedia air bersih	Ada ventilasi tapi jarang terbuka	Sering Pakai Jamban/Toilet	tbc

⇩

Nafsu Makan Kurang	Sering Jajan Diluar	Tersedia air bersih	Ada lalat/nyamuk di sekitar tumpukan sampah	Ada ventilasi disemua ruangan	Sering Pakai Jamban/Toilet	dbd	tbc	tifoid
0	0	1	0	0	1	0	1	0

Fig. 4. Sample data before and after Multilabel Binarizer

In data transformation, it will also be transformed into per de-sa form. Data normalization will also be carried out so that numeric data variables do not have the same scale or range. The data normalization method used is L2 Normalization (Euclidean Norm) to convert a vector of variable length into a vector of fixed length, which is 1 while maintaining the relative direction of the vector. As shown in Fig. 5 below displays the sample data after being grouped by address and as shown in Fig. 6 displays the results of data normalization using the L2 Normalization method.

	tbc	dbd	tifoid	Nafsu Makan Kurang	Sering Jajan Diluar	Tersedia air bersih	Ada lalat/nyamuk di sekitar tumpukan sampah
Bili Bili (Gowa)	11.0	9.0	14.0	26.0	19.0	33.0	1.0
Binangae (Gowa)	1.0	0.0	0.0	0.0	1.0	1.0	0.0
Bonto Bontoa (Gowa)	2.0	10.0	3.0	14.0	13.0	15.0	2.0
Bonto-Bonto (Gowa)	1.0	0.0	0.0	0.0	1.0	1.0	0.0
Bontomanai (Gowa)	36.0	29.0	54.0	103.0	73.0	119.0	5.0
Bontomarannu (Gowa)	5.0	1.0	2.0	5.0	6.0	8.0	0.0

Fig. 5. Patient data samples per village

	tbc	dbd	tifoid	Nafsu Makan Kurang	Sering Jajan Diluar	Tersedia air bersih	Ada lalat/nyamuk di sekitar tumpukan sampah
Bili Bili (Gowa)	0.275	0.169811	0.259259	0.228070	0.202128	0.239130	0.166667
Binangae (Gowa)	0.025	0.000000	0.000000	0.000000	0.010638	0.007246	0.000000
Bonto Bontoa (Gowa)	0.050	0.188679	0.055556	0.122807	0.138298	0.108696	0.333333
Bonto-Bonto (Gowa)	0.025	0.000000	0.000000	0.000000	0.010638	0.007246	0.000000
Bontomanai (Gowa)	0.900	0.547170	1.000000	0.903509	0.776596	0.862319	0.833333
Bontomarannu (Gowa)	0.125	0.018868	0.037037	0.043860	0.063830	0.057971	0.000000

Fig. 6. Data sample after normalization

3.3 Hybrid Sequential Clustering Algorithm

The K-means Clustering Algorithm

The *K-means* algorithm is an unsupervised learning algorithm that groups data into the nearest cluster. K-means performs data clustering by partitioning existing data into the form of one or more clusters/groups based on attributes into k partitions where k < n [10], so that data that has the same characteristics is grouped into the same cluster and data that has different characteristics is grouped into other groups [11]. This algorithm also aims to find groups in the data, with the number of groups represented by the variable K. The variable K itself is the desired number of clusters. In each cluster there is a center point (centroid) [12]. The process stages in the kmeans clustering algorithm are as follows [13]:

a Determine the number of clusters
b Randomly assign each cluster an initial center point value
c Calculate the distance of each village data in the dataset to the initial center point of each cluster using euclidean distance with the following formula [14]

$$D_e = \sqrt{(x_i - s_i)^2 + (y_i - t_i)^2} \tag{1}$$

where

D_e : *Euclidean Distance*
i : many objects
(x, y) : object coordinates
(s, t) : *centroid* coordinates

d Allocate each data to the nearest centroid i.e. by taking into account the minimum distance of the object
e Calculate the new center point in each cluster by calculating the average data in each cluster that has been formed
f Repeat the three processes above until the resulting centroid value is fixed, the cluster members do not move to other clusters, or have met the maximum iteration value specified

Hybrid With Particle Swarm Optimization Algorithm

Particle Swarm Optimization (PSO) algorithm is one of the evolutionary computing techniques that has similarities with Genetic Algorithm, namely this algorithm starts by generating a population randomly or randomly [15]. Particle Swarm Optimization is a very simple optimization technique to apply and modify several parameters [16].

a Determine the starting point position as the cluster center of k data points randomly.
b Group the data into clusters by inserting the data into one of the clusters that has the closest cluster center randomly.
c Calculate the Sum of Squared Error (SSE) and Silhouette coefficient values which become fitness functions because SSE and Silhouette will be searched for values in the clustering algorithm.
d Define the initial velocity of the particle (V_0).
e Update the velocity value and calculate particle x based on the velocity value obtained using the following equations [17]–[18]:

$$V_{ij}(t+1) = w * V_{ij}(t) + c_1 * rand_1 * \left(pbest_{ij}(t) - p_{ij}(t)\right) + c_2 * rand_2 * (gbest_{ij}(t) - p_{ij}(t)) \quad (2)$$

$$p_{ij}^{t} = p_{ij}^{t-1} + V_{ij}^{t} \quad (3)$$

where,

t: indicates the iteration counter.
V_{ij}: velocity of particle i in the jth dimension.
p_{ij}: position of particle i in the jth dimension.
$Pbest_{ij}$: pbest position of particle i in the jth dimension.
$Gbest_{ij}$: gbest position of the jth dimension.
w: inertia weight.
$rand_1$, $rand_2$: random function range [0, 1].
$c1$ *and* $c2$: positive acceleration coefficient.

f Update the cluster center position, by summing the cluster center with the velocity value.
g The SSE value obtained at the end is then compared with the previous SSE value to obtain the new cluster center
h Repeat the three processes above until it meets the maximum iteration value specified
i After reaching the stopping criteria, the new centroid value is obtained which will be the initial cluster center that will be used in the K-means clustering process.
j Repeat the clustering stage using K-means using the initial cluster center (centroid) obtained from PSO optimization

Evaluation Model

In this section, we will evaluate the clustering process to see whether the application of the PSO algorithm to K-means clustering is optimal or not by comparing the SSE, Silhouette coefficient, and Davies Bouldin Index (DBI) values between the clustering process using only K-means and the clustering process by applying the PSO algorithm.

3.4 WebGIS Visualization

Web-based GIS (WebGIS) is a Geographic Information System (GIS) application distributed in a computer network to integrate and disseminate geographic information visually on the World Wide Web [19]. WebGIS compared to desktop GIS offers several advantages such as cost efficiency, human resource workload efficiency for installation, maintenance and technical support, trimming the learning curve for end users and advantages in terms of integration of spatial and non-spatial data [20]. The WebGIS architecture consists of three layers, namely the user interface layer, the application server layer, and the database layer [21]. In this research, WebGIS visualization is made using NextJS for the user interface, Flask Python for the application server, and SQLAlchemy for the database. The clustering data that has been grouped using kmeans and PSO will be displayed in WebGIS so that users can easily read the results of grouping infectious disease data.

4 Result

4.1 Clustering Result

The final results of the clustering process are shown in the cluster result plots is shown in Fig. 7 and Fig. 8.

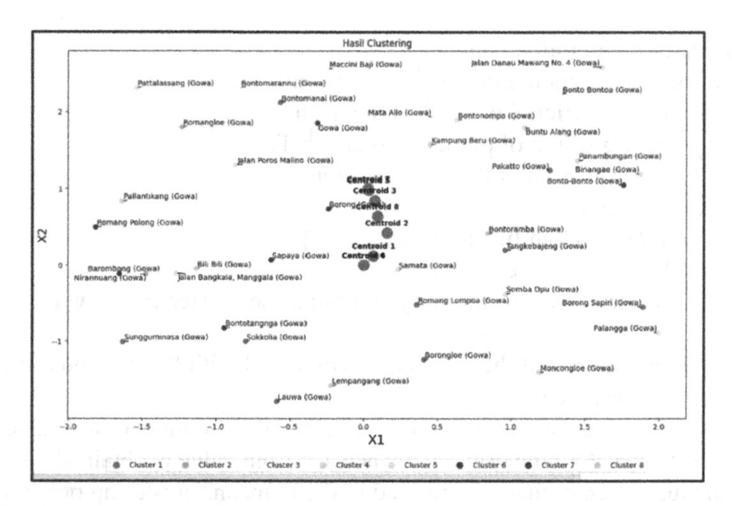

Fig. 7. Plot of cluster results by number of tuberculosis and e neighborhood conditions

Figure 7 displays a sample of cluster results of one infectious disease against environmental conditions where the x-axis is the infectious disease tuberculosis and the y-axis is one of the environmental conditions, namely the availability of clean water. The x-axis and y-axis values shown in the graph above are the centroid values of cluster results 1–8 and also the results of each cluster have been labeled with different colors.

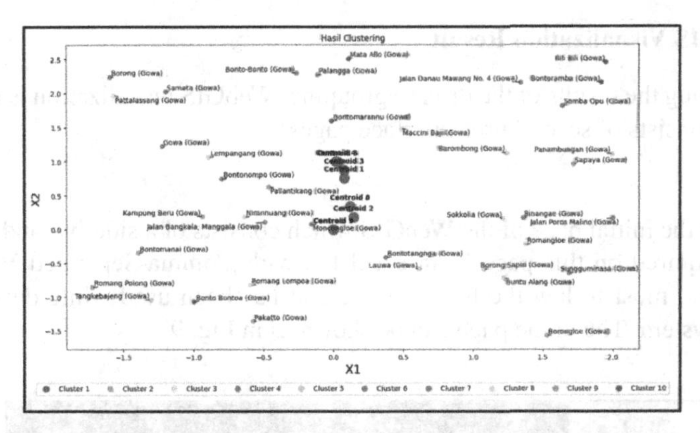

Fig. 8. Plot of cluster results by number of tuberculosis and lifestyle

Figure 8 displays a sample of cluster results of one infectious disease against lifestyle habits where the x-axis is the infectious disease tuberculosis and the y-axis is one of the lifestyle habits, namely poor appetite. The x-axis and y-axis values displayed in the graph above are the centroid values of cluster results 1–10 and the results of each cluster have been labeled with different colors.

4.2 Evaluation Result

Based on the results of K-means clustering optimization using the Particle Swarm Optimization (PSO) algorithm, it can be seen in Table 1 that a more optimal value is obtained with better SSE, Silhouette coefficient, and DBI values compared to using only K-means clustering.

Table 1. Evaluation Model Clustering Algorithm

Cluster	SSE		*Silhouette*		DBI	
	K-means	PSO *K-means*	*K-means*	PSO *K-means*	*K-means*	PSO *K-means*
1	1.116118	0.505705	0.502752	0.512179	0.608487	0.512963
2	0.858209	0.357805	0.535189	0.577407	0.576832	0.451819
3	1.043787	0.405219	0.517382	0.543259	0.582908	0.487524
4	1.180882	0.595071	0.842564	0.694180	1.087944	0.316235
5	1.056586	0.578959	0.861212	0.753807	0.439369	0.283324

4.3 WebGIS Visualization Result

After obtaining the results of the disease grouping, WebGIS visualization is then carried out which consists of several user interface pages.

Homepage

This page is the initial page of the WebGIS which consists of a side-bar and input page. The data required on this page is an excel file with Comma-Separated Values (.csv) extension and must follow the file template that has been used while developing the clustering system. The home page can be shown as in Fig. 9.

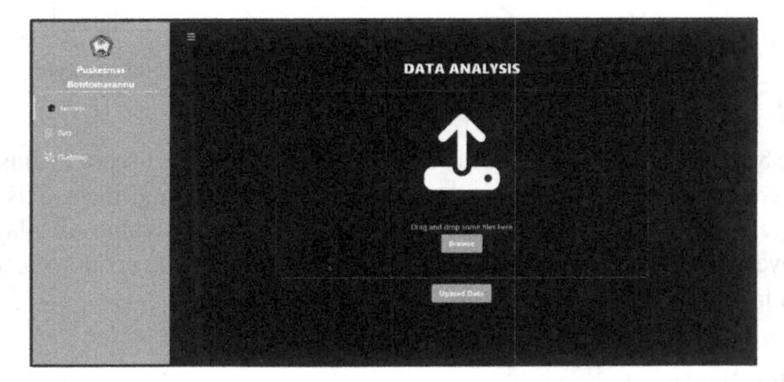

Fig. 9. WebGIS homepage view

Data Page

On this page, the user can see the details of the data that has been inputted on the previous home page. The generate cluster button at the top left serves to call the clustering program to execute the data that has been inputted by the user. The page to display the data that has been uploaded can be shown as in Fig. 10.

DATASET PASIEN PENYAKIT MENULAR

Umur	Alamat	Nafsu Makan	Kebiasaan Jajan	Kondisi Pembuangan Sampah	Ketersediaan Air Bersih	Sistem Ventilasi	Pemakaian Jamban/Toilet	Diagnosa
29	Maccini Baji (Gowa)	Nafsu Makan Normal	Sering Jajan Diluar	Bersih tanpa ada sampah	Tersedia air bersih	Ada ventilasi tapi hanya sedikit	Sering Pakai Jamban/Toilet	tbc
18	Maccini Baji (Gowa)	Nafsu Makan Kurang	Sering Jajan Diluar	Ada lalat/nyamuk di sekitar tumpukan sampah	Tersedia air bersih	Ada ventilasi disemua ruangan	Sering Pakai Jamban/Toilet	dbd
9	Maccini Baji (Gowa)	Nafsu Makan Kurang	Jarang Jajan Diluar	Ada sampah tapi tidak banyak	Tersedia air bersih	Ada ventilasi tapi hanya sedikit	Jarang pakai jamban/toilet	tifoid
28	Maccini Baji (Gowa)	Nafsu Makan Kurang	Sering Jajan Diluar	Ada sampah dan jarang dibersihkan	Tersedia air bersih	Ada ventilasi tapi jarang terbuka	Sering Pakai Jamban/Toilet	tbc

Fig. 10. Data Page view

Clustering Page

On this page, users can see the results of clustering infectious disease data. As shown in Fig. 11, the display of the clustering page.

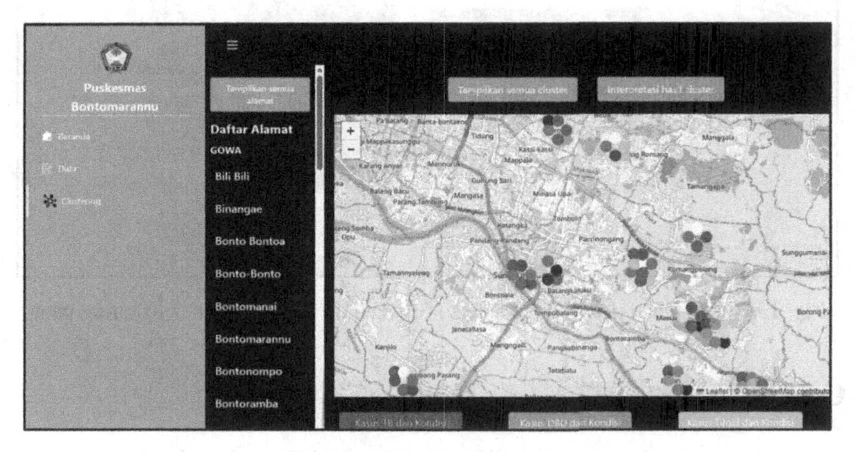

Fig. 11. Clustering page view

This page will display a digital geographic map that maps infectious disease clusters to each coordinate location point. In this map, there are several small circle symbols with different colors that show what clusters are included in a village in Gowa Regency. As shown in Fig. 12, the map view on the clustering page.

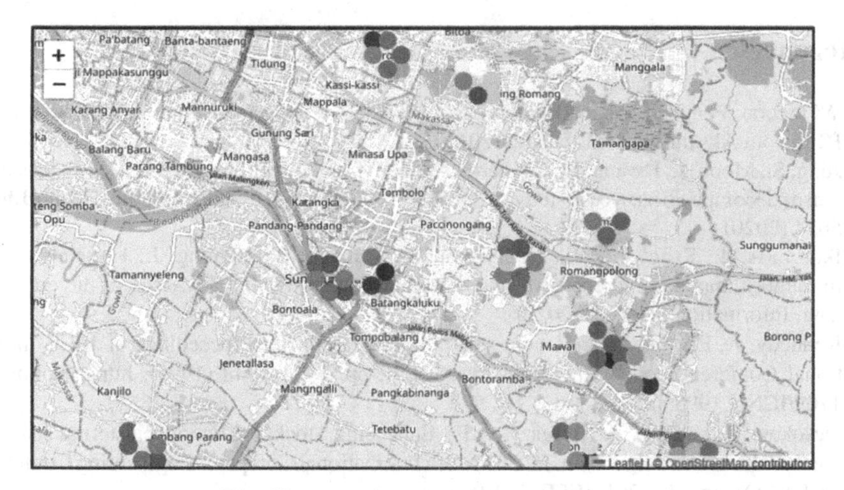

Fig. 12. Map view on the clustering page

If one of the circles is clicked on the map, it will display detailed information for that cluster as shown in Fig. 13.

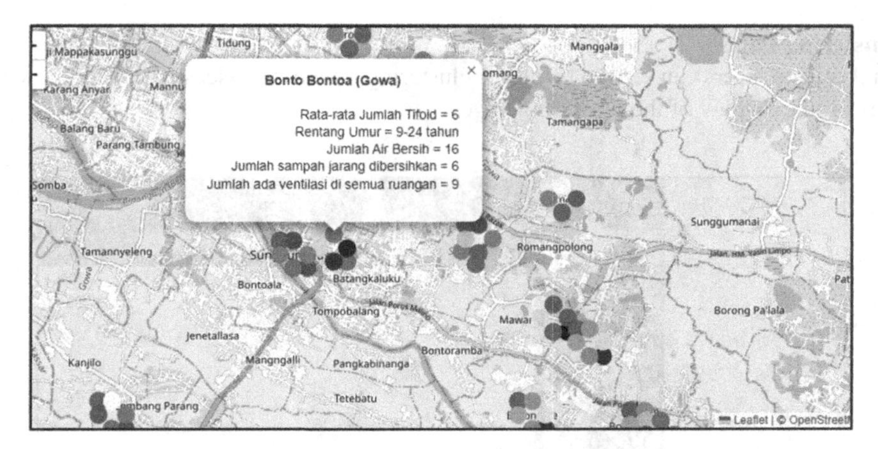

Fig. 13. View when one of the clusters is clicked on the map in the clustering page

5 Conclusion

This research aims to visualize WebGIS on infectious disease data that is clustered using a combination of algorithms, namely by optimizing K-means with PSO so as to obtain better cluster results. The results show that this WebGIS visualization has mapped 39 villages with different numbers of tuberculosis, dengue fever and typhoid fever infectious disease cases and with different environmental conditions and lifestyles. This research can be used as a reference or policy material by the health centers or even the government in handling and reducing cases of infectious diseases.

References

1. Widoyono: Penyakit Tropis: Epidemiologi, Penularan, Pencegahan dan Pemberantasannya. PT.Gelora Aksara Pratama, Jakarta (2011)
2. Jung, S., Moon, J., Hwang, E.: Cluster-based analysis of infectious disease occurrences using tensor decomposition: a case study of South Korea. Int. J. Environ. Res. Public Health **13**(21), 5634 (2020)
3. Ikotun, A.M., Ezugwu, A.E., Abualigah, L., Abuhaija, B., Heming, J.: K-means clustering algorithms: a comprehensive review, variants analysis, and advances in the era of big data. Information Sciences (2022)
4. Kennedy, J., Eberhart, R.: Particle swarm optimization. In: Proceedings of IEEE International Conference on Neural Networks, vol. 4, pp. 1942–1948 (1995). https://doi.org/10.1109/ICNN.1995.488968
5. Desokey, E.N., Badr, A., Hegazy, A.F.: Enhancing stock prediction clustering using K-means with genetic algorithm. In: 2017 13th International Computer Engineering Conference (ICENCO), pp. 256–261. IEEE (2017)
6. Kuo, R.J., Wang, M.J., Huang, T.W.: An application of particle swarm optimization algorithm to clustering analysis. Soft Comput. **15**, 533–542 (2011)
7. Arthur, D., Vassilvitskii, S.: K-means++ the advantages of careful seeding. In: Proceedings of the Eighteenth Annual ACM-SIAM Symposium on Discrete Algorithms, pp. 1027–1035 (2007)

8. Lu, X.: Web GIS based information visualization for infectious disease prevention. In: 2009 Third International Symposium on Intelligent Information Technology Application, vol. 1, pp. 148–151. IEEE (2009)
9. Randazzo, G., et al.: WebGIS Implementation for dynamic mapping and visualization of coastal geospatial data: a case study of BESS project. Appl. Sci. **11**(17), 8233 (2021)
10. Pu, Q., Gan, J., Qiu, L., Duan, J., Wang, H.: An efficient hybrid approach based on PSO, ABC and K-means for cluster analysis. Multimed. Tools App. **81**(14), 19321–19339 (2022)
11. Avanija, J., Ramar, K.: A hybrid approach using PSO and K-means for semantic clustering of web documents. J. Web Eng. 249–264 (2013)
12. Solaiman, B.: Energy optimization in wireless sensor networks using a hybrid K-means pso clustering algorithm. Turk. J. Electr. Eng. Comput. Sci. **24**(4), 2679–2695 (2016)
13. Peng, K., Leung, V.C., Huang, Q.: Clustering approach based on mini batch k means for intrusion detection system over big data. IEEE Access **6**, 11897–11906 (2018)
14. Nazeer, K. A., Sebastian, M.P.: Improving the accuracy and efficiency of the K-means clustering algorithm. In: Proceedings of the World Congress on Engineering, vol. 1, pp. 1–3. Association of Engineers London, London, UK (2009)
15. García, J., Martí, J.V., Yepes, V.: The buttressed walls problem: an application of a hybrid clustering particle swarm optimization algorithm. Mathematics **8**(6), 862 (2020)
16. Zheng, H., Hou, M., Wang, Y.: An efficient hybrid clustering-PSO algorithm for anomaly intrusion detection. J. Softw. **6**(12), 2350–2360 (2011)
17. Tan, L.: A clustering K-means algorithm based on improved PSO algorithm. In: 2015 Fifth International Conference on Communication Systems and Network Technologies, pp. 940–944. IEEE, April 2015
18. Prabha, K.A., Visalakshi, N.K.: Improved particle swarm optimization based K-means clustering. In: 2014 International Conference on Intelligent Computing Applications, pp. 59–63. IEEE, March 2014
19. Lu, X.: Web GIS based information visualization for infectious disease prevention. In: 2009 Third International Symposium on Intelligent Information Technology Application, (Vol. 1, pp. 148–151). IEEE, November 2009
20. Reed, R.E., Bodzin, A.M.: Using web GIS for public health education. Int. J. Environ. Sci. Educ. **11**(14), 6314–6333 (2016)
21. Lipeng, J., Xuedong, Z., Jianqin, Z., Zhijie, X., Shaocun, D.: Design and development of a visualization system for COVID-19 simulation based on WebGIS. In: 2020 International Conference on Public Health and Data Science (ICPHDS), pp. 278–282. IEEE, November 2020

Research on Security Assets Attention Networks for Temporal Knowledge Graph Enhanced Risk Assessment

Ying Cui[1], Xiao Song[1(✉)], Yancong Li[1], Wenxin Li[1], and Zuosong Chen[2]

[1] School of Cyber Science and Technology, Beihang University, Beijing, China
songxiao@buaa.edu.cn
[2] Qi An Xin Technology Group Inc., Beijing, China

Abstract. The rapid development and extensive application of cyberspace have brought numerous opportunities to Internet users. Due to the characteristics of virtual, and open nature, cybersecurity assets are highly susceptible to attacks. Therefore, security asset risk assessment is a challenging task in the field of cyberspace security. We present Security Asset Attention Network (SeAAN), a novel model that achieve risk assessment of asset node to capture temporal knowledge graph structural evolution. Specifically, SeAAN computes risk assessment of asset node through joint attention focus on both structural neighbor and temporal history, which assigns distinct snapshots to facts at various time stamps, capturing dynamic knowledge fluctuations effectively. Extensive experiments demonstrate that SeAAN achieves significant performance on a real-world benchmark dataset for temporal knowledge graph enhanced security asset risk assessment. Moreover, our ablation analysis confirms the efficacy of integrating structural attention and temporal self-attention in a joint manner. Empirical results on real-world datasets demonstrate that our model exhibits more substantial performance enhancements compared to conventional approaches.

Keywords: Attention Networks · Risk Assessment · Security Assets · Temporal Knowledge Graph

1 Introduction

The rapid development and extensive application of cyberspace have brought numerous opportunities to Internet users and organizations. However, due to the characteristics of connectivity, virtuality, and openness, security assets in cyberspace are highly susceptible to attacks [14]. Therefore, the real-time and accurate assessment of asset risks is an urgent problem that needs to be addressed. The risk assessment of cybersecurity assets in cyberspace has become an important task in protecting personal, organizational, and national security. Accurately assessing the risk level of cybersecurity assets is crucial for effectively formulating security strategies, optimizing resource allocation, and ensuring the sustainable operation of information systems.

© The Author(s), under exclusive license to Springer Nature Singapore Pte Ltd. 2024
F. Hassan et al. (Eds.): AsiaSim 2023, CCIS 1911, pp. 390–404, 2024.
https://doi.org/10.1007/978-981-99-7240-1_31

Asset risk assessment has been studied in numerous fields such as information networks [4], finance [24], smart grids [18], and the Internet of Things [3]. However, research on asset risk assessment in the field of cyberspace security is relatively limited. In addition, many research findings regarding risk assessment methods involve the use of Analytic Hierarchy Process (AHP) and Fuzzy Comprehensive Evaluation to assign weights to risk indicators for comprehensive risk assessment in terms of magnitude and probability regression [22,23]. There are also widely used machine learning methods such as logistic regression, XGBoost, support vector machines, random forests, and gradient boosting decision trees [7,25]. However, due to the complexity and dynamism of cyberspace, traditional risk assessment methods have become increasingly inadequate to meet the growing security demands.

In order to enhance the assessment of asset risks in the security domain, applying knowledge graphs to asset security risk assessment models enables the exploration of additional latent information through the analysis of relations between nodes. This approach can also provide a visual representation of the coupled risk elements within assets [21]. As a complex form of data, knowledge graphs present certain limitations for traditional machine learning methods in terms of feature representation and relation modeling. The deep learning method has significant advantages in the evaluation of the Knowledge graph. Through the deep learning method, information such as entities, relations and attributes in the Knowledge graph can be encoded into a low dimensional vector representation, so that more accurate reasoning and prediction can be made.

However, many asset risk graphs in the security field are dynamic. Graph structures undergo temporal evolution and are commonly depicted as sequences of graph snapshots at distinct time intervals. The complex temporal graph structure poses challenges in learning asset node risk assessment that changes over time [12]. Through dynamic security asset assessment, organizations can better identify and understand the potential risks and value of their security assets. They can then adjust security strategies and resource allocation in a timely manner to respond to the ever-changing threat and risk environment. In deep learning methods that handle dynamic data, traditional models such as RNN primarily focus on the temporal dependencies between sequence elements, while the structural information in the graph is often ignored [13]. However, in asset risk prediction, it is necessary to consider not only the temporal sequence but also the complex relations among assets. Therefore, we aim to capture both local and global temporal information in the graph structure by propagating information between nodes and relations.

For adeptly harnessing vital information like the topology's structure and dynamic temporal aspects, an innovative model named Security Asset Attention Network (SeAAN) is introduced, which can be designed to acquire node representations of assets, facilitating the capture of temporal knowledge graph structure evolution. By employing joint attention focus on both structural neighbor and temporal history, the model dynamically evaluates the risk of asset nodes, thereby enhancing the performance of the model. Specifically, structural attention network aggregates features extracted from local node neighborhoods collected from each snapshot using an attention mechanism. Attention weights are computed based on the relations between asset nodes. Temporal self-attention network dynamically focuses on the node information at key moments by flexibly weighting the historical representations based on the relative importance of each time snapshot with respect to other historical times. In order to adeptly capture the diverse fluctuations within the graph structure, we present a learning mechanism featuring multiple attention heads within both the structural and temporal attention layers. This enables the joint attention across different potential sub-spaces. We conducted experimental comparisons with several baselines of different scales, and the results demonstrate that SeAAN achieved significant performance benefits. Through ablation studies, we demonstrated the benefits of aggregating neighborhood information and tracking time history, which allows for a more comprehensive consideration of the relations between assets and temporal dependencies. As a result, it improves the accuracy and robustness of asset risk assessment. In summary, our contributions are shown as follows:

- We propose a Security Asset Attention Network(SeAAN), which dynamically evaluates the risk of assets based on temporal knowledge graph.
- The model consists of stacked structural attention network and temporal self-attention network, which model the interactions among assets to better understand and capture the trends of asset risks over time.
- We conducted extensive experiments on a security asset dataset and demonstrated the effectiveness of the model in risk assessment of cybersecurity assets.

2 Related Work

The open connectivity of information networks, along with the inherent vulnerabilities and design flaws of assets, poses significant challenges to security and management in various domains. Additionally, it introduces potential security risks in data transmission, storage, and processing. At present, asset risk assessment has been explored in various fields. Fu et al. [4] utilized interval intuitionistic fuzzy numbers for data processing and aggregation through a comprehensive evaluation model, reducing uncertainty and subjectivity, and increasing the goal-oriented risk assessment of safety assets. Utilizing mathematical and statistical methodologies, Zhang et al. [24] performed computations to analyze the economic and physical attributes of five stocks within China's A-share market. Priyanka et al. [18] proposed a framework that employs a multi-layered graph model to assist in a context-based risk assessment approach, which seeks to depict the asset interdependency model within a diverse smart grid environment, accounting for the distinct characteristics of assets in the context of risk assessment. Salim et al. [3] considered the system assets and potential threats of the IoT domain model, and identified and analyzed potential risks by defining security objectives. Arjun et al. [6] utilized digital assets of credit letters to facilitate logistics financing. But there is very little research on cyberspace security assets, and a similar article is about risk assessment of security assets in the field of information management systems. Mohammad et al. [19] proposed an asset-specific technological risk approach that is associated with the sensitivity of each asset, they analyzed the sensitivity of assets based on a network battlefield framework, determined the risks of each asset to enhance risk prediction in information management systems. Many studies on risk assessment methods are implemented using machine learning algorithms. Zhu et al. [25] formulated an empirical analysis model employing XGBoost, which was then compared with logistic regression, support vector machines, and random forest algorithms. This comparison yielded the optimal model and feature importance values, enabling the effective identification of collateral loan risks for emerging agricultural entities associated with biological assets.

Currently, in order to better assess the risk of security assets in various domains, knowledge graph is being applied to asset security risk assessment models to uncover more potential information. Yang et al. [21] utilized an aviation safety event knowledge graph, employed Correspondence Analysis to explore the relationships between event elements, which offers an additional decision-making instrument for evaluating the security risks in airspace. Li et al. [12] introduced a technique for learning dynamic interest sequences at multiple levels of granularity, which uses SEP2Vec for embedding representation and merges the entropy perception pool layer to obtain a user preference representation for learning dynamic user interest sequences. Nevertheless, numerous real-world graphs exhibit dynamic characteristics, with their structures evolving over time. Pushparaj et al. [2] processed the asset data of Industrial control system in a centralized manner to cope with dynamic changes. Amirhossein et al. [1] complete the temporal knowledge graph by combining the time aware relationship path and relationship context, this model can use neural network to improve the temporal knowledge graph completion method. Jia et al. [8] proposed to select a set of triples that are most affected by knowledge events to update by measuring the importance score of each triplet, so as to realize the embedded representation of the adaptive update temporal knowledge graph. Tang et al. [20] proposed a method for dynamically constructing a multi-hop knowledge graph, utilizing hierarchical graph attention to capture global features and mitigate semantic biases. Li et al. [11] proposed the Space Adaptation Network(SANe), which employs distinct latent spaces for time snapshots at various time stamps, enhancing the effectiveness of temporal knowledge graph completion.

In deep learning methods for processing temporal data, many studies focus on the temporal dependencies between sequence elements based on models such as RNN. Hong et al. [7] suggested the TKG-CRA credit risk assessment model, incorporating BiLSTM to encode temporally sorted sequences of neighboring nodes. The goal is to enhance the precision of enterprise credit risk evaluation. Liu et al. [13] achieved the visualization and quantitative assessment of railway accident network topology by integrating Hidden Markov Models, Conditional Random Fields algorithms, Bidirectional LSTM, and deep learning networks.

3 Methodology

We define the problem of asset risk assessment for temporal knowledge graph, with a focus on asset nodes. The research objective revolves around analyzing the relations between asset nodes to select the neighboring nodes that need to be aggregated.

The concept of the dynamic asset knowledge graph involves a sequence of captured instances of static graph snapshots, $\mathcal{G} = \{\mathcal{G}_1, \mathcal{G}_2, ..., \mathcal{G}_T\}$, $\mathcal{T}$ is the number of time steps. Every snapshot is an asset time series knowledge graph represented by quadruples, $\mathcal{G} = \{(u, r, v, \tau) \mid u, v \in \mathcal{V}, r \in \mathcal{R}, \tau \in \mathcal{T}\}$, where $\mathcal{V}$, $\mathcal{R}$, and $\mathcal{T}$ are collections of asset entities, relations, and time stamps. Every quadruple symbolizes a fact that varies with time, where asset entity u and asset entity v have a relationship r at time stamp τ. The key to risk assessment of asset temporal knowledge graph is to learn potential representations, $\mathbf{e}_v^\tau \in \mathbb{R}^d$. For each node $v \in \mathcal{V}$, and time step $\tau = \{1, 2, ..., \mathcal{T}\}$, $\mathbf{e}_v^\tau$ is desired to preserve both the local graph structure centered around asset v and its dynamic evolution, including the dynamic occurrence and disappearance of asset relations.

SeAAN is mainly composed of three modules: a Structural Attention Network (SAN), a Time Self-Attention Network (TSAN), and an Asset Risk Classful Network (ARCN). As shown in Fig. 1, SAN emphasizes the capture of local structural attributes in snapshots, TSAN emphasizes the capture of time evolution patterns, and ARCN focuses on comprehensively predicting asset risk based on structural neighborhood and time history. Dynamic risk assessment of cyberspace security assets can be achieved through layer stacking. Based on this, we propose the neural model SeAAN, which is constructed based on these three modules.

3.1 Structural Attention Network

The input to the network consists of a graph snapshot $\mathcal{G}_t$ and a sequence of asset node embeddings $\{\mathbf{x}_v \in \mathbb{R}^d, \forall v \in \mathcal{V}\}$, d is the representation dimension. The input is a learnable vector, and the output $\{\mathbf{z}_v \in \mathbb{R}^F, \forall v \in \mathcal{V}\}$ is a sequence of new asset node embedding with dimension F, which capture local structural attributes in the snapshot. Specifically, SAN aggregates the neighboring nodes of node v (in the snapshot $\mathcal{G}_t$) by computing the input node embeddings for node v. The definition of SAN is:

$$\mathbf{e}_{uv} = \sigma\left(\mathbf{a}^\top \left[\mathbf{W}^s \phi\left(\mathbf{x}_u, \mathbf{r}_{uv}\right) \| \mathbf{W}^s \mathbf{x}_v\right]\right), \tag{1}$$

$$\alpha_{uv} = \frac{\exp(\mathbf{e}_{uv})}{\sum_{w \in \mathcal{N}_v} \exp(\mathbf{e}_{wv})}, \tag{2}$$

$$\mathbf{z}_v = \sigma\left(\sum_{u \in \mathcal{N}_v} \alpha_{uv} \mathbf{W}^s \phi(\mathbf{x}_u, \mathbf{r}_{uv})\right), \tag{3}$$

$$\mathbf{h}_v = Concat(\mathbf{z}_v^1, \mathbf{z}_v^2, ..., \mathbf{z}_v^{H_S}), \forall v \in \mathcal{V}. \tag{4}$$

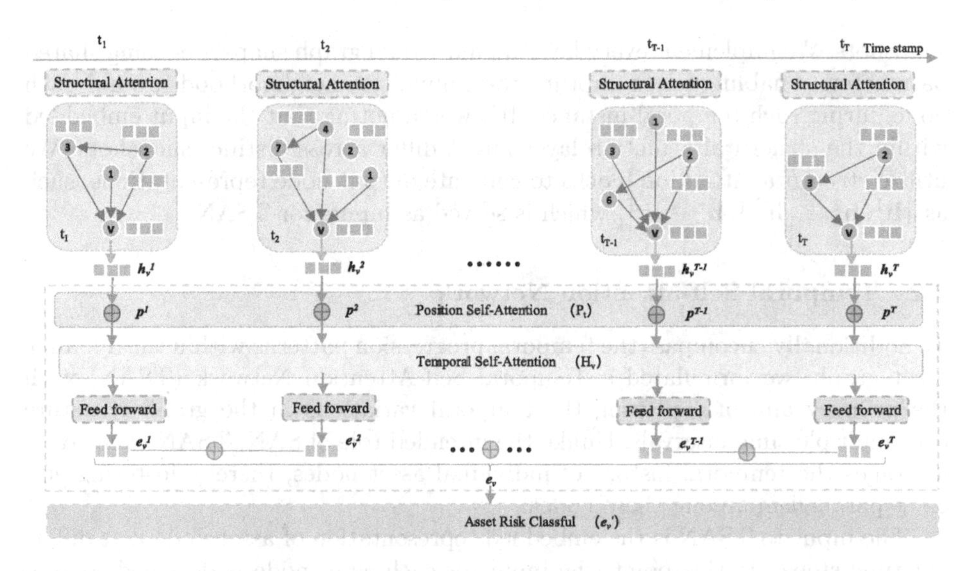

Fig. 1. The framework of our model SeAAN: we employ Structural Attention Network, Temporal Self-Attention Network followed by Asset Risk Classful Network.

Firstly, the similarity coefficients $\mathbf{e}_{uv}$ between the neighbors $\mathcal{N}_v$ of node v and itself are calculated one by one. $\mathcal{N}_v = \{u \in \mathcal{V} : (u, v) \in \mathcal{R}\}$ is the nearest neighbor sequence of asset node v in the snapshot $\mathcal{G}_\tau$. $\mathbf{W}^s \in \mathbb{R}^{F \times D}$ is a uniform weight transformation that is implemented on every node within the graph, serving as a prevalent technique for feature enhancement. $\mathbf{a} \in \mathbb{R}^{2F}$ is a weight vector, which is also an attention function implemented by parameterized feedforward layer. $\|$ refers to the concatenation operation, and $\sigma(\cdot)$ is a nonlinear activation function. $\phi(\mathbf{x}_u, \mathbf{r}_{uv})$ utilizes the asset-relation composition operation employed in knowledge graph embedding methods, embedding both the neighboring node u and the relation $\mathbf{r}_{uv}$ into the graph of relationships. It aggregates the messages $\mathcal{N}_v$ from all neighbors and combines them to get the updated embedding for the asset node v. Clearly, learning the correlation between nodes u and v is achieved through learnable parameters and mappings.

Then, we use LeakyReLU to compute the attention coefficients α_{uv} and apply exponential linear unit (ELU) activation to the asset embeddings. The learned coefficients α_{uv}, got by applying softmax to the neighborhoods of every asset node in $\mathcal{V}$, depicting the influence of node u on node v within the present snapshot. Next, using the computed attention coefficients, the updated representation of the asset node $\mathbf{z}_v$ is obtained by taking a weighted sum of the features of its neighboring nodes.

Finally, multi-head enhancement is employed to obtain multi-head representations of the asset nodes. H_S is the number of attention heads. $\mathbf{h}_v \in \mathbb{R}^F$ is the output after multi-head attention for node v. Please take into consideration that although structural attention is separately applied to each snapshot, the parameters governing the structural attention heads remain consistent across various snapshots. We implement every layer on individual graph snapshots using shared parameters, enabling us to capture the immediate neighborhood around each node during each temporal instance. It's worth noting that the input embedded within the structural attention layer might differ across distinct snapshots. We utilize structural attention blocks to generate output node representations, such as $\{\mathbf{h}_v^1, \mathbf{h}_v^2, ..., \mathbf{h}_v^{\mathcal{T}}\}, \mathbf{h}_v^{\tau} \in \mathbb{R}^F$, which is served as inputs for TSAN.

3.2 Temporal Self-attention Network

To additionally encompass the temporal progression patterns within the dynamic asset graph, we formulated a Temporal Self-Attention Network (TSAN) with the primary aim of capturing the temporal variations in the graph structure over multiple time intervals. Unlike the intended role of SAN, TSAN exclusively leverages the temporal history of individual asset nodes, thereby fostering efficient parallelism among these nodes.

The input to TSAN is the embedded representation of asset node v at different time steps. At this point, the input for each asset node is designed to adequately capture the local neighbor information at each time step. Specifically, for every asset node v, the input $\mathbf{X}_v \in \mathbb{R}^{\mathcal{T}\times F}$ is defined as $\{\mathbf{h}_v^1, \mathbf{h}_v^2, ..., \mathbf{h}_v^{\mathcal{T}}\}, \mathbf{h}_v^{\tau} \in \mathbb{R}^F$. The input embedding for asset node v at time step τ is $\mathbf{h}_v^{\tau}$, which encodes the immediate local structure surrounding the node v at the present moment. We use $\mathbf{h}_v^{\tau}$ as a query for attending to its historical embeddings, monitoring the changes in the nearby environment surrounding v. The output layer consists of a new sequence of embeddings for node v at each time step, denoted as $\mathbf{z}_v = \{\mathbf{z}_v^1, \mathbf{z}_v^2, ..., \mathbf{z}_v^T\}$, where $\mathbf{z}_v^{\tau} \in \mathbb{R}^F$, $\mathbf{z}_v \in \mathbb{R}^{T\times F}$. These representations are combined over time, with T being the number of historical snapshots used for prediction, and F being the dimension of the input representation.

To calculate the resulting representation of the asset node v at time τ, TSAN is defined as:

$$\mathbf{e}_{ij} = \sigma\left(\frac{\left(\left(\mathbf{X}_v\mathbf{W}_q\right)\left(\mathbf{X}_v\mathbf{W}_k\right)^{\top}\right)_{ij}}{\sqrt{F}}\right), \tag{5}$$

$$\beta_v^{ij} = \frac{\exp\left(e_v^{ij}\right)}{\sum\limits_{k=1}^{T}\exp\left(e_v^{ik}\right)}, \tag{6}$$

$$\mathbf{z}_v = \boldsymbol{\beta}_v\left(\mathbf{X}_v\mathbf{W}_v\right), \tag{7}$$

$$\mathbf{H}_v = Concat\left(\mathbf{z}_v^1, \mathbf{z}_v^2, ..., \mathbf{z}_v^{H_T}\right), \forall v \in \mathcal{V}. \tag{8}$$

First, we use the dot-product as the attention scoring function. Based on the dot-product calculation, we obtain the pairwise relationships between vectors. Initially, the query, key, and value undergo transformation into distinct spaces via linear projection matrices, these matrices are trainable and allow for learning, enhancing the model's fitting capability and acting as a buffer. We make predictions at each time step τ based on the previous T time steps to maintain the autoregressive nature.

Then, $\boldsymbol{\beta}_v \in \mathbb{R}^{T \times T}$ represents the attention weight matrix derived from the multiplication-based attention scoring function. At this point, we obtain the attention scores between asset nodes at a specific time relative to other historical moments, these scores indicate the level of attention that the asset node at this time should give to other moments.

Next, the important information $\mathbf{z}_v$ is extracted from these asset node vectors as the output of TSAN. All vectors are involved in the calculation, allowing for a global perspective. Indeed, the degree of involvement of each node vector in the calculation varies across different time steps. The attention scores determine the level of participation, with higher scores indicating greater involvement of the corresponding vectors. Consequently, the output vector becomes more similar to the vector with higher attention, allowing us to achieve a balance between a global perspective and focusing on key elements.

In addition, this paper employs an advanced version of self-attention mechanism known as multi-head self-attention. Since there can be multiple types of correlations between different time steps, a query can only capture one type of related key vector. Therefore, multiple query vectors and key vectors need to be introduced to capture multiple types of correlations. H_T is the number of temporal attention heads, satisfying $F = H_T \times F'$, $\mathbf{H}_v \in \mathcal{R}^F$ denotes the dimensionality of the output from the temporal multi-head attention.

Finally, independent position encoding is applied to decouple the position embeddings, making the model more lightweight. Because the self-attention network model lacks recursive or convolutional modules, it cannot perceive the position information of previous entities. Here, position information refers to historical interaction sequences, and it is necessary to add a position information embedding module to the model. Due to the lack of strong correlation between entities and absolute positions, this article does not adopt the common approach of adding entity embeddings and position embeddings for subsequent learning. Instead, it separately calculates the positional relevance.

First, the positional sequence is embedded to obtain a learnable positional embedding matrix $\mathbf{P} \in \mathbb{R}^{T \times F}$. Then, a linear transformation is applied to the initial positional embedding matrix. Here, $\mathbf{Y}_Q \in \mathbb{R}^{F \times F'}$, $\mathbf{Y}_K \in \mathbb{R}^{F \times F'}$, $\mathbf{Y}_V \in \mathbb{R}^{F \times F'}$ are all learnable parameter matrix. Next, the softmax function is employed to calculate the attention weights pertaining to positional information $\mathbf{A}_{pos} \in \mathbb{R}^{T \times T}$. The output, denoted as vector $\mathbf{p}$, is obtained, and multi-head are used to concatenate and obtain the final representation of positional embedding $\mathbf{P}_t$. Specifically,

$$\mathbf{P}_Q = \mathbf{P} \cdot \mathbf{Y}_Q, \tag{9}$$

$$\mathbf{P}_K = \mathbf{P} \cdot \mathbf{Y}_K, \tag{10}$$

$$\mathbf{A}_{pos} = Softmax\left(\frac{\mathbf{P}_Q \cdot (\mathbf{P}_K)^\top}{\sqrt{F}}\right), \tag{11}$$

$$\mathbf{p} = \mathbf{A}_{pos}\left(\mathbf{P} \cdot \mathbf{Y}_V\right), \tag{12}$$

$$\mathbf{P}_t = Concat(\mathbf{p}^1, \mathbf{p}^2, ..., \mathbf{p}^{H_t}), \forall t \in T. \tag{13}$$

Here, there is no embedding representation of any asset nodes. The attention weights for positional information are completely independent of the input information of the nodes and only need to be computed once per input batch. This helps to reduce computational costs.

By using the learnable positional embedding $\{\mathbf{p}_t^1, \mathbf{p}_t^2, ..., \mathbf{p}_t^T\}$, $\mathbf{p}_t^t \in \mathbb{R}^F$, we can obtain the temporal self-attention network sorting information for the asset nodes. Finally, the asset node output representation from the temporal self-attention network is obtained by combining the asset entity embeddings and positional embeddings, which are obtained separately using the self-attention method. The output representation $\mathbf{e}_v^t$ is of dimension F.

$$\mathbf{e}_v^t = \mathbf{H}_v + \mathbf{P}_t. \tag{14}$$

3.3 Asset Risk Classful Network

The ultimate layer of the model constitutes a node classification network. First, the asset node embeddings for T timestamps are averaged to obtain a comprehensive node representation. The output representation of the asset nodes $\mathbf{e}_v$, which has a dimension of F, is transformed into a representation $\mathbf{e}_v'$ with dimension C through a learned matrix and activation function. $\mathbf{W}_c \in \mathbb{R}^{C \times F}$, $\mathbf{W}_e \in \mathbb{R}^{F \times F}$ refer to the parameterized weight matrices between the node classification layers.

The feature dimension C is used as the risk category to be classified. Based on the risk level classification criteria, the risk categories are primarily categorized as high risk, medium risk, low risk, and no risk.

$$\mathbf{e}_v = \frac{1}{T} \sum_{t=1}^{T} \mathbf{e}_v^t, \tag{15}$$

$$\mathbf{e}_v' = \mathbf{W}_c \sigma(\mathbf{e}_v \mathbf{W}_e). \tag{16}$$

By applying the softmax function for normalization, we can obtain the probabilities of the risk levels for the asset node v.

$$\hat{y}_v = \frac{\exp(\mathbf{e}_v')}{\sum_{c=1}^{C} \exp(\mathbf{e}_c')}. \tag{17}$$

The model undergoes training through a supervised learning approach, employing the cross-entropy loss function for model evaluation and optimization. By adjusting the model's parameters, we aim to make the model's predictions as close as possible to the true labels. Throughout the training process, the loss function is systematically reduced, resulting in the progressive optimization of the model.

$$\mathcal{L} = -\sum_{l \in Y_L} \sum_{1}^{C} y_l \ln \hat{y}_l. \tag{18}$$

Table 1. The statistical information of the security asset dataset used for experiments.

Dataset	#Quadruple	#Timestamp	Risk-Free Ratio	Low-Risk Ratio	Medium-Risk Ratio	High-Risk Ratio
Train	887,562	20	0.65	0.16	0.11	0.08
Valid	131,554	3	0.69	0.13	0.11	0.07
Test	131,920	3	0.67	0.14	0.11	0.08

In the equation, Y_L represents the set of labeled asset nodes, y_l represents the true labels of the asset nodes, $\hat{y}_l$ represents the predicted probability values by the model.

4 Experiments

Within this section, the assessment of the proposed model on a dataset within the security domain is showcased. The dataset, baseline methods, evaluation metrics, and experimental details are introduced. A comparison is made between the model and the baselines. The experimental results are then analyzed and discussed. Furthermore, ablation studies are performed to assess the significance of various components within the model.

4.1 Experimental Setup

Dataset Description. We construct a security asset dataset by collecting data from a public information platform, which includes 10,719 devices and 78 types of relations. This dataset covers a vast number of interaction links between the devices, spanning 26 time periods. The dataset was divided into training, validation, and test sets based on timestamps, with proportions of 80%, 10%, and 10%, respectively. Table 1 summarizes the statistical information of the dataset.

Baseline Methods. Our proposed model is compared with several widely used methods in the field of risk assessment, which have demonstrated notable achievements. The benchmark models considered in this study include Logistic Regression (LR), K-Nearest Neighbors (KNN), Naive Bayes (NB), Support Vector Machine (SVM), Random Forest (RF), and Decision Tree (CART). These methods have been extensively studied in previous research, highlighting their efficacy in risk assessment [7,10,25]. To ensure consistency, all the aforementioned benchmark models are implemented using the Scikit-learn toolbox [17]. By evaluating our proposed model against these established methods, we aim to provide a comprehensive performance comparison and assess the effectiveness of our approach in the context of security asset risk assessment.

Evaluation Protocols. We report the evaluation protocols in this section, which include accuracy, precision, and F1 score as the evaluation protocols. These metrics are widely used in the evaluation of security asset risk assessment.

Table 2. Performance (in percentage) comparison of models in risk assessment.

Model	Accuracy	Precision	F1
LR	66.93	77.87	53.67
NB	66.94	77.87	53.69
KNN	66.66	58.34	57.63
SVM	67.21	72.02	56.42
RF	81.42	81.60	81.50
CART	81.45	81.64	81.54
SeAAN	87.28	87.18	87.22

Accuracy is calculated by assessing the ratio of correctly classified instances to the total number of instances. Precision evaluates the proportion of true positive predictions in relation to the total number of positive predictions. Lastly, the F1 score represents the harmonic mean of precision and recall, offering a balanced metric for gauging model performance.

For each evaluation protocol, we calculate its corresponding value based on the predictions made by our model. Subsequently, the outcomes are juxtaposed with the ground truth labels to evaluate the model's performance in terms of accuracy, precision, and F1 score.

Implementation Details. The hyperparameters' values are established through an evaluation of the F1 score on each respective validation set. The length of the historical graph sequence is set to 3, the number of attention heads is set to 8, and the embedding dimension is set to 200. The model parameters are initialized employing the Xavier initialization technique [5], and the Adam optimizer [9] with a learning rate of 0.001 is utilized for optimization. To optimize all the hyperparameters of the model, a grid search combined with early stopping is employed on the validation dataset. The training process is limited to 30 epochs, which is generally sufficient for convergence in most cases. The complete model is constructed using the PyTorch framework [16] and executed on an NVIDIA GeForce RTX 3090 GPU.

4.2 Experimental Results

The experimental outcomes in the risk assessment task demonstrate that our proposed model surpasses the performance of the baseline methods across all evaluation metrics, encompassing accuracy, precision, and F1 score. The detailed performance comparison is outlined in Table 2.

Our model demonstrates superior performance in accurately assessing security asset risks. It achieves high accuracy in classifying instances and exhibits excellent precision in minimizing false positives. Collectively, the F1 score serves as a testament to the resilience and efficacy of our proposed model in the realm

Table 3. Results (in percentage) by different variants of our model on the security asset dataset.

SAN	TSAN	ARCN	RNN	Accuracy	Precision	F1
✗	✗	✓	✗	66.57	67.26	66.76
✗	✓	✓	✗	69.68	68.99	69.24
✓	✗	✓	✗	85.72	85.58	85.64
✗	✗	✓	✓	68.48	67.49	67.27
✓	✗	✓	✓	86.12	86.10	86.09
✓	✓	✓	✓	87.28	87.18	87.22

of risk assessment. These findings underscore the superiority of our model when compared to baseline methods, reaffirming its capacity to precisely appraise and mitigate security asset risks.

4.3 Ablation Study

To scrutinize the individual contributions of various components within the SeAAN model, an ablation study is conducted. Variants of SeAAN are created by adjusting the utilization of model components, and the performance in risk assessment on the security asset dataset is compared. The results in Table 3 reveal insights into the importance of the structural attention network and temporal self-attention network in SeAAN.

Omitting the structural attention network results in a notable reduction of 26% in the F1 score, accompanied by substantial declines in other evaluation metrics. This indicates that aggregating local neighborhood information of nodes is beneficial for the risk assessment task. On the other hand, the exclusion of the temporal self-attention network results in a 2% drop in the F1 score, emphasizing the effectiveness of modeling historical information. These findings elucidate the promising performance of SeAAN, which stems from its ability to learn from both the structural aspects of the graph and the temporal evolution. Furthermore, even with only the asset risk classification network, the variant models achieve comparable performance, suggesting the effectiveness of the model in capturing essential risk-related patterns.

To further validate the importance of the TSAN, we introduce two variant models that replace the TSAN with a recurrent neural network (RNN) to capture the temporal evolution of security assets. The experimental findings offer valuable insights into the proficiency of RNNs in capturing temporal information for the purpose of risk assessment. However, it is observed that the performance of the RNN-based variants falls short compared to the models utilizing the TSAN. This finding underscores the superior efficacy of the TSAN in capturing and leveraging this temporal evolution of security assets for risk assessment.

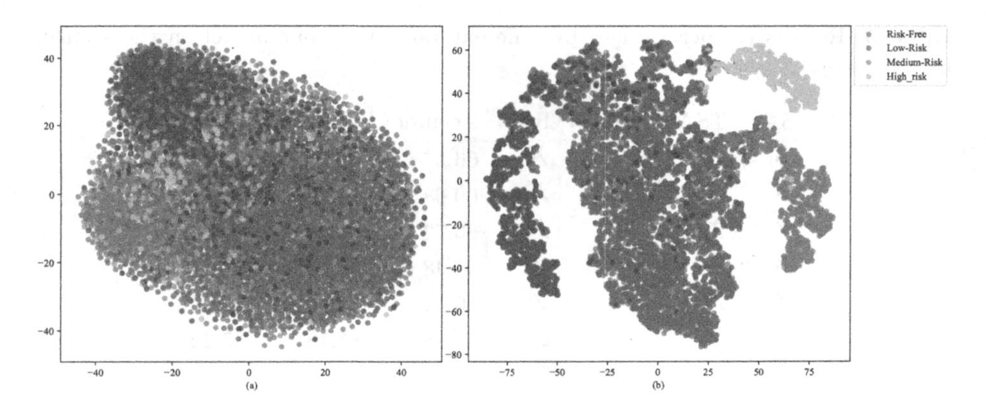

Fig. 2. t-SNE visualization of security asset node representations. (a) Visualization of security asset node representations without SeAAN processing. (b) Visualization of security asset node representations after SeAAN processing.

4.4 Visualization of Security Asset Node Representations

To gain visual insights into the efficacy of SeAAN in the risk assessment task, t-SNE [15] visualizations of the security asset nodes are provided, both before and after undergoing processing by SeAAN. These visualizations serve the purpose of intuitively observing the performance of SeAAN in distinguishing different classes of security assets.

As shown in Fig. 2, the t-SNE visualization of the nodes without SeAAN processing reveals a mixture of different classes, making it challenging to discern distinct patterns. In contrast, the t-SNE visualization of the nodes after SeAAN processing exhibits clearly separated clusters, facilitating risk classification. This visualization showcases the ability of SeAAN to effectively learn and represent the underlying structures and characteristics of security assets, enabling more accurate and meaningful risk assessment.

5 Conclusion

In order to better evaluate the risk of domain security assets, a Security Asset Attention Network named SeAAN is proposed, which is a comprehensive consideration of temporal knowledge graph topology and adjacent node sequences. Empirical assessments on real-world datasets demonstrate that our model achieves an accuracy of 87.28%, precision of 87.18%, and an F1 score of 87.22%. Notably, these results indicate a more significant performance enhancement compared to conventional methods.

Acknowledgements. This work was supported in part by the National Key Research and Development Program of China under Grant 2020YFB1712203, and in part by the fund from Qi An Xin Technology Group Inc.

References

1. Baqinejadqazvini, A., Tahery, S., Farzi, S.: Dynamic knowledge graph completion through time-aware relational message passing. In: 2023 28th International Computer Conference, Computer Society of Iran (CSICC), pp. 1–5. IEEE (2023)
2. Bhosale, P., Kastner, W., Sauter, T.: A centralised or distributed risk assessment using asset administration shell. In: 2021 26th IEEE International Conference on Emerging Technologies and Factory Automation (ETFA), pp. 1–4. IEEE (2021)
3. Chehida, S., et al.: Asset-driven approach for security risk assessment in IoT systems. In: Garcia-Alfaro, J., Leneutre, J., Cuppens, N., Yaich, R. (eds.) CRiSIS 2020. LNCS, vol. 12528, pp. 149–163. Springer, Cham (2021). https://doi.org/10.1007/978-3-030-68887-5_9
4. Fu, S., Zhou, H., Xiao, Y.: Research on information system assets risk assessment and defense decision-making. J. Amb. Intell. Human. Comput. **14**(2), 1229–1241 (2023)
5. Glorot, X., Bengio, Y.: Understanding the difficulty of training deep feedforward neural networks. In: Proceedings of the thirteenth international conference on artificial intelligence and statistics, pp. 249–256. JMLR Workshop and Conference Proceedings (2010)
6. Harish, A.R., Liu, X., Zhong, R.Y., Huang, G.Q.: Log-flock: a blockchain-enabled platform for digital asset valuation and risk assessment in e-commerce logistics financing. Comput. Indus. Eng. **151**, 107001 (2021)
7. Hong, C., Tan, M., Wang, S., Wang, J., Li, M., Qiu, J.: Small and medium-sized enterprises credit risk assessment based on temporal knowledge graphs. In: 2021 IEEE 20th International Conference on Cognitive Informatics & Cognitive Computing (ICCI* CC), pp. 166–172. IEEE (2021)
8. Jia, Z., Li, H., Chen, L.: Air: adaptive incremental embedding updating for dynamic knowledge graphs. In: Wang, X., et al. (eds.) Database Systems for Advanced Applications. DASFAA 2023. LNCS, vol. 13944, pp. 606–621. Springer, Cham (2023). https://doi.org/10.1007/978-3-031-30672-3_41
9. Kingma, D.P., Ba, J.: Adam: a method for stochastic optimization. arXiv preprint arXiv:1412.6980 (2014)
10. Lessmann, S., Baesens, B., Seow, H.V., Thomas, L.C.: Benchmarking state-of-the-art classification algorithms for credit scoring: an update of research. Eur. J. Oper. Res. **247**(1), 124–136 (2015)
11. Li, Y., Zhang, X., Zhang, B., Ren, H.: Each snapshot to each space: Space adaptation for temporal knowledge graph completion. In: Sattler, U., et al. (eds.) The Semantic Web. ISWC 2022. LNCS, vol. 13489, pp. 248–266. Springer, Cham (2022). https://doi.org/10.1007/978-3-031-19433-7_15
12. Li, Y., et al.: Learning dynamic user interest sequence in knowledge graphs for click-through rate prediction. IEEE Trans. Knowl. Data Eng. **35**(1), 647–657 (2021)
13. Liu, C., Yang, S.: Using text mining to establish knowledge graph from accident/incident reports in risk assessment. Expert Syst. Appl. **207**, 117991 (2022)
14. Lyu, M., Gharakheili, H.H., Sivaraman, V.: A survey on enterprise network security: Asset behavioral monitoring and distributed attack detection. arXiv preprint arXiv:2306.16675 (2023)
15. Van der Maaten, L., Hinton, G.: Visualizing data using t-SNE. J. Mach. Learn. Res. **9**(11), 1–27 (2008)
16. Paszke, A., et al.: Pytorch: an imperative style, high-performance deep learning library. In: Advances in Neural Information Processing Systems, vol. 32 (2019)

17. Pedregosa, F., et al.: Scikit-learn: machine learning in python. J. Mach. Learn. Res. **12**, 2825–2830 (2011)
18. Priyanka, A., Monti, A.: Towards risk assessment of smart grids with heterogeneous assets. In: 2022 IEEE PES Innovative Smart Grid Technologies Conference Europe (ISGT-Europe), pp. 1–6. IEEE (2022)
19. Shakibazad, M., Rashidi, A.J.: New method for assets sensitivity calculation and technical risks assessment in the information systems. IET Inf. Secur. **14**(1), 133–145 (2020)
20. Tang, C., Zhang, H., Loakman, T., Lin, C., Guerin, F.: Enhancing dialogue generation via dynamic graph knowledge aggregation. In: The 61st Annual Meeting of the Association for Computational Linguistics (2023)
21. Yang, Y., Huang, C., Zhang, H., Feng, C., Wang, Z., Cui, Z.: Research on airspace security risk assessment technology based on knowledge graph. In: 2021 IEEE 21st International Conference on Software Quality, Reliability and Security Companion (QRS-C), pp. 980–986. IEEE (2021)
22. Zandi, P., Rahmani, M., Khanian, M., Mosavi, A.: Agricultural risk management using fuzzy TOPSIS analytical hierarchy process (AHP) and failure mode and effects analysis (FMEA). Agriculture **10**(11), 504 (2020)
23. Zhang, J., et al.: Evaluating water resource assets based on fuzzy comprehensive evaluation model: a case study of Wuhan City, China. Sustainability **11**(17), 4627 (2019)
24. Zhang, Y., Liu, Y., Zhu, S.: Research on the dynamic assessment of comprehensive risk measurement and investment performance of financial assets. In: Proceedings of the 2023 6th International Conference on Computers in Management and Business, pp. 30–36 (2023)
25. Zhu, S., Chen, Y., Wang, W.: Risk assessment of biological asset mortgage loans of china's new agricultural business entities. Complexity **2020**, 1–12 (2020)

Parameter Study on the Use of Artificial Intelligence to Optimize Response to Unattended Bags to Increase Airport Security

Olaf Milbredt(✉), Andrei Popa, and Christina Draeger

German Aerospace Center (DLR), Institute of Transportation Systems, Lilienthalplatz 7, 38108 Braunschweig, Germany
{olaf.milbredt,andrei.popa}@dlr.de

Abstract. As a vulnerable transportation hub, the airport can become the target of an attack at any time. Airport security, therefore, touches a fundamental aspect of our society: moving without fear. The attackers' methods become increasingly sophisticated. It is therefore essential to react quickly and adequately through unforeseen events. Every piece of unattended baggage—a commonly occurring incident—potentially poses a threat. The digitalization of airports has been advanced over the last decade. This circumstance opens the possibility to a broader use of Artificial Intelligence. AI is already being successfully used in individual areas of an airport. These range from intelligent video surveillance through border control to monitoring high-security areas. In this work, an exemplary environment was examined depicting the actions necessary to neutralize an unattended piece of baggage (e. g. a suitcase). The AI method Reinforcement Learning, especially the Deep Q-Network method, was used to train an agent to solve the challenge of choosing an optimal sequence of actions. This special method gives rise to a set of parameters, namely learning rate, batch size and the number of iterations. By means of a parameter study, a set of parameters was searched for enabling the agent to adequately solve the challenge.

Keywords: Airport Security · Artificial Intelligence · Reinforcement Learning

1 Introduction

The air transportation system faces a growing number of challenges. The methodology of terrorists is getting more and more sophisticated. In the past, plots of the attack founded the base for the introduction of new security measures. Starting from metal detectors through the detection of explosive substances to the detection of liquids [1,2]. Besides ensuring each flight to be secure, the airport, as a point where many people are piled up together, has to put a

© The Author(s), under exclusive license to Springer Nature Singapore Pte Ltd. 2024
F. Hassan et al. (Eds.): AsiaSim 2023, CCIS 1911, pp. 405–414, 2024.
https://doi.org/10.1007/978-981-99-7240-1_32

huge effort in security. The attack on Brussels airport in 2016 shows how prone airports are to terrorist attacks [3].

An unattended item, for example a suitcase, is always seen as harassment. The items are regularly found at airports. At the airport of Munich, Germany, an unattended item is found up to 30 times each day. Mostly, the owner of the unattended item can be found. In contrast, in 8 out of 30 cases a sniffer dog is being used. If the canine hits or cannot be used due to an unpropitious area for using the sniffer dog, a bomb squad is needed to neutralize the situation. This happens in up to three of 30 cases [4].

To assist security officers with revealing potential danger, AI is being used at airports ranging from video surveillance through border control to surveillance of high-security areas [5–7]. These AI systems operate independently without the possibility of exchanging information [8]. To overcome this problem, the vision of a novel security concept was presented in [8], which is holistically interpreted by an AI suggesting adequate responses.

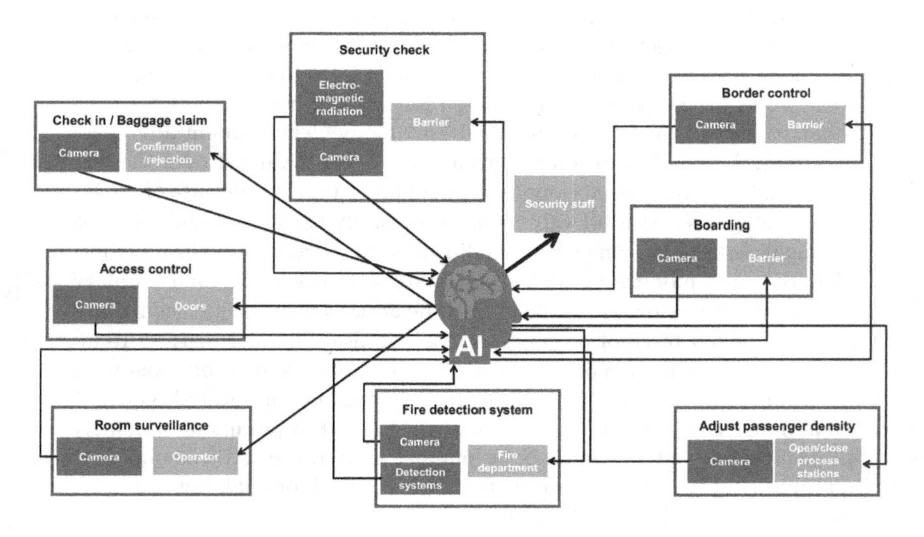

Fig. 1. Information flow within the vision of a holistic airport security including safety [8]. (Color figure online)

Figure 1 shows a scheme of the AI information flow. Different areas of the airport are shown with sensors (in blue) and actuators (in green). In the upper half are the security processes. By including safety (in the lower half), the AI is able to detect mutual ramifications.

The use of AI at the airport is primarily characterized by the use of images. This combination enables, for example, document- and contactless access to the security checkpoint and gate. Here, AI determines the identity of a passenger through facial recognition [9]. A similar approach is used for border control in the iBorderCtrl method [6]. Feature recognition in an image is also suitable for baggage. In [10], pieces of baggage are identified across airports based on their images.

Not every detected piece of baggage can be assumed to be an unattended item, since the primary characteristic of an unattended item is the non-traceability of the owner. For this reason, [11,12] have investigated the issue of how an object can be assigned to one or more persons and how the abandonment of this object can be detected. For example, there would be the possibility of cameras tracking the trajectory of an object moving away [11].

The system presented in this paper is not aimed at detection, but at controlling the processes for restoring security after the detection of an unattended item. It thus forms a complementary system to those presented above. As part of the AI system presented in [8] for a holistic control of security systems, it can be seen as interpreting the feedback from the systems described above.

In the following, a first methodological basis for the security concept mentioned in [8] is investigated. The method we used to train an AI is Reinforcement Learning. In this method, the agent learns to act autonomously in an environment. An action of the agent within the environment leads to a feedback from the environment in the form of a reward. In this paper, an exemplary environment was investigated representing the actions necessary to neutralize an unattended item.

The Deep Q-Network (DQN) method used here was presented in [13]. The authors tested the DQN method for the challenge to play a video game using solely the image data provided by the game. In order to check whether this method is also suitable for the application in the exemplary environment of this paper, a parameter study was performed on the parameters batch size, learning rate and number of iterations. Thus, on the one hand, it can be analyzed how the parameter settings affect the training results. On the other hand, it can be seen whether the specified goal—the fluctuation-free neutralization of the unattended item—can be achieved with the corresponding parameter setting. Fluctuation-free in this context refers to a standard deviation of zero of the mean of the maximum return of independent training runs of the agent at a given iteration. This means that a fluctuation-free neutralization of the unattended item is obtained if the fluctuation-freeness is fulfilled from a certain iteration onwards.

2 Methodology

2.1 Reinforcement Learning Method

This paper uses the Reinforcement Learning approach [14]. Here, an agent has to perform given actions in a given environment (see Fig. 2). For each action in a state of the environment, the agent receives feedback in the form of a new state and a numerical value (reward).

Here, the set of all actions and the set of all states are finite. Since the response of the environment depends only on the current state and the particular action of the agent, it is called a Marcov Decision Process [14].

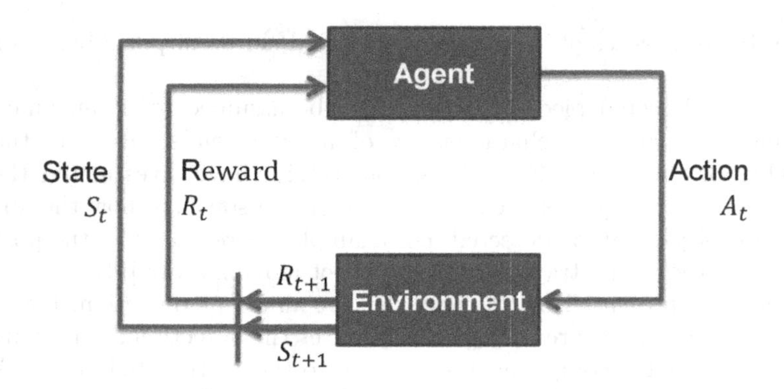

Fig. 2. Agent-environment interaction for Reinforcement Learning [14].

The task of an agent is to find a sequence of actions that maximizes the sum of all rewards ("return"). In the beginning, there is no experience for the agent to build on, so it chooses actions randomly.

2.2 Description of the Environment

The underlying idea behind the development of this environment is the neutralization of an unattended item at the airport. The unattended item used in this environment is 70% harmless baggage and 30% disguised explosive devices. This allocation was adopted because if the explosion rate would be orders of magnitude (power of ten) smaller, the algorithm would require many more iterations to experience an explosion. An explosion rate that is orders of magnitude smaller results in a much longer runtime, which was avoided for this initial methodological baseline study.

The environment is built on a framework for Python. This consists of a Reinforcement Learning structure as well as reward function and action execution. The actions in the implemented environment are based on the processes executed in reality when an unattended item is detected. These are here shutting off areas, deployment of the sniffer dog or calling the bomb squad.

The sniffer dog determines whether an active explosive device is present in the unattended item. The bomb squad can neutralize it. The shutting off results in an evacuation of the affected area so that no persons would be harmed. The reward function consists of two parts. The first part considers costs and the second part takes into account the extent of damage. Each action is associated with fixed costs, the magnitude of which corresponds to the effort of that action. In the case of detonation, the damage consists of property damage and personal injury. The former was determined using estimated costs for property damage per square meter. For personal damage, costs also had to be assumed in order to perform an addition. These were estimated to be three orders of magnitude (10^3) higher than the property damage.

The agent's task is to perform actions that result in a neutralization of the unattended item. To do this, the agent's return should be maximized. of a sentence.

2.3 Description of the Algorithm

In this paper, we use the Deep Q-Network (DQN) method, which had been successfully used several times [13]. It combines Q-learning with neural networks. The action-value function is a function depending on the particular state and the action performed in that state. It describes the return if the action is executed in the given state and a given strategy is followed. An optimal strategy is characterized by the best possible expected value of the return. Every optimal strategy has the same action-value function. If this function is known, an optimal strategy can in turn be reconstructed from it. Therefore, in this algorithm, the approximation of this function is refined step by step. Each optimal function satisfies the so-called Bellman equation, which was developed by Bellman in [15,16]. This can be used to construct an iteration rule. Such a sequence then converges to the optimal action-value function. In the case of DQN, a function with one parameter is used for the approximation [13].

DQN uses neural networks for function approximation due to the computer capacity now available allowing large networks with more than a thousand neurons in a layer. If a neural network is used to approximate the action-value function, instabilities can occur [17]. To counter these, the authors in [13] introduced the use of experience replay and periodically updated target values for the action-value function.

Various parameters are used in the DQN algorithm. In this paper, the parameters batch size, number of iterations and learning rate are varied. The batch size is the number of test cases randomly selected from the set of already experienced situations (experience replay). These selected experience cases serve as additional input to the learning process. The learning rate feeds into the Stochastic Gradient Descent (SGD) algorithm to minimize the error resulting from the Bellman equation. The gradient of the error is a vector pointing in the direction of the largest increase in the error function. A minimization is achieved by adjusting the point in the opposite direction.

The goal of this paper is to identify the best set of parameters. This means a result as free of fluctuations as possible, which can be achieved with a number of iterations as small as possible and results in the neutralization of the unattended item.

2.4 Realization of the Parameter Study

The following values were selected to perform the parameter study, namely

- batchsize = {64, 128, 256, 512},
- number of iterations = {5000, 10000, 20000, 40000} and
- learning rate = {0.005, 0.001, 0.0005, 0.0001}.

These cover the parameter space after previous test runs. Since any fluctuation behavior only occurred with a larger number of iterations, the smallest value specified here was used in this study. At greater iterations than the largest specified, similar fluctuation behavior was observed. A learning rate greater than 0.005 or a learning rate less than 0.0001 also resulted in behavior with larger fluctuation. With a batch size of 16, just as with an iteration number smaller than 2000, the desired goal was not achieved.

The outcome of a learning process was evaluated within 10 episodes—a sequence of states, actions, and rewards, concluding with a final state—and the mean was calculated. This was done in two identical environments to test generalizability. To decouple the results from the specific learning history, 20 runs were used. Of these, mean and standard deviation were calculated for each data point.

3 Results/Discussion

The range given above for the iterations was chosen because larger fluctuations can occur after a smaller number of iterations, as can be seen representatively in Fig. 3a.

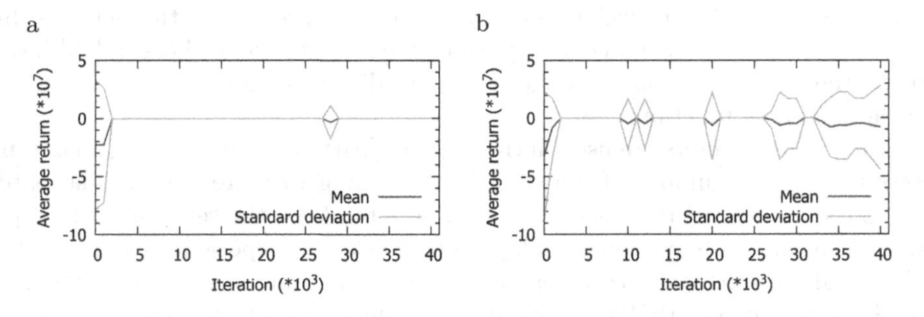

Fig. 3. Fluctuations of the average return during training of the agent with the parameters (a) batchsize: 256, iterations: 40,000, learning rate: 0.001; (b) batchsize: 256, iterations: 40,000, learning rate: 0.005.

For the largest learning rate of 0.005, the results fluctuated by more than 50% around the mean. For a batch size of 256, the result of the 40,000th iteration was more than 440% (see Fig. 3b).

At the smallest learning rate of 0.0001, on average at least 20,000 iterations were required to meet the target (neutralization of the unattended item) (see Fig. 4a).

It is difficult to see whether the fluctuation actually disappears. Additionally, we found that the fluctuation can disappear, but throughout the training increases again.

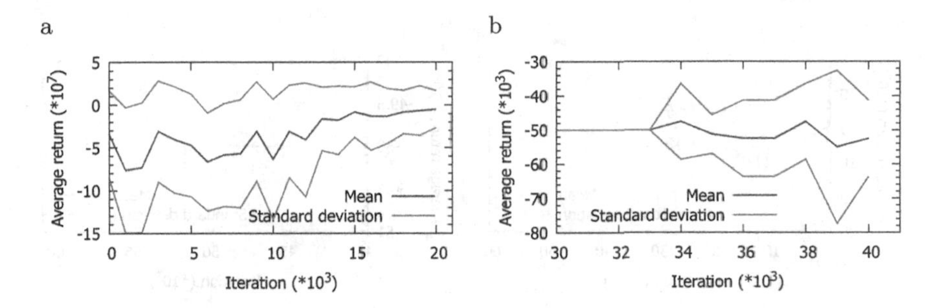

Fig. 4. Fluctuations of the average return during training of the agent with the parameters (a) batchsize: 128, iterations: 20,000, learning rate: 0.0001; (b) By scaling visible smaller fluctuations of the results during training of the agent with the parameters batchsize: 512, iterations: 40,000, learning rate: 0.0001;

In Fig. 4b, this behavior can be seen. In this case, the fluctuation of the results was about 30%. A faster and less fluctuating neutralization was seen for the learning rates 0.001 and 0.0005. At the learning rate 0.0005, the standard deviation was zero over several thousand iterations, but increased to 15–50% for larger iterations similar to Fig. 4b. For the learning rate 0.001, the standard deviation was at least 20% for the batch size 512. For batch sizes 64 and 256, the fluctuations were in the range of 15–50% with few exceptions.

In addition to the specified range of batch size, batch size 32 was used. Results from the previous training runs were used for this purpose. In order to get a broad picture, the additional training runs were performed with the iterations 10,000 and 40,000 as well as the two learning types 0.001 and 0.0001. The learning rate 0.001 reached the goal of neutralization at 40,000 iterations, but with fluctuations of at least 30%. For the smaller learning rate (0.0001), more than 10,000 iterations were required to achieve the goal of neutralizing the unattended item, as was evident in the previous training runs. For 40,000 iterations, the target was achieved without fluctuation from iteration 33,000.

To exclude the occurrence of fluctuation at a later iteration step, a training run with 60,000 iterations was performed (see Fig. 5a). Here, fluctuation-free target achievement was obtained from the 38,000th iteration (see Fig. 5b). The optimal parameter set for the given problem in this paper is represented by the batch size 32 and the learning rate 0.0001 at an iteration of at least 40,000 (see Fig. 5b).

In the following, the results are analyzed in their suitability. The largest learning rate 0.005 is unsuitable for the learning process, since fluctuations up to 440% can occur. For no batch size, the fluctuations drop below 50%. Decreasing the learning rate by one order of magnitude (0.0005; 0.0001) sometimes resulted in phases without fluctuation, but these occurred only temporarily in the learning process. The fluctuation was present for each batch size. With the given parameter combinations, no fluctuation-free neutralization of the unattended item was possible. Only with the additional tests with batch size 32 it turned out that a

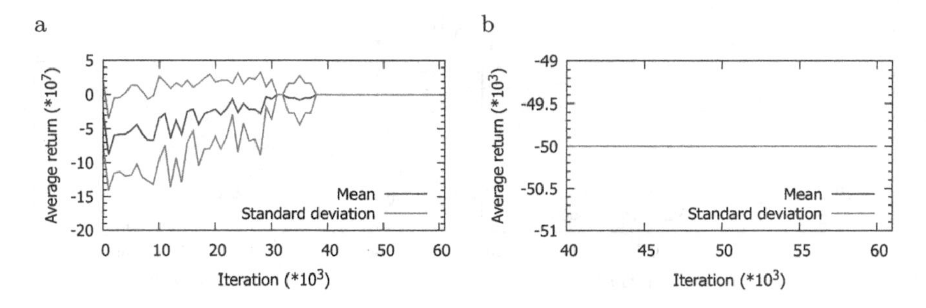

Fig. 5. (a) Evolution of the result during training of the agent with the parameters batchsize: 32, number of iterations: 60,000, learning rate: 0.0001; (b) Zoom of the left image for the iterations starting at 40,000/.

target achievement was possible. However, it was necessary to select a smaller learning rate (0.0001). A fluctuation-free solution to the problem was achieved with this batch size and a learning rate of 0.0001. This behavior was confirmed with an iteration count of 60,000.

During the execution of the training runs, it was observed that the duration of a training session increases with a larger batch size (number of iterations and learning rate fixed) on the one hand and with a smaller learning rate (number of iterations and batch size fixed) and a larger number of iterations (batch size and learning rate fixed) on the other hand. A variation-free solution to the problem was only achieved with a learning rate of 0.0001 and a batch size of 32 (see Fig. 5b).

4 Conclusion

In this paper, we present a parameter study for solving the challenge to choose the optimal actions to neutralize an unattended item. We used 20 independent runs to decouple the results from the specific learning history. The standard deviation over this 20 training runs for one parameter set is a measure for the quality of the results. This approach was necessary, since evaluating only a few runs may distort judging the performance of a parameter set.

For a parameter set beyond the boundaries of the chosen region for the parameters the results either the specified goal was not achieved or the calculation time was high and no better results were produced. A batch size of 32 has revealed to be an exception of the latter statement.

The results show that a large learning rate (0.005) did not result in fluctuation-free neutralization of the unattended item. The fluctuations ranged from 0.1% to 440%. At a learning rate of 0.001, a fluctuation of at least 15% was observed. A fluctuation-free solution to the task was obtained with a learning rate of 0.0001 and a batch size of 32.

The possibility of using AI as a new method of preventive monitoring of the airport terminal, especially in the selection of measures in case of an unattended item, is supported by the obtained results. The environment implemented here with the learning method DQN leads to the goal with the help of using the determined set of parameters—that is, making decisions to solve the problem without fluctuations, in this case, to neutralize an unattended item.

The implementation of the environment represents a small part of the actions when an unattended item is found. To approach the vision of the novel security concept from [8], the next step consists of an extension of the environment including the time dependency when neutralizing the unattended item. The spatial conditions as well as the position of the required resources and the personnel to be deployed are to be considered.

References

1. European Commission: Communication from the Commission to the European Parliament and the Council on the use of security scanners at EU airports (2010). https://eur-lex.europa.eu/legal-content/EN/TXT/?uri=CELEX
2. Olson, G.: Some European airports plan to end liquid limits for carry-ons, but when will the US? Thrifty Traveler (2022). https://thriftytraveler.com/news/travel/carry-on-liquid-restrictions/
3. Beake, N.: Brussels attacks: trial begins over 2016 attacks that killed 32. BBC News Europe (2022). https://www.bbc.com/news/world-europe-63834777
4. Moritz, H.: Das macht die Bundespolizei mit herrenlosem Gepäck. Merkur (2020). https://www.merkur.de/lokales/erding/flughafen-muenchen-ort60188/flughafen-muenchen-bundespolizei-herrenloses-gepaeck-so-reagiert-polizei-13311422.html
5. Donadio, F., Frejaville, J., Larnier, S., Vetault, S.: Artificial intelligence and collaborative robot to improve airport operations. In: Auer, M.E., Zutin, D.G. (eds.) Online Engineering & Internet of Things. LNNS, vol. 22, pp. 973–986. Springer, Cham (2018). https://doi.org/10.1007/978-3-319-64352-6_91
6. Jupe, L.M., Keatley, D.A.: Airport artificial intelligence can detect deception: or am I lying? Secur. J. **33**(4), 622–635 (2020)
7. Koroniotis, N., Moustafa, N., Schiliro, F., Gauravaram, P., Janicke, H.: A holistic review of cybersecurity and reliability perspectives in smart airports. IEEE Access **8**, 209802–209834 (2020)
8. Milbredt, O., Popa, A., Doenitz, F., Hellmann, M.: Neuartiges Konzept der Sicherheitsarchitektur eines Flughafens – Ganzheitliche Interpretation der Sicherheitsinfrastruktur am Flughafen mithilfe von KI. Internationales Verkehrswesen (3), 27–31 (2022)
9. Lufthansa: Star alliance biometrics. https://www.lufthansa.com/de/de/star-alliance-biometrics
10. Beck, A.: Neural Network for AutoID. PSI Logistics GmbH (2021). https://www.psilogistics.com/fileadmin/files/downloads/PSI_Logistics/PR_Media/Fachartikel/GATE_INSPIRE_Issue1_PSIairport.pdf
11. Arsić, D., Schuller, B.: Real time person tracking and behavior interpretation in multi camera scenarios applying homography and coupled HMMs. In: Esposito, A., Vinciarelli, A., Vicsi, K., Pelachaud, C., Nijholt, A. (eds.) Analysis of Verbal and Nonverbal Communication and Enactment. The Processing Issues. LNCS, vol. 6800, pp. 1–18. Springer, Heidelberg (2011). https://doi.org/10.1007/978-3-642-25775-9_1

12. Soh, Z.H.C., Kamarulazizi, K., Daud, K., Hamzah, I.H., Saad, Z., Abdullah, S.A.C.: Abandoned baggage detection & alert system via AI and IoT. In: Proceedings of the 2020 12th International Conference on Computer and Automation Engineering, ICCAE 2020, pp. 205–209. Association for Computing Machinery, New York, NY, USA (2020)
13. Mnih, V., et al.: Human-level control through deep reinforcement learning. Nature **518**, 529–533 (2015)
14. Sutton, R.S., Barto, A.G.: Reinforcement Learning, An Introduction. 2nd edn. MIT Press (2018)
15. Bellman, R.E.: Dynamic Programming. Princeton University Press, Princeton (2010)
16. Bellman, R.E.: A Markovian decision process. J. Math. Mech. **6**(5), 679–684 (1957). http://www.jstor.org/stable/24900506
17. Tsitsiklis, J., Van Roy, B.: An analysis of temporal-difference learning with function approximation. IEEE Trans. Autom. Control **42**(5), 674–690 (1997)

Revolutionizing Plant Disease Detection with CNN and Deep Learning

Fariha Tabassum, Imtiaj Ahmed(✉), Mahmud Hasan, Adnan Mahmud, and Abdullah Ahnaf

Department of Computer Science and Engineering, East West University, Dhaka 1212, Bangladesh
2018-3-60-076@std.ewubd.edu

Abstract. Recent advancements in plant leaf disease detection using image analysis have sparked a reevaluation of farming practices, especially in agriculture-dependent countries like Bangladesh. Automatic detection systems provide timely identification of roots, stems, leaves, and fruits, allowing farmers to capture and diagnose plant diseases through images easily. This approach saves time and reduces costs while enabling remote monitoring and support for farmers. Our paper introduces a CNN-based technology for plant disease detection and assesses its performance using various algorithms, including Alexnet, Googlenet, Resnet50, VGG16, VGG19, Darknet53, Shufflenet, Squzzenet, Mobilenetv2, Inceptionv3, Inceptionresnetv2, Efficientnetb0, Desnet201, Xception, Nasnetlarge, Yolov7, and Nasnetmobile. Among these, Yolov7 exhibited exceptional validation accuracy of 96.1%, showcasing its effectiveness in disease localization and identification. This study highlights the necessity of deep learning optimizers in improving picture classification outcomes, particularly for plant disease diagnosis, and recommends improvements in deep learning architectures and optimization techniques.

Keywords: Plant diseases · Deep learning · Automated Diagnosis · CNN algorithms · Alex net · Resnet50 · Yolov7

1 Introduction

Plant disease is well-known that the Asian nation has a difficult and costly method for diagnosing and categorizing plant diseases. Deep learning (DL) approaches for automating the categorization of plant diseases have been used by researchers to get over these obstacles. In addition to lowering the need for chemical sprays, DL-based models have shown they can increase the amount and quality of agricultural goods. The performance of DL-based techniques has been shown to be superior to that of conventional machine learning (ML) models. A number of strategies have been investigated to improve plant disease classification, including alterations to well-known DL models, various training methodologies, data augmentation techniques, cascaded versions of effective DL architectures, and visualization methods.

© The Author(s), under exclusive license to Springer Nature Singapore Pte Ltd. 2024
F. Hassan et al. (Eds.): AsiaSim 2023, CCIS 1911, pp. 415–425, 2024.
https://doi.org/10.1007/978-981-99-7240-1_33

The study highlights the value of using deep learning optimizers to complete picture classification jobs more effectively. It does point out that prior research has not specifically suggested DL architectural upgrades or the application of optimization techniques for plant disease categorization. DL architectures have been effectively used for a variety of agricultural applications, including leaf and fruit counting, identifying crop types, and identifying and categorizing plant diseases. This review article gives a brief summary of current advancements in visualization approaches and modified/cascaded versions of popular DL models for identifying plant diseases. It also identifies knowledge gaps and offers suggestions for more research.

The paper also covers the relevance of autonomous plant disease diagnosis and the many approaches used, such as spectral sensing, machine learning and image processing coupled, and deep learning using convolutional neural networks. Although these techniques have been used in several studies to identify different plant species that have diseases, further analysis is needed to see how well they hold up in less-than-perfect circumstances. The analysis highlights the promise of DL-based models for automating the categorization of plant diseases overall and identifies directions for future research to improve the precision and robustness of these models.

2 Literature Review

Arnab et al. [1] explored a deep-learning architecture based on convolutional neural networks (CNN) for plant disease detection. Although their model achieved high success rates, limitations were identified, indicating room for future improvements. In an experiment, the authors assessed their model's performance using images different from the training dataset, achieving an accuracy rate of 31%. They also evaluated the model in real-time conditions, attaining a 33% accuracy rate. Furthermore, we intend to advance this research by implementing novel deep-learning models such as ACNet, ViT, and MLP Mixer for plant disease identification, and assess their performance.

Narendra et al. [2] proposed the utilization of the Kaggle dataset, consisting of labeled healthy and leaf blast images, to train a CNN model.

MANOJ et al. [3] implemented an approach where they excluded papers that did not specifically address the detection and classification of Rice or other plant diseases using deep learning or CNN.

Bulent et al. [4] identified a fundamental problem in using CNN for plant disease detection and classification as the requirement for large datasets. Furthermore, external factors such as weather might have an influence on data gathering, making it a time-consuming operation that may take several days of effort. As previously reported, most CNN models for plant disease diagnosis used big-picture datasets. In their own investigation, the researchers used three different versions of the PlantVillage dataset to diagnose plant illnesses and evaluated how the picture backdrop affected disease diagnosis accuracy.

In their work, Arun et al. [5] used the suggested 14-DCNN model to categorize cherry healthy and strawberry leaf scorch classes. Table 5 displays the model's classification performance on individual classes. Figure 7 compares the performance of the proposed 14-DCNN model to that of various state-of-the-art classification algorithms, measuring accuracy, weighted average precision, weighted average recall, and weighted average F1 score.

Hsing et al. [6] highlighted the promise of artificial intelligence (AI) technology, particularly deep learning using convolutional neural networks (CNN), for disease diagnosis in tomato plants in their study. They suggested using CNN, especially the AlexNet architecture, in a smartphone application that allows tomato producers to diagnose illnesses based on changes in leaf appearance. The researchers hoped to give a viable and accessible solution for farmers in reliably identifying tomato illnesses using leaf pictures collected by smartphone cameras by utilizing AI and CNN technologies.

In their study, Gupta et al. [7] created a system for detecting plant diseases using a dataset comprising labelled leaf images. The focus was on Tomato leaves, and the dataset included samples for 5 types of Tomato diseases as well as healthy leaves, resulting in 4 classes for algorithm classification. Rather than utilizing pre-existing models, the researchers opted to develop a custom model from scratch, as depicted in the provided Figure. The dataset consisted of around 4500 images, with 60% allocated for training, 20% for validation, and 20% for testing purposes.

According to Ashiqul et al. [8], further research might broaden the scope of this work by integrating a broader range of paddy leaf diseases and investigating more sophisticated CNN models to obtain greater accuracy and faster diagnosis. To understand the different aspects influencing plant disease identification, such as dataset diversity, dataset size, learning rate, and lighting conditions, a rigorous and comprehensive analysis is required. Ultimately, overcoming these constraints will provide the groundwork for more accurately identifying additional plant leaf diseases, with the findings from this study serving as a stepping stone.

In their study, Murk et al. [9] successfully applied deep learning algorithms for automated plant disease identification. Their program achieved an excellent 98% testing accuracy on a publicly available dataset and demonstrated consistent performance on images of Sukkur IBA University plants. These findings imply that convolutional neural networks (CNN) can be used to automate the detection and diagnosis of plant diseases. The authors advise expanding the collection with fresh real-world photographs to improve the classification of various plant and disease types, particularly in real-world settings.

Muhammad et al. [10] examined the identification of tomato illnesses using SVM classifiers based on HSI in their study. They evaluated the effectiveness of the classifiers using measures such as F1-score, accuracy, specificity, and sensitivity. In another study, wheat disease diagnosis was achieved with an 89% accuracy utilizing a Random Forest (RF) classifier paired with multispectral imagery. SVM classifiers were also used to detect plant illnesses using hyperspectral data, with an accuracy rate of more than 86%. A 3D-CNN strategy was developed utilizing hyperspectral images to diagnose Charcoal rot disease in soybeans, and the CNN model's performance was tested based on accuracy.

According to Johan et al. [11], their suggested model had the lowest accuracy and loss values among the modified/improved deep learning models studied. It did, however, necessitate more training time. The MLCNN design, which included a dropout layer after each max pooling layer and lowered the number of filters in the original AlexNet's initial convolution layers, produced a respectable F1 score. They also looked at a hybrid version of the AlexNet and VGG architectures, which performed well in terms of validation accuracy and F1 score. Nonetheless, this hybrid model contained the most parameters, resulting in a longer training period for each epoch.

PENG et al.'s study [12] evaluated how well several deep convolution networks performed at recognizing objects on ALDD, including VGGNet-16 AlexNet, InceptionV3, ResNet-101, ResNet-50, ResNet-34, ResNet-18, ResNet-50 and GoogLeNet. For training, the stochastic gradient descent (SGD) technique was applied with randomly chosen batch sizes. They also presented a hybrid model dubbed VGG-INCEP, which combined VGGNet-16 with InceptionV3, and which outperformed the conventional networks in terms of accuracy and convergence speed.

This paper [13] diagnosis of plant diseases is essential for agriculture and the economy. Traditional techniques are time- and labor-intensive. A potential fix is provided by deep learning and machine learning. Using the examples of tomato, rice, potato, and apple as a case study, this article explores their potential for detecting plant illnesses. It includes illness analysis, detection procedures, datasets, current models, difficulties, and potential future avenues for study.

This paper [14] plant diseases can reduce productivity and quality. Early illness detection is critical. This research describes a unique approach for detecting plant leaf disease using picture segmentation and classification. The technology outperformed traditional approaches in terms of accuracy, detecting illnesses with a 75.59% success rate.

This paper [15] explores color-based techniques used in agriculture to identify plant diseases. Due to its capacity to isolate color information, the Lab color space performs better. Among other techniques, the K-Means algorithm is a frequently used threshold-based segmentation tool.

3 Proposed Methods

We use various CNN base pre-trained models to pre-train our data set, and for doing this we need to use various optimization algorithms and have to control the loss using the loss function algorithm or process to handle the loss of the data. First of all,

- we try to augment our trained images both vertically and horizontally.
- Then we also do the operation of the rotation and scale, shear both vertically and horizontally.
- Also, we do the translation here to augment the images
- Then we augmented the images with [227 227 1] size dimension images
- Then we use the CNN network model described above

- Then we optimize our model's using the above operations/methods.
- Here we set the Initial learning rate at 0.0001
- Then we complete the training process by following the rest of the functional tasks
- Then we detect the testing data and the training data
- then we detect the test accuracy, validation accuracy precision score, F1-score and confusion matrix.

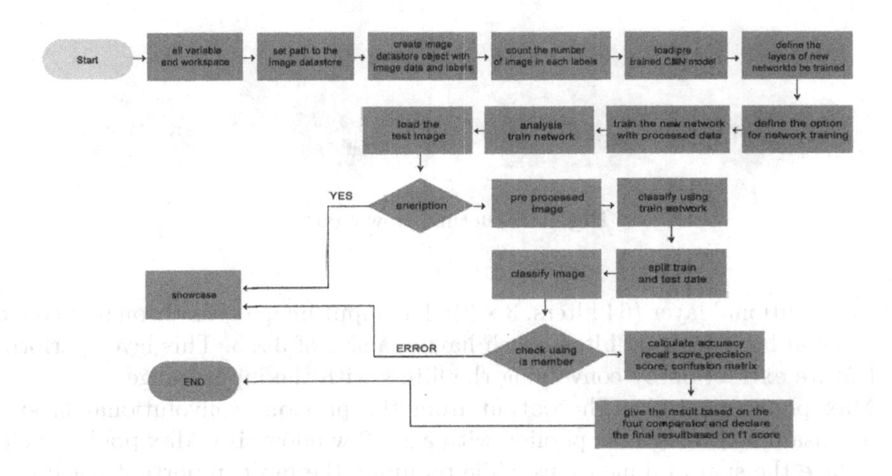

Fig. 1. Proposed Model.

The proposed plant disease prediction method takes input from the plant's leaves images. Figure 1 represents the block diagram of the proposed method. The working process described involves pre-training a dataset using various CNN base pre-trained models. Images are enhanced using augmentation techniques such as translation, scale and shear operations, rotation, and vertical and horizontal flipping. The photos are then downsized to [227 227 1] dimensions. The optimization process uses the CNN network model and is carried out with the techniques above, with an initial learning rate of 0.0001. The testing data accuracy, recall score, precision score, F1-score, and confusion matrix are established following the completion of the training procedure. The advantages of this strategy include enhanced data representation, higher model performance, and extensive assessment metrics.

4 Convolutional Neural Network

In Fig. 2, the architecture appears to be a convolutional neural network (CNN) for plant disease detection. Let's break it down step by step:

- Input: This is the input to the network, typically an image of a plant leaf or part of a plant.

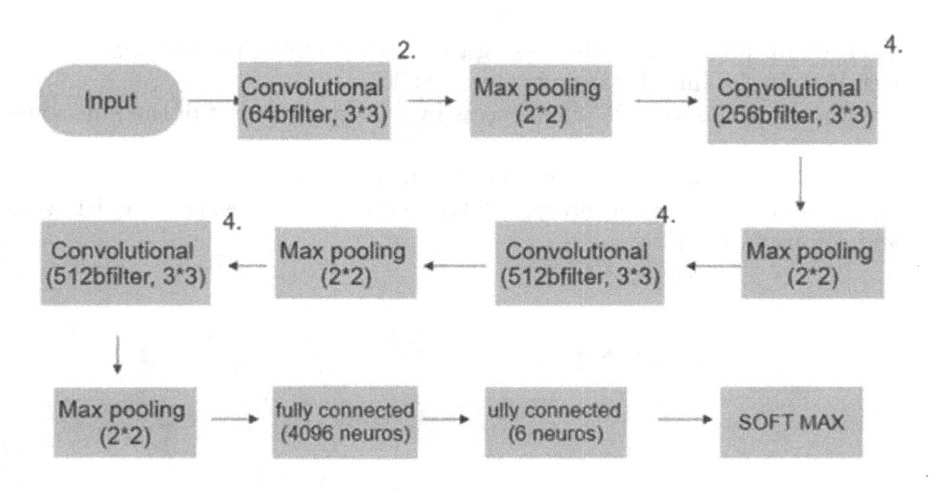

Fig. 2. Structural flow chart.

- Convolutional layer (64 filters, 3×3): The input image goes through a convolutional layer with 64 filters, each having a size of 3×3. This layer performs feature extraction by convolving the filters with the input image.
- Max pooling (2×2): The output from the previous convolutional layer is downsampled using max pooling with a 2×2 window size. Max pooling helps reduce the spatial dimensions while retaining the most important features.
- Convolutional layer (256 filters, 3×3): The downsampled output from the previous max pooling layer is passed through another convolutional layer with 256 filters of size 3×3. This layer continues to extract more complex features from the input.
- Convolutional layer (512 filters, 3×3): Another convolutional layer follows with 512 filters of size 3×3. This layer further captures higher-level features.
- Max pooling (2×2): The output from the previous convolutional layer is downsampled again using max pooling with a 2×2 window size.
- Convolutional layer (512 filters, 3×3): The downsampled output is passed through another convolutional layer with 512 filters of size 3×3. This layer aims to capture more abstract and intricate features.
- Max pooling (2×2): The output from the previous convolutional layer is downsampled using max pooling with a 2×2 window size.
- Max pooling (2×2): Another max pooling operation is applied, this time with a larger 2×2 window size.
- Fully connected layer (4096 neurons): The output from the last max pooling layer is flattened and passed through a fully connected layer with 4096 neurons. This layer helps in learning complex relationships between features.
- Fully connected layer (6 neurons): The previous fully connected layer is followed by another fully connected layer with 6 neurons, which corresponds to the number of classes or diseases you want to detect in plants.

- Softmax: The final layer applies a softmax activation function to produce probabilities for each class. Softmax ensures that the predicted probabilities sum up to 1, making it easier to interpret the results as class probabilities.

Overall, this architecture leverages convolutional layers for feature extraction and pooling layers for downsampling, followed by fully connected layers for classification. The softmax activation at the end provides the probabilities of each disease class for plant disease detection.

5 Experimental Result

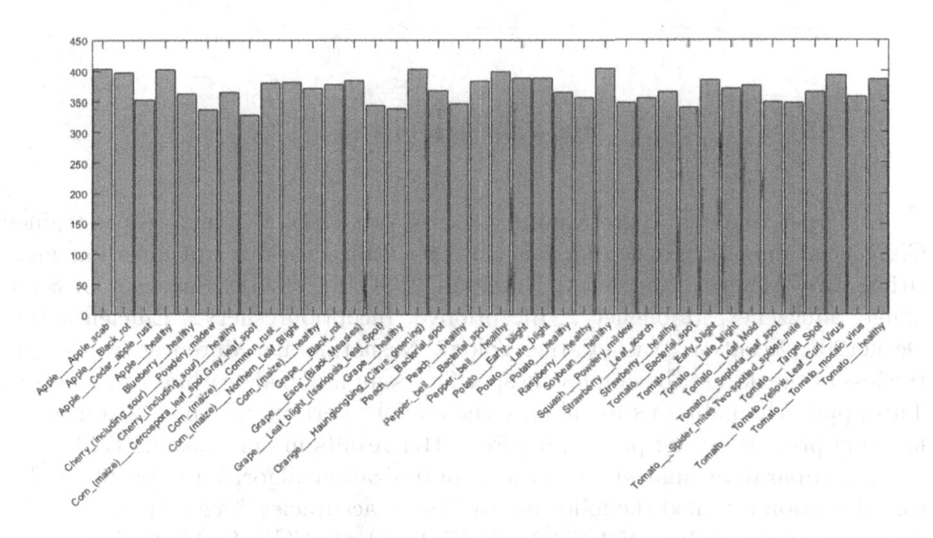

Fig. 3. Histogram of the detected plant disease

To enhance our plant disease detection system, we employed a comprehensive range of 17 optimization algorithms, including Alexnet, Googlenet, Resnet50, VGG16, VGG19, Darknet53, Shufflenet, Squzzenet, Mobilenetv2, Inceptionv3, Inceptionresnetv2, Efficientnetb0, Desnet201, Xception, Nasnetlarge, Yolov7, and Nasnetmobile. These algorithms were utilized to detect a total of 38 plant diseases, as illustrated in Fig. 3. We wanted to improve the accuracy and strength of our detection system by using such a broad range of optimization methods, ensuring accurate diagnosis and classification of plant illnesses for increased agricultural output and disease control.

Implementing automatic plant disease detection will save farmers time and money by enabling them to quickly diagnose and cure infections. The suggested CNN-based system, which uses a variety of algorithms, shows better illness recognition performance and accuracy. This technology revolutionizes how farmers combat crop diseases and raise agricultural output by enabling remote monitoring and prompt assistance.

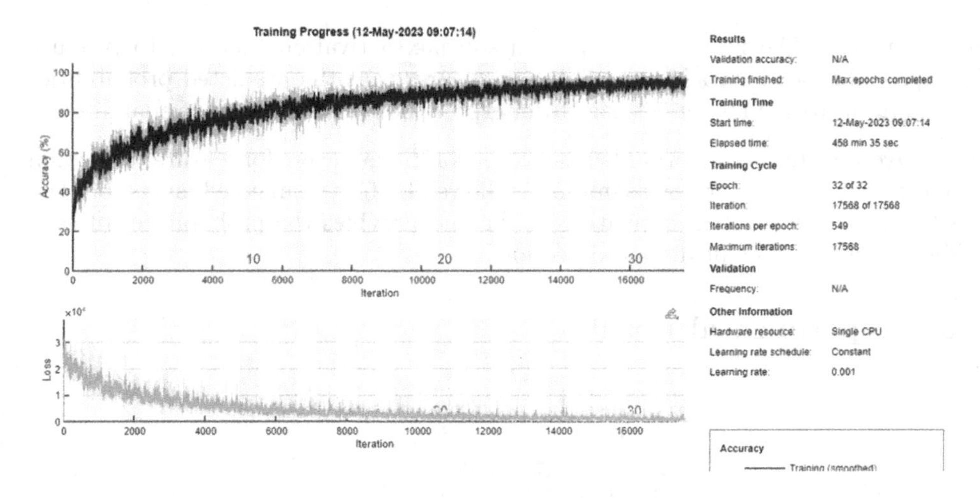

Fig. 4. Training Process

In Fig. 5, our data pretraining process, we employ multiple pre-trained CNN base models. To accomplish this, we utilize various optimization algorithms are Alexnet, Googlenet, Resnet50, VGG16, VGG19, Darknet53, Shufflenet, Squzzenet, Mobilenetv2, Inceptionv3, Inceptionresnetv2, Efficientnetb0, Desnet201, Xception, Nasnetlarge, Yolov7, Nasnetmobile. Moreover, we control the loss of the dataset by employing specific loss function algorithms or processes. This approach enables us to enhance the model's performance and optimize the learning process during pretraining for better results in our tasks in Table 1.

A comparative analysis of various optimization algorithms for plant disease detection revealed the following validation accuracies %age: Alexnet (94.8), Googlenet (92.8), Resnet50 (94.2), VGG16 (91.6), VGG19 (95.2), Darknet53 (89.8), Shufflenet (82.6), Squzzenet (94.8), Mobilenetv2 (87.1), Inceptionv3 (77.5), Inceptionresnetv2 (75.8), Efficientnetb0 (81.2), Desnet201 (85.8), Xception (86.5), Nasnetlarge (81.8), Yolov7 (96.1), and Nasnetmobile (81.0). Among these algorithms, Yolov7 exhibited the highest %age of validation accuracy of 96.1, showing its outstanding efficacy in diagnosing and localising plant diseases. This suggests that, when tested on a different validation dataset, Yolov7 outperformed the other algorithms in properly recognizing and localizing objects within pictures. It demonstrates that Yolov7's design and training procedure were successful in obtaining higher performance in object detection tests. These findings shed light on the efficacy of various algorithms for detecting plant diseases, leading to the selection of appropriate models to improve agricultural output and disease control techniques.

Table 1. Table of Models, % of their Accuracy, Precision and F1-score

Model	Training Accuracy	Validation Accuracy	Precision	F1-score
Alexnet	98.6	94.8	92.6	94.2
Googlenet	97.2	92.8	90.1	93.4
Resnet50	98.2	94.2	92.6	94.2
VGG16	96.6	91.6	89.5	92.4
VGG19	96.8	95.2	92.6	92.2
Darknet53	94.5	89.8	87.5	88.5
Shufflenet	89.1	82.6	81.6	81.8
Squzzenet	93.5	94.8	92.6	94.2
Mobilenetv2	90.2	87.1	87.2	87.4
Inceptionv3	82.5	77.5	76.8	76.9
Inceptionresnetv2	81.6	75.8	75.1	76.2
Efficientnetb0	86.6	81.2	79.6	80.2
Desnet201	90.5	85.8	83.6	84.4
Xception	91.1	86.5	84.6	95.2
Nasnetlarge	85.0	81.8	81.2	82.1
Yolov7	99.2	96.1	95.2	95.7
Nasnetmobile	85.4	81.0	79.9	80.6

The following figures provide an overview of the evaluation metrics used to assess our results, including accuracy, F1-score, recall score, precision score, and the confusion matrix. The accuracy of our detection method is measured by taking into account the average value and confusion avoidance inside the confusion matrix. The accuracy score detects real negatives whereas the recall score picks out false negatives. Precision and recall are combined in the F1-score, which strikes a compromise between the two.

This approach has the advantages of learning patterns of plant disease through feature extraction, spatial dimension reduction, capturing complex features, robustness to variations, accurate disease classification, probability-based predictions, flexibility for task adaptation, and scalability through adjustable parameters.

Figure 5 focuses on the importance of the F1-score in combining the accuracy and recall outcomes.

6 Limitation and Future Work

In our work, we analyze the effectiveness of several dataset identification techniques using some models as a benchmark. To improve the detection system's accuracy, we also expand our analysis by including more CNN models, using a

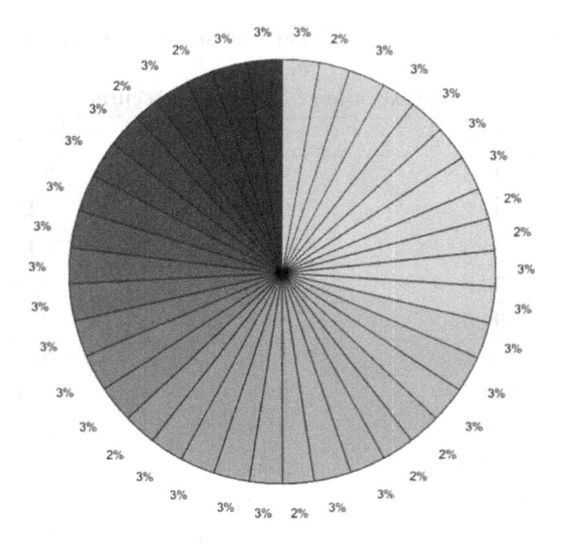

Fig. 5. F1-score pie chart

variety of optimization techniques, and using different loss functions. Additionally, we address the problem of undesirable and fuzzy photographs by putting in place methods to efficiently filter and process them. Our research focuses on enhancing the detection process over time by utilizing a variety of models, algorithms, and strategies to produce more reliable and precise outcomes.

7 Conclusion

Our model provides functionality with a broad range of CNN and deep learning-based image processing methods, allowing us to accurately recognize areas of damage and automatically detect plant diseases. The ability to offer focused ideas for handling illnesses improves this skill even further. To assure the dependability of our detection system, we combine several models and algorithms, utilizing their combined strength to reduce mistakes and improve overall accuracy. In research Yolov7 had an outstanding %age of validation accuracy of 96.1 among these algorithms, showing its superior efficacy in identifying and localizing plant diseases. Our goal is to provide a strong and adaptable system that can successfully handle the difficulties and complexities involved in detecting plant diseases by combining several methodologies. Our attention continues to be on constant development and the discovery of cutting-edge approaches to further hone and optimize our model for accurate and efficient illness diagnosis in the agricultural sector.

References

1. Hassan, S.M., et al.: Identification of plant-leaf diseases using CNN and transfer-learning approach. Electronics **10**(12), 1388 (2021)

2. Rathore, N.P.S., Prasad, L.: Automatic rice plant disease recognition and identification using convolutional neural network. J. Crit. Rev. **7**(15), 6076–6086 (2020)
3. Khatri, A., Agrawal, S., Chatterjee, J.M.: Wheat seed classification: utilizing ensemble machine learning approach. Sci. Programm. **2022**, 2626868 (2022)
4. Tugrul, B., Elfatimi, E., Eryigit, R.: Convolutional neural networks in the detection of plant leaf diseases: a review. Agriculture **12**(8), 1192 (2022)
5. Kalim, H., Chug, A., Singh, A.P.: Citrus leaf disease detection using hybrid CNN-RF model. In: 2022 4th International Conference on Artificial Intelligence and Speech Technology (AIST). IEEE (2022)
6. Chen, H.-C., et al.: AlexNet convolutional neural network for disease detection and classification of tomato leaf. Electronics **11**(6), 951 (2022)
7. Bhandari, G.B., et al.: Plant diseases detection system using deep learning (2021)
8. Islam, M.A., et al.: An automated convolutional neural network based approach for paddy leaf disease detection. Int. J. Adv. Comput. Sci. Appl. **12**(1), 0120134 (2021)
9. Chohan, M., Khan, A., Chohan, R., Katpar, S.H., Mahar, M.S.: Plant disease detection using deep learning. Int. J. Recent Technol. Eng. **9**(1), 909–914 (2020)
10. Saleem, M.H., Potgieter, J., Arif, K.M.: Plant disease detection and classification by deep learning. Plants **8**(11), 468 (2019)
11. Saleem, M.H., Potgieter, J., Arif, K.M.: Plant disease classification: a comparative evaluation of convolutional neural networks and deep learning optimizers. Plants **9**(10), 1319 (2020)
12. Jiang, P., Chen, Y., Liu, B., He, D., Liang, C.: Real-time detection of apple leaf diseases using deep learning approaches based on improved convolutional neural networks. IEEE Access **7**, 59069–59080 (2019)
13. Wani, J.A., Sharma, S., Muzamil, M., Ahmed, S., Sharma, S., Singh, S.: Machine learning and deep learning based computational techniques in automatic agricultural diseases detection: Methodologies, applications, and challenges. Arch. Comput. Methods Eng. **29**(1), 641–677 (2022)
14. Nanehkaran, Y.A., Zhang, D., Chen, J., Tian, Y., Al-Nabhan, N.: Recognition of plant leaf diseases based on computer vision. J. Ambient Intell. Human. Comput. (2020). https://doi.org/10.1007/s12652-020-02505-x

Comparative Analysis of Machine Learning Algorithms in Vehicle Image Classification

Nur Izzaty Muhammad Asry[1], Aida Mustapha[1](✉), and Salama A. Mostafa[2]

[1] Faculty of Applied Sciences and Technology, Universiti Tun Hussein Onn Malaysia, KM 1, Jalan Panchor, 84600 Pagoh, Johor, Malaysia
aidam@uthm.edu.my

[2] Faculty of Computer Science and Information Technology, Universiti Tun Hussein Onn Malaysia, Parit Raja, 86400 Batu Pahat, Johor, Malaysia

Abstract. Object identification in the area of Computer Vision essentially focus on how robots are able to perceive in comparison of human eyes. The main task for vehicle detection is detecting vehicles from image sources. This study utilises Weka to categorise vehicle photos into 3 vehicle groups based on data mining approach. In feature extraction, three image filtering algorithms are used, which are Colour Layout, Edge Histogram, and Pyramid Histogram of Oriented Gradients (PHOG). The features extracted are then fed into five classification algorithms for a comparative analysis: Multilayer Perceptron (MLP), Simple Logistic, Sequential Minimal Optimization (SMO), Logistic Model Tree (LMT), and Random Forest. The experimental results showed that PHOG filtering algorithm produced the best set of features to accurately classify the vehicle images. During classification, the SMO classifier achieved the best accuracy of 46.3158%.

Keywords: Vehicle detection · Image filtering · Data mining

1 Introduction

Vehicle detection categorization is a significant operation in the Intelligent Transportation System (ITS) since it enables calculation of a traffic measurement known as vehicles count by category. For efficient traffic operations in an ITS, a precise classification of vehicles into distinct categories is critical and they all depend on input video feeds from the security or surveillance cameras. Identification of vehicle types are important to not only safety community [14], but security, law enforcement, and traffic monitoring agencies.

Machine learning techniques for visual traffic monitoring are known to be easy to deploy, non-invasive, and cost-effective [11]. The data from traffic monitoring systems are usually acquired from a security camera or Closed-Circuit Television (CCTV) in a specific location. Installation of CCTV in the highway and on the roadways have increasingly become the highest priority in the communities. CCTVs were discovered to be more efficient and less expensive than sensors.

© The Author(s), under exclusive license to Springer Nature Singapore Pte Ltd. 2024
F. Hassan et al. (Eds.): AsiaSim 2023, CCIS 1911, pp. 426–436, 2024.
https://doi.org/10.1007/978-981-99-7240-1_34

CCTV feed coupled with machine learning and big data analytics might reveal significant information about the characteristics of built-up car traffic in roads and highways [8].

In machine learning approach to vehicle detection, various algorithms are used to first convert the video images into image feed, extract features from the images, and identify the type of vehicle on the road. Popular algorithms include the Support Vector Machine (SVM), K-Nearest Neighbor, Logistic Regression, Decision Tree, and Random Forest [6]. There have been various vehicle classification techniques developed. Earliest innovations in sensing and machine learning have produced lots of new alternative vehicle classification systems. Although these new classification systems improve vehicle accuracy rate, they have various features and requirements, including sensor selections, produced, and stored, as well as the planning process [8]. Meanwhile, deep learning approaches have also been used to perform vehicle object detection, which is an important piece of research in traffic controlling. For example, organizations that operate road tolls use deep learning techniques to detect fraud caused by different fees structure imposed on different vehicle types [13].

In the realm of vehicle tracking, the accuracy of vehicle detection algorithm is highly crucial due to the broad range in vehicle sizes. This, in turn, will affect the precision of vehicle counts when not accurately detected. From a angled camera in high position, vehicle image sizes varies significantly based on their distance to the camera. Therefore, it is very important to investigate the effectiveness of image filtering algorithms which will extract the features before the classification process begins. To address this issue, the main objective of this project is to compare the detection performance between five machine learning algorithms to classify images of vehicles into three categories, which are buses, cars and trucks. In order to achieve the objective, three main processes will be carried out.

First, to implement three feature extractions algorithms or feature descriptors that extract the features from the input vehicle images. The feature descriptors or also known as image filters include the Color Layout, Edge Histogram, and Pyramid Histogram of Oriented Gradients (PHOG). Second, to implement five machine learning algorithms, which are Multilayer Perceptron (MLP), Sequential Minimal Optimization (SMO), Logistic Model Trees, Simple Logistic, and Random Forest (RF) to classify the category of vehicles based on each feature descriptors. Third, to compare the performance of the machine learning algorithms based on the accuracy metric and Kappa statistics.

The remainder of this paper proceeds as follows. Section 2 presents the materials and methods to conduct the comparative vehicle classification experiments along with description on the dataset, algorithms, and evaluation metrics. Section 3 presents and discusses the results and finally Sect. 4 concludes the paper with some direction for future research.

2 Materials and Methods

This paper reports a comparative experiment for classifying vehicles from an image dataset into categories such as bus, car, and truck. The research methodology is shown in Fig. 1. Based on this figure, the image dataset for this investigation is obtained from [15] with a total of 380 images of buses, cars, and trucks.

The feature extraction steps are carried out using three different filtering algorithms, which are Color Layout, Edge Histogram and Pyramid Histogram of Oriented Gradients (PHOG). The features extracted are then used in five classification algorithms to compare their classification accuracy in classifying the categories of bus, car, and truck. The classification algorithms include the Multilayer Perceptron (MLP), Simple Logistic, Sequential Minimal Optimization (SMO), Logistic Model Tree (LMT), and Random Forest.

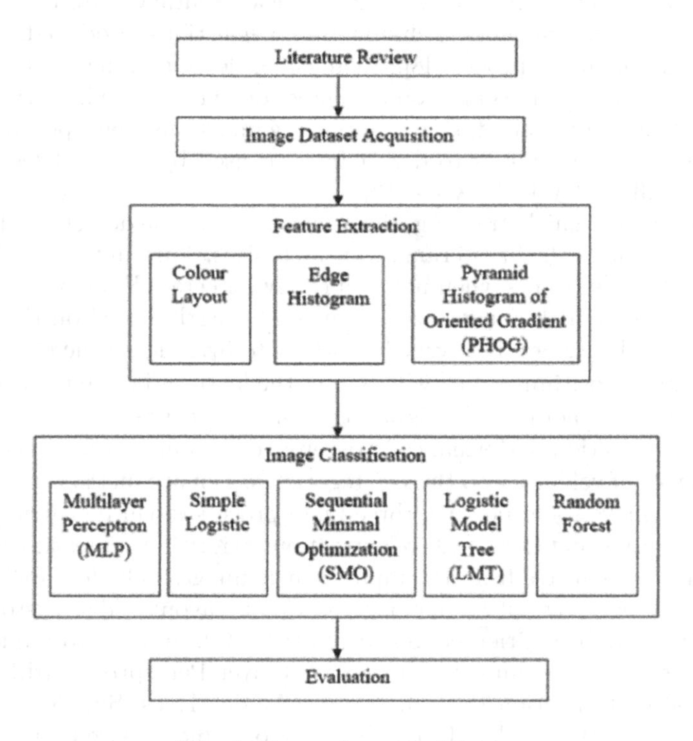

Fig. 1. Research methodology.

The comparative experiments are carried out using the Waikato Environment for Knowledge Analysis (WEKA) [4] with 10-fold cross validation method for training and testing the data. This cross-validation method systematically generates ten equal parts ("folds), analyzes one at a time, while train upon that remaining nine sections simultaneously.

2.1 Data Description

The vehicle image dataset used in this study is sourced from [15]. The dataset consists of 7,143 photos of buses, cars, motorcycles, and trucks. Table 1 shows the summary of instances in the image dataset. All images as shown in Fig. 2 are labelled and assigned with a class of bus, car or truck.

Table 1. Vehicle image dataset.

Type of Vehicle	Number of Images
Bus	128
Car	125
Truck	127
Total	380

Fig. 2. Sample images.

2.2 Feature Extraction

In the comparative experiments, three feature extraction algorithms or feature descriptors are used, which are Color Layout, Edge Histogram, and Pyramid Histogram of Oriented Gradients (PHOG) to extract the features.

- **Color Layout:** A color layout filter is utilized for image data and video retrieval to capture spatial distribution of color in an image. Because color is the most fundamental aspect of visual content, it is feasible to use it to describe and represent an image [5]. The MPEG-7 specification is one of the most efficient method known for their best results.
- **Edge Histogram:** An image's edge is a key minimal characteristic to distinguish between shape features details, which are important for accurate image classification. The edges are one of the most used picture aspects in evidence-based image analysis, demanding a clear procedure for its description because of its importance. The edge histogram identifies the image's five edge types: horizontal, vertical, two diagonal, and non-directional [5].
- **Pyramid Histogram of Oriented Gradients (PHOG):** Pattern classifiers for object detection are histograms of directed gradients. The method counts the number of times a gradient orientation is computed on a linear pattern of symmetrical cells in a picture. The theory behind this approach is that the distribution of edge directions can be used to characterize the local appearance of objects in an image [12].

2.3 Vehicle Image Classification

Following the literature [6], five classification algorithms are chosen for the comparative experiment in classifying vehicle images into categories such as buses, cars, and trucks. The algorithms include the Multilayer Perceptron (MLP), Simple Logistic, Sequential Minimal Optimization (SMO), Logistic Model Tree (LMT), and Random Forest.

- **Multilayer Perceptron (MLP):** The network of classifier in Multilayer Perceptron (MLP) includes three layers: input layer, hidden layer, and output layer. Every node conducts a weight value among its inputs and limits its outcome, because each link does have a weight (amount). The nodes are frequently referred to as "neurons" using a sigmoid function [3].
- **Simple Logistic:** A binary classification algorithm does seem to be logistic regression. All independent variables are generally thought to have a bell-shaped Gaussian distribution and be numerical in nature. The second cause is indeed not necessarily valid; logistic regression may still give accurate findings regardless of whether the dataset is indeed not gaussian [2]. Throughout the dispersion dataset, some output characteristics receive a Gaussian-like distribution whereas the majority do not [9].
- **Sequential Minimal Optimization (SMO):** Seems to be an approach towards handling an optimization problem which exists throughout support vector machine training. SMO seems to be an effective algorithm towards tackling the above-mentioned optimization method. SMO divides the issue through as many specific subs as practical, which subsequently obtained from the analysis [10].
- **Logistic Model Tree (LMT):** LMT tree structural attribute the logistic learning algorithm combining linked features in addition to serving the classifier. If nominal attributes were binarized using regression applications, this regression function considers a selection regarding all the features within this data [7].
- **Random Forest:** One classification tree machine learning method which extends reducing. The fact of categorized decision trees being built and use a different algorithm which takes majority optimum ways of generating at every phase within this tree-building method would be a disadvantage [1].

2.4 Evaluation Metrics

Two main evaluation metrics are used, which are classification accuracy and Kappa statistics between the actual vehicle class and the predicted class.

- **Accuracy:** Accuracy is the measure of percentage of correct predictions (all true positives and true negatives) from the total number of predictions (sum of true positives, false positives, true negatives, and false negatives). In WEKA, this percentage is calculated as shown in Eq. 1.

$$\text{Accuracy} = \frac{\text{Number of correct predictions}}{\text{Total number of predictions}} \tag{1}$$

- **Kappa Statistics:**. Another measurement which contrasts current precision with predicted precision is based on probability and is measured using the Kappa Statistics as value shown in Eq. 2. A kappa value that is significantly lower than 0.4 is negative. There is absolutely no differentiation between the observations and random chance whenever the Kappa value is 0.

$$k = \frac{(p_0 - p_e)}{(1 - p_e)} \tag{2}$$

Based on Eq. 2, p_0 is the relative observed agreement among rates and p_e is the hypothetical probability of chance agreement. Table 2 shows the interpretation for Kappa statistics value.

Table 2. Interpretation for values of Kappa Statistics.

Kappa Statistics	Interpretation
0	No agreement
0.10–0.20	Slight agreement
0.21–0.40	Fair agreement
0.41–0.60	Moderate agreement
0.61–0.80	Substantial agreement
0.81–0.99	Near perfect agreement
1	Perfect agreement

In addition to the measurement of accuracy percentage (correctly classified instances) and the Kappa statistics between the actual vehicle class and the predicted class, a Receiver Operating Characteristic (ROC) is also calculated. An ROC value contrasts the true positive and the false positive for different classification parameters. In an ROC chart, y-axis has mainly been used to visualise actual genuine positive value while the x-axis have generally been employed to visualise the fake particular value. Equation 3 is the formula to calculate true positive rate (TPR) and Eq. 4 is the formula for false positive rate (FPR).

$$\mathrm{TPR} = \frac{\mathrm{TP}}{\mathrm{TP} + \mathrm{FN}} \tag{3}$$

$$\mathrm{FPR} = \frac{\mathrm{FP}}{\mathrm{TN} + \mathrm{FP}} \tag{4}$$

where TP is true positive, FN is false negative value, FP is false positive value, and TN is true negative value.

3 Results and Discussion

The main goal of this study is to compare the performance of five image classification algorithms using different feature extraction algorithms or feature descriptors, which are Color Layout, Edge Histogram, and Pyramid Histogram of Oriented Gradients (PHOG). The algorithms are as follows:

- Multilayer Perceptron (MLP)
- Simple Logistic
- Sequential Minimal Optimization (SMO)
- Logistic Model Trees (LMT)
- Random Forest (RF)

The comparison results between the five image classification algorithms are presented based on specific feature descriptor. The ROC values are also compared in a similar manner, based on specific feature descriptors. In this way, the results the experiments will show the impact of each feature descriptor towards each vehicle image classification algorithms.

3.1 Color Layout Feature Descriptor

Using the Color Layout feature descriptor, the overall classification accuracy for the images is compared in Table 3. This feature descriptor was used to add 33 image features to the dataset. The results revealed that image classification experiment achieved the best accuracy of 43.94745% by the Random Forest classifier utilising only the colour layout elements of the images.

Table 3. Results for the image classification using Color Layout Filter.

Classifiers	Correctly Classifiers Instances (%)	Incorrectly Classified Instances (%)	Kappa Statistics (%)
Multilayer Perceptron	35.5263	64.4737	0.0329
Simple Logistic	39.4737	60.5263	0.0920
Sequential Minimal Optimization	40.7895	59.2105	0.1116
Logistic Model Tree	39.4737	60.5263	0.0920
Random Forest	**43.9474**	56.0526	**0.1590**

This result is also consistent to the Kappa statistics, where the value of 0.1590 is the highest among other classifiers. Note that in a certain cross-evaluation outcomes with readily obvious behaviours, Kappa statistic values under 0.70 will be deemed as unsatisfactory.

3.2 Edge Histogram Feature Descriptor

With Edge Histogram features used as the key differentiating factors in classifying all images, the classification accuracy is shown in Table 4. The results showed that some of the classifiers had accuracy of 41.5789% and that's very much like the prior features applied.

Table 4. Results for the image classification using Edge Histogram.

Classifiers	Correctly Classified Instances (%)	Incorrectly Classified Instances (%)	Kappa Statistics (%)
Multilayer Perceptron (MLP)	36.0526	63.9474	0.0406
Simple Logistic	34.2105	65.7895	0.0130
Sequential Minimal Optimization	35.7895	64.2105	0.0364
Logistic Model Tree	33.1579	66.8421	−0.0026
Random Forest	**41.5789**	58.4211	**0.1231**

While using Edge Histogram information like the size and direction of edges in images, the Random Forest classifier delivers the maximum accuracy. According to Kappa Statistics data, Random Forest is also the best classifier because the value of 0.1231% was the highest when compared to the other classifiers. This resulting score of 0.1231% shows that there was agreement among the assessments.

3.3 Pyramid Histogram Oriented Gradient (PHOG) Feature Descriptor

When Pyramid Histogram of Oriented Gradients (PHOG) features were applied for classification, the results of the various classifiers are shown in Table 5. PHOG feature descriptors allows classifiers to detect car images with a 46.3158% accuracy, with Sequential Minimal Optimization (SMO) classifiers providing the best results in this test.

Table 5. Results for the image classification using PHOG.

Classifiers	Correctly Classified Instances (%)	Incorrectly Classified Instances (%)	Kappa Statistics (%)
Multilayer Perceptron	43.6842	56.3158	0.1549
Simple Logistic	44.2105	55.7895	0.1630
Sequential Minimal Optimization	**46.3158**	53.6842	**0.1943**
Logistic Model Tree	43.6842	56.3158	0.1551
Random Forest	45.7895	54.2105	0.1863

This finding is consistent with the results for Kappa statistics, which also showed that Sequential Minimal Optimization (SMO) achieved the highest Kappa value at 0.1943. Due to its simplification in proportion to predicted accuracy, the Kappa statistic may be used to comparing classifiers developed and assessed on datasets with various class probabilities very effectively instead of accuracy alone.

3.4 Receiver Operating Characteristic (ROC)

Note that the ROC is a probability curve that measure the degree of separability, or how much a classification model is capable of distinguishing between the classes. Table 6 shows the comparison of ROC values based on separate feature descriptors, which are the Color Layout, Edge Histogram, and PHOG.

Table 6. ROC values based on specific feature descriptor.

Feature Descriptor	Classifiers	Class		
		Bus	Car	Truck
Color Layout	Multilayer Perceptron	0.564	0.51	0.381
	Simple Logistic	0.544	0.59	0.577
	Sequential Minimal Optimization	0.534	0.565	0.544
	Logistic Model Tree	0.544	**0.59**	0.577
	Random Forest	**0.635**	0.579	**0.647**
Edge Histogram	Multilayer Perceptron	0.553	0.575	0.438
	Simple Logistic	0.551	0.561	0.439
	Sequential Minimal Optimization	0.545	**0.595**	0.476
	Logistic Model Tree	0.543	0.553	0.451
	Random Forest	**0.614**	**0.595**	**0.517**
PHOG	Multilayer Perceptron	0.677	0.653	0.549
	Simple Logistic	**0.715**	**0.68**	0.521
	Sequential Minimal Optimization	0.649	0.64	0.566
	Logistic Model Tree	0.699	0.678	0.524
	Random Forest	**0.715**	0.65	**0.589**

Based on Table 6, it is shown that Random Forest has consistently achieved the highest ROC value for most vehicle classes. A value of lower than 0.5 suggests no capability of discrimination or unable to categorize the classes independently or worst than random classification. A value between 0.6 to 0.7 is considered acceptable, 0.8 to 0.9 is considered excellent, and more than 0.9 is considered outstanding. The majority of classifiers provide a value that is quantitative models to determine the categorization. It is typical to view something above 0.5 as favorable whenever a classifier provides a value within 0.0 which is certainly negative and 1.0 which is obviously positive.

4 Conclusions

This paper discusses the role of feature descriptors in image classification, which is for extracting features from the image datasets. In order to extract the image features from the dataset during feature extraction, three image filters or feature descriptors were used, which are Color Layout, Edge Histogram, and Pyramid Histogram of Oriented Gradients (PHOG). The overall findings demonstrated that the Pyramid Histogram of Oriented Gradients (PHOG) feature descriptor produced the most useful features, which in turn resulting the classifiers to accurately categorise the images up to 46.3158% accuracy with SMO algorithm. This is because PHOG encodes key to identifying to the direction of gradients in intensity across an image while SMO breaks down a big problem into smaller ones that can be tackled analytically. When using SMO, all missing values are replaced, nominal attributes are converted to binary ones, and all attributes are normalised. In the future, this project is set to investigate more feature descriptors to assess the impact of different feature sets to the classification algorithms using the PHOG as the baseline of feature extraction. As for the image classification algorithm, SMO will be used as the benchmark results to be compared with future deep learning vehicle detection models.

Acknowledgements. This study is funded by Universiti Tun Hussein Onn Malaysia.

References

1. Chen, Z., et al.: Random forest based intelligent fault diagnosis for PV arrays using array voltage and string currents. Energy Convers. Manage. **178**, 250–264 (2018)
2. Chu, J., Lee, T.H., Ullah, A.: Component-wise AdaBoost algorithms for high-dimensional binary classification and class probability prediction. In: Vinod, H.D., Rao, C. (eds.) Financial, Macro and Micro Econometrics Using R, Handbook of Statistics, vol. 42, pp. 81–114. Elsevier (2020). https://doi.org/10.1016/bs.host.2018.10.003. https://www.sciencedirect.com/science/article/pii/S0169716118300932
3. Dalatu, P.I., Fitrianto, A., Mustapha, A.: A comparative study of linear and nonlinear regression models for outlier detection. In: Herawan, T., Ghazali, R., Nawi, N.M., Deris, M.M. (eds.) SCDM 2016. AISC, vol. 549, pp. 316–326. Springer, Cham (2017). https://doi.org/10.1007/978-3-319-51281-5_32
4. Hall, M., Frank, E., Holmes, G., Pfahringer, B., Reutemann, P., Witten, I.H.: The WEKA data mining software: an update. SIGKDD Explor. Newsl. **11**(1), 10–18 (2009). https://doi.org/10.1145/1656274.1656278
5. Janković, R.: Machine learning models for cultural heritage image classification: comparison based on attribute selection. Information **11**(1) (2020). https://doi.org/10.3390/info11010012. https://www.mdpi.com/2078-2489/11/1/12
6. Khan, A.A., Laghari, A.A., Awan, S.A.: Machine learning in computer vision: a review. EAI Endorsed Trans. Scalable Inf. Syst. **8**(32), e4–e4 (2021)
7. Kumar, M.N., Koushik, K., Deepak, K.: Prediction of heart diseases using data mining and machine learning algorithms and tools. Int. J. Sci. Res. Comput. Sci. Eng. Inf. Technol. **3**(3), 887–898 (2018)

8. Peppa, M.V., Bell, D., Komar, T., Xiao, W.: Urban traffic flow analysis based on deep learning car detection from CCTV image series. Int. Arch. Photogrammetry Remote Sens. Spatial Inf. Sci. **XLII-4**, 499–506 (2018). https://doi.org/10.5194/isprs-archives-XLII-4-499-2018. https://isprs-archives.copernicus.org/articles/XLII-4/499/2018/
9. Ray, S.: A quick review of machine learning algorithms. In: 2019 International Conference on Machine Learning, Big Data, Cloud and Parallel Computing (COMIT-Con), pp. 35–39. IEEE (2019)
10. Rehman Javed, A., Jalil, Z., Atif Moqurrab, S., Abbas, S., Liu, X.: Ensemble AdaBoost classifier for accurate and fast detection of botnet attacks in connected vehicles. Trans. Emerging Telecommun. Technol. **33**(10), e4088 (2022)
11. Shokravi, H., Shokravi, H., Bakhary, N., Heidarrezaei, M., Rahimian Koloor, S.S., Petrů, M.: A review on vehicle classification and potential use of smart vehicle-assisted techniques. Sensors **20**(11) (2020). https://doi.org/10.3390/s20113274. https://www.mdpi.com/1424-8220/20/11/3274
12. Sugiharto, A., Harjoko, A., Suharto, S.: Indonesian traffic sign detection based on HAAR-PHOG features and SVM classification. Int. J. Smart Sens. Intell. Syst. **13**(1), 1–15 (2020). https://doi.org/10.21307/ijssis-2020-026
13. Suguitan, A.S., Dacaymat, L.N.: Vehicle image classification using data mining techniques. In: Proceedings of the 2nd International Conference on Computer Science and Software Engineering, CSSE 2019, pp. 13–17. Association for Computing Machinery, New York, NY, USA (2019). https://doi.org/10.1145/3339363.3339366
14. Won, M.: Intelligent traffic monitoring systems for vehicle classification: a survey. IEEE Access **8**, 73340–73358 (2020). https://doi.org/10.1109/ACCESS.2020.2987634
15. Zuraimi, M.A.B., Zaman, F.H.K.: Vehicle detection and tracking using YOLO and DeepSORT. In: 2021 IEEE 11th IEEE Symposium on Computer Applications & Industrial Electronics (ISCAIE), pp. 23–29. IEEE (2021)

Evaluation of Linear Imputation Based Pediatric Appendicitis Detection System Using Machine Learning Algorithm

Md Al-Imran[(✉)], Nafisha Nower Juthi, Tasnima Sabrina Mahi, and Safayet Hossain Khan

East West University, Dhaka, Bangladesh
al.imran@ewubd.edu,
{2020-1-60-200,2020-1-60-121,2020-1-60-122}@std.ewubd.edu

Abstract. Appendicitis is a condition that can be crucial, and lack of proper treatment among the children may cause unwanted death. The primary objective of this study is to develop a model specifically trained for diagnosing pediatric appendicitis. Detailed study has been performed by following specific steps and techniques involved in data preprocessing and model development phase. The experiment found the challenges associated with diagnosing appendicitis in children, including the presence of nonspecific symptoms, missing symptoms, and variations in clinical presentations. The researchers acknowledged the importance of addressing these challenges associated with nonspecific symptoms, missing symptoms, and variations in clinical presentations, the developed model demonstrates the potential to improve diagnostic accuracy in children. In learning that the experiment delved into the model development process, encompassing the selection of suitable machine-learning algorithms or statistical techniques like interpolation for handling missing values. The rationale behind these choices is explained, along with insights into how the model was trained and evaluated using the available dataset. The linear interpolation algorithm has been found to achieve optimal results with logistic model accuracy of 87% for the prediction of appendicitis, 95% for the prediction of appendicitis treatment, and 91% for the prediction of complications of appendicitis. However, an estimated prediction model based on ML has been built to predict pediatric appendicitis.

Keywords: Pediatric Appendicitis · Imputation · Machine Learning · Medical Data · Child health

1 Introduction

The inflammation of the appendix is an acute process known as appendicitis. This paper emphasizes the importance of accurate diagnosis and timely treatment of pediatric appendicitis [1]. It highlights the significance of data preprocessing in improving the accuracy of diagnosing appendicitis in children. By

© The Author(s), under exclusive license to Springer Nature Singapore Pte Ltd. 2024
F. Hassan et al. (Eds.): AsiaSim 2023, CCIS 1911, pp. 437–450, 2024.
https://doi.org/10.1007/978-981-99-7240-1_35

effectively handling missing data and transforming the dataset, the study aims to enhance the precision and effectiveness of managing pediatric appendicitis. The objective is to develop an advanced model that surpasses existing approaches in accuracy and complexity. The successful integration of existing knowledge and innovative techniques can lead to better patient outcomes and improved healthcare decision-making [2].

Chapter two provides an overview of the literature review. Chapter three provides an overview of the dataset, including its variables. Chapter four focuses on the data methodology and workflow employed in the study. Chapter five details the data preprocessing steps, encompassing imputation, transformation, and cleaning and also presents the exploratory data analysis conducted on the preprocessed data. Chapter six describes the experimental setup used for developing the prediction model. Chapter seven discusses the results and performance evaluation of the developed model. The conclusion highlights the study's limitations and potential biases.

2 Literature Review

For children who have suspected appendicitis, the authors of [1] have developed an accessible online appendicitis prediction tool using logistic regression, random forests, and gradient boosting machines. Also previously published study by [2] who have also used Random Forest, Logistic Regression, Naïve Bayes, Generalized Linear, Decision Tree, Support Vector Machine, and Gradient Boosted Tree to develop such model.

3 Dataset Overview

The dataset [1] contains survey of 0 to 18 years children who were admitted to the pediatric surgery department with suspected appendicitis.

3.1 Properties of Pediatric Appendicitis Dataset

The 430 patients who were admitted to a pediatric surgical unit provided the current dataset [1]. Table 1 contains details information about the dataset.

Table 1. Dataset Information

Public availability	Yes
Number of Categorical Features	30
Number of Numerical Features	11
Number of records	430
Number of features	41
Total number of missing values	25.09%

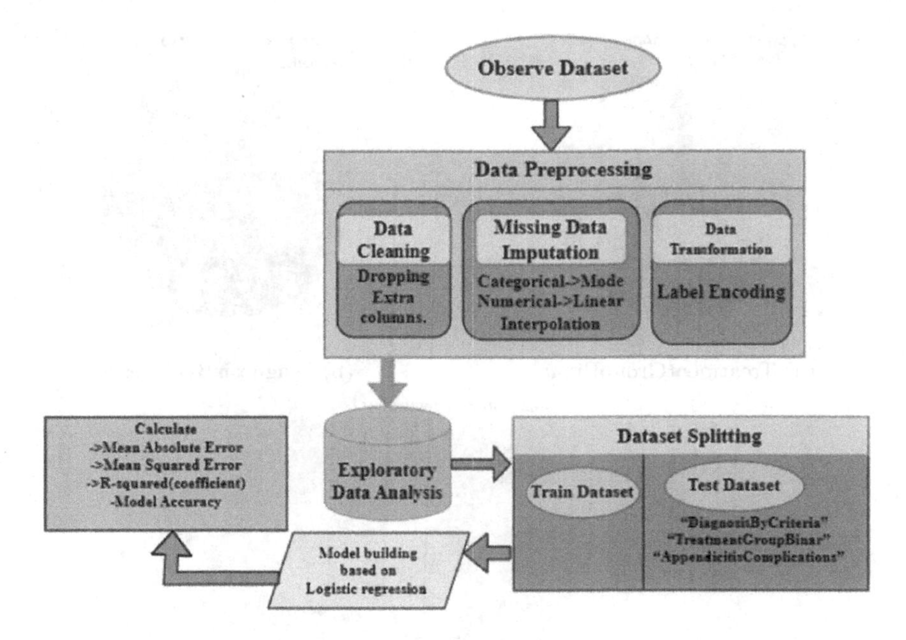

Fig. 1. Work Flow Diagram.

3.2 Response Feature Information

The pie charts shown in Fig. 2 will help us to understand the dependencies of the target variables on independent features.

3.3 Features Exploration

The dataset [1] has total of 41 features among which three features are target features and others are independent. Among 38 independent variables, 11 are numerical such as 'AlvaradoScore', 'PediatricAppendicitisScore' etc., and 27 features are categorical such as 'MigratoryPain', 'LowerAbdominalPainRight', 'ReboundTenderness' etc. Rigorous illustration of the dataset can be found at [1].

4 Methodology

The aim of the methodology is to provide sufficient detail for readers to understand the study. Figure 1 shows the steps of this experiment

4.1 Data Preprocessing

Data preprocessing deals with irrelevant data, missing values noise, null value of a dataset and makes the dataset suitable to work with.

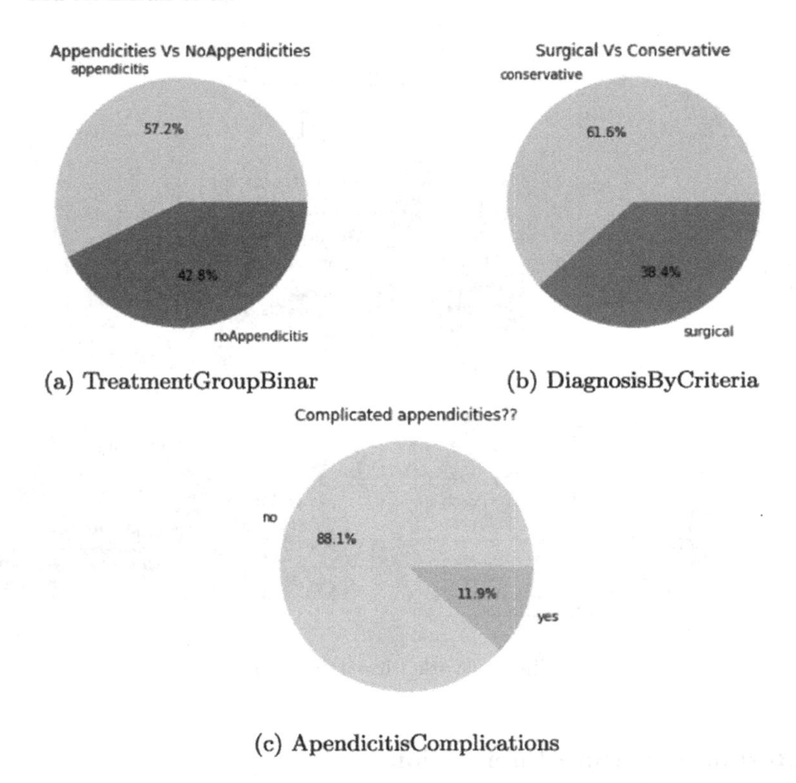

(a) TreatmentGroupBinar

(b) DiagnosisByCriteria

(c) ApendicitisComplications

Fig. 2. Class variable distribution with pie charts

Data Cleaning. Data cleaning involves identifying and eliminating errors in a dataset, which is challenging but it ensures high data quality and consistency [3]. The dataset had some irrelevant data features like 'Age', 'Height', 'Weight', and 'Sex' features which were inessential to the model; so these features were dropped. Those features which had more than 75% missing values like 'TissuePerfusion', 'BowelWallThick', 'Ileus', 'FecalImpaction', 'Meteorism' and 'Enteritis' were dropped from the dataset.

Missing Value Imputation. Missing data refers to the absence of recorded data for a particular variable in a given observation [4]. Missing values can be classified into three types [5]:

- Missing Completely At Random (MCAR)
- Missing At Random (MAR)
- Missing Not At Random (MNAR)

MAR refers to types of missing data that are dependent on observed values but not related to unobserved values. Since clinical symptoms may depend on the known values of other variables in the data but not on the missing variable itself, the missing values in our dataset are of the second type (MAR). Our dataset contains both numerical and categorical missing values.

■ Categorical missing value imputation:

To fill the categorical missing values "mode method" has been used. Mode imputation is the easiest method because in this process most frequent value of the entire feature (mode) is used to replace the missing values. [6].

$$Mode = l + h\frac{fm - f1}{fm - f1 + fm - f2} \tag{1}$$

In Eq. 1 [7], l is considered as the lower limit and h is the size of observed data. fm is the iteration count.

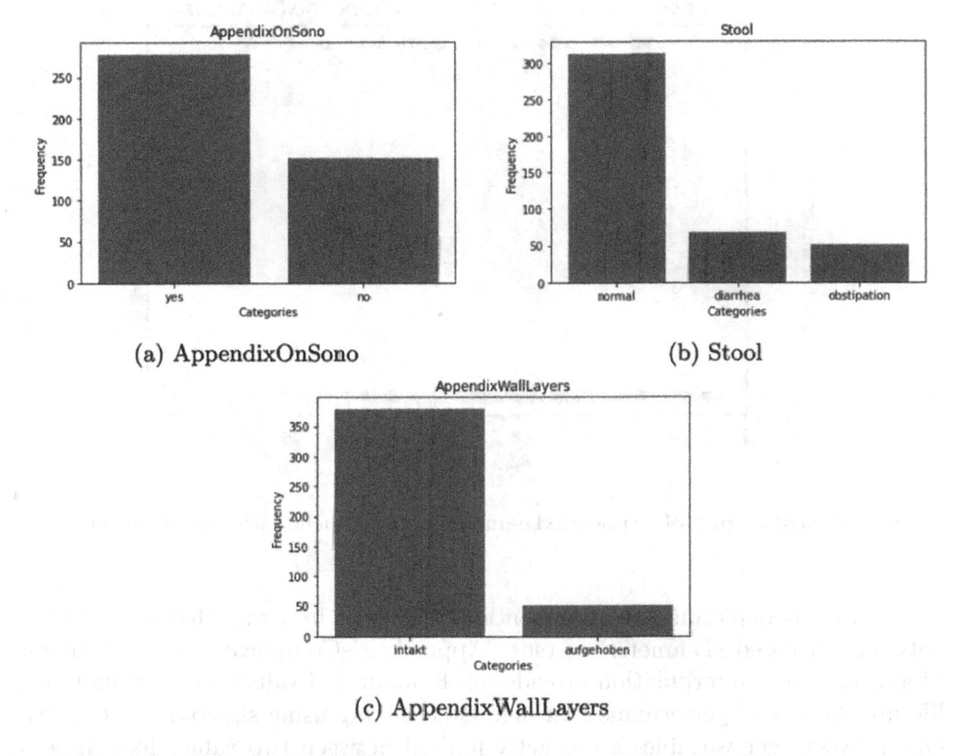

(a) AppendixOnSono (b) Stool

(c) AppendixWallLayers

Fig. 3. Feature Skewness using bar chart which is showing right skewness of the feature values

The dataset contains clinical symptoms of pediatric appendicitis, so categorical values of this dataset is skewed as shown in Fig. 3 as most of the patients will have the most common symptoms. When data is skewed and the data type is a string (object), the mode can be reasonable approximation of the central tendency of the data than the mean or the median.

■ **Numerical missing value imputation:**

To impute the numerical missing values of dataset Linear Interpolation is a suitable method for regression models and it is a flexible method [7].

⟶ **Reasons behind choosing Linear Interpolation:**

"AppendixDiameter" has 38% missing values which is needed to impute. Three target variables of the dataset are "AppendicitisComplications", "TreatmentGroupBinar", and "DiagnosisByCriteria" which is a binary data type. If these target variables share linear relation with "AppendixDiameter", then their regression plot should share two different lines which can be fitted using sigmoid function.

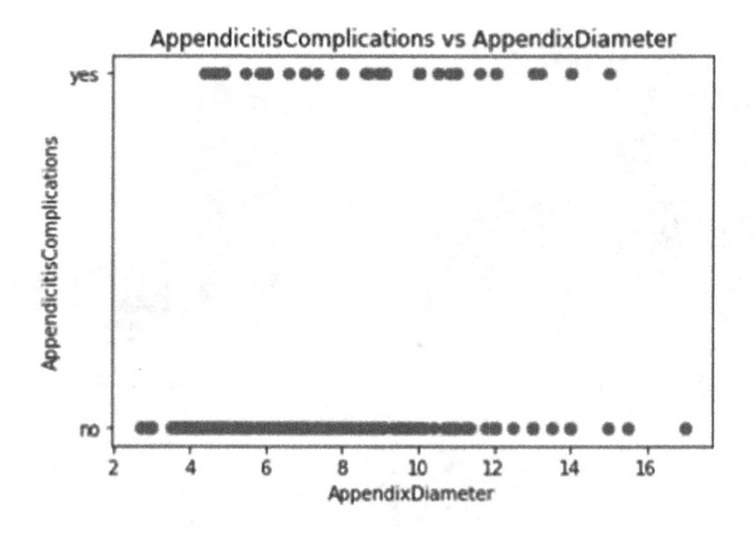

Fig. 4. Scatter plot of "AppendixDiameter" and "AppendicitisComplications"

Figure 4 is depicting the aforementioned concept in a way that the relation between "AppendixDiameter" versus "AppendicitisComplications" is sigmoid. Moreover, linear interpolation provides such imoutated values which would proliferate the model performance through curve fitting using sigmoid function [8]. Other two target variables are exactly limited between two values like 'appendicitis' or 'noAppendicitis', 'Surgical' or 'Conservative' and 'yes' or 'no'. For this reason, it is an assumption that linear interpolation is the best way to fill the missing values which will escalate training of the model as it will be discussed in the result and discussion part.

⟶ **Reasons behind not choosing other interpolations:**

• **Cubic Interpolation:**

Cubic interpolation estimates missing values by fitting a cubic polynomial curve between adjacent data points which will be a smooth curve but clearly not matching with plot of this work.

- **Spline Interpolation:**

Spline interpolation is similar to cubic interpolation, but it is a technique that constructs a smooth curve by breaking down the data into smaller sections and fitting each section with a polynomial function [9].

- **Nearest Neighbor Interpolation:**

The nearest neighbor or proximate interpolation method estimates missing values by finding the closest data point to the missing value and using this value to interpolate the missing value [9].

4.2 Data Transforming

Most of the features in the dataset are categorical, so it was necessary to convert them to numbers, as most machine learning algorithms and deep learning programs accept numbers as inputs [8].

We have used the "Factorization" method to convert those categorical features to numerical ones for model calculation. The factorization method is also known as the "Label encoding" or "Ordinal encoding" method. Label encoding is simply assigning an integer value to every possible value of a categorical variable [10]. The advantage of factorization is that it can result in a smaller dataset because factorization creates only one new column for each categorical column. This can be particularly useful when dealing with large datasets or datasets with many categorical columns.

The 'pd.factorize()' function from pandas takes each column as input and returns new columns with name columnName+"F" and assigns the integer codes returned by the pd.factorize() function to them. The original categorical column will be still present on the dataset unless we drop it.

4.3 Exploratory Data Analysis (EDA)

Exploratory Data Analysis (EDA) is a crucial preliminary stage which substantially help to achieve reliability of the model [11].

After completing all the steps of data preprocessing as shown in Table 2, the dataset is ready for model building.

5 Feature Selection and Model Implementation

In this section model building method and feature selection have been explained.

Table 2. Dataset information after preprocessing

Type	Amount
Numeric	8
Categorical encoded to numerical	20
Number of dropped columns	10
Total features	31
Target features	3
Total number of missing values	0.00%

5.1 Model Building

All three dependent variables of our dataset are binary so we have used 'Logistic Regression' to build our prediction model. Logistic regression is a statistical analysis technique that explores the association between multiple independent variables and a dependent variable [12]. In addition, logistic regression has only two categories for the dependent variable. The occurrence of an event is encoded as 1, and its absence is encoded as 0 [13].

Logistic regression is a suitable research method when the focus is on determining the presence or absence of an event, rather than its timing [14]. If a model involves disease state then particularly logistic regression is appropriate for that model, and therefore it is widely used in studies of health sciences [14]. Logistic regression helps to study medical disease by predicting the presence or absence [15]. The functional formula of logistic regression is [12]:

$$Y = \frac{e^{\alpha+\beta}}{1 + e^{\alpha+\beta}} \tag{2}$$

Here α and β determine logistic intercept and slope.

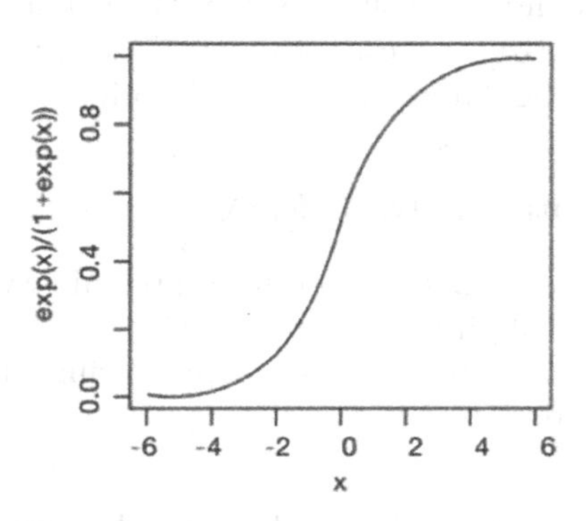

Fig. 5. Graph of the logistic curve where $\alpha = 0$ and $\beta = 1$ [12]

The curve on Fig. 5 is a s-shaped curve that builds the relation between 0 and 1 values. Clearly logistic regression is a perfect fit for our model.

5.2 Feature Selection:

After dropping all 10 features that have more than 75% missing values, we have worked on 28 independent features (predictors). From the whole clinical dataset to determine the target variable we have worked on questions like:

1. What are the dependent variables?
2. what is the target of our model building?
3. which features have complete binary data with no missing value?

After finding solutions to these questions, we have got three features that match all the criteria. So we marked 'AppendicitisComplications', 'Treatment-GroupBinar', and 'DiagnosisByCriteria' as our target variables.

6 Experimental Setup

In this section, we described data splitting and all the formulas that have been used for evaluating the prediction model.

6.1 Data Splitting

Data splitting is the act of partitioning available data into two portions named training dataset and test dataset. By utilizing the training data, the model is trained and fitted with specific parameters to capture the underlying patterns in the data using stratified sampling [16]. The dataset is split up by dividing the dataset into 70% train data and 30% test data. Since our dataset has three responsive variables and every time we have split one of those response variables as test data and the other variables as train data. We have built three different regression models to gain the accuracy for three different response variables respectively.

6.2 Mean Absolute Error (MAE) and Mean Squared Error (MSE)

The Mean Absolute Error (MAE) is widely used in evaluating models, and it provides valuable information about model accuracy [17] and so as Mean Squared Error (MSE) also do. Because MAE provides a more comprehensible measurement of the model's average error, it is frequently chosen [18]. MSE means to square the error so that the result is positive.

$$MAE = \frac{1}{N}\sum_{i=1}^{N}|y_i - \hat{y}| \tag{3}$$

$$MSE = \frac{1}{N}\sum_{i=1}^{N}(y_i - \hat{y})^2 \tag{4}$$

Equations 3 and 4 is from [19].

Here $\hat{y}$ is referred to as the predicted value and yi as the actual value. A regression model's accuracy is higher when the MAE and MSE values are lower. Both MAE and MSE referred to error so the lower MSE and MAE value the more accurate the model is.

6.3 R-squared

R squared is actually the coefficient of determination. R squared displays the percentage of the dependent variable's variance that can be accounted for by the independent variables [19]. A high R-squared value denotes a more precise estimation of the dependent variable from the explanatory factors when the purpose of the model is to make a prediction. R squared can be calculated as [20]:

$$R^2 = 1 - \frac{\sum(y_i - \hat{y})^2}{\sum(y_i - \bar{y})^2} \tag{5}$$

Here $\hat{y}$ is referred to as the predicted value and yi as the actual value. Another way to calculate R-squared with the value of MSE [21]:

$$R^2 = 1 - \frac{SSR}{SST} \tag{6}$$

Here SSR is the sum of squared residuals and SST is the total sum of squares. R-squared and MSE shares negative monotonic relations. It is desirable to have a higher R-squared value because it indicates a good accuracy of the model. Best value of R-squared is 1 and the worst value is -infinity [19].

7 Result and Discussion

The model accuracy and experimental results are discussed in this section.

7.1 Result Analysis:

To check the fitness of the model that we built, we have calculated the mean squared error, mean absolute error, and R-squared. From the Table 3 and Fig. 6, we can see that the value of MAE and MSE is the same for individually all three variables. It indicates that the errors do not have a particular bias or systematic deviation from the true values. It indicates that the model's predictions are, on average, equally close to the actual values in terms of both magnitude (MAE) and squared magnitude (MSE). The value of R-square > 0 for all three target variables which means the model has some kind of prediction power.

Table 3. Obtained MAE and MSE and R-Squared

Variable	MAE	MSE	R-Squared
"DiagnosisByCriteria"	0.13	0.13	0.45
"TreatmentGroupBinar"	0.05	0.05	0.80
"AppendicitisComplications"	0.09	0.09	0.01

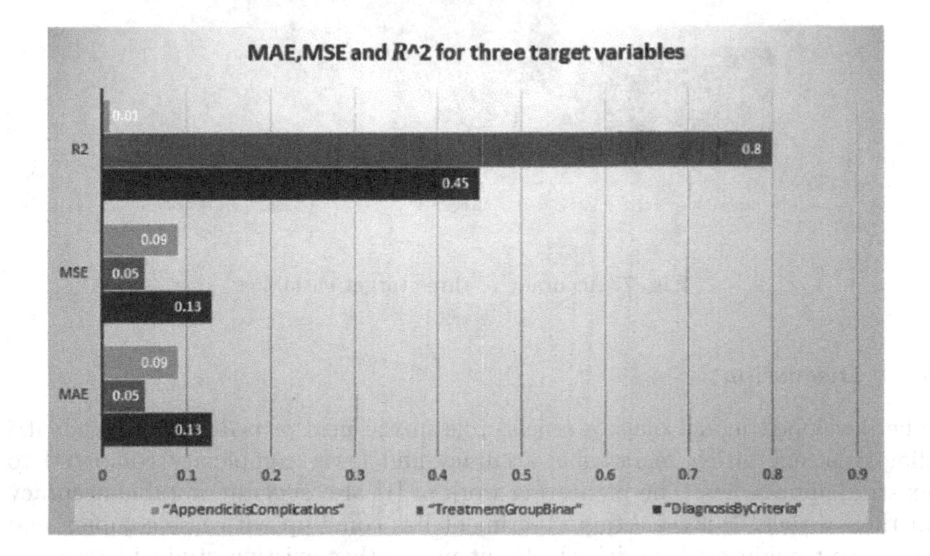

Fig. 6. Comparison between R-Square, MAE, MSE Values of three target variables.

Table 4. Accuracy of three target variable.

Variables	Accuracy
DiagnosisByCriteria	0.87
TreatmentGroupBinar	0.95
AppendicitisComplications	0.91

Table 4 and Fig. 7 shows the accuracy of three target variables.

It seems that "TreatmentGroupBinar" has the highest accuracy that is 0.95. It means our model correctly predicts the class label for 95% of the data points in the test set. The lowest accuracy has "DiagnosisByCriteria" which is 0.87 implies that the model makes correct predictions of the class label for 87% of the test set data points. "AppendicitisComplications" has 0.91 (91%) accuracy.

A well-fit model should have good predictive accuracy on the test set. The developed model has pretty good accuracy for all three response variables which proves that our logistic regression model is a well-fit regression model.

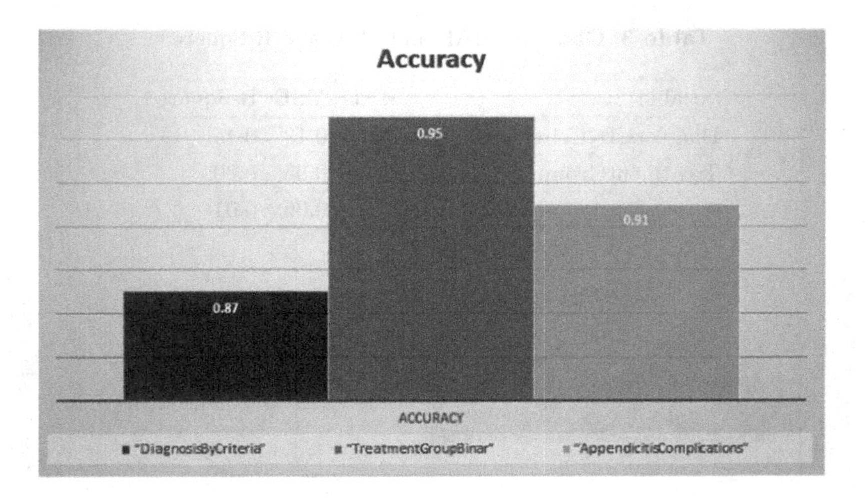

Fig. 7. Accuracy of three target variable

7.2 Discussion

The developed model plays a crucial role in the field of pediatric appendicitis diagnosis, exhibiting remarkable accuracy and lower complexity compared to existing approaches. The pioneering work of [1] also suggests notable accuracy in their article. However, this experiment has come up with less complex and significant engineered model which outsmart other existing study in terms of accuracy.

8 Conclusion

The conducted experiment surpasses others in the field of pediatric appendicitis diagnosis due to several distinctive factors. Firstly, our model achieves higher accuracy and lower complexity compared to existing approaches, resulting in improved diagnostic outcomes. This Paper provides a clear and detailed explanation of the steps taken to clean and transform the dataset. The paper has provided justifications for the methods has used. The dataset's small patient population, particularly those with severe appendicitis, can be noted as a significant limitation of this analysis. In summary, future efforts in this field should focus on the synergy of machine learning and image processing techniques. The work will be performed with the help of physicians, image processing, and extracting information from image data to extend pediatric appendicitis diagnosis to the highest trustworthiness among the stakeholders which will eventually automate the disease detection.

References

1. Marcinkevics, R., Wolfertstetter, P.R., Wellmann, S., Knorr, C., Vogt, J.E.: Using machine learning to predict the diagnosis, management and severity of pediatric appendicitis. Front. Pediatrics, 360 (2021)
2. Mijwil, M.M., Aggarwal, K.: A diagnostic testing for people with appendicitis using machine learning techniques. Multimedia Tools Appl. **81**, 7011–7023 (2022). https://doi.org/10.1007/s11042-022-11939-8
3. Wangikar, V.C., Deshmukh, R.R.: Data cleaning: current approaches and issues. In: IEEE International Conference on Knowledge Engineering (2011)
4. Kang, H.: The prevention and handling of the missing data. Korean J. Anesthesiol. **64**(5), 402–406 (2013)
5. Feng, S., Hategeka, C., Grépin, K.A.: Addressing missing values in routine health information system data: an evaluation of imputation methods using data from the democratic Republic of the Congo during the COVID-19 pandemic. Popul. Health Metrics **19**(1), 1–14 (2021)
6. Munguía, J.A.T.: Comparison of imputation methods for handling missing categorical data with univariate pattern. Revista de Métodos Cuantitativos para la Economía y la Empresa **17**, 101–120 (2014)
7. Jegadeeswari, K., Ragunath, R., Rathipriya, R.: Missing data imputation using ensemble learning technique: a review. Soft Comput. Secur. Appl. Proc. ICSCS **2022**, 223–236 (2022)
8. Al-Imran, Md., Ripon, S.H.: Network intrusion detection: an analytical assessment using deep learning and state-of-the-art machine learning models. Int. J. Comput. Intell. Syst. **14**, 1–20 (2021)
9. Morelli, D., Rossi, A., Cairo, M., Clifton, D.A.: Analysis of the impact of interpolation methods of missing RR-intervals caused by motion artifacts on HRV features estimations. Sensors **19**(14), 3163 (2019)
10. Hancock, J.T., Khoshgoftaar, T.M.: Survey on categorical data for neural networks. J. Big Data **7**(1), 1–41 (2020). https://doi.org/10.1186/s40537-020-00305-w
11. MIT Critical Data, Komorowski, M., Marshall, D.C., Salciccioli, J.D., Crutain, Y.: Exploratory data analysis. In: Secondary Analysis of Electronic Health Records, pp. 185–203 (2016)
12. Park, H.-A.: An introduction to logistic regression: from basic concepts to interpretation with particular attention to nursing domain. J. Korean Acad. Nurs. **43**(2), 154–164 (2013)
13. Fernandes, A.A.T., Filho, D.B.F., da Rocha, E.C., da Silva Nascimento, W.: Read this paper if you want to learn logistic regression. Revista de Sociologia e Política **28**, e006 (2021)
14. Boateng, E.Y., Abaye, D.A.: A review of the logistic regression model with emphasis on medical research. J. Data Anal. Inf. Process. **7**(4), 190–207 (2019)
15. Araveeporn, A.: Comparison of logistic regression and discriminant analysis for classification of multicollinearity data. WSEAS Trans. Math. **22**, 120–131 (2023)
16. Birba, D.E.: A comparative study of data splitting algorithms for machine learning model selection (2020)
17. Chai, T., Draxler, R.R.: Root mean square error (RMSE) or mean absolute error (MAE)?-Arguments against avoiding RMSE in the literature. Geosci. Model Dev. **7**(3), 1247–1250 (2014)
18. Robeson, S.M., Willmott, C.J.: Decomposition of the mean absolute error (MAE) into systematic and unsystematic components. PLOS One **18**(2), e0279774 (2023)

19. Chicco, D., Warrens, M.J., Jurman, G.: The coefficient of determination R-squared is more informative than SMAPE, MAE, MAPE, MSE and RMSE in regression analysis evaluation. PeerJ Comput. Sci. **7**, e623 (2021)
20. Moksony, F., Heged, R.: Small is beautiful. The use and interpretation of R2 in social research. Szociológiai Szemle, Special issue, pp. 130–138 (1990)
21. Tjur, T.: Coefficients of determination in logistic regression models-a new proposal: the coefficient of discrimination. Am. Stat. **63**(4), 366–372 (2009)

MMoE-GAT: A Multi-Gate Mixture-of-Experts Boosted Graph Attention Network for Aircraft Engine Remaining Useful Life Prediction

Lu Liu[1], Xiao Song[2(✉)], Bingli Sun[2], Guanghong Gong[1], and Wenxin Li[2]

[1] School of Automation Science and Electrical Engineering, Beihang University, Beijing 100191, China
[2] School of Cyber Science and Technology, Beihang University, Beijing 100191, China
songxiao@buaa.edu.cn

Abstract. Accurately estimating remaining useful life (RUL) is critical to reducing unplanned downtime, lowering maintenance costs, and improving safety and reliability in the field of prognostics and health management (PHM). At present, most of the data-driven RUL estimation methods are single-task learning models, i.e., the auxiliary tasks related to RUL are neglected, resulting in limited prediction accuracy. In this case, this study presents a multi-task learning (MTL) framework composed of a structure of multi-gate mixture-of-experts (MMoE) and a graph attention network (GAT) model, aiming to utilize the health state (HS) evaluation task to improve the prognostics accuracy. Specifically, GAT was employed to extract the intrinsic spatial information from the sensor network. A gating mechanism in the MMoE was utilized to adjust the parameters based on the distinctive features of different tasks. Moreover, we applied a learnable regularization term to deal with the fusion of HS loss and RUL loss. Experiments on the aircraft engine datasets reveal that the RUL prediction performances of MMoE-GAT are superior to those of available state-of-the-art (SOTA) methods.

Keywords: Aircraft engine · Remaining useful life prediction · Health state evaluation · Multi-task learning · Multi-gate mixture-of-experts

1 Introduction

PHM has been essential to equipment operation and maintenance over the years. It helps to realize effective fault forecasting and maintenance planning by monitoring, diagnosing, and predicting the status of a system or equipment in real time to improve equipment reliability, availability, and performance. This type of technology is employed in many industries, including aerospace, military industry, petrochemical industry, wind power, transportation, etc. As one of the main tasks of PHM, RUL estimation aims to forecast the residual service lifetime of equipment on the basis of condition monitoring (CM) data and historical information by employing various methods such as mechanistic models,

© The Author(s), under exclusive license to Springer Nature Singapore Pte Ltd. 2024
F. Hassan et al. (Eds.): AsiaSim 2023, CCIS 1911, pp. 451–465, 2024.
https://doi.org/10.1007/978-981-99-7240-1_36

statistical models, or artificial intelligence models, so that appropriate maintenance measures can be taken before failures occur, thereby reducing downtime and maintenance costs, and ultimately achieving predictive maintenance. Aircraft engines are the source of electricity and power for the aircraft, and their dependability and safety are vital to air travel. Estimating the RUL of aircraft engines helps to make efficient repair plans and decisions, thus guaranteeing proper engine operations and preventing probable breakdowns. With the rapid development of big data and artificial intelligence technologies, data-driven methods for remaining life prediction have gained widespread attention. Researchers have employed various machine learning-based methods such as support vector regression (SVR) models [1], extreme learning machines (ELM) [2], hidden Markov models (HMMs) [3]. To extract relevant features, traditional machine learning algorithms typically rely on feature engineering which is a time-consuming and specialized procedure that necessitates domain expertise and experience. In practice, multi-source CM data are collected through multiple sensors in order to extensively track the status of complex machinery. Although multi-sensor signals provide more information resources, it turns out to be a technical challenge to fuse them and explore the hidden features using conventional machine learning models. On the contrary, deep learning (DL) performs admirably when dealing with multi-source data processing and fusion. It can automatically derive more detailed and abstract feature representations through the multi-layer structure from large-scale, high-dimensional data. Consequently, an increasing amount of DL-based RUL prediction approaches have emerged. Babu et al. [4] presented a convolutional neural network (CNN)-based model for aircraft engine prognostics. Li et al. [5] combined CNN and long short-term memory (LSTM) networks to acquire the spatial and temporal characteristics from the measured data. Jin et al. [6] introduced a position encoding-based CNN to capture sequential information and enhance the performance of prognostics. Lin et al. [7] proposed an attention-based gate recurrent unit (ABGRU) prognostics model for aero-engines. The enhanced RUL prediction model combined an encoder-decoder architecture with GRU networks for efficient feature fusion. Liang et al. [8] developed an RUL estimation framework, which integrated a graph attention network and deep adaptive transformer (GAT-DAT), to improve the forecasting accuracy.

Through a review of the above literature, it can be concluded that some representative popular DL models such as CNN, LSTM, and Transformer have been broadly applied to machinery prognostics. Nevertheless, most of those models are designed to mine information related to RUL prediction and ignore other information about tasks related to the RUL prediction task, which can lead to non-maximized use of the data and limited model representation learning ability. With the emergence of the multi-task learning (MTL) paradigm, more researchers tend to integrate the RUL prediction and health state (HS) assessment as a dual-task learning problem and utilize DL-based models to enhance the prognostics performance. Zhang et al. [9] established a multi-task learning-based maintenance framework using a stacked autoencoder long short-term memory

(SAE-LSTM) network to execute cooperative training of the classification and regression tasks to create prognostics outcome for the degrading equipment. To highlight the importance of determining the health status to the process of RUL estimation, Kim et al. [10] developed a CNN-based multi-task learning framework. In general, MTL-based prognostics methods extensively adopt the shared-bottom architecture. In this instance, the hidden layers at the bottom are accessible for different tasks. This structure can essentially reduce the risk of overfitting, while the model performance may be affected by task differences and data distribution. In addition, the traditional loss function of MTL is comprised of the weighted summation of losses from single-task learning, and choosing the appropriate weights requires time-consuming attempts.

Therefore, in order to better model task relationships and extract the equipment degradation information, we propose a multi-task learning framework based on a multi-gate mixture-of-experts and graph attention network (MMoE-GAT) to enhance RUL prediction performance. The MMoE model follows the ensemble learning design of MoE that employs expert modules to explicitly solve a subtask-based prediction challenge and balances the difference between tasks through the mechanism of gated networks. Moreover, an adaptive and flexible design of loss function was employed to achieve automatic learning of multi-task weights. The main contributions of this paper are summarized below.

1. An integration of MMoE and GAT enhances the attention-based ensemble learning ability of MMoE and the feature fusion capacity of GAT.
2. A MMoE was utilized to model relationships between RUL prediction and HS assessment tasks, and automatically adjust shared and task-specific information by regulating gated networks.
3. We employed the multi-source data to construct a sensor graph from which a GAT model was used to extract the degradation features of the aero-engine.
4. A loss function for MTL with a learnable positive regularization item was suggested to overcome the problem of manual tuning of weights.

The remainder of this work is organized as follows. Section 2 introduces the basic theories of MMoE and GAT. Section 3 describes the proposed multi-task learning model. An aircraft engine dataset is provided to verify the presented MMoE-GAT prognostics framework in Sect. 4. Meanwhile, some comparative experiments are also conducted to show the model's superiority. Finally, Sect. 5 concludes this study.

2 Preliminary

2.1 MMoE

MMoE was first introduced by Ma et al. [11] to explore the relationship between multiple tasks in the process of multi-task learning. As shown in Fig. 1, the MMoE was built on the popular shared-bottom design. Several network layers are stacked and shared by all tasks. After that, each task has a separate network

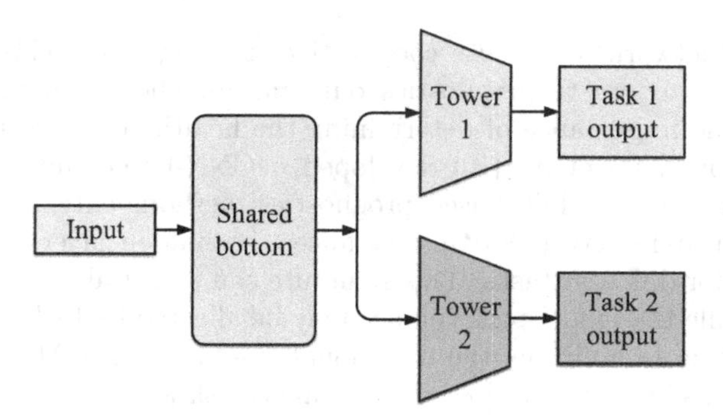

Fig. 1. Structure of the shared-bottom model.

"Tower" to regulate respective feature representation. Nevertheless, the shared-bottom architecture is limited in the face of complex tasks and heterogeneous data from multiple sources. By combining multiple expert modules, MoE is able to provide greater expressive ability. Each expert module is allowed to have varied parameters and structure so that the diverse and complex input data will be processed appropriately. Besides, MoE with a gating module can dynamically assign weights to different experts, thereby adaptively extracting the common and comprehensive features and providing better overall performance. Given M tasks and N expert networks, the single-gate MoE model is illustrated in Fig. 2 and formulated by

$$f(x) = \sum_{i=1}^{N} g(x)_i e_i(x), \tag{1}$$

$$y_m = t^m(f(x)). \tag{2}$$

where $f(x)$ represents the comprehensive outcome of expert networks, $g(x)_i$ is the ith logit of the gating network output $g(x)$, which is assigned to the ith expert e_i. Note that $\sum_{i=1}^{n} g(x)_i = 1$. y_m and t^m indicate the output of task m and function of tower m.

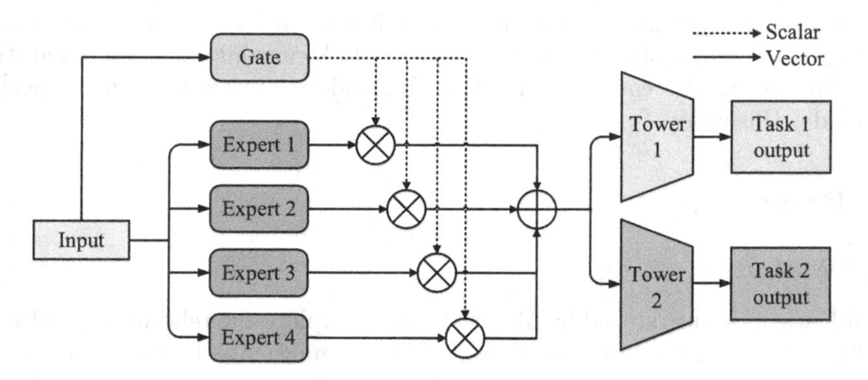

Fig. 2. Structure of the single-gate MoE model.

Inspired by the gating mechanism of MoE, MMoE further developed a task-specific gating network to optimize each task adaptively. As depicted in Fig. 3, similar to the MoE, the structure of MMoE involves a group of expert networks. Different from the single-gate network, MMoE employs individual gating networks for each task allowing different tasks to take expert modules in various ways. In this way, the output of task m obtained from MMoE can be expressed as follows.

$$y_m = t^m(f^m(x)), \tag{3}$$

$$f^m(x) = \sum_{i=1}^{N} g^m(x)_i e_i(x), \tag{4}$$

$$g^m(x) = \text{softmax}(W_{g^m} x), \tag{5}$$

where $W_{g^m} \in \mathbb{R}^{N \times D}$ represents a learnable parameter matrix. N indicates the number of expert networks. D refers to the input dimension.

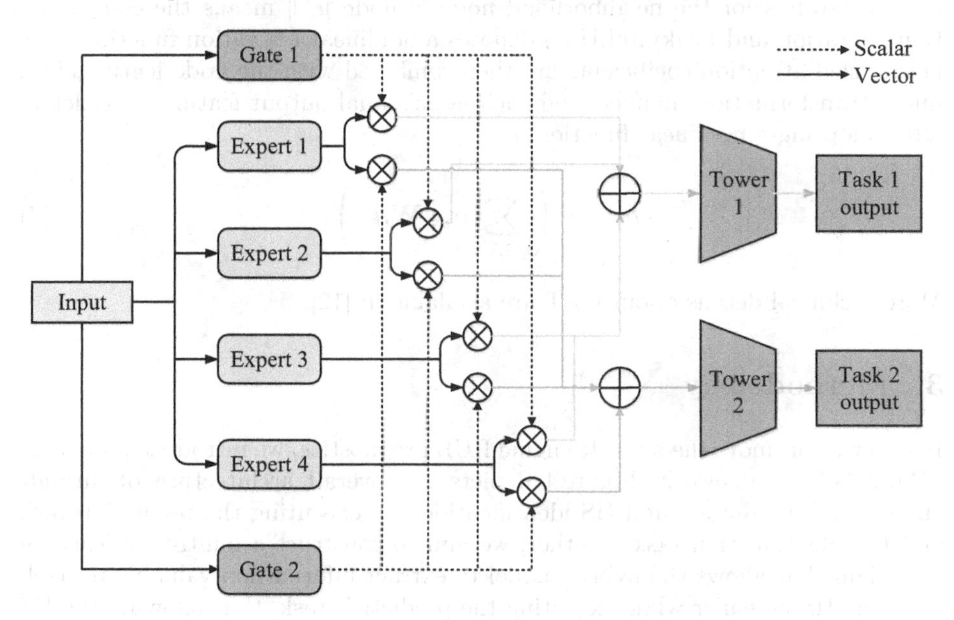

Fig. 3. Structure of the MMoE model.

2.2 GAT

The majority of current deep learning-based techniques rely on grid-structured data or sequence-structured data, even sequence-grid-structured data to complete the prediction task, resulting in an inability of extracting non-Euclidean spatial characteristics from multi-source measurements. Recently, GAT is gaining popularity in processing graph-structured data, which takes into account the relationships among multi-sensor signals. GAT utilizes an attention mechanism to perform a weighted summation of adjacent node representations, and the weights of adjacent node characteristics depend exclusively on the node features.

Assume that a graph consists of K nodes and each node has F features. The set of node features can be expressed as $\mathbf{h} = \{\boldsymbol{h}_1, \boldsymbol{h}_2, ..., \boldsymbol{h}_K\}, \boldsymbol{h}_k \in \mathbb{R}^F$. Through the operation of the graph attention convolutional (GATConv) layer, a new set of node features $\mathbf{h}' = \{\boldsymbol{h}'_1, \boldsymbol{h}'_2, \ldots, \boldsymbol{h}'_K\}, \boldsymbol{h}'_k \in \mathbb{R}^{F'}$ is obtained. In general, the GATConv layer includes the steps of linear transformation, attention coefficient computation, and attention coefficient normalization. The normalized attention coefficient between node u and v can be formulated as

$$\alpha_{uv} = \frac{\exp\left(\mathrm{LeakyReLU}\left(\mathbf{a}^T[\mathbf{W}\boldsymbol{h}_u \| \mathbf{W}\boldsymbol{h}_v]\right)\right)}{\sum_{w \in \mathcal{N}_u} \exp\left(\mathrm{LeakyReLU}\left(\mathbf{a}^T[\mathbf{W}\boldsymbol{h}_u \| \mathbf{W}\boldsymbol{h}_w]\right)\right)}, \tag{6}$$

where $\mathbf{W}$ is a trainable matrix, $\mathbf{a}^T$ is a single-layer feed-forward neural network, $w \in \mathcal{N}_u$ stands for the neighborhood node of node u, $\|$ means the concatenation operation and LeakyReLU $(\cdot)$ denotes a nonlinear activation function. The normalized attention coefficients are then combined with the node features in a linear transformation manner, and each node's final output features will derive after adopting a nonlinear function, σ.

$$\boldsymbol{h}'_u = \sigma\left(\sum_{w \in \mathcal{N}_u} \alpha_{uw} \mathbf{W}\boldsymbol{h}_w\right). \tag{7}$$

More technical details about GAT are available in [12].

3 Methodology

In order to promote the aircraft engine RUL prognostics, we introduced a hybrid MMoE-GAT framework. Figure 4 depicts the overall architecture of aircraft engine RUL prediction and HS identification. By executing the prediction task and the identification task together, we aim to construct a multi-task learning paradigm that allows the hybrid model to extract information valuable to itself from the HS classifier while executing the prediction task, thus allowing the HS assessment task to be used as a auxiliary task to improve the accuracy of the prognostics task. The whole prognostics framework can be roughly divided into four parts: data acquisition, data selection and normalization, sliding window processing, and model construction. Firstly, CM data are collected from various components of aircraft engines including compressors, fan spool, core spool, and turbines. Not all of these signals provide valid degradation information, and some of them have constant values that do not change over time, so it is essential to analyze and filter the signals collected by the sensors. At the same time, normalized inputs have a variety of effects on neural networks, such as accelerating training convergence, enhancing gradient propagation, and improving resilience and generalization capacities. Therefore, it is necessary to normalize the selected

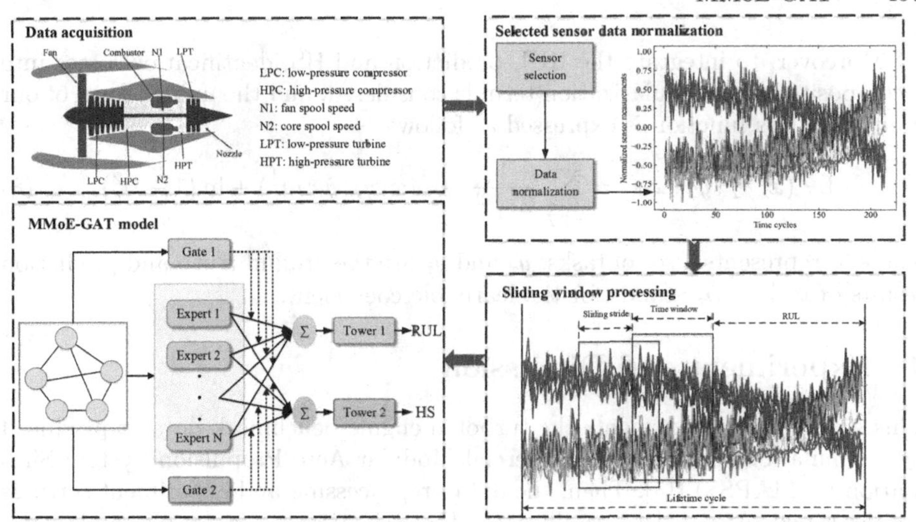

Fig. 4. Proposed overall flowchart of aircraft engine prognostics and HS assessment.

sensor data into the same range. Then, in order to get more training samples, we used sliding window operation to process the sequence signals. Because the proposed MMoE-GAT model works with graph data, we adapted the cosine similarity to analyze the correlation of any two sensor data and construct the CM data-based sensor graph. In this study, we employed four GAT-based expert modules, each of which is illustrated in Fig. 5. The input graph data can be distilled into higher-level features after multiple hidden layers. The combination of batch normalization (BN) and Gaussian error linear units (GELU) helps to enhance the representation of neural networks and improve network robustness and training stability. The role of the pooling and readout layers is to reduce the number of nodes, summarize the information, and accelerate computation. We utilized EdgePool [13] and global mean pool methods as pooling and readout layers, respectively.

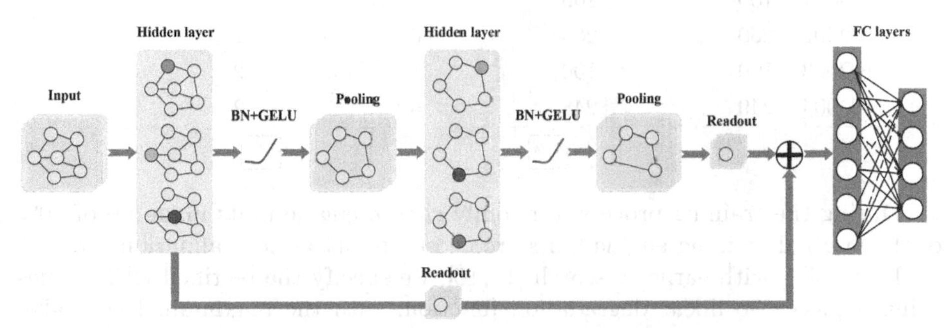

Fig. 5. Proposed GAT-based expert module.

Moreover, to integrate the RUL prediction and HS identification losses into a composite loss, a regularization term is considered, and the unified form of our multi-task loss function is expressed as follows.

$$\mathrm{L}_{\mathcal{T}}\left(\boldsymbol{x}, y_{\mathcal{T}}, \hat{y}_{\mathcal{T}}; \boldsymbol{\omega}_{\mathcal{T}}\right)=\sum_{\tau \in \mathcal{T}} \frac{1}{2 \cdot \beta_\tau^2} \cdot \mathrm{L}_\tau\left(\boldsymbol{x}, y_\tau, \hat{y}_\tau; \boldsymbol{\omega}_\tau\right)+\ln\left(1+\beta_\tau^2\right), \tag{8}$$

where $\mathcal{T}$ represents a set of tasks, y_τ and $\hat{y}_\tau$ are the ground truth and prediction results of task τ, β_τ stands for the learnable coefficient.

4 Experiment and Discussion

This section initially presents the turbofan engine benchmark dataset produced by a simulation tool called Commercial Modular Aero-Propulsion System Simulation (C-MAPSS) [14]. Then, the data preprocessing and experiment settings are described. Afterwards, we illustrate the RUL prognostics results and evaluate them against existing SOTA methods. The experiments based on single-task single-gate and multi-task single-gate models are eventually conducted to evaluate the impact of the task number and gate number.

4.1 Dataset Description

The C-MAPSS dataset contains four subsets named FD001 to FD004. They differ from each other by operation conditions and fault modes. In contrast to FD002 and FD004, which comprise six operating situations, FD001 and FD003 capture engine sensor measurements acquired under a single operating condition. FD001 and FD002 have one failure mode, but two fault modes exist in FD003 and FD004. Table 1 lists the details of the C-MAPSS dataset.

Table 1. Description of C-MAPSS dataset.

Subsets	Training Engine units	Testing Engine units	Operation conditions	Fault modes
FD001	100	100	1	1
FD002	260	259	6	1
FD003	100	100	1	2
FD004	249	248	6	2

During the training process, randomly chosen engine units made up of 10% of the original training engine units are chosen to act as the validation set.

In keeping with earlier research [15,16], we specify the rectified RUL values using a piecewise linear degradation function, with the maximum RUL value set to 125. Meanwhile, on the basis of the rectified RUL values, we divide the engine lifespan into three HS phases: healthy, slight degeneration, and severe degeneration. The correlation between the HS and RUL of engines is shown in Table 2.

Table 2. Relation between the HS and RUL.

HS	Phase 1	Phase 2	Phase 3
RUL	[20, 125]	[10, 20)	[0, 10)

4.2 Sensor Selection and Data Normalization

21 kinds of sensor values were collected simultaneously around the low-pressure compressor (LPC), high-pressure compressor (HPC), low-pressure turbine (LPT), and high-pressure turbine (HPT), including several physical quantities such as temperature, pressure, and speed. It should be noted that, following the shifting patterns, the 21 sensor measurements can be categorized into three types: ascending, descending, and irregular. Sensors with indices of 1, 5, 6, 10, 16, 18, and 19 always have irregular values and are unable to deliver crucial information concerning degradation. As a result, we made use of the rest 14 sensor measurements to train our model and forecast the engine RUL values. Subsequently, the min-max normalization operation is indispensable for accelerating model convergence and improving prediction performance, which is formulated as follows.

$$x_{norm}^{i,j} = \frac{2\left(x^{i,j} - x_{\min}^{j}\right)}{x_{\max}^{j} - x_{\min}^{j}} - 1, \forall i, j \tag{9}$$

where $x^{i,j}$ is the raw measurement value collected from ith data point and jth sensor, $x_{norm}^{i,j}$ denotes the normalized sensor data. $x_{\max}^{j}$ and $x_{\min}^{j}$ are the maximum and minimum values from the j-th sensor, respectively.

4.3 Sliding Window Processing and Sensor Graph Construction

In the field of time series analysis, the relationship between data from neighboring moments is very crucial. Sliding window processing captures these dependencies by using the time window to gather several consecutive moments of sensor data. In addition, it is seen as a technique for data augmentation. In this work, in order to obtain more training samples, the sliding step size was set to 1 and the window sizes of FD001 and FD003 were specified as 25 and 30, respectively. After that, it is essential to convert the time series data into graph data, that is, sensor graph construction. Here, we employed the cosine similarity method to determine if there is a connection between two sensor nodes. When the cosine similarity between them is larger than 0, we recognized that they are related. Conversely, if the cosine similarity between them is less than 0, we consider them to be uncorrelated. The constructed sensor network graph is illustrated in Fig. 6.

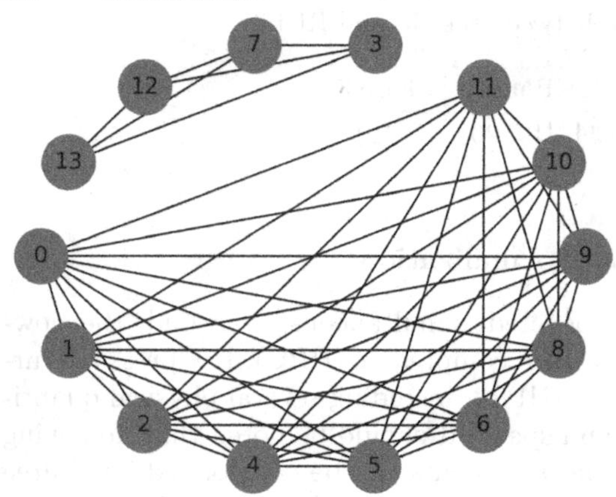

0: Total temperature at LPC outlet (T24)
1: Total temperature at HPC outlet (T30)
2: Total temperature at LPT outlet (T50)
3: Total pressure at HPC outlet (P30)
4: Physical fan speed (Nf)
5: Physical core speed (Nc)
6: Static pressure at HPC outlet (Ps30)
7: Ratio of fuel flow to Ps30 (phi)
8: Corrected fan speed (NRf)
9: Corrected core speed (NRc)
10: Bypass Ratio (BPR)
11: Bleed Enthalpy (htBleed)
12: HPT coolant bleed (W31)
13: LPT coolant bleed (W32)

Fig. 6. Diagram of the FD001 sensor network.

4.4 Evaluation Indicators

For the purpose of comparing our prediction results with those of existing studies, it is important to specify the unified metrics. Herein, we employ two generalized metrics, the root mean square error (RMSE) and the score functions, formulated as

$$\mathrm{RMSE} = \sqrt{\frac{1}{M_T} \sum_{i=1}^{M_T} (\hat{y}_i - y_i)^2} \tag{10}$$

$$S = \begin{cases} \sum_{i=1}^{M_T} \left[exp\left(-\frac{\hat{y}_i - y_i}{13}\right) - 1\right], & \text{for } \hat{y}_i < y_i \\ \sum_{i=1}^{M_T} \left[exp\left(\frac{\hat{y}_i - y_i}{10}\right) - 1\right], & \text{for } \hat{y}_i \geq y_i \end{cases} \tag{11}$$

where $\hat{y}_i$ and y_i are the estimated and actual RUL for the i-th testing sample, respectively. M_T represents the amount of testing samples. Notably, smaller RMSE and S values mean better prediction performance. As shown in Eq. (11), the score function penalizes delayed prediction more because it may bring downtime and cause more serious accidents.

4.5 Hyperparameters

We employed the Adam optimizer as the backpropagation algorithm and set the training epoch to 100. Table 3 lists the hyperparameter details.

Table 3. Hyperparameters of the proposed method.

Hyperparameter	Value
Number of nodes	14
Number of experts	4
Feature dimension of expert out	128
Hidden layer size of tower	32
GATconv out channels	1024
Batchsize	128
Pooltype	Edgepool
Readout	Global mean pool
Epoch	100
Learning rate	0.005
Dropout	0.2

4.6 RUL Prognostics Results

In this work, we adopted the CM data from FD001 and FD003 to analyze the RUL prediction performance. Figure 7 illustrates the RUL curves over time from four randomly chosen engines. As observed in the figures, the prediction curves gradually approach the true RUL value as time goes by, indicating that the prediction accuracy becomes higher. This is because in the later stages of the engine lifespan, the degradation or failure information progressively increases so that the proposed MMoE-GAT model can better capture and extract features.

Furthermore, all the engine samples in the FD001 and FD003 test sets were evaluated and the predicted results were quantified to compare with those of existing studies. The prognostic results of the SOTA approaches as well as our proposed method are illustrated in Table 4. As can be seen from the table, in the FD001 and FD003 subsets, our suggested MMoE-GAT model performs best in terms of RMSE and score function. Moreover, it is clear in Table 4 that the MS (Multi-task single-gate) model performs the second best on most of the metrics and datasets. In fact, the single-gate MoE in Fig. 2 is the prototype of the MS model. Compared with the MMoE-GAT model, the MS model lacks one gating module, and two towers of the dual-task share the common input from the expert modules. This will result in two tasks not being able to selectively acquire features from the expert modules.

The Single-task single-gate (SS) model has only one gated network and one target task, i.e., RUL prediction. As mentioned earlier, the advantage of multi-task learning is that it taps into the correlations between tasks and provides complementary information for each task, thus improving the performance of both sides. Both MS and MMOE-GAT models consider the two tasks of RUL prediction and HS evaluation, both of which happen to be correlated, which will undoubtedly improve the prediction performance. Moreover, we suggested a loss function strategy that avoids manually adjusting the weights between subtask loss functions, thereby facilitating the training process. Figure 8 illustrates the

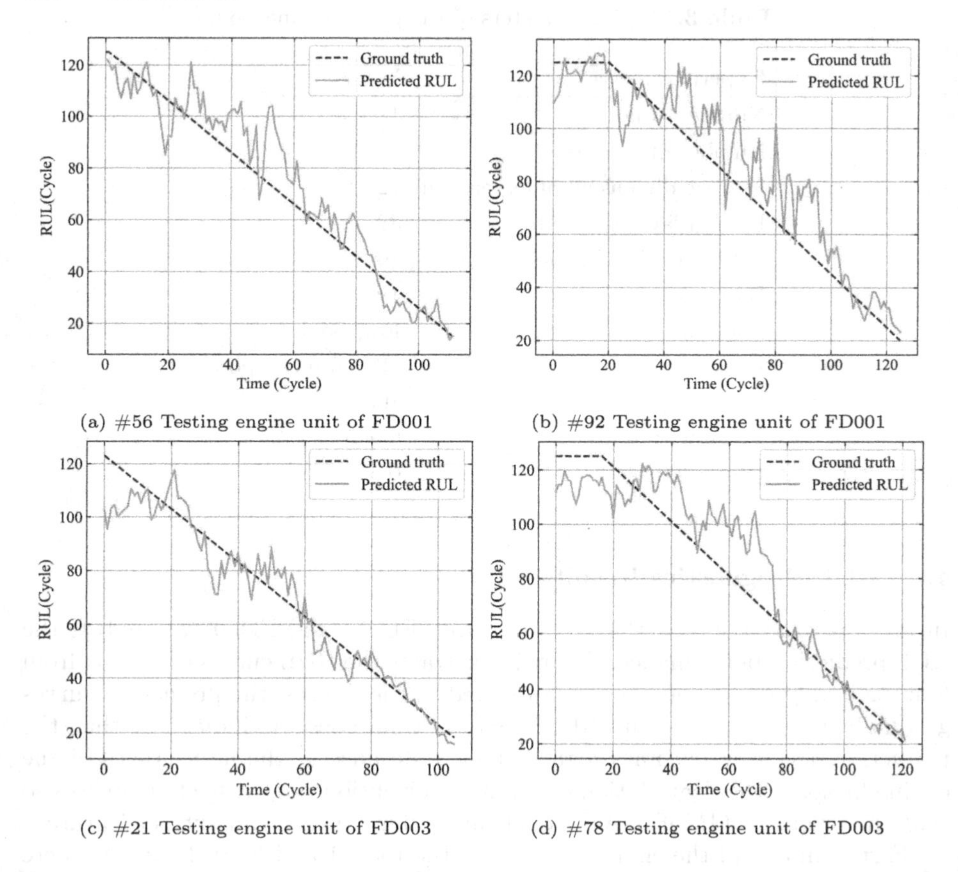

(a) #56 Testing engine unit of FD001

(b) #92 Testing engine unit of FD001

(c) #21 Testing engine unit of FD003

(d) #78 Testing engine unit of FD003

Fig. 7. RUL prognostic performances for the random testing engine units from subsets FD001 and FD003.

Table 4. Comparison with other SOTA approaches. Red text indicates the best and **blue** text indicates the second best result. SS: Single-task single-gate; MS: Multi-task single-gate.

Methods	Year	FD001 RMSE	Score	FD003 RMSE	Score
GatBoost [17]	2020	15.8	398.7	16.0	584.2
BiGRU-AS [18]	2021	13.68	284	15.53	428
GAT-EdgePool [19]	2022	13.53	309.6	14.66	470.7
VAE-RNN [20]	2022	15.81	326	14.88	722
GAT-DAT [8]	2023	13.83	318.6	14.85	438.5
ABGRU [7]	2023	12.83	**221.5**	13.23	279.2
SS	2023	12.81	235.7	12.98	300.1
MS	2023	**12.58**	231.8	**12.67**	**257.4**
MMoE-GAT (Proposed)	2023	11.85	206.5	12.29	253.2

RUL prediction results provided by MM, MS, and SS methods. As can be seen in Fig. 8, the prediction results of these models are all acceptable. The estimation errors of the three techniques, particularly in the later phases of the engine operating cycle, are not considerable, which results from the expert modules that they all possess. The expert modules can be considered as an ensemble learning and it will enhance the model's generalization and robustness abilities.

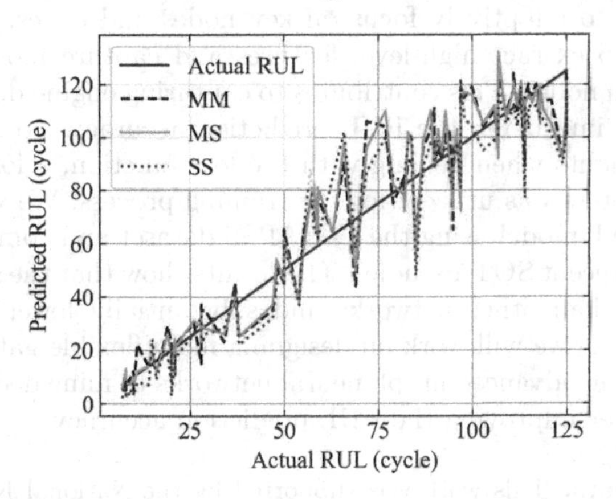

(a) Comparison of the MM, MS, and SS models on FD001 subset.

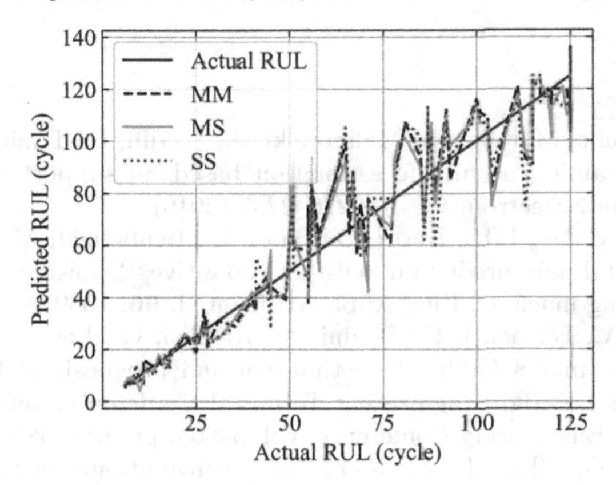

(b) Comparison of the MM, MS, and SS models on FD003 subset.

Fig. 8. Visualization of RUL prediction results obtained by our proposed models. MM: Multi-task multi-gate, i.e., MMoE-GAT.

5 Conclusion

In this paper, we propose an aircraft engine RUL prediction model that incorporates MMoE and GAT networks. MMoE allows multiple related tasks to be learned simultaneously in a single model. While all tasks share the expert networks, an independent gating network allows task-specific filtering and correction of information from the expert networks, controlling the degree of expert attention to features required for different tasks.

In addition, we applied a GAT network to model the relationships between engine sensors and learn their interactions. With the self-attention mechanism, the GAT is able to adaptively focus on key nodes and edges. The multi-layer structure helps to extract high-level features and capture more complex relationships between nodes. This contributes to capturing engine degradation information and thus improving the RUL prediction accuracy. To avoid artificially adjusting the weights when dealing with the loss function, a loss function with learnable parameters was utilized for the training process. We validate the proposed MMoE-GAT model using the C-MAPSS dataset and compare the prediction results with recent SOTA studies. The results show that the suggested model performs better than other networks and substantially lowers the prediction error. In the future, we will work on designing more flexible gating mechanisms while incorporating advanced graph neural networks to mine deeper degradation features for further improving the RUL prediction accuracy.

Acknowledgements. This work was supported by the National Key Research and Development Program of China under Grant 2020YFB1712203.

References

1. Khelif, R., Chebel-Morello, B., Malinowski, S., Laajili, E., Fnaiech, F., Zerhouni, N.: Direct remaining useful life estimation based on support vector regression. IEEE Trans. Ind. Electron. **64**(3), 2276–2285 (2016)
2. Berghout, T., Mouss, L.H., Kadri, O., Saïdi, L., Benbouzid, M.: Aircraft engines remaining useful life prediction with an adaptive denoising online sequential extreme learning machine. Eng. Appl. Artif. Intell. **96**, 103936 (2020)
3. Giantomassi, A., Ferracuti, F., Benini, A., Ippoliti, G., Longhi, S., Petrucci, A.: Hidden Markov model for health estimation and prognosis of turbofan engines. In: International Design Engineering Technical Conferences and Computers and Information in Engineering Conference, vol. 54808, pp. 681–689 (2011)
4. Sateesh Babu, G., Zhao, P., Li, X.-L.: Deep convolutional neural network based regression approach for estimation of remaining useful life. In: Navathe, S.B., Wu, W., Shekhar, S., Du, X., Wang, X.S., Xiong, H. (eds.) DASFAA 2016. LNCS, vol. 9642, pp. 214–228. Springer, Cham (2016). https://doi.org/10.1007/978-3-319-32025-0_14
5. Li, J., Jia, Y., Niu, M., Zhu, W., Meng, F.: Remaining useful life prediction of Turbofan engines using CNN-LSTM-SAM approach. IEEE Sensors J. **23**, 10241–10251 (2023)
6. Jin, R., Wu, M., Wu, K., Gao, K., Chen, Z., Li, X.: Position encoding based convolutional neural networks for machine remaining useful life prediction. IEEE/CAA J. Autom. Sin. **9**(8), 1427–1439 (2022)
7. Lin, R., Wang, H., Xiong, M., Hou, Z., Che, C.: Attention-based gate recurrent unit for remaining useful life prediction in prognostics. Appl. Soft Comput. **143**, 110419 (2023)

8. Liang, P., Li, Y., Wang, B., Yuan, X., Zhang, L.: Remaining useful life prediction via a deep adaptive transformer framework enhanced by graph attention network. Int. J. Fatigue **174**, 107722 (2023)
9. Zhang, L., Zhang, J.: A data-driven maintenance framework under imperfect inspections for deteriorating systems using multitask learning-based status prognostics. IEEE Access **9**, 3616–3629 (2020)
10. Kim, T.S., Sohn, S.Y.: Multitask learning for health condition identification and remaining useful life prediction: deep convolutional neural network approach. J. Intell. Manuf. **32**, 2169–2179 (2021)
11. Ma, J., Zhao, Z., Yi, X., Chen, J., Hong, L., Chi, E.H.: Modeling task relationships in multi-task learning with multi-gate mixture-of-experts. In: Proceedings of the 24th ACM SIGKDD International Conference on Knowledge Discovery & Data Mining, pp. 1930–1939 (2018)
12. Veličković, P., Cucurull, G., Casanova, A., Romero, A., Lio, P., Bengio, Y.: Graph attention networks. arXiv preprint arXiv:1710.10903 (2017)
13. Diehl, F.: Edge contraction pooling for graph neural networks. arXiv preprint arXiv:1905.10990 (2019)
14. Frederick, D.K., DeCastro, J.A., Litt, J.S.: User's guide for the commercial modular aero-propulsion system simulation (c-mapss). Tech. Rep. (2007)
15. Li, H., Zhao, W., Zhang, Y., Zio, E.: Remaining useful life prediction using multi-scale deep convolutional neural network. Appl. Soft Comput. **89**, 106113 (2020)
16. Liu, L., Song, X., Zhou, Z.: Aircraft engine remaining useful life estimation via a double attention-based data-driven architecture. Reliab. Eng. Syst. Safety **221**, 108330 (2022)
17. Deng, K., et al.: A remaining useful life prediction method with long-short term feature processing for aircraft engines. Appl. Soft Comput. **93**, 106344 (2020)
18. Duan, Y., Li, H., He, M., Zhao, D.: A BiGRU autoencoder remaining useful life prediction scheme with attention mechanism and skip connection. IEEE Sens. J. **21**(9), 10905–10914 (2021)
19. Li, T., Zhou, Z., Li, S., Sun, C., Yan, R., Chen, X.: The emerging graph neural networks for intelligent fault diagnostics and prognostics: a guideline and a benchmark study. Mech. Syst. Signal Process. **168**, 108653 (2022)
20. Costa, N., Sánchez, L.: Variational encoding approach for interpretable assessment of remaining useful life estimation. Reliab. Eng. Syst. Safety **222**, 108353 (2022)

Integrating Dynamic Limiting Mechanism in Electric Scooters to Mitigate Range Anxiety via SoC-Tracking Range-Enhancement Attachment (STRETCH)

Khairul Azhar Che Kamarudin[1], Saiful Azrin M. Zulkifli[1,2](✉), and Ghulam E Mustafa Abro[1,2]

[1] Electrical & Electronics Engineering, Universiti Teknologi PETRONAS, Seri Iskandar, Malaysia
saifulazrin_mz@utp.edu.my
[2] Center for Automotive Research & Electric Mobility (CAREM), Universiti Teknologi PETRONAS, 32610 Seri Iskandar, Perak, Malaysia

Abstract. Electric vehicles face a remarkable challenge in the form of range anxiety, as their limited operational range per charge has hindered widespread adoption. This concern is further amplified by consumers' apprehensions about running out of battery and the scarcity of charging infrastructure, coupled with the time-consuming recharging process. Surmounting this barrier will not only revolutionize transportation but also empower individuals to embrace the sustainable future of electric mobility, transforming the way we move and live. This work develops a dynamic current-limiting mechanism for existing personal electric mobility (PEM) equipment such as electric scooters. Based on onboard battery state-of-charge (SoC) and pre-set limit levels, the throttle signal is bypassed by an added microcontroller (STRETCH) to reduce battery depletion and hence prolong the scooter's range. This study performs modeling and simulation of the e-scooter using MATLAB-Simulink, integrating the dynamic limiting algorithm. Simulation results are compared and validated with field test data of the enhanced e-scooter, using an FTDI serial adapter for serial communication with the user interface. The acquired data is meticulously analyzed to obtain crucial operating parameters, facilitating a comprehensive performance comparison between the real scooter and the simulation model.

Keywords: Electric Vehicle · Personal Electric Mobility · E-Scooter · Range Anxiety · State of Charge · Dynamic Limiting

1 Introduction

As the world faces climate change issues such as rising of sea levels [1] and global warming [2], many initiatives have been taken to reduce CO_2 emissions, most significant of which is promotion of green transportation. Electric vehicles (EV) have increased in popularity as alternative choice to conventional internal combustion engine vehicle. For

© The Author(s), under exclusive license to Springer Nature Singapore Pte Ltd. 2024
F. Hassan et al. (Eds.): AsiaSim 2023, CCIS 1911, pp. 466–482, 2024.
https://doi.org/10.1007/978-981-99-7240-1_37

instance, China has been proactive in promoting EV usage under the Belt and Road initiative by investing in more charging station along the highway [3]. However, one of the major issues restricting vehicle users from adopting EV according to research by [4] is a phenomenon called range anxiety.

Range anxiety can be defined as the feeling of worrying of whether the EV will be able to reach charging station before running out of power. This can also apply to smaller and personal electric mobility transport (PEM) such as electric scooters. In addition to the smaller battery pack due to the size and purpose of usage, the low range of electric scooters may also prevent user from using it for daily purpose. Recent technological developments, particularly greater charging rate and battery capacity, have expanded the acceptance of electric mobility and PEM devices. Depending on the vehicle, PEM can go up to 30 km or more at speeds above 40 km/h. PEM examples include electric scooter, electric unicycle, and electric skateboard, as illustrated in Fig. 1.

Fig. 1. Different types of personal electric mobility (PEM)

This research will focus on extending the range of a PEM device using smart energy management via an add-on device, to implement signal limiter from the user input to the motor controller, with the following objectives in mind:

- To perform modelling and simulation of electric scooter with current-limiting mechanism using Simscape Electrical Toolbox.
- To design and analyze the control algorithm in the simulation program to vary parameters for dynamic limiting to extend the range.
- To perform datalogging and data analysis on PEM for field-testing and modelling validation.

2 Literature Review

Two-wheeled electric skateboards have grown in acceptance as a viable alternative to personal transportation vehicles in recent years. The control approach being studied by [4] aims to increase the energy effectiveness of personal electric vehicles. Fuzzy-PID control algorithms have been created and used to regulate the vehicle motion [4]. To address the system-related uncertainty, fuzzy algorithms are applied. The fuzzy method is used to update the PID controller so that it can adjust to changing load amounts. The Fuzzy-PID method has the benefit of not requiring prior knowledge of the systems model, making it very flexible [4]. The findings show that compared to the standalone PID controller, the suggested technique is able to reduce energy consumption by up to 2.03%. This work also shows the importance of including the various insignificant variables and uncertainties into the control algorithm to improve the performance.

A technique to maximize the energy efficiency of electric vehicles using a distributed drive system has been proposed by the research work [5]. It achieves this by carefully balancing the torque sent to the traction and regenerative braking systems. Minimizing energy waste during vehicle operation and enhancing the overall performance of the electric vehicle are the goals of the design. To guarantee that the torque allocation is adjusted for optimal energy efficiency, the design takes into account a number of variables, including battery state-of-charge, road conditions, and driving style [5]. The use of this approach results in less energy being used and longer battery life, making electric mobility more environmentally friendly and sustainable. The advantage of the suggested design is that it uses less calculations thanks to the offline optimization process. The proposed design has the advantage of low calculation usage by utilizing the offline optimization procedure and online application by simple interpolation [5]. The technique also included safety mechanisms for both the driver safety as well as the battery health condition. This signifies the importance of the safety mechanism to be considered in the design of a control mechanism.

As this project also requires the execution of the hardware to the PEM device, this research work [6] can give an insight on how to design and implement the hardware for an electric vehicle (EV) with regenerative braking capabilities. The system focuses on controlling speed and torque while utilizing regenerative braking to increase energy efficiency. The hardware includes components such as electric motors, inverters, and batteries, while the control system uses algorithms to manage the interactions between these components and the vehicle brake system. The selections of the required sensor and MOSFET are shown with the details. The goal is to improve the overall performance and energy efficiency of the EV for personal mobility purposes.

The research paper is a comprehensive examination of various methods used for estimating the state of charge (SOC) of a battery. SOC refers to the amount of energy stored in a battery, expressed as a percentage of its maximum capacity [7]. Accurately estimating SOC is important for ensuring the proper operation of battery-powered systems and for maximizing battery life. The review provides an overview of traditional SOC estimation methods, such as Coulomb counting, and more advanced methods, including open-circuit voltage, neural networks, and Kalman filters [7]. It also discusses the challenges associated with each method, such as the difficulty in obtaining accurate models of battery behaviour and the limitations of hardware-based methods. The review concludes by highlighting the need for further research in the field, particularly regarding developing more accurate and reliable SOC estimation methods for different types of batteries.

Through eco-routing, eco-driving, and energy consumption prediction, this study offers a comprehensive strategy for increasing the range of electric cars (EVs) [8]. The strategy proposed considers several elements, including as driving style, road conditions, and energy usage, that have an impact on the range of EVs [8]. Based on elements including road gradient, speed limits, and traffic circumstances, the eco-routing component chooses the most effective route for the EV. The eco-driving component focuses on enhancing the driver behaviour while operating the vehicle to reduce energy consumption, for instance by recommending the ideal speed and acceleration. Utilizing machine learning techniques, the energy consumption prediction component forecasts the energy that the vehicle will use based on a variety of variables, including road conditions, driving style, and weather conditions. The best route and driving style to extend the range of the vehicle can then be decided using this information. The research work also assesses the effectiveness of the suggested technique and comparing it with current methods. The result has demonstrated that their strategy outperforms current approaches in terms of range maximisation.

The research work focuses on producing a fast velocity trajectory planning and control technique for a four-wheel drive (4WD) electric vehicle [9]. The method used is a time-based model predictive control (MPC) strategy to create the vehicle optimal velocity profiles while taking into account a multiple constraint, including the road gradient, the vehicle top speed, and the battery capacity. The researcher uses a vehicle dynamics model to first simulate the dynamic behaviour of the vehicle, and then they include this model into the MPC framework. The MPC algorithm predicts future vehicle states using the vehicle model and determines the best acceleration/deceleration inputs to reduce energy consumption while meeting the limitation [9]. The suggested approach surpasses conventional methods in terms of energy efficiency and velocity tracking precision, according to experimental data. The outcomes also show that the algorithm can successfully handle various driving styles and road circumstances, such as uphill and downhill grades. The work concludes by showing how the time-based MPC approach may produce velocity profiles for 4WD electric vehicles that are energy efficient. Using the suggested method, it is possible to create workable energy-saving control schemes for electric vehicles.

MATLAB-Simulink provides a unified platform for quick assessment of design concepts, simulation of large-scale systems utilising reusable components and libraries, and integration of specialised third-party modelling tools [19]. It also makes model deployment for desktop simulation and real-time testing easier [14]. In this project, MATLAB-Simulink is used to develop a control algorithm or limiting mechanism that takes into account the battery state-of-charge to provide a limited signal for motor control.

The FT232RL adaptor, which makes use of the dependable FTDI FT232RL chipset [15], is ideal for data logging tasks. Its ability to handle 3.3V and 5.5V voltage levels provides interoperability with a wide range of devices and sensors. It enables effective real-time monitoring with data rates of up to 3MBaud. The adapter USB power supply eliminates the need for additional sources, increasing mobility, and its small dimension of 16mm x 34mm makes it easier to fit into tight spaces. Furthermore, its broad interoperability with microcontrollers and Arduinos allows smooth integration. Overall, the high-quality chipset, wide voltage range, quick data rates, USB power supply, and extensive compatibility of the FT232RL converter make it a dependable and efficient solution for data logging.

RealTerm is a popular terminal programme noted for its ability to capture, manipulate, and debug binary data. Its user-friendly interface and capability make it useful for datalogging, particularly in development, reverse engineering, and automated testing. The ability of the programme to produce log and trace files assists in the debugging of complicated serial problems, and its specialised design for processing tough data streams, particularly in binary format, makes it extremely beneficial for engineers and anyone dealing with problematic data. RealTerm strong terminal capabilities, such as the ability to transfer precise bytes, facilitate accurate datalogging and data analysis jobs, making it a valuable tool for technical applications [16].

Python flexibility and extensive data processing and analysis capabilities have led to its widespread use in dealing with numbers and texts in a variety of forms, including binary, octal, decimal, and hexadecimal [17]. This literature review investigates several Python methods for translating hexadecimal data to decimal. The hex() method converts an integer number to its equivalent hexadecimal form, whereas the int() function does the opposite, turning a hexadecimal string to an integer. Iterative conversion entails dividing the decimal number by 16 until a quotient of zero is reached, then deriving the hexadecimal equivalent from the sequence of remainders. Another way uses the int() function in conjunction with a for loop, iterating over each letter in the hexadecimal string and computing its decimal equivalent to achieve the final decimal result. Python libraries provide efficient methods for converting hexadecimal data to decimal, which contributes to the language success in data manipulation and analysis tasks.

3 Methodology

The research process begins with modelling the electric scooter in MATLAB-Simulink. The weight, battery capacity, motor power, and motor peak current of the electric scooter models could all be customized. This is referred to as the modelling phase of our planned work. The second task is to compare the MATLAB-Simulink electric scooter modelling to a real-world scooter to determine the difference in performance between the Simulink and real-world models.

If the performance is comparable to that of a real-world scooter, the dynamic limiting mechanism control block or subsystem was modelled in order to extend the electric scooter model range. The dynamic limiting mechanism uses state-of-charge of the battery as trigger to limit the throttle input according to the available limiting percentage on the electric scooter. There are 3 mode available on the user interface which limit the throttle input up to certain percentage. In this project, a crucial aspect involves the comprehensive analysis of communication between the motor controller and the user interface.

A meticulous wire-tapping technique is employed, using an FTDI serial adapter, to intercept and read transmitted code without disrupting normal operations. Datalogging is conducted on a flat road surface to avoid slope-induced acceleration variations. The throttle input is consistently set to the maximum position due to the difficulty in maintaining precise control within small percentage gaps. The obtained data is logged for examination, enabling a deep understanding of communication flow and message exchange. Deciphering the logged data involves analyzing patterns, structures, and encoding mechanisms to extract meaningful information about communication protocols and data exchange formats.

The knowledge derived from the analyzed data provides valuable insights into the electric scooter behavior and interaction between motor controller with user interface. These insights are pivotal in assessing system performance, identifying issues, and making informed decisions for potential optimizations or improvements.

MATLAB-Simulink was used to create a simulation model of an electric scooter. After the control method is implemented into the model, the simulation will allow one to investigate the effectiveness of the existing restricting mechanism. As minimal needed parameters for the simulation, the PEM model must be capable of producing battery SoC, current, voltage, distance travelled, and velocity graph. Figure 2 depicts an electric scooter model created in MATLAB-Simulink. The simulation model is divided into four major subsystems: the driver input subsystem, the motor and controller subsystem, the vehicle body and tyre subsystem, and the battery subsystem, as seen in Figs. 4, 5, 6, and 7. Each subsystem contains its own set of blocks and components that allow the subsystems to function as a complete integrated electric scooter system.

The longitudinal driver block from the powertrain block library delivers normalised accelerating and braking instructions depending on the reference and feedback velocity for the driver input subsystem. It has been configured for the reference velocity to be able to switch between a drive cycle source block and a signal builder block. The motor and controller subsystem has an H-bridge driver block for applying acceleration and braking to the subsystem's DC motor block. The battery powers the motor, which converts electrical energy into mechanical energy. The motor rated speed, rated voltage, and many other characteristics of the DC Motor block may be customized.

Following the modelling of the PEM on MATLAB-Simulink, the control method or dynamic limiting mechanism for the PEM simulation's range expansion will be developed. The control algorithm or limiting mechanism will regulate battery depletion, hence extending the PEM's driving range. It will take the throttle input from the user or drive cycle and dynamically limit using limiting percentage of 40% and 80% available on the electric scooter namely mode of operation. The state-of-charge will be used as a trigger to automatically change the mode of operation with different limiting percentages. The dynamic limiting algorithm is implemented using If block and If-Action block in the Simulink as shown in Fig. 3. The 3 modes of the electric scooter is represented using the If-Action block with limiting percentage of 40%, 80% and 100%. The transition point of the mode is pre-selected using SOC% as the trigger prior to simulation run.

The vehicle body and tire subsystem are where the vehicle parameters are specified; here, the scooter weight, rider weight assumption, number of wheels per axle, and many other options are available. Because this is an electric scooter model, the model has been set up to use two tires, front and rear, as a real-world 2-wheel electric scooter. Finally, there is the battery subsystem, which houses the battery pack that powers the motor. The battery pack could be swapped out for different types of batteries and battery parameters utilized in the system. The simulation yielded the SoC% graph, current and voltage graphs, distance in km graph, and lastly the velocity graph.

The user interface, controller, motor drive, BLDC motor, and battery pack are depicted in Fig. 8 The controller and motor drive are integrated onto a single circuit board in the electric scooter, which is located near the BLDC motor. The battery pack is directly attached to the motor drive and will be activated.

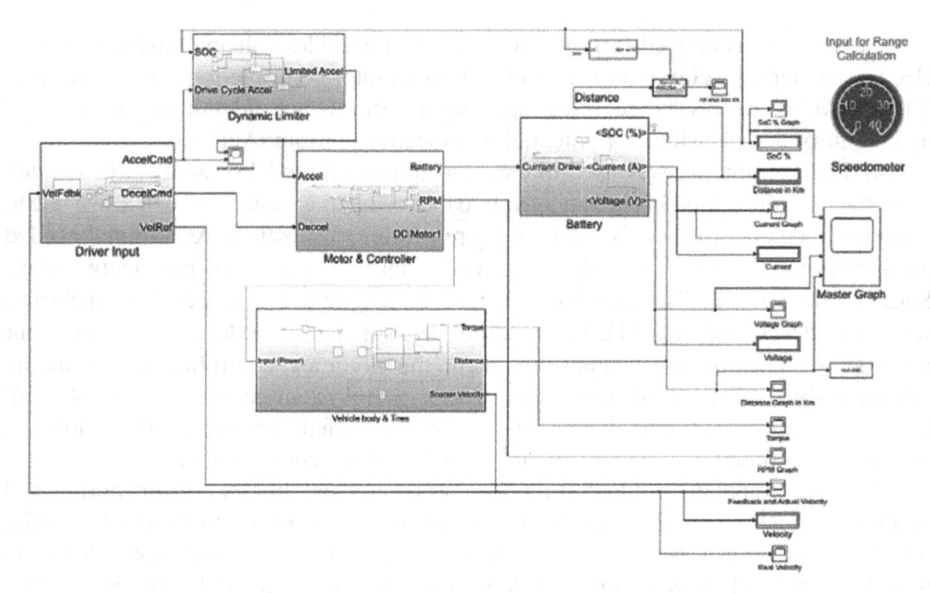

Fig. 2. PEM modelling with all subsystems

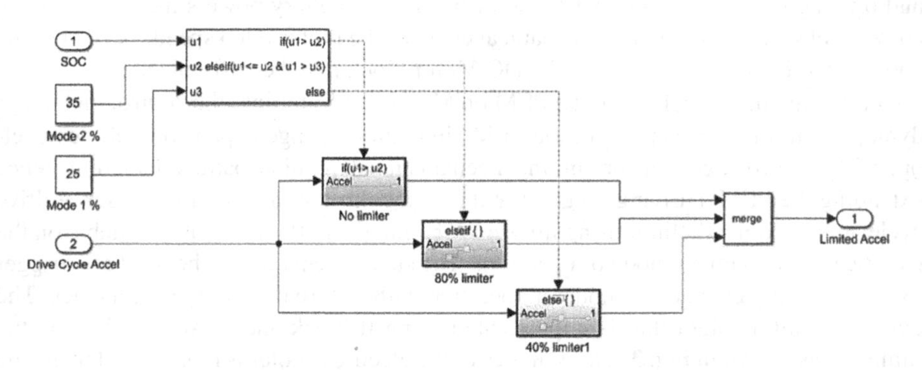

Fig. 3. Dynamic limiting algorithm subsystem

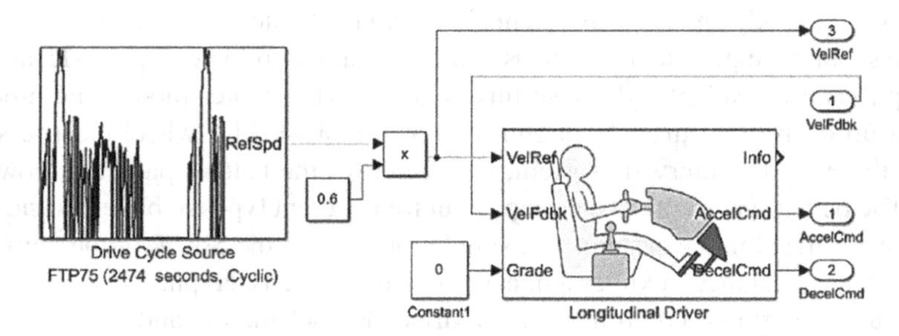

Fig. 4. Drive cycle subsystem

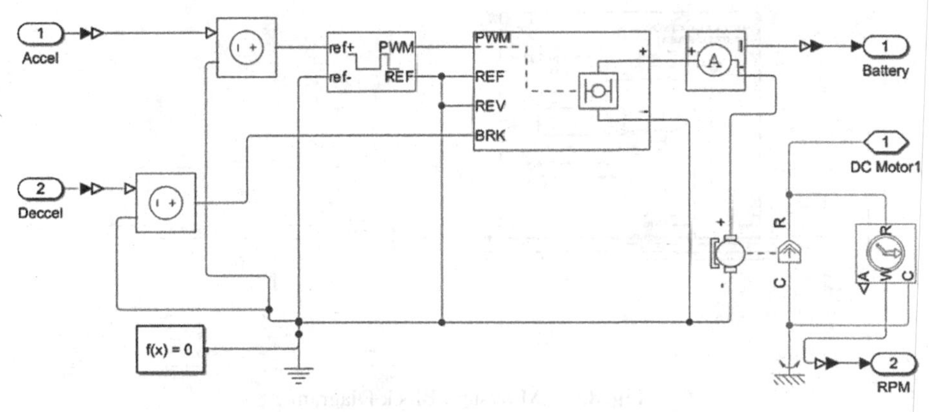

Fig. 5. Motor controller and DC motor subsystem

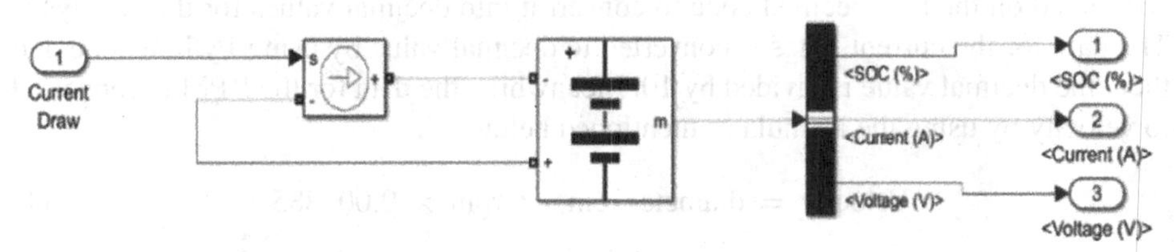

Fig. 6. Battery subsystem

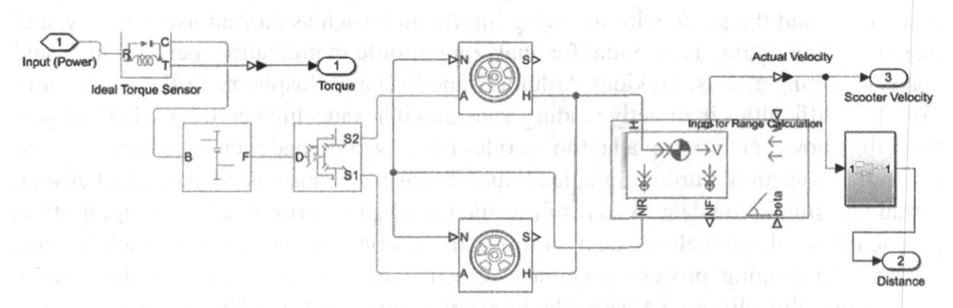

Fig. 7. Physical tire and body subsystem

4 Results and Discussion

4.1 Datalogging

For the validation of the PEM modelling on MATLAB-Simulink, the datalogging is conducted to gather data on the performance of the actual electric scooter. This method is called characterization of the system. The data logging is done using the receiving wire from the motor controller to the user interface. Upon investigation, it is found that the receiving wire contains data on current usage and RPM of the motor. The total byte for the receiving wire is 14 byte and the data is logged in hexadecimal code. Realterm software is used to capture and save the data in a text file. Data post processing is

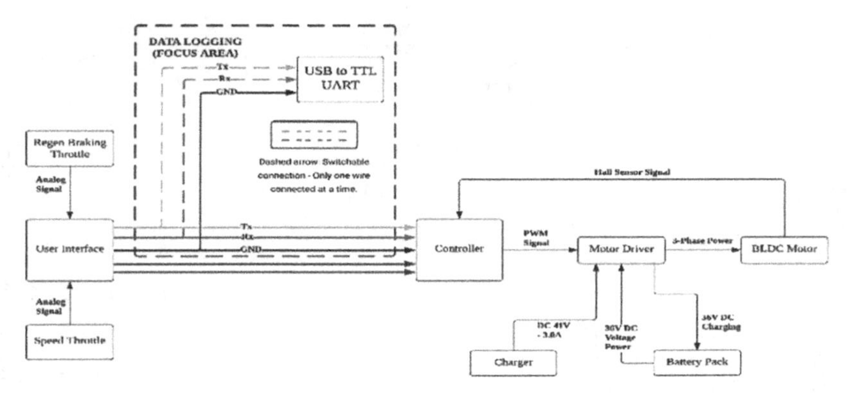

Fig. 8. PEM System Block Diagram

conducted on the hexadecimal code to convert it into decimal values for data analysis. The data for the current bytes is converted to decimal value by using Python code and then, the decimal value is divided by 10. Meanwhile, the data for the RPM is converted to velocity by using the formula as mentioned below:

$$\text{Velocity} = \text{diameter (cm)} \times \text{rpm} \times 0.001885 \quad (1)$$

The electric scooter's user interface board is linked to the BLDC controller through five wires, with the yellow wire delivering throttle input, brake input, and selected speed mode data, and the green wire receiving information such as current use, velocity, and distance. Datalogging is essential for analyzing throttle modification, performance, and current limiting effects. Previous Arduino Nano [22] and Raspberry Pi Pico configurations had difficulties in directly reading transmitted hexadecimal code. A FTDI adaptor fixed this, however carrying a laptop on rides added weight and increased the chance of cable disconnection. Furthermore, load fluctuations may cause inconsistencies between actual and simulation data. As a result, while the adapter permits data logging, it offers practical obstacles as well as the possibility of data inconsistencies owing to load changes.

The datalogging process generated velocity and current graphs for the electric scooter, revealing different drive cycles based on various modes and limiting percentages. Although validation was done on a low-traffic public road, adjustments are necessary to avoid incidents, making it impractical to conduct runs with the same distance and time. Nonetheless, the achieved steady state of the scooter allows characterization. Steady state is identified by constant velocity with minimal changes, enabling determination of current and velocity values at the same timestep for each mode. The throttle was set to maximum due to limited control at lower inputs (25%, 50%, and 75%). System characterization of the electric scooter and MATLAB modelling after changes made according to the validation results are shown in Table 1 and Table 2.

Figures 9 and 10 exhibit distinctive characteristics, primarily revealing the raw data depicting both instability and steady-state velocity of each mode. The objective of datalogging is to attain the steady-state velocity of the actual scooter across different modes, facilitating a comprehensive comparison with the steady-state velocity and current derived from the simulation model.

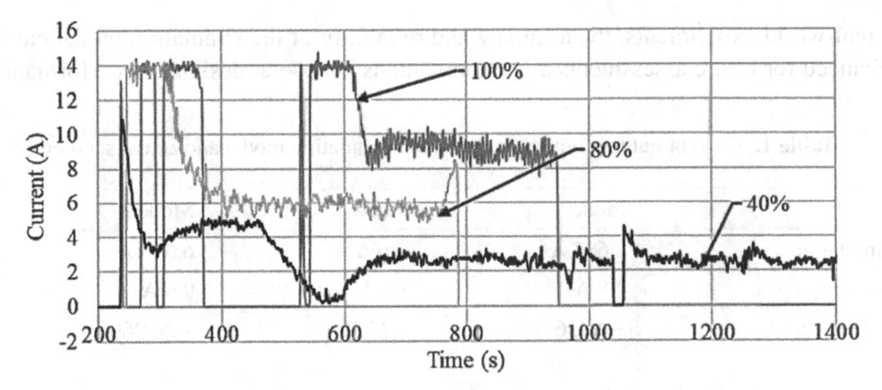

Fig. 9. Current vs. time profiles from datalogging for different modes

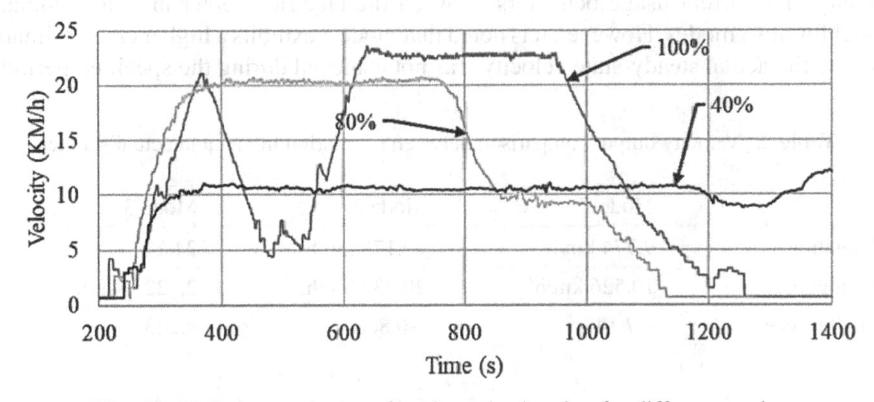

Fig. 10. Velocity vs. time profiles from datalogging for different modes

Notably, a significant event in the data is observed during mode 3 (100%) between the time intervals of 300s to 500s. This event corresponds to a sudden drop in velocity, which can be attributed to the need to decelerate the scooter for traffic avoidance. However, it is noteworthy that the steady-state velocity is subsequently achieved during periods extending beyond 700s.

It is important to note that conducting such experiments in a public open road environment introduces inherent variability. Consequently, each run may exhibit inconsistencies concerning the predefined run plan, particularly during the phases of acceleration and deceleration. Findings from this datalogging exercise provide valuable insights into the behavior of the scooter under different modes, shedding light on its performance characteristics and response to real-world conditions. It is imperative to consider the impact of external factors, such as traffic, during data interpretation, which inevitably contributes to the observed fluctuations in the raw data.

To ensure the credibility of the comparative analysis with the simulation model, it is essential to carefully account for these real-world variables and thoroughly validate the simulation model's accuracy against the observed data. Thus, by incorporating the lessons learned from the datalogging process and considering the uncertainties inherent in real-world experiments, the accuracy and reliability of the simulation model can be enhanced for future assessments and optimizations in scooter design and performance.

Table 1. Current output comparison between simulation model and actual scooter

	Mode 1	Mode 2	Mode 3
Simulation	2.687 A	5.166 A	6.695 A
Datalogging	2.8 A	5.6 A	9.6 A
% Difference	−4.036	−7.750	−30.260

Tables 1 and 2 illustrate consistently low errors, indicating strong similarity in the velocity and current usage behaviors between the electric scooter and the simulated model for most modes. However, it is noted that mode 3 exhibits a higher error, primarily because the actual steady-state velocity was not achieved during the specified period.

Table 2. Velocity output comparison between simulation model and actual scooter

	Mode 1	Mode 2	Mode 3
Simulation	9.774 km/h	20.175 km/h	24.276 km/h
Datalogging	10.526 km/h	20.339 km/h	22.222 km/h
% Difference	−7.147	−0.806	9.243

Several external factors contribute to the discrepancy in mode 3 data. Wind-induced drag is one such factor that can affect the scooter's performance, especially at higher velocities, leading to variations in the recorded data. Additionally, gradual battery depletion during operation can influence the scooter's overall performance, impacting its velocity and current usage over time. Moreover, uncertainties related to the actual load carried by the scooter during the experiment can also introduce variations in the measured data.

4.2 Hard Limiting Discussion

To comprehensively understand the current graph in Fig. 13 behavior, the relationship between velocity and throttle input demand is crucial. Throttle input demand, ranging from 0 to 1, follows a standard drive cycle velocity using a PID controller in the driver input subsystem, evident from the orange line data in all graphs. Steeper acceleration leads to a faster increase in throttle input, ensuring the electric scooter quickly reaches the target velocity. Higher throttle input results in increased acceleration, torque demand, and subsequently higher current usage.

Next, the effect of limiters on throttle input is discussed, with limiters set at 40%, 80%, and 100%, determined after electric scooter characterization to validate the MATLAB-Simulink model. The throttle input is restricted below 0.4 in 40% mode and below 0.8 in

80% mode, affecting current usage, torque, and velocity. The 40% limiter significantly deviates from the reference velocity, flattening at 14 km/h. When the reference velocity is not reached, the PID controller increases throttle input, limited by the limiter to 0.4 and below, impacting current usage accordingly. As acceleration decreases, reaching the maximum velocity as per throttle input, torque demand lowers resulting in constant current during nearly constant velocity.

The current and torque graphs show the same pattern, confirming the close relationship between torque and current according to formula below. Another intriguing behavior in limiter mode is delayed deceleration compared to the reference velocity, observed in Figs. 12 and 13 during timestamps 300 to 350, where throttle input and current usage remain constant despite reduced reference values at time step 300. The electric scooter maintains high throttle input until the referenced and limited velocities match, prompting deceleration.

$$\tau = \mathrm{kT} \times \mathrm{I} \tag{2}$$

where τ is the torque, kT is the torque constant (specific to the motors) and I is the current through the windings [23] (Figs. 11, 14 and 15).

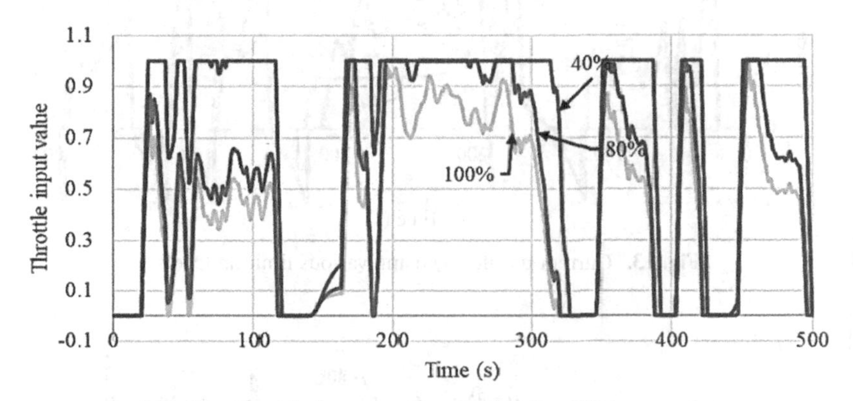

Fig. 11. Throttle input vs. time profiles for differenr modes

4.3 Dynamic Limiting Algorithms

The drive cycle data presented in Fig. 16 is obtained using a Proportional-Integral-Derivative (PID) controller integrated within the drive cycle subsystem. The PID controller is utilized to generate throttle input, ensuring that the electric scooter closely follows the predefined velocity profile of the standard drive cycle. The throttle input from the PID controller is constrained within the range of 0 to 1, representing the percentage of throttle demand, ensuring consistency and comparability across experimental runs for reliable modeling and analysis. The methodology guarantees robustness and reproducibility, facilitating accurate analysis and performance evaluation of the electric scooter.

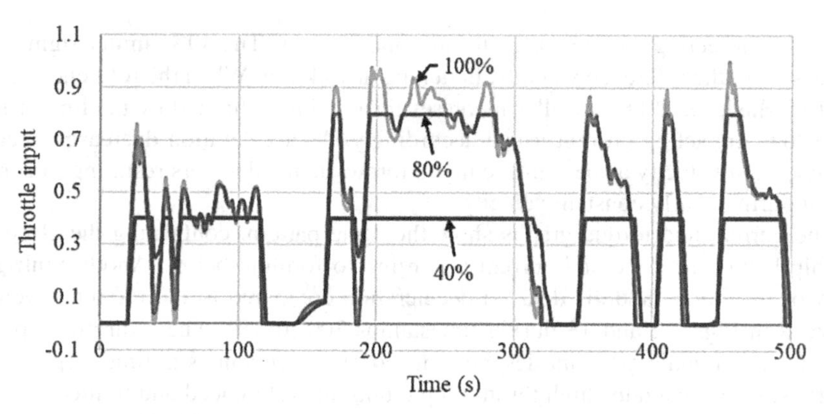

Fig. 12. Limited throttle input by hard limiting algorithm

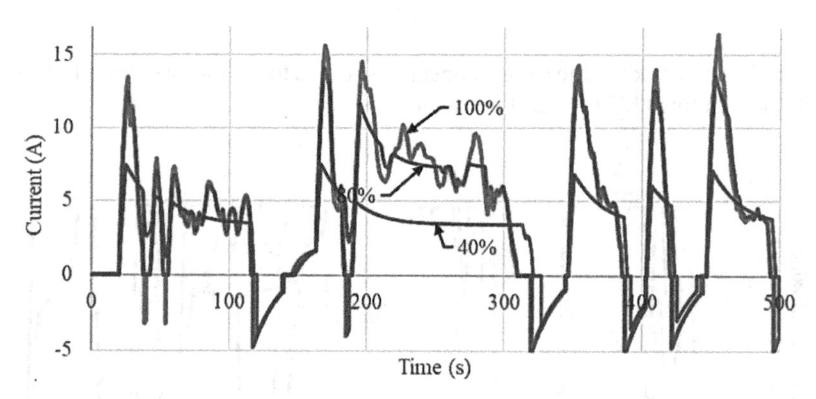

Fig. 13. Current profiles by using various limiting levels

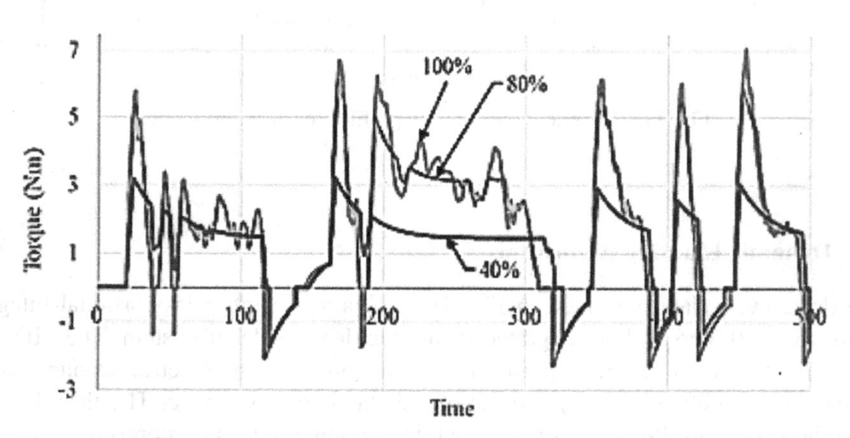

Fig. 14. Torque profiles by using various limiting levels

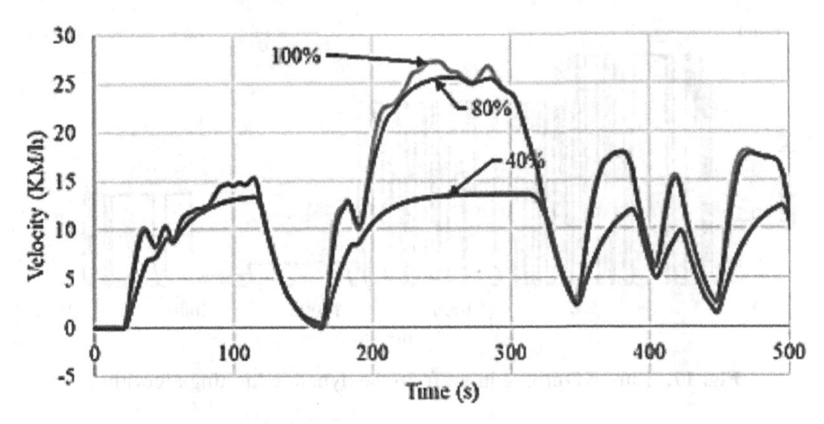

Fig. 15. Velocity profiles by using various limiting levels

Figure 16 shows the real-world throttle demand by the user during electric scooter operation. However, since the throttle input is not inherently limited, a dynamic limiting algorithm is applied to restrict the throttle input before transmitting it to the motor controller and DC motor subsystem, as shown in Fig. 17. The dynamically limited throttle input exhibits three distinct maximum limits, with the throttle input remaining unrestricted from 0 s to 600 s, limited to 80% from 600 s to 1400 s, and further limited to 40% after 1400 s. The duration of the transition period depends on the battery's SOC, as specified in the dynamic limiting algorithm.

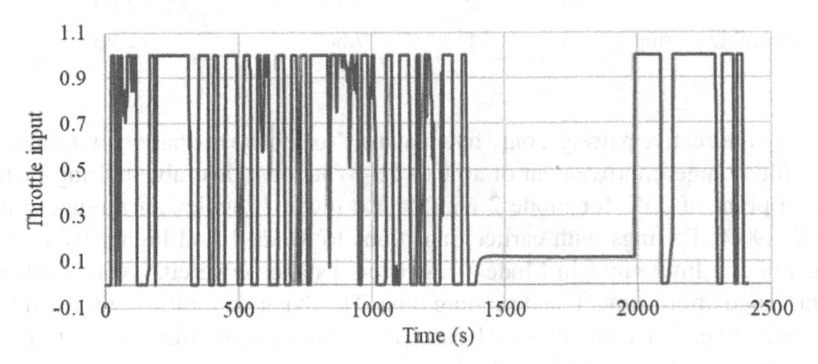

Fig. 16. Throttle input from the drive cycle

Through a meticulous iterative process, various combinations for the transition point between modes were tested to achieve the objective of attaining a 0% state of charge (SOC) within the designated drive cycle duration. Two parameters were adjusted: the initial state-of-charge set to 40% for a suitable starting point, and the drive cycle down-scale constant increased from 30% to 60% to amplify power consumption and stimulate higher velocities as the target. Based on the data in Table 3, the benchmark run in mode 3 achieved a total range of 7.39 km.

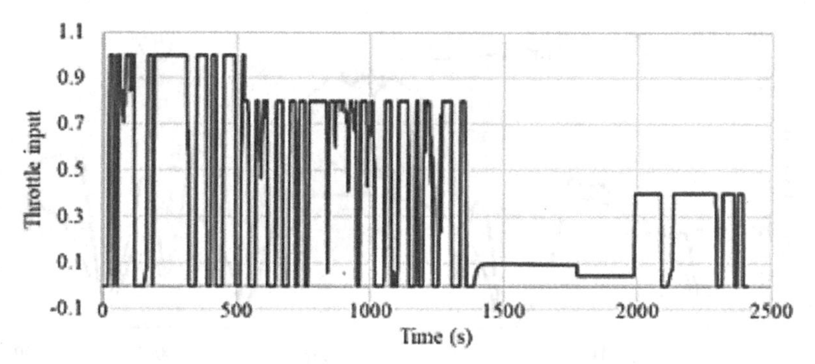

Fig. 17. Limited throttle input from the dynamic limiting algorithm

Table 3. Range improvement for different SoC% trigger limiter level

	SoC Trigger Level (%)		Total Distance (km)	Range Improvement (%)
	Mode 2	Mode 1		
Benchmark using Mode 3	-	-	7.39	-
Range Improvement (km)	30	10	7.599	2.83
	25	10	7.531	1.91
	35	5	7.547	2.12
	30	5	7.538	2.00
Best comparable pairing	25	5	7.604	2.90

Among alternative pairings, only five managed to deplete the battery within the drive cycle, with a range improvement of at least 2%. The most favorable pairing featured a transition point of 25% for mode 2 and 5% for mode 1, achieving a range improvement of 2.90%. Pairings with earlier transitions to Mode 1 had higher balance SOC, but the velocity limitations in Mode 1 restricted distance covered, creating a trade-off between range improvement, and running time. The dynamic limiting control algorithm has potential to enhance range, but AI-based algorithms could offer more optimized and adaptable solutions across various driving scenarios.

Hence, based on the obtained results, it can be concluded that the dynamic limiting control algorithm has the potential to enhance the range when the optimal pairing for the transition point between modes is applied. However, it is important to acknowledge the limitations of the trial-and-error method employed to determine the transition point in this study. This approach does not test every point within the state of charge (SOC) range, and the pairing that exhibits the highest range improvement may not necessarily be the best choice for different driving cycles.

To address these limitations and further improve the algorithm, future research can consider implementing a control algorithm, such as artificial intelligence (AI), to regulate the transition point. This AI-based control algorithm can leverage parameters such as changing load, throttle input profiles, and power consumption to determine the most effective transition point dynamically.

5 Conclusion

This research has successfully completed the datalogging process, validating the simulation modeling and providing a comprehensive understanding of the wire data, which supports the implementation of the proposed dynamic limiting control algorithm for electric scooters. This validation enhances the credibility of the algorithm's real-world application. The research primarily focused on understanding wire information, limiting exploration of other system aspects.

The performance analysis of the dynamic limiting control algorithm showcased its potential for range enhancement in electric scooters by dynamically adjusting throttle based on battery SOC. However, determining optimal transition points for each mode remains a challenge, necessitating iterative trial and error. Despite this limitation, the algorithm represents a significant improvement over the previous hard current limiting control, offering a more adaptable and responsive approach that optimizes power consumption and enhances the scooter's overall range.

Acknowledgment. The authors would like to acknowledge financial support provided through International Collaborative Research Fund (ICRF), with grant no. 015ME0–287. The authors would also like to express gratitude to the support of Center for Graduate Studies (CGS) of Universiti Teknologi *PETRONAS*, Malaysia for providing research facilities to carry out this work.

References

1. Quinn, R. K.: The coastal squeeze: rising seas and upland plant invasions differentially affect vertical exchange of greenhouse gases, vol. 2016 (2016)
2. Houghton, J.: Global warming. Rep. Progress Phys. **68**(6), 1343–1403 (2005). https://doi.org/10.1088/0034-4885/68/6/r02
3. Ouyang, X., Min, X.: Promoting green transportation under the belt and road initiative: locating charging stations considering electric vehicle users' travel behavior. Transp. Policy **116**, 58–80 (2022). https://doi.org/10.1016/j.tranpol.2021.11.023
4. Sumantri, B. et al.: Fuzzy-PID controller for an energy efficient personal vehicle: two-wheel electric skateboard. Int. J. Electrical Comput. Eng. (IJECE), **9**(6), 5312 (2019). https://doi.org/10.11591/ijece.v9i6.pp5312-5320
5. Zhang, X., Gohlich, D., Li, J.: Energy-efficient toque allocation design of traction and regenerative braking for Distributed Drive Electric Vehicles. IEEE Trans. Veh. Technol. **67**(1), 285–295 (2018). https://doi.org/10.1109/TVT.2017.2731525
6. A Pulamarin 2020 Design and implementation of hardware and speed and torque control with regenerative braking system for an electric vehicle for personal mobility. In: 2020 IEEE MIT Undergraduate Research Technology Conference (URTC). https://doi.org/10.1109/urtc51696.2020.9668888
7. Chang, W.-Y.: The state of charge estimating methods for battery: a review. ISRN Appl. Math. **2013**, 1–7 (2013). https://doi.org/10.1155/2013/953792

8. Thibault, L., De Nunzio, G., Sciarretta, A.: A unified approach for electric vehicles range maximization via eco-routing, eco-driving, and energy consumption prediction. IEEE Trans. Intell. Veh. **3**(4), 463–475 (2018). https://doi.org/10.1109/TIV.2018.2873922
9. Wu, D., et al.: Fast velocity trajectory planning and control algorithm of intelligent 4WD electric vehicle for energy saving using time-based MPC. IET Intell. Trans. Syst. **13**(1), 153–159 (2018). https://doi.org/10.1049/iet-its.2018.5103
10. Collins, D.: What's the relationship between current and DC motor output torque?, Motion Control Tips. https://www.motioncontroltips.com/faq-whats-the-relationship-between-current-and-dc-motor-output-torque/. Accessed 17 Mar 2023
11. Oldvetteguy, enzom32, 79galinakorczak, SteveThackery, HillmanImp: Best way to conver hall effect sensor singal into km/h. Arduino Forum (2021). https://forum.arduino.cc/t/best-way-to-conver-hall-effect-sensor-singal-into-km-h/860977/15?page=4
12. Sadiq, M.M., Abro, G.E.M., Yazsid, A.Z.S.M., Zulkifli, S.A.M.: Development of range enhancement device for personal electric mobility. In: 2022 7th International Conference on Electric Vehicular Technology (ICEVT), Bali, Indonesia, 2022, pp. 144–152. https://doi.org/10.1109/ICEVT55516.2022.9924865
13. Renesas Solutions for BLDC Motor Control (no date) Renesas. https://www.renesas.com/us/en/support/engineer-school/brushless-dc-motor-03-rssk. Accessed 17 Jul 2023
14. Chaturvedi, D.K.: Modelling and Simulation of Systems Using MATLAB® and Simulink®. CRC Press, Boca Raton (2017)
15. Data logging. FTDI. https://ftdichip.com/product-category/applications/data-logging/. Accessed 20 Jul 2023
16. Realterm: Serial terminal. Terminal Software. https://realterm.sourceforge.io/. Accessed 20 Jul 2023
17. Python program to convert decimal to hexadecimal. GeeksforGeeks https://www.geeksforgeeks.org/python-program-to-convert-decimal-to-hexadecimal/. Accessed 20 Jul 2023

Atrial Fibrillation Detection Based on Electrocardiogram Features Using Modified Windowing Algorithm

Kong Pang Seng, Farah Aina Jamal Mohamad, Nasarudin Ahmad, Fazilah Hassan, Mohamad Shukri Abdul Manaf, Herman Wahid, and Anita Ahmad(✉)

Faculty of Electrical Engineering, Universiti Teknologi Malaysia, 81310 UTM Skudai, Johor, Malaysia

anita@utm.my

Abstract. The most prevalent kind of heart disease, Atrial Fibrillation (AF), is said to increase the risk of stroke, heart failure, and other health concerns. Clinical observation of the electrocardiogram (ECG) waveform needs an experienced person to observe and takes long hours. We provide an approach to identify AF in the MIT-BIH Arrhythmia and AF databases in this study. Several parameters, including QRS complex, RR Interval, heart rate, coefficient of variance (CV), normalized root mean square of successive difference (nRMSSD) and peak frequency are calculated from the features of the ECG signal. With holdout validation, we analyse the AF classification performance of various classifiers. With the input parameters mentioned, the greatest outcome in AF classification is obtained by the weighted KNN classifier using the DWT algorithm and modified windowing algorithm in holdout validation, with sensitivity, specificity, and accuracy of 90%, 100%, and 92.31%, accordingly. On this basis, it is recommended that classifying ECG signals using machine learning methods will assist in improving the research's accuracy. Further research is needed to test the proposed algorithm on a large database for better accuracy. This algorithm will be involved in hardware and implemented in the detection of real-time AF ECG patients.

Keywords: Atrial Fibrillation · RR Intervals · Machine Learning

1 Introduction

A variety of cardiac conditions affect people, including atrial fibrillation (AF). A malfunction of one of the heart's atria causes AF. 33.5 million people worldwide had AF in 2013. [1]. In the general population, it affects 1% to 2% of people. Furthermore, by 2050, its estimated mileage would triple, posing a serious danger to fitness [2]. Its prevalence rises with age, from 0.5% in those between the ages of 40 and 50 to 5% to 15% in those between the ages of 80 and older [3]. AF may be caused by atrial regions responsible for maintaining the arrhythmia [4]. AF continues to be the main contributor to heart disease, stroke, cardiac arrest, and early mortality worldwide. Therefore, AF patients have a much higher chance of having a stroke than others in the population [5]. Framingham Heart Research estimates that the lifetime risk of AF is roughly 25% [6]. So, early detection and treatment of AF are crucial.

© The Author(s), under exclusive license to Springer Nature Singapore Pte Ltd. 2024
F. Hassan et al. (Eds.): AsiaSim 2023, CCIS 1911, pp. 483–495, 2024.
https://doi.org/10.1007/978-981-99-7240-1_38

A wave-based signal called an electrocardiogram (ECG) signal records and presents the electrical rhythm of the beating heart. It is a useful clinical technique for identifying unusual cardiac activity. AF happens when the electrical activity in the atria spreads erratically [7]. However, clinical ECG waveform observation requires a specialist to examine large volumes of ECG data and takes a lot of time. To speed up the process of detecting AF, an automated AF detection method is needed to automatically analyse the ECG signals [8]. The risk of consequences from later discovery can be greatly reduced if AF can be quickly and automatically recognised in the early stages of its pathogenesis.

AF happens when the electrical activity in the atria spreads erratically. The two features of the AF signal are irregular RR intervals and the disappearance of P waves forming into f waves. The difficulty in establishing the PR interval is as follows, as f waves are of very low amplitude and the absence of a P wave is easily affected by signal waves [7]. Furthermore, the accuracy of AF classification using the threshold values of ECG signals' features is low [9].

P waves, QRS complexes, and T waves compose the ECG pattern. Each of these waves correlates to a distinct heart motion (repolarization or depolarization). To calculate the placement of PQRST waves, one must determine the R peak sites of every QRS complex. The QRS complexes were then separated from the ECG signal using the reference R points. Due to the different ECG signal structures, the locations of PQST waves from each segmented QRS complex are also developed utilising local minima and local maxima detection methods [10].

To implement machine learning in signal classification, a summary of machine learning methods is discussed. The decision tree (DT) classifier is organised like a tree, with a root node that is further separated into child nodes and leaf nodes depending on if-else conditions [11]. The linear discriminant (LD) is one of the classifiers used in machine learning. A diagnostic project is taught to enhance the between-class distance and decrease the within-class distance to improve classification accuracy [12]. The KNN classifier is a supervised learning algorithm with an excellent balance of computing speed and classification accuracy. The majority decision and separation from the "K" closest samples are used to classify a new test sample [12]. Moreover, SVM is a typical classifier that creates the largest possible margin between training and testing data. The support vectors are the samples that are nearest to the decision border [12]. SVM handles problems such as tiny sample sizes, large dimensions, regional minima, and others, particularly for the individual testing set used for training. Its benefit is that it finds a solution to the nonlinear classification and short sample size issues [13].

Shrikanth et al. [11] suggest a method for identifying AF that combines QRS detection, PCA, energy computation, and a DT-based classifier with ten-fold cross validation for classification with an overall average accuracy of 85.1%. On the MIT-BIH Arrhythmia database, Szymon et al. [14] demonstrated AF classification using a fine decision tree with 10-fold cross validation and attained an accuracy result of 95.8% with time domain features.

Adiwijaya et al. [15] suggested study to distinguish between AF ECG signals and regular ECG signals utilising RR interval approaches and a KNN classifier has a 91.75% accuracy rate. In the MIT-BIH Arrhythmia database, P.Kora et al. [16] proposed a KNN classifier-based DWT approach that achieved 99.5% accuracy, 96.97% sensitivity, and

96.97% specificity and included features including R peak, inverted T wave logic, and ST segment elevation. In comparison to DT, LD, and KNN approaches, Areej Almazroa et al. [9] presented AF classification using input features (QRS duration and heart rate) with 80% of training data and 20% of testing data from the MIT-BIH Arrhythmia database obtained 96.7% accuracy.

This study aspires to enhance PQRST point recognition in ECG signals and apply machine learning techniques to enhance the precision of AF signal detection. This study is done by using the software MATLAB 2021a and the classification toolbox, datasets from the MIT-BIH Arrhythmia and AF databases, respectively.

2 Materials and Method

2.1 Proposed Study

Figure 1 shows the summary flow process of the study. First, the ECG signals are downloaded online from Physionet and saved in the local database. Next, the ECG signal preprocessing is applied to the ECG signal because the signals are complicated. After the preprocessing stage, the features are extracted from the signal. Lastly, the signals are classified as AF signals or normal signals by using different types of machine learning algorithms and comparing their performance.

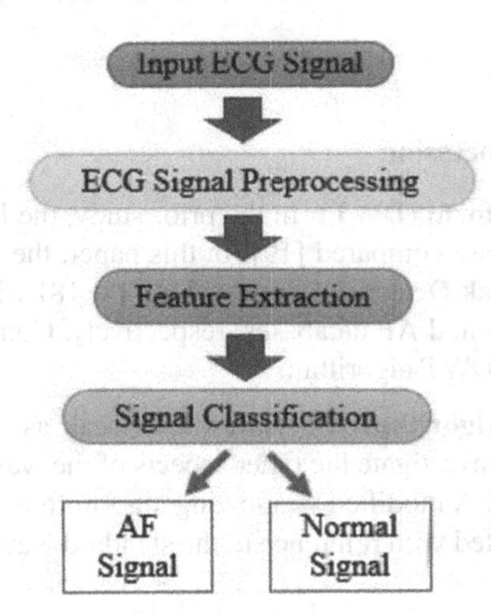

Fig. 1. Flow Process of the Study

2.2 Database

The two types of datasets used are MIT-BIH Arrhythmia Database and the MIT-BIH Atrial Fibrillation Database. The signals are downloaded from the physionet in ".mat" format.

MIT-BIH Arrhythmia Database There are 48 recordings of heartbeats at a sampling frequency of 360 Hz in the MIT-BIH Arrhythmia Database. Each of the 48 records for the various patients is 10 s long and has two leads, designated Lead A and Lead B [17]. The Lead A datasets in this study are used for detection, but the "102m" and "104m" datasets are missing Lead A data and are therefore not used for detection.

MIT-BIH Atrial Fibrillation Database There are 25 long-term heartbeat recordings of people with atrial fibrillation in this MIT-BIH AF database. Each record lasts for 10 s and has a resolution of 12 bits with a 10 mV range. The sampling frequency is 250 Hz. The "ECG1" and "ECG2" ECG signals are included in each record. The only signals used in this study are "ECG1" signals [18].

Table 1 shows 19 records of normal signal and 48 records of AF signal with a 10 s long each and resampled at 360Hz from both databases mentioned. These 67 records of signal are investigated in this study.

Table 1. Records of Normal Signal and AF Signal

Signal	Database	Records
Normal	MIT-BIH Arrhythmia	19
AF	MIT-BIH Arrhythmia and MIT-BIH AF	48

2.3 ECG Signal Preprocessing

Discrete Wavelet Transform (DWT) In the prior study, the Pan-Tompkins algorithm and the DWT algorithm are compared [19]. In this paper, the DWT algorithm achieve better sensitivity of R Peak Detection compared to [17, 18] with 99.86% and 100% in the MIT-BIH Arrhythmia and AF databases, respectively. Consequently, the following investigation utilises the DWT algorithm.

Modified Windowing Algorithm By using the R peak as a reference, the window analysis is carried out to investigate the other aspects of the wave, such as the P peak, Q peak, S peak, and T peak. A modified windowing algorithm is used to detect the peaks, and this algorithm is created with reference to the standard waveform of the ECG signal as shown in Fig. 2.

The modified windowing algorithm is stated accordingly with some adjustments from the previous study [20]. Initially, ECG signal is preprocessed using DWT algorithm. Then, ECG signal is reconstructed to find the R peak and R location (Rloc). The Rloc for the first one and the last one is eliminated because the first and last QRS complexes do not show all the characteristics of the peaks. The Rloc depicts the reference for detecting PQST points. For finding the P peak of the ECG signal by finding the highest number in the window range from the left side of Rloc: Rloc – 60 to Rloc – 10. For identifying the Q peak, a window is created on the left edge of the Rloc in the region of Rloc – 50 to Rloc – 5 by discovering the lowest value of the point. Furthermore, finding the smallest

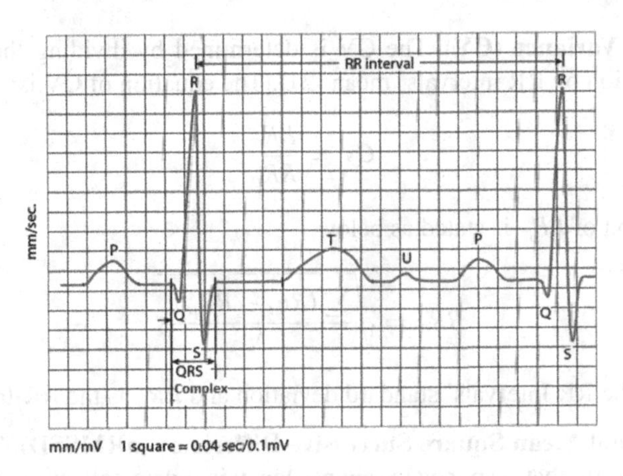

Fig. 2. Standard ECG Waveform

value from the right side of Rloc in the range of Rloc + 0 to Rloc + 50 yields the S peak value. The highest value inside the window range of Rloc + 10 to Rloc + 50 is used to classify T peak. The positions of the P, Q, R, S, and T peaks are then plotted and indicated.

2.4 Feature Extraction

Six characteristics of the ECG signal sequence are computed in this study.

QRS Complex The distance in seconds between the Q and S locations is known as the QRS complex and is calculated as given by Eq. (1), where f_s indicates sampling frequency, S_{loc} indicates S peak locations and Q_{loc} indicates Q peak locations. This equation produces an array of values for the QRS complex, which are subsequently averaged to produce a single QRS complex.

$$QRS\ Complex(i) = \frac{S_{loc}(i) - Q_{loc}(i)}{f_s} \tag{1}$$

RR Interval and Heart Rate Since R peaks are found, the period between one R-spike and the next R-spike (successive R's found) is calculated to get the RR interval in seconds. The calculation of the RR interval is in Eq. (2), which R_{loc} denotes R peak locations. A single RR interval is obtained by averaging the values of the array of RR intervals produced by this equation. The heart rate can then be determined with 60 divided by the value of a single RR interval. The heart rate formula is in Eq. (3).

$$RR\ Interval(i) = \frac{R_{loc}(i+1) - R_{loc}(i)}{f_s} \tag{2}$$

$$Heart\ Rate(bpm) = \frac{60}{RRInterval} \tag{3}$$

Coefficient of Variance (CV) The CV is determined by dividing the RR intervals' standard deviation by RR intervals' mean [21]. The equation of CV is:

$$CV = \frac{RR_\sigma}{RR_\mu} \tag{4}$$

The equation of RR_σ is stated as below:

$$RR_\sigma = \sqrt{\frac{\sum (RR_i - RR_\mu)^2}{N}} \tag{5}$$

where RR_σ is the RR Intervals' standard deviation and RR_μ is the RR Intervals' mean.

Normalised Root Mean Square Successive Difference (nRMSSD) The RMSSD is a descriptive metric that responds to inconsistency in a data collection. RMSSD stands for the square root of the normal of the corresponding squares of differences between RR Intervals [21]. RMSSD is calculated as below:

$$RMSSD = \sqrt{\frac{\sum_{i=1}^{N-1} (RR_{i+1} - RR_i)^2}{N-1}} \tag{6}$$

where $RR_i = i^{th}$ RR interval in the component with length N, where $i = 1, 2, \ldots, N$, and N = length of RR Interval. Since the progressions in heartrate take additional time and unexpected ventricular compressions, RMSSD is partitioned by mean RR Interval which is called nRMSSD. The formula of nRMSSD is:

$$nRMSSD = \frac{RMSSD}{RR_\mu} \tag{7}$$

Peak Frequency The peak frequency of ECG signal is found by drawing a graph of power spectral density. To draw the graph of power spectral density, the welch method is used. The initial signal section of length N is divided by the welch method into K sub-segments, each of which has a length of L = N/K. The overlapping K sub-segments, which are the K sub-segments of length L, are used to generate the modified periodogram. To reduce the signal energy loss caused by the windowing method, the periodogram is normalised by the factor U [22]. Then, the equation is

$$\widehat{P}_w\left(e^{jw}\right) = \frac{1}{KLU} \sum_{i=0}^{K-1} \left| \sum_{n=0}^{L-1} w(n)x(n+iD)e^{-jnw} \right|^2 \tag{8}$$

where D is the distance between two successive sub segments and U is given as

$$U = \frac{1}{L} \sum_{n=0}^{L-1} |w(n)|^2 \tag{9}$$

An "event" is classified as a signal segment (N) with a duration of 10 s. It is split every two seconds of overlapping sub-segments. The percentage of overlap (D) stayed at 50%. The modified periodogram of each subsegment is captured using Hanning window w(n) [22]. The MATLAB syntax used in this part is Hs = spectrum.welch (WindowName,

SegmentLength, OverlapPercent). The WindowName used in the syntax is the Hanning window, the SegmentLength is the length of each of the time-based segments into which the input signal is divided and the default OverlapPercent is 50%. A red circle will be plotted in the figure to point out the peak value of the power spectrum, which represents the peak frequency of the signal. The red circle plotted in the power spectrum in Fig. 3 states that the peak frequency of the (04015) ECG signal ID is 12.70 Hz.

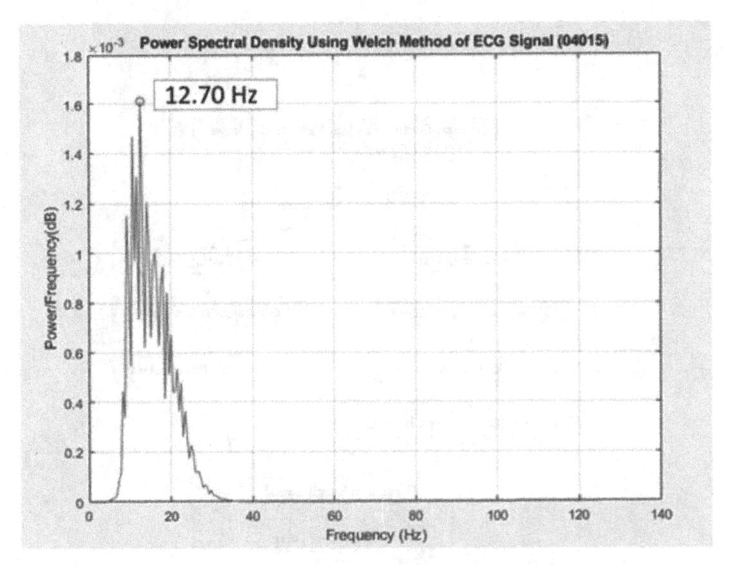

Fig. 3. Power Spectral Density using Welch Method of ECG Signal ID (04015)

2.5 Signal Classification

Holdout Validation The signal classification between normal signal and AF signal uses features such as QRS complex, RR interval, heart rate, CV, nRMSSD, and peak frequency from the PSD method. These features are input parameters for machine learning classifiers. Data is divided into two sets for the holdout validation, with the first set being used for training with 80% of the data and the second set being used for testing with 20% of the data.

Figure 4 displays the flowchart of holdout validation. First, the database used is from Table 1 with six types of stated features above. Then, the control random number generator is inserted with the syntax rng('default'). This function is used to set the settings to their default state and is used for repeatability. The data set goes through the process of holdout validation using the MATLAB syntax cvpartition(). Both training data and testing data undergo the process of normalization which converts the table data to an array. Additionally, machine learning classifiers like Decision Tree (DT), K-Nearest Neighbour (KNN), Linear Discriminant (LD), and Support Vector Machine (SVM) are trained to create the model from the processed training data.

The signal outcome is predicted using the trained and processed test data along with the function predict. The confusion matrix, which is generated using the test data that has been processed and the predicted data, assesses the performance of each classifier for each model. Finally, each model's computation of sensitivity, specificity, and accuracy is listed and compared to see which classifier performs the best.

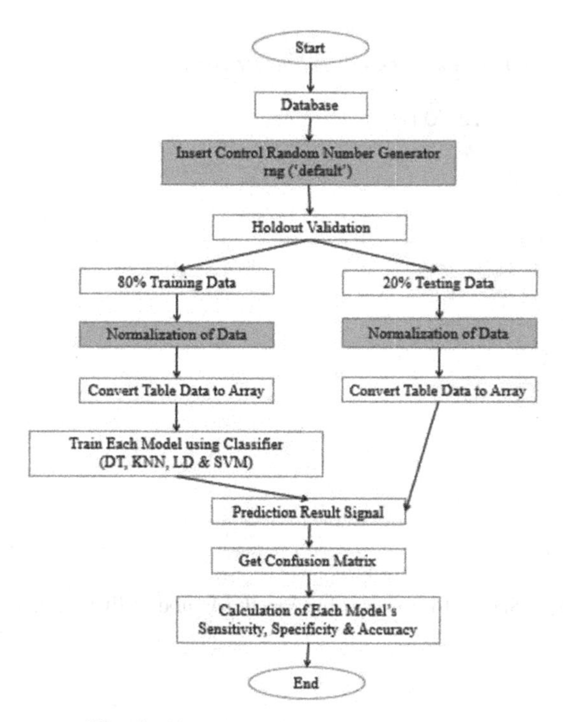

Fig. 4. Flowchart of Holdout Validation

3 Results and Discussion

3.1 Result Modified Windowing Algorithm

In Fig. 5, the result for the normal ECG signal ID (100) from the MIT-BIH Arrhythmia database is displayed. The R peak is denoted by an inverted triangle '∇' in red, the P peak by a star '*' in red, the Q peak by a circle 'o' in green, the S peak by as a square in blue and the T peak by a triangle 'Δ' in magenta, in that sequence. The result for AF ECG signal ID (04015) from the MIT-BIH AF database is shown in Fig. 6. The irregularity of R peaks can be seen Fig. 6.

The proposed method's output is contrasted with that of other strategies already in use. It should be emphasized that the methods are based on linear prediction (LP), independent component analysis (ICA), wavelet transform (WT), eigenvector, and fast

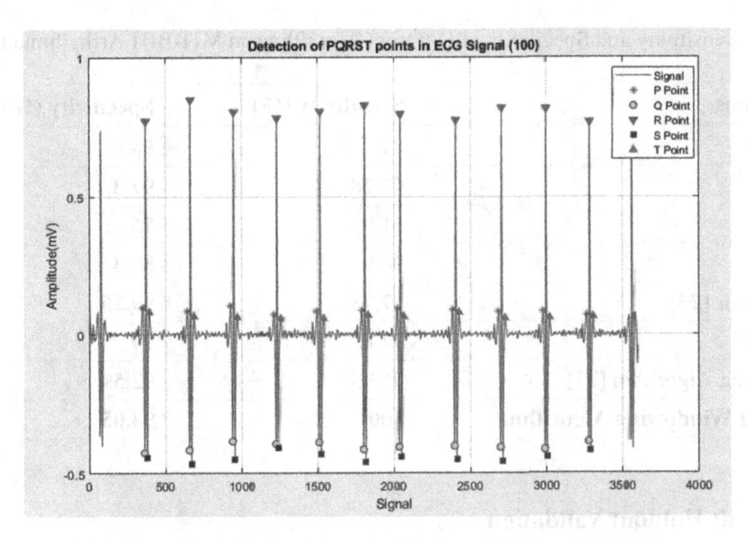

Fig. 5. Detection of PQRST Points in ECG Signal ID (100)

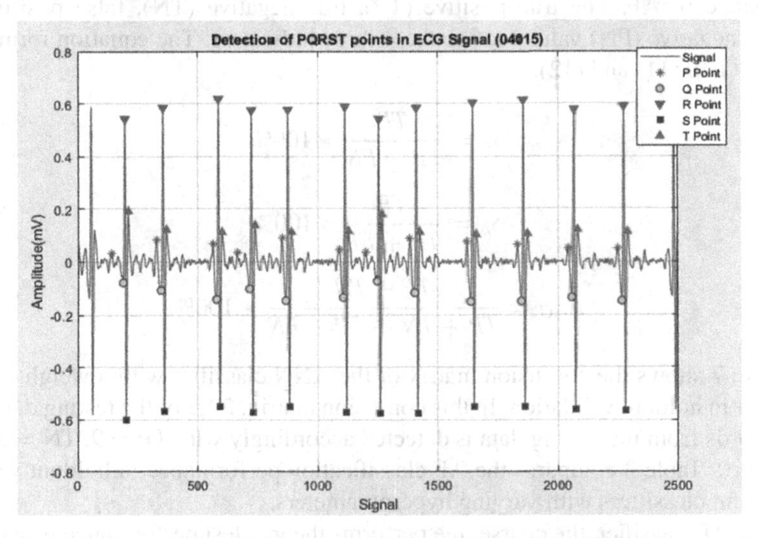

Fig. 6. Detection of PQRST Points in ECG Signal ID (04015) (Color figure online)

fourier transform (FFT) [23]. Table 2 indicates the sensitivity and specificity values of different types of algorithms in the MIT-BIH Arrhythmia database. The sensitivity of the modified windowing algorithm is the highest compared to the other algorithms which is 100% while its specificity value is 94.05%.

Table 2. Sensitivity and Specificity of Different Algorithms in MIT-BIH Arrhythmia Database

Algorithms	Sensitivity (%)	Specificity (%)
FFT [23]	81	98
AR [23]	97.28	97.3
WT [23]	61	75
LP [23]	96.9	80.4
Eigenvector [23]	97.78	99.25
ICA [23]	97.8	99
Windowing Algorithm [23]	96.95	92.59
Modified Windowing Algorithm	**100**	**94.05**

3.2 Result Holdout Validation

Sensitivity (Se), specificity (Sp), and accuracy (Acc) are used to assess the classification performance of AF. The true positive (TP), true negative (TN), false positive (FP), and false negative (FN) values define these three indicators. The equation formulas are formulas (10), (11) and (12).

$$S_e = \frac{TP}{TP + FN} * 100\% \tag{10}$$

$$S_p = \frac{TP}{TP + FP} * 100\% \tag{11}$$

$$A_{CC} = \frac{TP + TN}{TP + TN + FP + FN} * 100\% \tag{12}$$

Figure 7 shows the confusion matrix of the KNN classifier with a weighted hyperparameter in holdout validation. In this confusion matrix, 20% of the testing data, which is 13 records from the testing data is detected accordingly with TP = 9, TN = 3, N = 1 and FP = 0. Table 3 compares the AF classification performance in holdout validation for different classifiers with varying hyperparameters.

In the DT classifier, the coarse tree performs the greatest performance in sensitivity, specificity and accuracy of 90%, 66.67% and 84.62%. Weighted KNN achieves the best performance in the KNN classifier with Se = 90%, Sp = 100% and Acc = 92.31%. The linear discriminant classifier establishes the performance of Se = 80%, Sp = 100% and Acc = 84.62%. Linear SVM is the hyperparameter in the SVM classifier that achieves the highest performance with Se = 80%, Sp = 100% and Acc = 84.62%. Lastly, the weighted KNN classifier achieves the best performance in holdout validation compared to other classifiers.

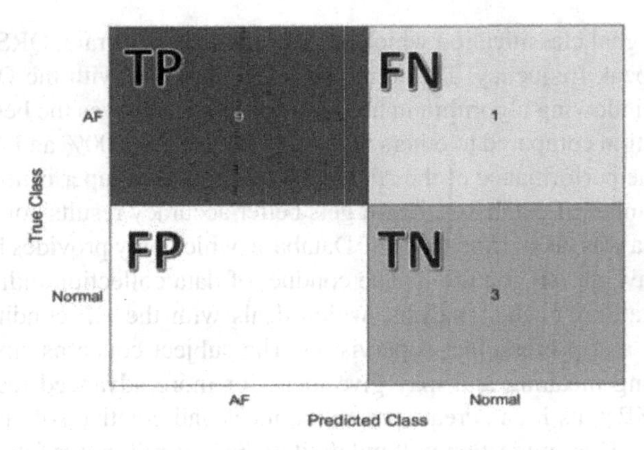

Fig. 7. The Confusion Matrix in Holdout Validation

Table 3. AF Classification Performance in Holdout Validation

Classifier	Hyperparameter	Se (%)	Sp (%)	Acc (%)
Decision Tree (DT)	Fine	80	66.67	76.92
	Medium	80	66.67	76.92
	Coarse	90	66.67	84.62
K-Nearest Neighbour (KNN)	Fine	80	100	84.62
	Medium	90	66.67	84.62
	Coarse	100	0	76.92
	Cosine	90	66.67	84.62
	Cubic	90	66.67	84.62
	Weighted	**90**	**100**	**92.31**
Linear Discriminant (LD)		80	100	84.62
Support Vector Machine (SVM)	Linear	80	100	84.62
	Quadratic	90	66.67	84.62
	Cubic	70	66.67	69.23
	Fine Gaussian	100	0	76.92
	Medium Gaussian	90	66.67	84.62
	Coarse Gaussian	100	0	76.92

4 Conclusion

In this study, we used the MIT-BIH Arrhythmia and AF databases to provide a unique approach for detecting AF signals. In comparison to the windowing algorithm and other PQRST detection techniques, the modified windowing algorithm exhibits considerable gains in terms of sensitivity and specificity. We have extracted several features to be the parameters in signal classification which are RR interval, heart rate, QRS complex, CV, nRMSSD and peak frequency. The weighted KNN classifier with the DWT algorithm and modified windowing algorithm in holdout validation achieves the best performance in AF classification compared to others with Se = 90%, Sp = 100% and Acc = 92.31%, respectively. The performance of the AF classification in this paper is high. Although it uses a small number of databases, it still gets better accuracy results compared to other studies. Our data was taken from the MIT Database, which only provides limited data for patients who have an AF condition. The conduct of data collection individually offers numerous limitations to this analysis, which deals with the AF condition, including ethical concern and professional supervision. The subject concerns simulation in the digital processing meaning and may give ideas for more advanced techniques using both AI and DSP, thus it can create pragmatic ideas and creative solutions in medical application area. This suggestion will indefinitely be covered in our future work. In the future, the proposed algorithms will be investigated in a large number of databases for better accuracy. This algorithm will be performed in hardware and implemented for detecting real-time AF ECG patients.

Acknowledgements. This research is fully supported by UTM Fundamental Research Grant, Q.J130000.2551.21H43. The authors fully acknowledged Ministry of Higher Education (MOHE) and Universiti Teknologi Malaysia for the approved fund which makes this important research viable and effective.

References

1. M.G. Foundation: World Health Organization Study: Atrial Fibrillation is a Growing Global Health Concern, pp. 2–4 (2021)
2. Young, M.: Atrial Fibrillation. Crit. Care Nurs. Clin. North Am. **31**(1), 77–90 (2019)
3. Naccarelli, G.V., Varker, H., Lin, J., Schulman, K.L.: Increasing prevalence of atrial fibrillation and flutter in the United States. Am. J. Cardiol. **104**(11), 1534–1539 (2009)
4. Narayan, S.M., Krummen, D.E., Clopton, P., Shivkumar, K., Miller, J.M.: Direct or coincidental elimination of stable rotors or focal sources may explain successful atrial fibrillation ablation: on-treatment analysis of the CONFIRM trial (conventional ablation for af with or without focal impulse and rotor modulation). J. Am. Coll. Cardiol. **62**(2), 138–147 (2013)
5. Wang, T.J., et al.: Temporal relations of atrial fibrillation and congestive heart failure and their joint influence on mortality: the Framingham heart study. Circulation **107**(23), 2920–2925 (2003)
6. Lloyd-Jones, D.M., et al.: Lifetime risk for development of atrial fibrillation: the Framingham heart study. Circulation **110**(9), 1042–1046 (2004)
7. Zhang, Y., Zhang, L.: Research on algorithm for detecting atrial fibrillation using RR intervals. In: 2018 International Conference on Audio, Language and Image Processing (ICALIP), Shanghai, China, pp. 205–209 (2018)
8. Peimankar, A., Puthusserypady, S.: Ensemble learning for detection of short episodes of Atrial fibrillation. In: 26th European Signal Processing Conference (EUSIPCO). Rome, Italy, pp. 66–70 (2018)
9. Almazroa, A., Sun, H.: An Internet of Things (IoT) management system for improving homecare - a, case study. In: International Symposium on Networks, Computers and Communications (ISNCC). Istanbul, Turkey, pp. 1–5 (2019)

10. Dissanayake, T., Roshan Ragel, Y.R., Nawinne, I.: An ensemble learning approach for electrocardiogram sensor based human emotion recognition. Sensors, **19**(20), 4495 (2019)
11. Shrikanth Rao, S.K., Martis, R.J.: Machine learning based decision support system for atrial fibrillation detection using electrocardiogram. In: 2020 IEEE International Conference on Distributed Computing, VLSI, Electrical Circuits and Robotics (DISCOVER), Udupi, India, pp. 263–266 (2020)
12. Bashar, S.K., et al.: Atrial fibrillation detection in ICU patients: a pilot study on MIMIC III data. In: 2019 41st Annual International Conference of the IEEE Engineering in Medicine and Biology Society (EMBC), Berlin, Germany, pp. 298–301 (2019)
13. Deng, M., Qiu, L., Wang, H., Shi, W., Wang, L.: Atrial fibrillation classification using convolutional neural networks and time domain features of ECG sequence. In: 2020 IEEE 19th International Conference on Trust, Security and Privacy in Computing and Communications (TrustCom), Guangzhou, China, pp. 1481–1485 (2020)
14. Sieciński, S., Kostka, P.S., Tkacz, E.J.: Comparison of atrial fibrillation detection performance using decision trees, SVM and artificial neural network. In: Rocha, Á., Ferrás, C., Paredes, M. (eds.) Information Technology and Systems. ICITS 2019. Advances in Intelligent Systems and Computing, vol. 918. Springer, Cham (2019). https://doi.org/10.1007/978-3-030-11890-7_65
15. Resiandi, K., Resiandi, A., Utama, D.Q.: Detection of atrial fibrillation disease based on electrocardiogram signal classification using RR interval and K-nearest neighbor. In: 2018 6th International Conference on Information and Communication Technology (ICoICT), Bandung, Indonesia, pp. 501–506 (2018)
16. Kora, P., Kumari, C.U., Swaraja, K., Meenakshi, K.: Atrial fibrillation detection using discrete wavelet transform. In: 2019 IEEE International Conference on Electrical, Computer and Communication Technologies (ICECCT), Coimbatore, India, pp. 1–3 (2019)
17. Fariha, M.A.Z., Ikeura, R., Hayakawa, S., Tsutsumi, S.: Analysis of Pan-Tompkins algorithm performance with noisy ECG signals. J. Phys. Conf. Ser. Bristol **1532**(1), 012022 (2020)
18. Lai, D., Zhang, X., Bu, Y., Su, Y., Ma, C.S.: An automatic system for real-time identifying atrial fibrillation by using a lightweight convolutional neural network. IEEE Access **7**, 130074–130084 (2019)
19. Kong, P.S., Ahmad, N., Hassan, F., Ahmad, A.: The sensitivity of the heart rate detection for atrial electrograms signal. Elektr. J. Electr. Eng. **20**(2), 33–41 (2021)
20. Patro, K.K., Kumar, P.R., Viswanadham, T.: An efficient signal processing algorithm for accurate detection of characteristic points in abnormal ECG signals. In: 2016 International Conference on Electrical, Electronics, and Optimization Techniques (ICEEOT), Chennai, India, pp. 1476–1479 (2016)
21. Seng, K.P., Ahmad, N., Hassan, F., Manaf, M.S.A., Wahid, H., Ahmad, A.: Signal detection based on atrial fibrillation detection algorithms using RR interval measurements. In: Wahab, N.A., Mohamed, Z. (eds.) Control, Instrumentation and Mechatronics: Theory and Practice. Lecture Notes in Electrical Engineering, vol. 921. Springer, Singapore (2022). https://doi.org/10.1007/978-981-19-3923-5_51
22. Sheikh, S.A., Majoka, A.Z., Rehman, K.U., Razzaq, N.: Nonparametric spectral estimation technique to estimate dominant frequency for atrial fibrillation detection. J. Sig. Inf. Process. **6**, 266–276 (2015)
23. Umer, M., Bhatti, B.A., Tariq, M.H., Zia-ul-hassan, M., Khan, M.Y., Zaidi, T.: Electrocardiogram feature extraction and pattern recognition using a novel windowing algorithm. Adv. Biosci. Biotechnol. **5**, 886–894 (2014)

Author Index

© The Editor(s) (if applicable) and The Author(s), under exclusive license to Springer Nature Singapore Pte Ltd. 2024
F. Hassan et al. (Eds.): AsiaSim 2023, CCIS 1911, pp. 497–500, 2024.
https://doi.org/10.1007/978-981-99-7240-1

Printed in the United States
by Baker & Taylor Publisher Services

Printed in the United States
by Baker & Taylor Publisher Services